W0080972

Mathematics Study Resources

Volume 10

Series Editors

Kolja Knauer, Departament de Matemàtiques Informàtic, Universitat de Barcelona, Barcelona, Barcelona, Spain

Elijah Liflyand, Department of Mathematics, Bar-Ilan University, Ramat-Gan, Israel

This series comprises direct translations of successful foreign language titles, especially from the German language.

Powered by advances in automated translation, these books draw on global teaching excellence to provide students and lecturers with diverse materials for teaching and study.

Norbert Henze

Asymptotic Stochastics

An Introduction with a View towards Statistics

 Springer

Norbert Henze
Institut für Stochastik
Karlsruher Institut für Technologie (KIT)
Karlsruhe, Germany

ISSN 2731-3824   ISSN 2731-3832   (electronic)
Mathematics Study Resources
ISBN 978-3-662-68922-6   ISBN 978-3-662-68923-3   (eBook)
https://doi.org/10.1007/978-3-662-68923-3

English translation of the 2nd original German edition published by Springer-Verlag Heidelberg, 2024

This book is a translation of the original German edition "Asymptotische Stochastik: Eine Einführung mit Blick auf die Statistik," 2nd edition, by Norbert Henze, published by Springer-Verlag GmbH, DE in 2024. The translation was done with the help of an artificial intelligence machine translation tool. A subsequent human revision was done primarily in terms of content, so that the book will read stylistically differently from a conventional translation. Springer Nature works continuously to further the development of tools for the production of books and on the related technologies to support the authors.

Translation from the German language edition: "Asymptotische Stochastik: Eine Einführung mit Blick auf die Statistik" by Norbert Henze, © Der/die Herausgeber bzw. der/die Autor(en), exklusiv lizenziert an Springer-Verlag GmbH, DE, ein Teil von Springer Nature 2024. Published by Springer Berlin Heidelberg. All Rights Reserved.

This Springer imprint is published by the registered company Springer-Verlag GmbH, DE, part of Springer Nature.
The registered company address is: Heidelberger Platz 3, 14197 Berlin, Germany

If disposing of this product, please recycle the paper.

To Edda, Martin, Michael, and Matthias

Preface

This book is at the beginning graduate level. It is based on notes of the one-semester course *Asymptotic Stochastics* (two double hours of lecture and one double hour of exercise), which I gave several times at the Karlsruhe Institute of Technology (KIT). The main target group are students of mathematics, business mathematics, or technomathematics who have completed a bachelor's degree in one of these subjects and are at the beginning of their master's degree. These students often have only the basic knowledge of mathematical stochastics acquired in introductory courses. As a rule, they have experienced at least sporadically that asymptotic methods and procedures are of considerable importance. A textbook that is supposed to give these students a quick insight into asymptotic stochastics must take this limited prior knowledge into account.

Existing books on asymptotic stochastics are often written only for researchers in this field and are therefore not suitable for the target group, and often a claim to completeness leads to a presentation that lacks clarity and hardly shows a thread. Other books, in turn, occasionally omit proofs, or focus on a particular aspect of asymptotic stochastics. With this work I would like to fill a gap in the literature on asymptotic stochastics in this respect, and in particular to teach material in the area of the interface between probability theory and mathematical statistics in a purposeful way, deliberately omitting some interesting details. The aforementioned students who want to familiarize themselves with the subject or lecturers who are planning a lecture on it will find orientation in this work and will get an overview of important methods and procedures of asymptotic stochastics.

The book assumes a basic knowledge of stochastics and of measure and integration theory, as usually taught in an undergraduate course. It is deliberately designed for self-study. This purpose is served not only by the 193 problems, whose solutions can be found at the end of the book, but also by the 138 self-questions interspersed throughout the text. Answers to these self-questions can be found at the end of each chapter. What is striking is a noticeable redundancy that is conducive to the learning process. For example, the Portmanteau theorem about equivalent criteria for convergence in distribution is first proved for the finite-dimensional case. In the framework of the generalization of the theory to convergence in distribution in general metric spaces one recognizes which parts of the proof can be taken over unchanged and which not.

The individual chapters of this work differ in part significantly from each other, both in terms of length and in terms of the degree of difficulty. Chapter 1 has an introductory character in that it essentially provides basic concepts and results from probability theory and measure theory that will be referred to in later chapters. Chapters 2– 4 treat topics that are independent of each other, and which are also already suitable for undergraduate seminars. Chapter 2 deals with a necessary and sufficient condition for the convergence in distribution of independent non-negative integer random variables to a Poisson distribution. Furthermore, the importance of Poisson convergence for extreme value stochastics is emphasized. Chapter 3 is devoted to the method of moments, and Chap. 4 provides a central limit theorem for m-dependent stationary sequences of random variables. Chapter 5 deals with the multivariate normal distribution. With respect to the generalization to separable Hilbert spaces in Chap. 17, this distribution is introduced via the (possibly degenerate) univariate normal distributions of all one-dimensional projections.

The topic of Chap. 6 is convergence in distribution of d-dimensional random vectors. Main results are the Portmanteau theorem as well as multivariate central limit theorems and the delta method. Statistical applications concern the chi-square test and variance-stabilizing transformations. Chapter 7 deals with empirical distribution functions and the Glivenko–Cantelli theorem, an interface between probability theory and mathematical statistics. The same applies to Chap. 8, which is devoted to U-statistics. Central results are the convergence in distribution of non-degenerate as well as singly-degenerate one-sample U-statistics and a central limit theorem for two-sample U-statistics. Statistical applications concern the Mann–Whitney U test and the Cramér–von Mises goodness-of-fit test.

Chapter 9 provides a setting for asymptotic statistics and discusses estimators and confidence intervals. Chapter 10 is devoted to the maximum likelihood method and thus to a basic estimation principle. Main results are on the one hand the almost sure convergence of maximum likelihood estimators if the parameter set is compact, and on the other hand the asymptotic normal distribution of the estimation error, where the Fisher information matrix occurs. In Chap. 11 we discuss the method of moments as another important design principle for estimators, as well as a comparison of the quality of estimators with a large sample size. Topics include BAN estimators and asymptotic relative Pitman efficiency. Chapter 12 deals with the testing of hypotheses within parametric models. The main result is the limiting distribution of the generalized likelihood ratio test statistic under the null hypothesis. Applications concern, among others, tests of independence in contingency tables. The chapter concludes with a proof of the asymptotic validity of a parametric bootstrap procedure.

Chapter 13 leaves the finite-dimensional framework, because the topics are probability measures on the Borel σ-field in general metric spaces. The main focus is on the space C[0, 1] of continuous real-valued functions defined on the unit interval. The subject of Chap. 14 is the weak convergence of probability measures and the convergence in distribution of random variables that take on values in metric spaces. Keywords are the portmanteau theorem, the mapping theorem, the Prokhorov theorem on tightness and relative compactness and a criterion for convergence in distribution in the space C[0, 1] with the help of tightness and convergence of finite-dimensional distributions. Topics of Chap. 15 are the Wiener process, Donsker's theorem, the functional central limit theorem, and the Brownian bridge.

In Chap. 16 we are concerned with convergence in distribution in the *Càdlàg space* D[0, 1] of all right-continuous real-valued functions on [0, 1] with existing left-hand limits. Statistical applications are the goodness-of-fit tests of Kolmogorov–Smirnov and Cramér–von Mises as well as the nonparametric Kolmogorov–Smirnov two-sample test. The concluding Chap. 17 deals with random elements that take on values in a separable Hilbert space $\mathbb{H}$ and with convergence in distribution of such random elements. Basic notions are the normal distribution in $\mathbb{H}$ as well as expectation, covariance operator, and the characteristic functional of a $\mathbb{H}$-valued random element. The main result is a central limit theorem for a triangular array of $\mathbb{H}$-valued random elements. Statistical applications concern weighted L^2 statistics. This last chapter is certainly the most challenging because it introduces topics of current research. For example, the publications [DE1], [DE2], [HEK], and [HMA] are based on master's theses written by students directly after attending the lecture *Asymptotic Stochastics*.

Pfinztal, Germany   Norbert Henze
January 2024

Acknowledgments

At this point I would like to thank all those who have been an invaluable help during the development phase of this book. Yakov Yu. Nikitin (1947–2019) and Andreas Rüdinger of Springer Verlag encouraged me to tackle this work. I am also indebted to Andreas Rüdinger for numerous tips. Bruno Ebner, Bernhard Klar, Celeste Mayer, and Bianca Alton from Springer Verlag were able to uncover various deficiencies in the manuscript through meticulous proofreading. I also owe valuable advice to my colleagues Ludwig Baringhaus, Lutz Dümbgen, Hajo Holzmann, and Mathias Trabs. In particular, I would like to thank Lutz Mattner, whose numerous comments have contributed to a substantial improvement of this book.

Reading Notes

Chapter 1 shall serve as an introduction. Chapters 2, 3, and 4 can be read independently of each other. Except for Problem 7.4, which connects to the multivariate central limit theorem, the same is true for Chap. 7.

Chapters 5 and 6 are fundamental for all other chapters except Chaps. 7 and 9. The statistics Chaps. 9–12 build on each other; however, they are not needed for an understanding of the further chapters.

Chapters 13 and 14 are prerequisites for each of the following chapters. Chapters 15 and 16 build on each other, but the last chapter can be read directly after Chap. 14.

Contents

List of Symbols

Symbol	Meaning	Page
$\xrightarrow{\text{a.s.}}$	almost sure convergence	1
$\xrightarrow{\mathbb{P}}$	convergence in probability	1
$\|\cdot\|$	norm (meaning depends on chapter)	2
$\xrightarrow{\mathcal{L}^p}$	convergence in mean of order p ($\mathcal{L}^p$-convergence)	2
$\mathbb{E}(Y)$	expectation of a random variable Y	2
$1\{A\}$, $\mathbf{1}_A$	indicator function of an event A	3
$\mathbb{V}(Y)$	variance of a random variable Y	4
C_G	set of continuity points of a function G	5
$\xrightarrow{\mathcal{D}}$	convergence in distribution	5
φ_X	characteristic function of X	8
$\overset{\mathcal{D}}{=}$	equality in distribution	8
$\top$	transposition (of vectors and matrices)	9
$N(\mu, \sigma^2)$	normal distribution with expectation μ and variance σ^2	13
$\mathbb{E}[X|\mathcal{G}]$	conditional expectation of X given a sub-σ-field $\mathcal{G}$	16
$\mathcal{B}$	σ-field of Borel sets of $\mathbb{R}$	17
$\overline{\mathbb{R}}$	$\mathbb{R} \cup \{\infty, -\infty\}$	17
$\overline{\mathcal{B}}$	σ-field of Borel sets of $\overline{\mathbb{R}}$	17
$\nu \ll \mu$	the measure ν is absolutely continuous with respect to μ	18
$\mu_1 \otimes \mu_2$	product of the measures μ_1 and μ_2	19
$\mathcal{A}_1 \otimes \mathcal{A}_2$	product of the σ-fields $\mathcal{A}_1$ and $\mathcal{A}_2$	19
$\text{Po}(\lambda)$	Poisson distribution with expectation λ	23
$x \wedge y$	minimum of x and y	24
$X \sim ?$	X has the distribution ?	25
0_d	zero vector in $\mathbb{R}^d$	57
$\delta_{i,j}$	Kronecker's symbol	59

(continued)

Symbol	Meaning	Page
$N_d(\mu, \Sigma)$	d-dimens. normal distr. with expect. μ and cov. matrix Σ	60
I_d	unit matrix of order d	61
$0_{r \times s}$	zero matrix with r rows and s columns	62
χ_k^2	Chi square distribution with k degrees of freedom	65
$\mathcal{B}^d$	σ-field of Borel sets of $\mathbb{R}^d$	70
$\mathcal{O}^d$	class of open subsets of $\mathbb{R}^d$	71
$\mathcal{A}^d$	class of closed subsets in $\mathbb{R}^d$	71
B°	interior of a set B	71
$\overline{B}$	closure of a set B	71
$\partial B := \overline{B} \setminus B^\circ$	boundary of a set B	71
$\mathcal{C}_b(\mathbb{R}^d)$	class of continuous bounded functions $f : \mathbb{R}^d \to \mathbb{R}$	71
$A \uplus B$	$A \cup B$, if $A \cap B = \emptyset$	75
$X_n = O_{\mathbb{P}}(a_n)$	the sequence (X_n/a_n) is tight	79
$X_n = o_{\mathbb{P}}(a_n)$	the sequence (X_n/a_n) converges to zero in probability	79
δ_x	Dirac measure centered on x	96
$\text{Exp}(\lambda)$	exponential distribution with density $\lambda \exp(-\lambda x)$, $x \geq 0$	144
$(\mathcal{X}, \mathcal{B}, \mathbb{P}_\vartheta)_{\vartheta \in \Theta}$	statistical space	145
t_k-distribution	Student's t-distribution with k degrees of freedom	151
$L_{n,\mathbf{x}}(\vartheta)$	likelihood function	162
$I_{KL}(\vartheta : \vartheta')$	Kullback–Leibler information of $f(\cdot, \vartheta)$ w.r.t. $f(\cdot, \vartheta')$	168
$I_1(\vartheta)$	Fisher information matrix	174
(S, ρ)	metric space	230
$B(x, \varepsilon)$	$\{y \in S : \rho(x, y) < \varepsilon\}$ (open ball centered at x with radius ε)	230
$\mathcal{O}$	class of open subsets of a metric space	230
$\mathfrak{A}$	class of closed subsets of a metric space	230
$\rho(x, M)$	distance of x to the set M	231
M^ε	$\{y \in S : \rho(y, M) < \varepsilon\}$ (parallel set w.r.t. M at distance ε)	231
$\mathcal{B}(S)$	σ-field of Borel sets in a metric space (S, ρ)	233
$\mathcal{C}_b(S)$	class of continuous bounded functions $f : S \to \mathbb{R}$	234
$\mathcal{C}_b^0(S)$	class of uniformly continuous bounded functions $f : S \to \mathbb{R}$	234
$C = C[0, 1]$	class of continuous functions $x : [0, 1] \to \mathbb{R}$	235
$w_x(\delta) := w(x, \delta)$	modulus of continuity	237
$\pi_{t_1,\ldots,t_k}$	$\pi_{t_1,\ldots,t_k}(x) := (x(t_1), \ldots, x(t_k))$ (coordinate projection)	239
$\mathcal{C}_f$	class of finite-dimensional sets	239
$\mathbb{R}^\infty$	set of sequences $x = (x_j)_{j \geq 1}$ of real numbers	240
$\mathcal{R}_f^\infty$	system of finite-dimensional sets in $\mathbb{R}^\infty$	241

(continued)

Symbol	Meaning	Page
$\mathcal{C}(P)$	$\{B \in \mathcal{B}(S) : P(\partial B) = 0\}$ (class of P-continuity sets)	249
$X_n \xrightarrow{\mathcal{D}_{\text{fidi}}} X$	conv. of finite-dimensional distributions (fidi-convergence)	258
$\mathrm{D} = \mathrm{D}[0, 1]$	Càdlàg space	308
$w'_x(\delta)$	Càdlàg modulus of continuity	310
$\mathcal{G}$	continuous strictly increasing functions $g : [0, 1] \to [0, 1]$	311
d_S	Skorokhod metric	311
$\mathcal{L}^+_{tr}(\mathbb{H})$	positive operators of trace class on $\mathbb{H}$	363
$\mathcal{N}_d$	class of nondegenerate d-variate normal distributions	377

In this book it is assumed that the reader has some familiarity with basic concepts and results of probability. Useful texts include [BI3], [DUR], [SHI], and [KLE]. This chapter contains a brief review of those notions and results that are relevant in the sequel.

For real-valued random variables $X, X_1, X_2, \ldots$ that all are defined on the same probability space $(\Omega, \mathcal{A}, \mathbb{P})$, $\mathbb{P}$-*almost sure convergence* of X_n to X, abbreviated with $X_n \xrightarrow{\text{a.s.}} X$, is defined by

$$X_n \xrightarrow{\text{a.s.}} X :\Longleftrightarrow \mathbb{P}\left(\left\{\omega \in \Omega : \lim_{n \to \infty} X_n(\omega) = X(\omega)\right\}\right) = 1.$$

Here, the symbol $:\Longleftrightarrow$ denotes definition, and each limit is taken as $n \to \infty$, unless stated otherwise. Mostly, *almost sure convergence* is shorthand for $\mathbb{P}$-almost sure convergence. Notice that, in probability theory the notion of pointwise convergence $X_n(\omega) \to X(\omega)$ *for each* $\omega \in \Omega$ is too restrictive, since sets having probability 0 are often negligible. Thus, almost sure convergence is pointwise convergence on a set having probability 1.

Since

$$X_n \xrightarrow{\text{a.s.}} X \Longleftrightarrow \lim_{n \to \infty} \mathbb{P}\left(\left\{\sup_{k \geq n} |X_k - X| > \varepsilon\right\}\right) = 0 \text{ for each } \varepsilon > 0,$$

almost sure convergence implies *convergence in probability* or, equivalently, *stochastic convergence*, which is defined by

$$X_n \xrightarrow{\mathbb{P}} X :\Longleftrightarrow \lim_{n \to \infty} \mathbb{P}(|X_n - X| > \varepsilon) = 0 \text{ for each } \varepsilon > 0.$$

The reverse implication does not hold in general (Problem 1.2 a)).

N. Henze, *Asymptotic Stochastics*, Mathematics Study Resources 10, https://doi.org/10.1007/978-3-662-68923-3_1

Self-question 1: Why does $X_n \xrightarrow{\mathbb{P}} X \Longrightarrow X_n \xrightarrow{\text{a.s.}} X$ hold if Ω is a countable set?

1.1 Theorem (Subsequence Criterion for Convergence in Probability)
The following statements are equivalent:

(a) $X_n \xrightarrow{\mathbb{P}} X$.

(b) *Each subsequence* (X_{n_k}) *of* (X_n) *contains a sub-subsequence* $(X_{n'_k})$ *such that* $X_{n'_k} \xrightarrow{a.s.} X$ *as* $n'_k \to \infty$.

The notions of almost sure convergence and convergence in probability can immediately be generalized to random vectors. If $X = (X^{(1)}, \ldots, X^{(d)})$ and $X_n = (X_n^{(1)}, \ldots, X_n^{(d)})$, $n \geq 1$, are d-dimensional random vectors, then almost sure convergence $X_n \xrightarrow{\text{a.s.}} X$ can be adopted without change. Since the intersection of finitely many sets of probability 1 has probability 1, $X_n \xrightarrow{\text{a.s.}} X$ holds if and only if each of the sequences of components converges almost surely, that is, if $X_n^{(j)} \xrightarrow{\text{a.s.}} X^{(j)}$ for each $j \in \{1, \ldots, d\}$. Using any norm $\| \cdot \|$ on $\mathbb{R}^d$, one defines *convergence in probability* of X_n to X by

$$X_n \xrightarrow{\mathbb{P}} X :\Longleftrightarrow \lim_{n \to \infty} \mathbb{P}(\|X_n - X\| > \varepsilon) = 0 \quad \text{for each } \varepsilon > 0.$$

Self-question 2: Why is the special choice of the norm $\| \cdot \|$ irrelevant?

By analogy with almost sure convergence, we have

$$X_n \xrightarrow{\mathbb{P}} X \Longleftrightarrow X_n^{(j)} \xrightarrow{\mathbb{P}} X^{(j)} \text{ for each } j \in \{1, \ldots, d\}$$

(Problem 1.1).

Given $p \in (0, \infty)$, let $\mathcal{L}^p$ be the collection of real-valued random variables X on Ω that are integrable to the p-th power with respect to $\mathbb{P}$, that is, for which $\mathbb{E}|X|^p < \infty$. For $X, X_1, X_2, \ldots \in \mathcal{L}^p$, we define

$$X_n \xrightarrow{\mathcal{L}^p} X :\Longleftrightarrow \lim_{n \to \infty} \mathbb{E}|X_n - X|^p = 0$$

and say that X_n *converges to* X *in the mean of order* p. In the special cases $p = 1$ and $p = 2$, we speak of *convergence in mean* and *convergence in quadratic mean*, respectively.

Since $\mathbf{1}\{|X_n - X| > \varepsilon\} \leq \varepsilon^{-p}|X_n - X|^p$, convergence in the mean of order p implies convergence in probability. However, $X_n \xrightarrow{\text{a.s.}} X$ does not generally follow from $X_n \xrightarrow{\mathcal{L}^p} X$ (Problem 1.2 b)).

1.2 Theorem (Strong Law of Large Numbers)
Let $X_1, X_2, \ldots$ be independent and identically distributed (in short: i.i.d.) random variables. Then the following statements are equivalent:

(a) There is a random variable X such that $\frac{1}{n}\sum_{j=1}^{n} X_j \xrightarrow{\text{a.s.}} X$.
(b) $\mathbb{E}|X_1| < \infty$.

If (a) or (b) holds, then $\frac{1}{n}\sum_{j=1}^{n} X_j \xrightarrow{\text{a.s.}} \mathbb{E}(X_1)$.

This classical result by A. N. Kolmogorov[1] immediately generalizes to random vectors (Problem 1.3). It also holds for random variables that take on values in certain infinite-dimensional spaces (see Theorem 17.15).

The next two results are important tools for proving almost sure convergence. Here, writing := for definition,

$$\limsup_{n\to\infty} A_n := \bigcap_{n=1}^{\infty} \bigcup_{k=n}^{\infty} A_k$$

denotes the *limit superior* of a sequence (A_n) of events.

1.3 Theorem (Borel[2]–Cantelli[3]-Lemma)
Let $A_1, A_2, \ldots$ be events in a probability space $(\Omega, \mathcal{A}, \mathbb{P})$.

(a) If $\sum_{n=1}^{\infty} \mathbb{P}(A_n) < \infty$ then $\mathbb{P}\left(\limsup_{n\to\infty} A_n\right) = 0$.
(b) If $A_1, A_2, \ldots$ are independent and $\sum_{n=1}^{\infty} \mathbb{P}(A_n) = \infty$ then $\mathbb{P}\left(\limsup_{n\to\infty} A_n\right) = 1$.

---

[1] Andrej Nikolajewitsch Kolmogorov (1903–1987), professor at Moscow University (from 1930), one of the most proliferous mathematicians of the twentieth century. Kolmogorov made fundamental contributions to probability theory, mathematical statistics, mathematical logic, topology, measure and integration theory, functional analysis, information theory, and algorithm theory.

[2] Émile Borel (1871–1956), from 1909 professor at the Sorbonne in Paris. Main areas of work: Complex analysis, set theory, measure theory, probability theory, game theory.

[3] Francesco Paolo Cantelli (1875–1966), Italian mathematician, professor of actuarial mathematics at the universities of Catania, Naples and Rome. He founded the Italian Actuarial Association *Istituto Italiano degli Attuari*. In probability theory, he is primarily known for the Borel–Cantelli lemma, the Cantelli inequality and the Glivenko–Cantelli theorem.

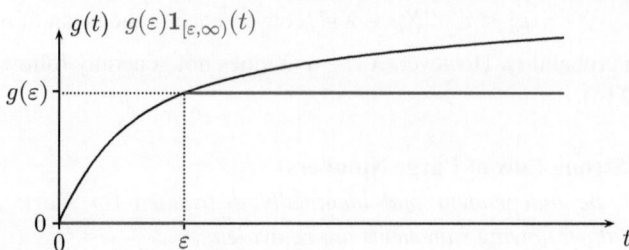

Fig. 1.1 Illustrating Markov's inequality

1.4 Theorem (Maximal Inequality, Kolmogorov)

Suppose $X_1, \ldots, X_n$ are independent random variables with $\mathbb{E}(X_j^2) < \infty$ for each $j \in \{1, \ldots, n\}$, and put $S_k := \sum_{j=1}^{k}(X_j - \mathbb{E}[X_j])$, $k \in \{1, \ldots, n\}$. Then

$$\mathbb{P}\left(\max_{1 \leq k \leq n} |S_k| \geq \varepsilon\right) \leq \frac{\mathbb{V}(S_n)}{\varepsilon^2} \quad \text{for each } \varepsilon > 0.$$

For the case $n = 1$, if X is a random variable with $\mathbb{E}(X^2) < \infty$, then this inequality boils down to *Chebyshev's[4] inequality*

$$\mathbb{P}(|X - \mathbb{E}(X)| \geq \varepsilon) \leq \frac{\mathbb{V}(X)}{\varepsilon^2} \quad \text{for each } \varepsilon > 0. \tag{1.1}$$

A further generalization of Chebyshev's inequality is the following result, which traces back to the Russian mathematician A. A. Markov.[5]

1.5 Theorem (Markov's Inequality)

If $g : [0, \infty) \rightarrow [0, \infty)$ is an increasing function and X is any random variable with $\mathbb{E}g(|X|) < \infty$, then

$$\mathbb{P}(|X| \geq \varepsilon) \leq \frac{\mathbb{E}g(|X|)}{g(\varepsilon)} \quad \text{for each } \varepsilon > 0$$

such that $g(\varepsilon) > 0$.

Markov's inequality immediately follows from the inequality $g(\varepsilon)\mathbf{1}_{[\varepsilon,\infty)}(|X|) \leq g(|X|)$, which holds pointwise on the underlying probability space, and taking expectations (see Fig. 1.1).

---

[4] Pafnuti Lwowitsch Chebyshev (1821–1894), from 1850 professor at the University of St. Petersburg. Main areas of work: Number theory, complex analysis, integration theory, probability theory.

[5] Andrey Andreyevich Markov (1856–1922), professor at the University of St. Petersburg (from 1893). Markov was politically progressive. Among other things, he protested by returning all orders and honors when the election of Maxim Gorky as a member of the Academy was rejected on the orders of the Tsar. Main area of work: Probability theory (*Markov chains*).

The most important special case of Markov's inequality applies to the function $g(t) :=$ t^p for some $p > 0$. If we set $p = 2$ and consider instead of X the centered random variable $X - \mathbb{E}(X)$, Markov's inequality boils down to Chebyshev's inequality (1.1).

Upon combining the Borel–Cantelli lemma and Chebyshev's inequality, one obtains the following series criterion for almost sure convergence.

1.6 Theorem (Series Criterion for Almost Sure Convergence)
Let $X, X_1, X_2, \ldots$ be random variables. If

$$\sum_{n=1}^{\infty} \mathbb{P}\big(|X_n - X| > \varepsilon\big) < \infty \quad \text{for each } \varepsilon > 0$$

then $X_n \xrightarrow{a.s.} X$.

Regarding inequalities, the following result is often used.

1.7 Theorem (Jensen's[6] Inequality)
Let $I \subset \mathbb{R}$ be an interval and $g : I \to \mathbb{R}$ a convex function, that is, $\lambda g(x) + (1 - \lambda) g(y) \geq g(\lambda x + (1 - \lambda) y)$ for all $\lambda \in (0, 1)$ and $x, y \in I$. If X is an I-valued random variable with $\mathbb{E}|X| < \infty$ and $\mathbb{E}|g(X)| < \infty$, then

$$\mathbb{E}\big(g(X)\big) \geq g\big(\mathbb{E}(X)\big).$$

In the sequel, C_G denotes the set of continuity points of a function $G : \mathbb{R}^n \to \mathbb{R}^k$, that is, the set of all x in $\mathbb{R}^n$ with the property that G is continuous at x.

If $X, X_1, X_2, \ldots$ are random variables with distribution functions $F, F_1, F_2, \ldots$, respectively, *convergence in distribution of X_n to X* is defined by

$$X_n \xrightarrow{\mathcal{D}} X :\Longleftrightarrow \lim_{n \to \infty} F_n(x) = F(x) \text{ for each } x \in C_F. \tag{1.2}$$

Equivalent notations for this notion of convergence are $F_n \xrightarrow{\mathcal{D}} F$, $\mathbb{P}^{X_n} \xrightarrow{\mathcal{D}} \mathbb{P}^X$ or $X_n \xrightarrow{\mathcal{D}} \mathbb{P}^X$.

We call $\mathbb{P}^X$ the *limit distribution* or *asymptotic distribution* of the sequence (X_n) or the sequence $(\mathbb{P}^{X_n})$. Convergence in probability implies convergence in distribution. The converse implication holds if $\mathbb{P}^X$ is degenerate, that is, if $\mathbb{P}(X = t) = 1$ for some $t \in \mathbb{R}$. If F is continuous, then the convergence of F_n to F in (1.2) is not only pointwise,

---

[6] Johann Ludvig Valdemar Jensen (1859–1925), essentially self taught Danish mathematician and engineer. Jensen never held an academic appointment but worked with the Copenhagen Telephone Company. Main fields of interest: Riemann hypothesis, convex functions.

Fig. 1.2 Hierarchy of notions
of convergence for random
variables

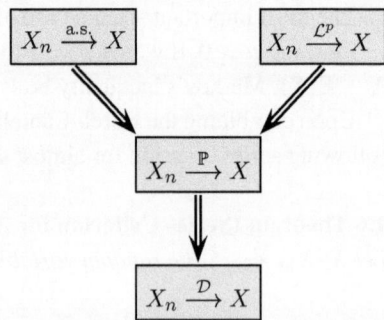

but even uniform. Figure 1.2 illustrates the implications between the different notions of convergence.

In what follows, let

$$\mathcal{C}_b := \{f : \mathbb{R} \to \mathbb{R} : f \text{ is bounded and continuous}\},$$

$$\mathcal{C}_b^{(0)} := \{f \in \mathcal{C}_b : f \text{ is uniformly continuous}\}.$$

Moreover, denoting $f^{(j)}$ the jth derivative of a function f, put for $r \in \mathbb{N}$

$$\mathcal{C}_b^{(r)} := \{f \in \mathcal{C}_b^{(0)} : f \text{ is } r\text{-times differentiable}, \ f^{(j)} \in \mathcal{C}_b^{(0)} \text{ for } j \in \{1, \dots, r\}\}.$$

The next result illustrates the notion of convergence in distribution.

1.8 Theorem (Characterization of Convergence in Distribution)
For each integer r, the following statements are equivalent:

(a) $X_n \xrightarrow{D} X$.
(b) $\lim_{n \to \infty} \mathbb{E} f(X_n) = \mathbb{E} f(X)$ *for all* $f \in \mathcal{C}_b$.
(c) $\lim_{n \to \infty} \mathbb{E} f(X_n) = \mathbb{E} f(X)$ *for all* $f \in \mathcal{C}_b^{(r)}$.

This result is not surprising, since (1.2) is the same as

$$\lim_{n \to \infty} \mathbb{E} \mathbf{1}_{(-\infty, x]}(X_n) = \mathbb{E} \mathbf{1}_{(-\infty, x]}(X) \quad \text{for all } x \in C_F,$$

and since the indicator function $\mathbf{1}_{(-\infty, x]}$ can be approximated as accurately as required by smooth functions (see Fig. 1.3 for an approximation by a uniformly continuous function).

The fact that conditions (a) and (b) of Theorem 1.8 are equivalent gives rise to a definition of convergence in distribution for random variables that take on values in general metric spaces (see Definition 14.6).

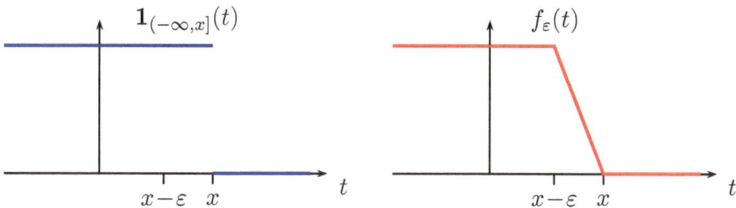

Fig. 1.3 The function f_ε approximates the indicator function $\mathbf{1}_{(-\infty,x]}$

The following notions are ubiquitous in the context of convergence in distribution. They will be revisited in much greater generality in Chaps. 6 and 14. For the second part, if $Q, Q_1, Q_2, \ldots$ are probability measures on the σ-field (or: σ-algebra) $\mathcal{B}$, we write $Q_n \xrightarrow{\mathcal{D}} Q$ if $F_n \xrightarrow{\mathcal{D}} F$, where $F_n(x) = Q((-\infty, x])$, $n \geq 1$, and $F(x) = Q((-\infty, x])$, $x \in \mathbb{R}$.

1.9 Definition (Tightness and Relative Compactness)
A non-empty collection $\mathcal{Q}$ of probability measures on the σ-field $\mathcal{B}$ of Borel sets is called

(a) *tight*, if for every $\varepsilon > 0$ there is a compact set $K \subset \mathbb{R}$ such that $Q(K) \geq 1 - \varepsilon$ for every Q in $\mathcal{Q}$,
(b) *relatively compact*, if for every sequence (Q_n) in $\mathcal{Q}$ there exists a subsequence (Q_{n_k}) and a probability measure Q on $\mathcal{B}$ such that $Q_{n_k} \xrightarrow{\mathcal{D}} Q$ as $k \to \infty$.

The next result also holds in much greater generality (see Theorem 14.22).

1.10 Theorem   *In the setting of Definition 1.9, we have:*

$$\mathcal{Q} \text{ is tight } \iff \mathcal{Q} \text{ is relatively compact.}$$

1.11 Corollary

(a) *If $X_n \xrightarrow{\mathcal{D}} X$ then the set $\{\mathbb{P}^{X_n} : n \geq 1\}$ is tight.*
(b) *Suppose $\{\mathbb{P}^{X_n} : n \geq 1\}$ is tight. If there is a probability measure Q such that $X_{n_k} \xrightarrow{\mathcal{D}} Q$ as $k \to \infty$ for every subsequence (X_{n_k}) of (X_n) that converges in distribution, then $X_n \xrightarrow{\mathcal{D}} Q$.*

The following concept provides a basic tool for analytical methods in probability theory. Among other things, it has proved useful for the characterization of probability distributions and for deriving limit theorems. A generalization is first made for random vectors (Definition 1.14), and in Chap. 17 under the naming *characteristic functional* for random variables that take on values in infinite-dimensional Hilbert spaces (Definition 17.22).

1.12 Definition (Characteristic Function)

If X is a random variable X, the function $\varphi_X : \mathbb{R} \to \mathbb{C}$, defined by

$$\varphi_X(t) := \mathbb{E}\left[e^{itX}\right] = \mathbb{E}[\cos(tX)] + i\,\mathbb{E}[\sin(tX)], \quad t \in \mathbb{R},$$

is called the *characteristic function (of the distribution) of* X.

An immediate corollary of this definition is that, for $a, b \in \mathbb{R}$, we have

$$\varphi_{aX+b}(t) = e^{itb}\,\varphi_X(at), \quad t \in \mathbb{R}. \tag{1.3}$$

With regard to proofs of limit theorems for sums of independent random variables, the *multiplication rule*

$$X, Y \text{ independent} \implies \varphi_{X+Y} = \varphi_X\,\varphi_Y$$

is important. If a and b are points of continuity of the distribution function F of X and $a < b$, then

$$F(b) - F(a) = \lim_{T \to \infty} \frac{1}{2\pi} \int_{-T}^{T} \frac{e^{-ita} - e^{-itb}}{it}\,\varphi(t)\,dt. \tag{1.4}$$

As a consequence, we obtain the *uniqueness theorem*

$$X \overset{\mathcal{D}}{=} Y \iff \varphi_X = \varphi_Y.$$

Here and in what follows, $X \overset{\mathcal{D}}{=} Y$ is shorthand for equality in distribution, that is, $X \overset{\mathcal{D}}{=} Y :\iff \mathbb{P}^X = \mathbb{P}^Y$.

Self-question 3: Why does the implication "$\impliedby$" follow from (1.4) ?

If X has the normal distribution $N(\mu, \sigma^2)$, then the characteristic function of X is given by

$$\varphi_X(t) = e^{i\mu t}\,\exp\left(-\frac{\sigma^2 t^2}{2}\right), \quad t \in \mathbb{R} \tag{1.5}$$

(Problem 1.4).

The next result is fundamental for deriving convergence in distribution. It may be used to prove the central limit Theorems 1.18–1.20.

1.13 Theorem (Continuity Theorem, Lévy[7]–Cramér[8])

Let $X, X_1, X_2, \ldots$ be random variables with characteristic functions $\varphi, \varphi_1, \varphi_2, \ldots,$ respectively. Then

$$X_n \xrightarrow{\mathcal{D}} X \iff \lim_{n \to \infty} \varphi_n(t) = \varphi(t) \quad \text{for every } t \in \mathbb{R}.$$

The generalization of the notion of a characteristic function for random vectors is as follows.

1.14 Definition (Characteristic Function of a Random Vector)

If $X = (X_1, \ldots, X_d)^\top$ is a d-dimensional random vector (written as a column vector), the function $\varphi_X : \mathbb{R}^d \to \mathbb{C}$, defined by

$$\varphi_X(t) := \mathbb{E}\left[e^{it^\top X}\right] = \mathbb{E}\left[\cos(t^\top X)\right] + i\,\mathbb{E}\left[\sin(t^\top X)\right], \quad t \in \mathbb{R}^d,$$

is called the *characteristic function (of the distribution) of* X.

Note that $t^\top X$ is the inner product of t and X. This perspective motivates a generalization of the concept of a characteristic function for random variables that take on values in a Hilbert space (see Definition 17.22).

The next result generalizes Eq. (1.4), and it justifies the naming *characteristic function* also for random vectors.

1.15 Theorem *Let* $X = (X_1, \ldots, X_d)$ *be a d-dimensional random vector. In addition, suppose* $B := [a_1, b_1] \times \ldots \times [a_d, b_d] \subset \mathbb{R}^d$ *is a d-dimensional rectangle and* $-\infty < a_j < b_j < \infty$ *for* $j \in \{1, \ldots, d\}$. *If*

$$\mathbb{P}\big(X_j \in \{a_j, b_j\}\big) = 0 \quad \text{for each } j \in \{1, \ldots, d\}, \tag{1.6}$$

then

$$\mathbb{P}^X(B) = \lim_{T \to \infty} \frac{1}{(2\pi)^d} \int_{-T}^{T} \cdots \int_{-T}^{T} \prod_{j=1}^{d} \frac{e^{-it_j a_j} - e^{-it_j b_j}}{it_j} \, \varphi_X(t) \, dt. \tag{1.7}$$

---

[7] Paul Lévy (1886–1959), 1919–1959 professor at the École Polytechnique in Paris. Alongside A.N. Kolmogorov and A.J. Khinchin, Lévy can be considered one of the main founders of modern measure-theoretically founded probability theory.

[8] Harald Cramér (1893–1985), professor at the University of Stockholm (1929–1958). Main areas of work: Stochastics, Actuarial Mathematics. His influential book [CRA] is considered the first mathematically accurate and at the same time readable textbook of statistics.

Proof We follow the proof in the case $d = 1$ (see, e.g., [LUK], pp. 31–33) and recall the *Dirichlet[9]-Integral*

$$\lim_{t \to \infty} \int_0^t \frac{\sin(x)}{x} \, dx = \frac{\pi}{2} \tag{1.8}$$

(see, e.g., [ZR2], p. 424). Furthermore, we write $\int_{C_T}$ for the d-fold integral occurring in (1.7) and set

$$I(T) := \frac{1}{(2\pi)^d} \int_{C_T} \prod_{j=1}^d \frac{e^{-it_j a_j} - e^{-it_j b_j}}{it_j} \, \varphi_X(t) \, dt_1 \cdots dt_d, \quad T > 0.$$

In view of

$$\left| \frac{e^{-it_j a_j} - e^{-it_j b_j}}{it_j} \right| = \left| \int_{a_j}^{b_j} e^{-it_j \xi} \, d\xi \right| \le b_j - a_j, \quad j \in \{1, \dots, d\},$$

Fubini's theorem (Theorem 1.34) can be applied, and using symmetry arguments we obtain

$$I(T) = \frac{1}{(2\pi)^d} \int_{C_T} \prod_{j=1}^d \frac{e^{-it_j a_j} - e^{-it_j b_j}}{it_j} \int_{\mathbb{R}^d} e^{it^\top x} \, \mathbb{P}^X(dx) \, dt_1 \cdots dt_d$$

$$= \int_{\mathbb{R}^d} \frac{1}{(2\pi)^d} \int_{C_T} \prod_{j=1}^d \frac{e^{-it_j a_j} - e^{-it_j b_j}}{it_j} \prod_{j=1}^d e^{it_j x_j} \, dt_1 \cdots dt_d \, \mathbb{P}^X(dx)$$

$$= \int_{\mathbb{R}^d} \frac{1}{(2\pi)^d} \int_{C_T} \prod_{j=1}^d \frac{e^{it_j(x_j - a_j)} - e^{it_j(x_j - b_j)}}{it_j} \, dt_1 \cdots dt_d \, \mathbb{P}^X(dx)$$

$$= \int_{\mathbb{R}^d} \prod_{j=1}^d \frac{1}{2\pi} \int_{-T}^T \frac{e^{it_j(x_j - a_j)} - e^{it_j(x_j - b_j)}}{it_j} \, dt_j \, \mathbb{P}^X(dx)$$

$$= \int_{\mathbb{R}^d} \prod_{j=1}^d \frac{1}{\pi} \int_0^T \frac{\sin(t_j(x_j - a_j)) - \sin(t_j(x_j - b_j))}{t_j} \, dt_j \, \mathbb{P}^X(dx).$$

---

[9] Peter Gustav Lejeune Dirichlet (1805–1859), 1831–1855 professor at the University of Berlin, 1855 appointment to the chair of C.F. Gauss in Göttingen. Main areas of research: Number theory, analysis, mathematical physics.

Putting

$$S(T) := \int_0^T \frac{\sin x}{x} \, dx,$$

we have

$$\int_0^T \frac{\sin t\vartheta}{t} \, dt = \operatorname{sgn}(\vartheta) \, S(T \, |\vartheta|), \quad T \geq 0, \ \vartheta \in \mathbb{R}.$$

Moreover, letting

$$g_j(z, T) := \frac{1}{\pi} \left(\operatorname{sgn}(z - a_j) S(T \, |z - a_j|) - \operatorname{sgn}(z - b_j) S(T \, |z - b_j|) \right), \ z \in \mathbb{R},$$

$j \in \{1, \ldots, d\}$, it follows that

$$I(T) = \int_{\mathbb{R}^d} \prod_{j=1}^d g_j(x_j, T) \, \mathbb{P}^X(dx).$$

The function g_j is bounded, and due to (1.8) we obtain

$$\lim_{T \to \infty} g_j(z, T) = \begin{cases} 0, & \text{if } z < a_j \text{ or } z > b_j, \\ \frac{1}{2}, & \text{if } z = a_j \text{ or } z = b_j, \\ 1, & \text{if } \quad a_j < z < b_j. \end{cases}$$

Since, by assumption, a_j and b_j are continuity points of the marginal distribution function of X_j for each $j \in \{1, \ldots, d\}$, we have $\mathbb{P}^X(\partial B) = 0$ and ∂B is the boundary of B. Lebesgue's dominated convergence theorem (Theorem 1.31) now yields

$$\lim_{T \to \infty} I(T) = \int_{\mathbb{R}^d} \prod_{j=1}^d \lim_{T \to \infty} g_j(x_j, T) \, \mathbb{P}^X(dx)$$

$$= \int_{\mathbb{R}^d} \prod_{j=1}^d \mathbf{1}\{x_j \in (a_j, b_j)\} \, \mathbb{P}^X(dx)$$

$$= \mathbb{P}^X(B \setminus \partial B) = \mathbb{P}^X(B).$$

$\square$

Since the set of continuity points of a distribution function is dense in $\mathbb{R}$, the collection of d-dimensional rectangles B satisfying (1.6) meets the conditions of the uniqueness theorem for measures (Theorem 1.28). We thus obtain the following result.

1.16 Corollary *If X and Y are d-dimensional random vectors with characteristic functions φ_X and φ_Y, respectively, then*

$$X \stackrel{\mathcal{D}}{=} Y \Longleftrightarrow \varphi_X = \varphi_Y.$$

The next theorem, which lays the basis for computer tomography, follows readily from Corollary 1.16.

1.17 Theorem (Herglotz[10]–Radon[11]–Cramér–Wold[12])
If X and Y are d-dimensional random vectors, then

$$X \stackrel{\mathcal{D}}{=} Y \Longleftrightarrow t^\top X \stackrel{\mathcal{D}}{=} t^\top Y \quad \text{for every } t \in \mathbb{R}^d.$$

Proof To prove the implication "$\Longleftarrow$", we write φ_Z for the characteristic function of a random variable or a random vector Z. Then, for each $t \in \mathbb{R}^d$,

$$\varphi_X(t) = \mathbb{E}\big[e^{it^\top X}\big] = \mathbb{E}\big[e^{i \cdot 1 \cdot t^\top X}\big] = \varphi_{t^\top X}(1)$$
$$= \varphi_{t^\top Y}(1) = \mathbb{E}\big[e^{i \cdot 1 \cdot t^\top Y}\big] = \mathbb{E}\big[e^{it^\top Y}\big]$$
$$= \varphi_Y(t).$$

Here, the fourth equal sign is justified by the assumption $t^\top X \stackrel{\mathcal{D}}{=} t^\top Y$. The claim now follows from Corollary 1.16.   □

The following limit theorems for sums of independent random variables are the core of classical probability theory.

---

[10] Gustav Herglotz (1881–1953), professor at the universities of Göttingen (1907), Vienna (1908), Leipzig (1909–1925) and Göttingen (1925–1947). Main areas of work: Complex analysis, potential theory, differential geometry, number theory, differential equations, flow problems.

[11] Johann Karl August Radon (1887–1956), professor at the universities of Hamburg (from 1919), Greifswald (from 1922), Erlangen (from 1925), Breslau (from 1928), and Vienna (from 1947). Main areas of work: calculus of variations, differential geometry, absolutely additive set functions.

[12] Herman Herman Ole Andreas Wold (1908–1992), professor of Mathematical Statistics at the Universities of Uppsala (1942–1970) and Gothenburg (1970–1975). Main areas of work: Mathematical Statistics, econometrics.

1.18 Theorem (Central Limit Theorem, Lindeberg[13]–Lévy)

Let $X_1, X_2, \ldots$ *be i.i.d. random variables with* $\mathbb{E}(X_1^2) < \infty$ *and* $\mathbb{V}(X_1) > 0$, *and put* $S_n := X_1 + \ldots + X_n, n \geq 1$. *Then*

$$\frac{S_n - \mathbb{E}(S_n)}{\sqrt{\mathbb{V}(S_n)}} \xrightarrow{\mathcal{D}} N(0,1) \quad as \; n \to \infty.$$

1.19 Theorem (Central Limit Theorem, Lindeberg–Feller[14]) *For each* $n \geq 1$, *let* $X_{n,1}, X_{n,2}, \ldots, X_{n,r_n}$ *be independent random variables with* $\mathbb{E}(X_{n,j}^2) < \infty$ *for each* $j \in \{1, \ldots, r_n\}$. *Put* $S_n := X_{n,1} + \ldots + X_{n,r_n}, n \geq 1$, *and assume* $\sigma_n^2 := \mathbb{V}(S_n) > 0$. *Furthermore, let*

$$L_n(\varepsilon) := \frac{1}{\sigma_n^2} \sum_{k=1}^{r_n} \mathbb{E}\left[\left(X_{n,k} - \mathbb{E}X_{n,k}\right)^2 \mathbf{1}\{|X_{n,k} - \mathbb{E}X_{n,k}| > \varepsilon\sigma_n\}\right], \quad \varepsilon > 0.$$

If the so-called Lindeberg condition

$$\lim_{n\to\infty} L_n(\varepsilon) = 0 \quad for \; every \; \varepsilon > 0 \tag{1.9}$$

holds, then

$$\frac{S_n - \mathbb{E}(S_n)}{\sqrt{\mathbb{V}(S_n)}} \xrightarrow{\mathcal{D}} N(0,1) \quad as \; n \to \infty. \tag{1.10}$$

1.20 Theorem (Central Limit Theorem, Lyapunov[15]) *Assume the situation of Theorem 1.19. If, for some* $\delta > 0$,

$$\lim_{n\to\infty} \frac{1}{\sigma_n^{2+\delta}} \sum_{k=1}^{r_n} \mathbb{E}|X_{n,k} - \mathbb{E}X_{n,k}|^{2+\delta} = 0, \tag{1.11}$$

then (1.10) *holds.*

Self-question 4: Why does the Lindeberg condition (1.9) follow from (1.11)?

---

[13] Jarl Waldemar Lindeberg (1876–1932), Finnish farmer and mathematician.

[14] William Feller (1906–1970), 1928 private lecturer at the University of Kiel, 1933 emigration to Copenhagen and 1934–1939 to Stockholm, professor at Brown University Providence (1939–1945), Cornell University (1945–1059), Princeton University (1950–1970). Main areas of work: Analysis, geometry, functional analysis, measure theory, probability theory.

[15] Alexander Mikhailovich Lyapunov (1857–1918), 1885 lecturer and 1892 professor at the University of Kharkov, from 1901 professor at the University of St. Petersburg. Main areas of work: Stability theory of mechanical systems, probability theory, potential theory.

1.21 On the Speed of Convergence in the Central Limit Theorem

There are numerous studies on the speed of convergence of the distribution function G_n (say) of $S_n^* := (S_n - \mathbb{E}(S_n))/\sqrt{\mathbb{V}(S_n)}$ to the distribution function Φ of the standard normal distribution. The theorem of Berry[16]–Esseen[17] (see [BER] and [ESS]), discovered independently of each other, states that in the situation of the Lindeberg–Lévy central limit theorem, under the additional assumption $\mathbb{E}|X_1|^3 < \infty$, there exists a universal positive constant C that does not depend on the distribution of X_1, such that

$$\sup_{x \in \mathbb{R}} \left| G_n(x) - \Phi(x) \right| \leq \frac{C}{\sqrt{n}} \cdot \mathbb{E} \left| \frac{X_1 - \mathbb{E}(X_1)}{\sqrt{\mathbb{V}(X_1)}} \right|^3. \tag{1.12}$$

It is known that

$$0.4097 \cdots = \frac{\sqrt{10} + 3}{6\sqrt{2\pi}} \leq C \leq 0.4690.$$

The upper bound for C has been improved over the course of 80 years from originally 7.59 in [ESS] (see, e.g., [ZO1], [ZO2], and [VBE]) to 0.4690 (see [SHE]).

There are also upper bounds for the supremum distance between G_n and Φ in the more general situation of the Lindeberg–Feller theorem, wherein we may w.l.o.g. let $\sigma_n^2 = 1$ and $\mathbb{E}(X_{n,j}) = 0$ for each $j \in \{1, \ldots, n\}$. If, in addition, we assume that third moments exist, then

$$\sup_{x \in \mathbb{R}} \left| G_n(x) - \Phi(x) \right| \leq 6 \sum_{j=1}^{n} \mathbb{E}|X_j|^3 \tag{1.13}$$

(see, e.g., [FEL], Section XVI.5, p. 544). The right-hand sides in (1.12) and (1.13) are comprehensible due to a so-called *Edgeworth expansion* (see, e.g., [FEL], Section XIV.4.6). For further results see, e.g., [PET].

The book [CGS], which promotes the method tracing back to C. Stein[18] for error estimation in approximations of distributions by the normal distribution (see [STE]), contains a weaker result (see their Theorem 3.6 and the constant 9.4), without mentioning the better estimate (1.13). The strength of Stein's method lies primarily in the fact that it can often be better adapted to analogous situations with *dependent* summands.

---

[16] Andrew Campbell Berry (1906–1998), American mathematician, from 1941 professor at Lawrence College, Appleton, Wisconsin. Main areas of work: Fourier transformation, probability theory.

[17] Carl-Gustav Esseen (1918–2001), Swedish mathematician, professor at the Royal Technical College in Stockholm and at the University of Uppsala. Main area of work: Probability theory.

[18] Charles M. Stein (1920–2016), American statistician, from 1953 professor at the University of Stanford, California. Named after him are the *Stein's method*, the *Stein's lemma* and the *Stein's paradox*, among others. Stein was a declared pacifist. Due to protests against apartheid politics, he was the first member of his university to be arrested.

As the following result shows, convergence in distribution is inherited under continuous mappings.

1.22 Theorem (Mapping Theorem)

Let $X_n \xrightarrow{\mathcal{D}} X$. If $h : \mathbb{R} \to \mathbb{R}$ is a continuous function, then $h(X_n) \xrightarrow{\mathcal{D}} h(X)$.

We will see that the statement of the mapping theorem remains valid in a much more general framework under a weaker assumption (Theorems 6.6 and 14.4). The same applies to the following result, which is revisited in Chap. 6 and in Chap. 14.

1.23 Theorem (Slutsky's[19] Lemma)

If $X_n \xrightarrow{\mathcal{D}} X$ and $Y_n \xrightarrow{\mathbb{P}} a$, where $a \in \mathbb{R}$, then

(a) $X_n + Y_n \xrightarrow{\mathcal{D}} X + a$,

(b) $X_n \cdot Y_n \xrightarrow{\mathcal{D}} a X$.

Part (a) of Slutsky's lemma conveys the following special message: If one wants to prove convergence in distribution of a sequence (Z_n) of random variables Z_n, and if one can write Z_n as a sum $X_n + Y_n$, where Y_n converges to 0 in probability, then one only needs to worry about the sequence (X_n) (which may be much easier to handle than Z_n).

Although, at least at first sight, convergence in distribution seems to have little in common with almost sure convergence, there is a direct connection between these two notions of convergence, as the following theorem, which is attributed to the Ukrainian mathematician A. V. Skorokhod,[20] shows. We will use this result in Chap. 3.

1.24 Theorem (Skorokhod)

Let $X, X_1, X_2, \ldots$ be random variables on a probability space $(\Omega, \mathcal{A}, \mathbb{P})$ with $X_n \xrightarrow{\mathcal{D}} X$. Then there exists a probability space $(\widetilde{\Omega}, \widetilde{\mathcal{A}}, \widetilde{\mathbb{P}})$ and random variables $Y, Y_1, Y_2, \ldots$ on $\widetilde{\Omega}$ with

$$\widetilde{\mathbb{P}}^Y = \mathbb{P}^X, \quad \widetilde{\mathbb{P}}^{Y_n} = \mathbb{P}^{X_n} \text{ for every } n$$

(and thus, in particular, $Y_n \xrightarrow{\mathcal{D}} Y$) and $\lim_{n\to\infty} Y_n = Y$ $\widetilde{\mathbb{P}}$-almost surely.

---

[19] Evgeny Evgenyevich Slutsky (1880–1948), frome 1926 scientist at the Center for Statistics and from 1931 at the Central Institute for Meteorology in Moscow. From 1934 he worked at the Mathematical Institute of the Academy of Sciences. Main areas of work: Statistics, especially time series analysis.

[20] Anatoliy Volodymyrovych Skorokhod (1930–2011), Ukrainian mathematician, professor at the University of Kiev (from 1964) and at Michigan State University (from 1993), member of the National Academy of Sciences of Ukraine from 1985. Main areas of work: Stochastic differential equations, limit theorems for stochastic processes.

The following concept and the subsequent Theorems 1.26 and 1.27 will be needed in Chap. 8.

1.25 Definition (Conditional Expectation)

Let $(\Omega, \mathcal{A}, \mathbb{P})$ be a probability space, $X \in \mathcal{L}^1(\Omega, \mathcal{A}, \mathbb{P})$ and $\mathcal{G}$ a sub-σ-field of $\mathcal{A}$.

A random variable Y is called *conditional expectation of X given $\mathcal{G}$*, and we write $Y = \mathbb{E}[X|\mathcal{G}]$, if the following conditions hold:

(a) $\mathbb{E}|Y| < \infty$,
(b) Y is $\mathcal{G}$-measurable,
(c) $\mathbb{E}(Y\mathbf{1}_A) = \mathbb{E}(X\mathbf{1}_A)$ for all $A \in \mathcal{G}$.

A conditional expectation $\mathbb{E}[X|\mathcal{G}]$ in the above sense exists, and it is uniquely determined $\mathbb{P}$-almost surely. Due to the latter property, $\mathbb{E}[X|\mathcal{G}]$ is, strictly speaking, the *set* of all random variables Y with properties (a)–(c). Therefore, each of the following equations is to be understood to hold $\mathbb{P}$-almost surely.

1.26 Theorem (Properties of Conditional Expectations)

Let $(\Omega, \mathcal{A}, \mathbb{P})$ be a probability space, $\mathcal{G}$ a sub-σ-field of $\mathcal{A}$ and X and Y random variables on Ω with $\mathbb{E}|X| < \infty$ and $\mathbb{E}|Y| < \infty$.

(a) $\mathbb{E}\left(\mathbb{E}[X|\mathcal{G}]\right) = \mathbb{E}X$ (law of iterated expectation).
(b) If X is $\mathcal{G}$-measurable then $\mathbb{E}[X|\mathcal{G}] = X$.
(c) $\mathbb{E}[aX + bY|\mathcal{G}] = a\,\mathbb{E}[X|\mathcal{G}] + b\,\mathbb{E}[Y|\mathcal{G}], \quad a, b \in \mathbb{R}$.
(d) If $\mathbb{E}|XY| < \infty$ and Y is $\mathcal{G}$-measurable then $\mathbb{E}[XY|\mathcal{G}] = Y\,\mathbb{E}[X|\mathcal{G}]$.
(e) If $\mathcal{F} \subset \mathcal{G}$ is a sub-σ-field of $\mathcal{G}$ then $\mathbb{E}[X|\mathcal{F}] = \mathbb{E}\left[\mathbb{E}[X|\mathcal{G}]|\mathcal{F}\right]$.
(f) If X and $\mathcal{G}$ are independent then $\mathbb{E}[X|\mathcal{G}] = \mathbb{E}(X)$.

1.27 Theorem (Factorization Theorem)

Let $(\Omega', \mathcal{A}')$ be a measurable space (that is, a set Ω', equipped with a σ-field $\mathcal{A}'$). If $\mathcal{G} = \sigma(Z) = Z^{-1}(\mathcal{A}')$ for an $(\mathcal{A}, \mathcal{A}')$-measurable mapping $Z : \Omega \to \Omega'$, then there is a measurable function $h : \Omega' \to \overline{\mathbb{R}}$ such that

$$\mathbb{E}[X|\mathcal{G}] = \mathbb{E}[X|\sigma(Z)] =: \mathbb{E}[X|Z] = h(Z).$$

We finally list some important results from measure theory and integration theory. To this end, we fix a measure space $(\Omega, \mathcal{A}, \mu)$. A class $\mathcal{M}$ of subsets of Ω is called a *π-system* if it is closed under intersection, that is, if $A, B \in \mathcal{M}$ then $A \cap B \in \mathcal{M}$. The following result is needed at several places.

1.28 Theorem (Uniqueness Theorem for Measures)

Let $(\Omega, \mathcal{A})$ be a measurable space and $\mathcal{M} \subset \mathcal{A}$, such that $\sigma(\mathcal{M}) = \mathcal{A}$ and $\mathcal{M}$ is a π-system. Furthermore, let μ_1 and μ_2 be measures on $\mathcal{A}$ with $\mu_1(M) = \mu_2(M)$ for each

$M \in \mathcal{M}$. If there is an increasing sequence (M_n) of sets in $\mathcal{M}$ (which means that $M_n \subset M_{n+1}$ for each $n \geq 1$) with $\Omega = \cup_{n=1}^{\infty} M_n$ and $\mu_1(M_n) < \infty$, $n \geq 1$, then $\mu_1 = \mu_2$.

To state three important theorems from the theory of integration, an $\mathcal{A}$-measurable *numerical* function on Ω is a function $f : \Omega \to \overline{\mathbb{R}} := \mathbb{R} \cup \{\infty, -\infty\}$, which is measurable with respect to the σ-field $\overline{\mathcal{B}} := \{B \cup E : B \in \mathcal{B}, E \subset \{\infty, -\infty\}\}$ of Borel sets on $\overline{\mathbb{R}}$.

1.29 Theorem (Fatou's Lemma[21])

If $(f_n)_{n \geq 1}$ is a sequence of non-negative $\mathcal{A}$-measurable numerical functions on Ω then

$$\int \liminf_{n \to \infty} f_n \, d\mu \leq \liminf_{n \to \infty} \int f_n \, d\mu. \qquad (1.14)$$

Self-question 5: Does (1.14) also hold if $f_n \geq c$, $n \geq 1$, for some constant $c < 0$?

1.30 Theorem (Monotone Convergence, Beppo Levi[22])

If $(f_n)_{n \geq 1}$ is an increasing sequence of non-negative $\mathcal{A}$-measurable numerical functions on Ω (that is, $f_n \leq f_{n+1}$ for each $n \geq 1$), then

$$\int \lim_{n \to \infty} f_n \, d\mu = \lim_{n \to \infty} \int f_n \, d\mu.$$

Self-question 6: Can the condition $f_n \leq f_{n+1}$, $n \geq 1$, be weakened?

Let E be a property that, for each ω in Ω, either holds or does not hold. We say that E holds μ-everywhere (in short: μ-a.e.), if it holds for each ω in the complement of a set $N \in \mathcal{A}$ with $\mu(N) = 0$ (a so-called μ-*null set*). If $\mu = \mathbb{P}$ is a probability measure, then E holds $\mathbb{P}$-almost surely (in short: $\mathbb{P}$-a.s.), if it holds for each ω in a set $\Omega_0 \in \mathcal{A}$ with $\mathbb{P}(\Omega_0) = 1$.

---

[21] Pierre Joseph Louis Fatou (1878–1929), from 1901 at the astronomical observatory in Paris. Main areas of work (in addition to astronomical research): Complex analysis, functional equations.

[22] Beppo Levi (1875–1961), professor at the universities of Piacenza (from 1901), Cagliari (from 1906), Parma (from 1910) and Bologna (from 1928), from 1939 honorary professor at the University of Rosario. Main areas of work: Algebraic geometry, analysis, number theory, partial differential equations.

1.31 Theorem (Dominated Convergence, H. Lebesgue[23])
Let $f, f_1, f_2, \ldots$ be $\mathcal{A}$-measurable numerical functions on Ω with

$$f = \lim_{n \to \infty} f_n \quad \mu\text{-a.e.}$$

If there is a μ-integrable numerical function g with $|f_n| \leq g$ μ-a.e. for each $n \geq 1$, then f is μ-integrable, and

$$\int f \, d\mu = \lim_{n \to \infty} \int f_n \, d\mu.$$

In the special case of a probability space $(\Omega, \mathcal{A}, \mathbb{P})$, one faces a different notation, inasmuch as the functions f_n are termed *random variables* and are denoted X_n. Moreover, integration is written using the expectation operator $\mathbb{E}(\cdot)$. For non-negative random variables X_n, Fatou's lemma then takes the form

$$\mathbb{E}\left(\liminf_{n \to \infty} X_n \right) \leq \liminf_{n \to \infty} \mathbb{E}(X_n),$$

and the monotone convergence theorem states that, if $0 \leq X_n \leq X_{n+1}$ $\mathbb{P}$-a.s. for each n, then

$$\mathbb{E}\left(\lim_{n \to \infty} X_n \right) = \lim_{n \to \infty} \mathbb{E}(X_n).$$

If the sequence (X_n) converges $\mathbb{P}$-a.s. to a random variable X, and if there is a random variable Y with $|X_n| \leq Y$ $\mathbb{P}$-a.s. for each n, the dominated convergence theorem states that $\mathbb{E}|X| < \infty$ and

$$\mathbb{E}(X) = \lim_{n \to \infty} \mathbb{E}(X_n).$$

A measure μ on $\mathcal{A}$ is said to be *σ-finite*, if there is an *increasing* sequence (A_n) of sets in $\mathcal{A}$ with $\cup_{n=1}^{\infty} A_n = \Omega$ and $\mu(A_n) < \infty$ for each $n \geq 1$. Suppose ν is another measure on $\mathcal{A}$. ν is called *absolutely continuous* with respect to μ (in short: $\nu \ll \mu$), if, for each $A \in \mathcal{A}$, $\mu(A) = 0$ implies $\nu(A) = 0$. An equivalent wording is that *the measure μ dominates the measure ν.* In this situation, there is the following fundamental theorem (which, in particular, guarantees the existence of conditional expectations).

---

[23] Henri Léon Lebesgue (1875–1941), 1919 professor at the Sorbonne, from 1921 professor at the Collège de France. Main areas of work: Real analysis, measure and integration theory, topology.

1.32 Theorem (Radon–Nikodým[24])

If the measure μ is σ-finite, then the following assertions are equivalent:

(a) $\nu \ll \mu$.
(b) There exists a non-negative measurable function f on Ω with

$$\nu(A) = \int_A f \, d\mu, \qquad A \in \mathcal{A}.$$

The function f is called a *density of ν with respect to μ*. If $\nu(\Omega) < \infty$, f is uniquely determined μ-a.e.

If $(\Omega_1, \mathcal{A}_1, \mu_1)$ and $(\Omega_2, \mathcal{A}_2, \mu_2)$ are measure spaces, where μ_1 and μ_2 are σ-finite, then there exists exactly one measure $\mu =: \mu_1 \otimes \mu_2$ on the product σ-field $\mathcal{A}_1 \otimes \mathcal{A}_2 =: \mathcal{A}$ on $\Omega := \Omega_1 \times \Omega_2$ with $\mu(A_1 \times A_2) = \mu_1(A_1)\mu_2(A_2)$ for any choice of $A_j \in \mathcal{A}_j$ ($j \in \{1, 2\}$). The following theorems refer to the integration with respect to this so-called *product measure*.

1.33 Theorem (Tonelli[25])

If $f : \Omega \to \overline{\mathbb{R}}$ is a non-negative $\mathcal{A}_1 \otimes \mathcal{A}_2$-measurable function, then

$$\int_\Omega f \, d(\mu_1 \otimes \mu_2) = \int_{\Omega_2} \left(\int_{\Omega_1} f(\omega_1, \omega_2)\mu_1(d\omega_1) \right) \mu_2(d\omega_2) \qquad (1.15)$$

$$= \int_{\Omega_1} \left(\int_{\Omega_2} f(\omega_1, \omega_2)\mu_2(d\omega_2) \right) \mu_1(d\omega_1). \qquad (1.16)$$

Here, the inner integrals are $\mathcal{A}_2$- and $\mathcal{A}_1$-measurable functions, respectively.

1.34 Theorem (Fubini[26])

If $f : \Omega \to \overline{\mathbb{R}}$ is an arbitrary $\mu_1 \otimes \mu_2$-integrable $\mathcal{A}_1 \otimes \mathcal{A}_2$-measurable function, then the function $\Omega_2 \ni \omega_2 \mapsto f(\omega_1, \omega_2)$ is μ_2-integrable for almost all $\omega_1 \in \Omega_1$ with respect to μ_1, and the function $\Omega_1 \ni \omega_1 \mapsto f(\omega_1, \omega_2)$ is μ_1-integrable for almost all $\omega_2 \in \Omega_2$

---

[24] Otton Martin Nikodým (1887–1974), lecturer at the universities of Warsaw and Krakow (1927–1939), professor at Kenyon College in Gambier (USA) (1947–1965). Main areas of work: functional analysis, measure theory, set-theoretic topology.

[25] Leonida Tonelli (1885–1946), professor at the universities of Cagliari (from 1913), Parma (from 1915), Pisa (from 1930 with a break in Rome (1939–1942)). Main area of work: Calculus of variations.

[26] Guido Fubini (1879–1943), professor at the universities of Turin (from 1910) and Princeton (from 1943). Main areas of work: Projective differential geometry, automorphic functions, discontinuous groups.

with respect to μ_2. The μ_1-a.e. and μ_2-a.e. defined functions $\omega_1 \mapsto \int f(\omega_1, \cdot)d\mu_2$ and $\omega_2 \mapsto \int f(\cdot, \omega_2)d\mu_1$ are μ_1- and μ_2-integrable, respectively, and (1.15) and (1.16) hold.

Answers to the Self-Questions

Answer 1 Let $\omega_0 \in \Omega$ with $\mathbb{P}(\{\omega_0\}) > 0$ and $\varepsilon > 0$ be arbitrary. By assumption, $\mathbb{P}(A_n(\varepsilon)) \to 0$, with $A_n(\varepsilon) := \{\omega \in \Omega : |X_n(\omega) - X(\omega)| > \varepsilon\}$. For sufficiently large n, $\omega_0 \notin A_n(\varepsilon)$, that is, $|X_n(\omega_0) - X(\omega_0)| \leq \varepsilon$, and thus $X_n \xrightarrow{\text{a.s.}} X$.

Answer 2 Because any two norms $\|\cdot\|_1$ and $\|\cdot\|_2$ are equivalent, that is, there exist positive constants K and L with $\|x\|_1 \leq K\|x\|_2 \leq L\|x\|_1$ for all $x \in \mathbb{R}^d$ (see, e.g., [WEI], Theorem 3.7.2).

Answer 3 The set $\mathcal{C}(F)$ of continuity points of F is dense in $\mathbb{R}$. Bearing this in mind, we let $a = a_n$ in (1.4) be a sequence in $\mathcal{C}(F)$ with $a_n \to -\infty$. As a consequence, $F(b)$ is known for each $b \in \mathcal{C}(F)$ and thus, in view of continuity from the right, for every $b \in \mathbb{R}$.

Answer 4 The reason is the inequality

$$(x - a)^2 \mathbf{1}\{|x - a| > \varepsilon\sigma\} \leq \frac{|x - a|^{2+\delta}}{(\varepsilon\sigma)^\delta}, \qquad x, a \in \mathbb{R}, \ \varepsilon, \sigma > 0.$$

Answer 5 Not in general, but when μ is a finite measure, since in that case one can apply Fatou's lemma to the functions $f_n - c$.

Answer 6 The inequality $f_n \leq f_{n+1}$ only needs to hold μ-a.e. for each n, since the μ-integral does not change if the integrand is modified on a μ-null set.

Problems

1.1 Problem Let $X_n = (X_n^{(1)}, \ldots, X_n^{(d)})$, $n \geq 1$, and $X = (X^{(1)}, \ldots, X^{(d)})$ be d-dimensional random vectors. Prove:

$$X_n \xrightarrow{\mathbb{P}} X \iff X_n^{(j)} \xrightarrow{\mathbb{P}} X^{(j)} \text{ for each } j \in \{1, \ldots, d\}.$$

1.2 Problem Almost sure convergence does not necessarily follow from

(a) convergence in probability,
(b) $\mathcal{L}^p$- convergence for some $p > 0$.

Provide counterexamples.

Hint Use the probability space $(\Omega, \mathcal{A}, \mathbb{P})$, in which $\Omega := [0, 1]$, $\mathcal{A} := \Omega \cap \mathcal{B}$, $\mathbb{P} := \lambda^1_{|[0,1]}$, and employ suitable indicator functions.

1.3 Problem State and prove a strong law of large numbers for random vectors.

1.4 Problem Show that the characteristic function of a random variable with the normal distribution $N(\mu, \sigma^2)$ is given by (1.5).

Hint Use (1.3) and the fact that the density f of the standard normal distribution satisfies the differential equation $f'(x) = -xf(x)$, $x \in \mathbb{R}$.

1.5 Problem Assume $X_n \xrightarrow{\mathcal{D}} X$, and suppose t is a point of continuity of the distribution function F of X. Prove the following: If (t_n) is a sequence and $\lim_{n \to \infty} t_n = t$, then

$$\lim_{n \to \infty} \mathbb{P}(X_n \le t_n) = F(t).$$

1.6 Problem For random variables $Z_n \sim \text{Bin}(n, p_n)$, in which $0 < p_n < 1$ for each $n \ge 1$ and $\lim_{n \to \infty} p_n =: p \in (0, 1)$, show that

$$\frac{Z_n - np_n}{\sqrt{np_n(1 - p_n)}} \xrightarrow{\mathcal{D}} N(0, 1) \quad \text{as } n \to \infty.$$

1.7 Problem Suppose $X_n \sim N(\mu_n, \sigma_n^2)$, $n \ge 1$, are normally distributed random variables, X is a random variable with a non-degenerate distribution, and $X_n \xrightarrow{\mathcal{D}} X$. Prove:

(a) The sequences (μ_n) and (σ_n^2) are bounded.
(b) The limits $\mu := \lim_{n \to \infty} \mu_n$ and $\sigma^2 := \lim_{n \to \infty} \sigma_n^2$ exist, and $X \sim N(\mu, \sigma^2)$.

1.8 Problem Let $(\Omega, \mathcal{A}, \mu)$ be a measure space, $(f_n)_{n \ge 1}$ a sequence of *non-negative* $\mathcal{A}$-measurable numerical functions on Ω, and $g : \Omega \to \overline{\mathbb{R}}$ an $\mathcal{A}$-measurable μ-integrable function satisfying $f_n \le g$ for each $n \ge 1$. Prove the following variant of Fatou's lemma:

$$\limsup_{n \to \infty} \int f_n \, d\mu \le \int \limsup_{n \to \infty} f_n \, d\mu.$$

Problems

Note: Also throughout the proof spaces $(\Omega, \mathcal{F}, P)$ as σ-algebras. $\mathbb{R} = \{x_1, x_2, \ldots\} \subset (\Omega, \mathcal{F})$...
and easily verify reflexive and transitive.

4.5 Problem Compute the σ-algebra $\mathcal{F}$ of all Borel ... Borel subsets ... for random vectors.

4.6 Problem Show that the σ-algebras are linear transformations measurable with the general definition of ...

Hint: ... and that the range σ of the Borel set ... of the discrete random distribution variable ...
interval ... integral $\int_{\mathbb{R}} f \, dP$ has a ...

4.7 Problem ... Let X, Y ... and Z ... a joint probability density from the distribution ... σ-algebra $\mathcal{F}$ ... integrals ...

$$E[f(X)] = \int_{\mathbb{R}} f \, dP$$

4.8 Problem Let the random variables $X_1, X_2, \ldots, X_n$ be i.i.d. ... μ_X ... so that $P(\Omega) = 1$ ... for each $x > 0$ and let $P(x_i = a, x_i = b) = 0$ for all i ... for ...

$$\int_{\Omega} \ldots = \lim_{n \to \infty} \frac{1}{n} \sum_{i=1}^{n} ...$$

4.9 Problem Suppose $X_n \to X$ a.s. (i.e. ...) with ... with probability distributed random variables ...
$X, X_1, X_2, \ldots$ and that $E[X_n] \to E[X]$ if ... $X_n \leq X$ a.s. and ...

(a) Let $\sup_n E[x_n]$ and ... $\leq a$ a density ...

... random variables ... $\lim$ ... $X_n \to X$ ... $E[X_n] \to E[X]$ ...

By construction $p(x) = \frac{d}{dx} P(X \leq x)$. Prove the following verification: an expectation

$$\lim_{n \to \infty} \int_{\Omega} ... = \int_{\Omega} \lim ...$$

For each $n \geq 1$, let X_n be a random variable with the binomial distribution $\text{Bin}(n, p_n)$, that is, $\mathbb{P}(X_n = k) = \binom{n}{k} p_n^k (1 - p_n)^{n-k}$, $k \in \{0, \ldots, n\}$. In an introductory course on probability theory, one learns that, if

$$\lim_{n \to \infty} n p_n = \lambda \tag{2.1}$$

for some $\lambda \in (0, \infty)$, then X_n converges in distribution to a random variable with the Poisson[1] distribution $\text{Po}(\lambda)$, that is,

$$\lim_{n \to \infty} \mathbb{P}(X_n = k) = e^{-\lambda} \frac{\lambda^k}{k!}, \quad k = 0, 1, 2, \ldots \tag{2.2}$$

This limit theorem is often referred to as the *law of rare events*. This naming is due to the equality in distribution $X_n \overset{\mathcal{D}}{=} \mathbf{1}\{A_{n,1}\} + \ldots + \mathbf{1}\{A_{n,n}\}$, in which $A_{n,1}, \ldots, A_{n,n}$ are independent events that have the same probability p_n and, in view of (2.1), this probability is of the order n^{-1} as n increases. In what follows, we ask whether such a law of rare events continues to hold under weaker conditions on $A_{n,1}, \ldots, A_{n,n}$. We will maintain the assumption of independence, but weaken the condition $\mathbb{P}(A_{n,1}) = \ldots = \mathbb{P}(A_{n,n})$. In addition, the random variable X_n can take the form

$$X_n = X_{n,1} + \ldots + X_{n,n}$$

---

[1] Siméon Denise Poisson (1781–1840), from 1806 professor at the École Polytechnique. Poisson made important contributions, especially to mathematical physics and analysis. In 1827, he was appointed as the geometer of the Bureau of Lengths in place of the deceased P.S. Laplace.

© The Author(s), under exclusive license to Springer-Verlag GmbH, DE, part of Springer Nature 2024
N. Henze, *Asymptotic Stochastics*, Mathematics Study Resources 10,
https://doi.org/10.1007/978-3-662-68923-3_2

with independent and $\mathbb{N}_0$-valued random variables $X_{n,1}, \ldots, X_{n,n}$. So, as with the Lindeberg–Feller central limit theorem (Theorem 1.19), we assume a *triangular array* $(X_{n,j} : 1 \leq j \leq n)_{n \geq 1}$ of row-wise independent random variables. Whereas in that limit theorem the Lindeberg condition (1.9) ensures that none of the summands has an overwhelming influence on the sum $X_{n,1} + \ldots + X_{n,n}$, the following condition is appropriate in the present situation.

2.1 Definition (Asymptotic Negligibility)

Let $\Delta := (X_{n,j} : 1 \leq j \leq n)_{n \geq 1}$ be a triangular array of real-valued random variables.
    Δ is called *asymptotically negligible* or a *null array*, if

$$\lim_{n \to \infty} \max_{1 \leq j \leq n} \mathbb{P}(|X_{n,j}| > \varepsilon) = 0 \text{ for every } \varepsilon > 0. \tag{2.3}$$

Condition (2.3) is equivalent to

$$X_{n,k_n} \xrightarrow{\mathbb{P}} 0 \text{ for each subsequence } (k_n) \text{ with } k_n \in \{1, \ldots, n\} \text{ for all } n \geq 1 \tag{2.4}$$

(Problem 2.1).
    In what follows, we get to know equivalent conditions to (2.3). For the first one, we use the common notation $x \wedge y := \min(x, y)$ for the minimum of two real numbers x and y.

2.2 Proposition *The following statements are equivalent:*

(a) $\lim_{n \to \infty} \max_{1 \leq j \leq n} \mathbb{P}(|X_{n,j}| > \varepsilon) = 0$ *for each $\varepsilon > 0$.*
(b) $\lim_{n \to \infty} \max_{1 \leq j \leq n} \mathbb{E}[|X_{n,j}| \wedge 1] = 0.$

Proof

"(b) $\Longrightarrow$ (a)":   Since the probability $\mathbb{P}(|X_{n,j}| > \varepsilon)$ increases when ε decreases, it is
    sufficient to consider the case $0 < \varepsilon \leq 1$. For such ε, the inequality

$$\mathbf{1}\{|X_{n,j}| > \varepsilon\} \leq \frac{1}{\varepsilon}(|X_{n,j}| \wedge 1)$$

    holds, and thus the claim follows by first taking expectations and then the maximum
    over j.
"(a) $\Longrightarrow$ (b)":   Fix $\varepsilon > 0$. Since $|X_{n,j}| \wedge 1 \leq 1$, we obtain

$$\mathbb{E}[|X_{n,j}| \wedge 1] = \mathbb{E}[(|X_{n,j}| \wedge 1)\mathbf{1}\{|X_{n,j}| > \varepsilon\}] + \mathbb{E}[(|X_{n,j}| \wedge 1)\mathbf{1}\{|X_{n,j}| \leq \varepsilon\}]$$
$$\leq \mathbb{P}(|X_{n,j}| > \varepsilon) + (\varepsilon \wedge 1),$$

    and the claim follows by taking the maximum over j.

$\square$

The main result of this chapter is the following Poisson limit theorem for triangular arrays. This theorem is remarkable in the sense that, under the assumption of asymptotic negligibility, it provides a *necessary and sufficient condition* for convergence in distribution to a Poisson distribution. Among other things, it is a generalization of the law of rare events (2.2) (Problem 2.2 (a) and Problem 2.3). Problems 2.6 and 2.7 contain applications to extreme value stochastics.

2.3 Theorem (Poisson Limit Theorem, cf. [KAL], Theorem 6.7)
Let $\left(X_{n,j} : n \geq 1, \ 1 \leq j \leq n\right)_{n \geq 1}$ be a null array of row-wise independent $\mathbb{N}_0$-valued random variables and $X \sim Po(\lambda)$. We then have:

$$X_{n,1} + \ldots + X_{n,n} \xrightarrow{\mathcal{D}} X \iff \text{(i)} \lim_{n \to \infty} \sum_{j=1}^{n} \mathbb{P}(X_{n,j} > 1) = 0,$$

$$\text{(ii)} \lim_{n \to \infty} \sum_{j=1}^{n} \mathbb{P}(X_{n,j} = 1) = \lambda.$$

We will prove this theorem using generating functions. For a $\mathbb{N}_0$-valued random variable X, the *generating function* of X is defined by the power series g_X, with

$$g_X(s) := \sum_{k=0}^{\infty} \mathbb{P}(X = k)s^k, \quad |s| \leq 1. \tag{2.5}$$

The distribution of X is uniquely determined by g_X, and if X has the Poisson distribution $Po(\lambda)$, then

$$g_X(s) = e^{\lambda(s-1)}, \quad s \in \mathbb{R}.$$

Furthermore, the *multiplication rule for generating functions* applies: If X and Y are independent $\mathbb{N}_0$-valued random variables, then $g_{X+Y}(s) = g_X(s)g_Y(s), |s| \leq 1$.

Self-question 1: Can you prove this multiplication rule?

The proof of the Poisson limit theorem uses the following *continuity theorem for generating functions*.

2.4 Theorem (Continuity Theorem for Generating Functions)
Let $X_0, X_1, \ldots$ be $\mathbb{N}_0$-valued random variables with generating functions $g_0, g_1, \ldots,$ respectively. Then the following statements are equivalent:

(a) $X_n \xrightarrow{\mathcal{D}} X_0$.

(b) $\lim\limits_{n\to\infty} \mathbb{P}(X_n = k) = \mathbb{P}(X_0 = k)$ *for each* $k \in \mathbb{N}_0$.

(c) $\lim\limits_{n\to\infty} g_n(s) = g_0(s)$ *for each* $s \in [0, 1]$.

Proof Try to prove this theorem yourself (Problem 2.4)!

To prepare the proof of Theorem 2.3, we need a condition that is equivalent to (2.3) and is stated with the help of generating functions.   □

2.5 Lemma *For each* $n \geq 1$, *let* $X_{n,1}, \ldots, X_{n,n}$ *be* $\mathbb{N}_0$-*valued random variables with generating functions* $g_{n,1}, \ldots, g_{n,n}$, *respectively. Then the following are equivalent:*

(a) $\left(X_{n,j} : n \geq 1,\ 1 \leq j \leq n\right)_{n \geq 1}$ *is a null array.*

(b) $\lim\limits_{n\to\infty} \max\limits_{1\leq j\leq n} \left(1 - g_{n,j}(s)\right) = 0$ *for each* $s \in [0, 1]$.

Proof The assertion follows from

(a) $\Longleftrightarrow \lim\limits_{n\to\infty} \max\limits_{1\leq j\leq n} \mathbb{P}\left(|X_{n,j}| > \varepsilon\right) = 0$ for each $\varepsilon > 0$

$\Longleftrightarrow X_{n,k_n} \xrightarrow{\mathbb{P}} 0$ for each subsequence (k_n) with $k_n \in \{1, \ldots, n\}$

$\Longleftrightarrow X_{n,k_n} \xrightarrow{\mathcal{D}} \delta_0$ for each subsequence (k_n) with $k_n \in \{1, \ldots, n\}$

$\Longleftrightarrow g_{n,k_n}(s) \to 1,\ 0 \leq s \leq 1$, for each subsequence (k_n) with $k_n \in \{1, \ldots, n\}$

$\Longleftrightarrow 1 - g_{n,k_n}(s) \to 0,\ 0 \leq s \leq 1$, for each subsequence (k_n) with $k_n \in \{1, \ldots, n\}$

$\Longleftrightarrow$ (b).

The second double arrow used Problem 2.1. The third double arrow holds since the notions of convergence in distribution and convergence in probability coincide if the limit distribution is a Dirac measure and thus degenerate.   □

Proof of Theorem 2.3 We first prove the implication "$\Longleftarrow$": Let $g_{n,j}$ be the generating function of $X_{n,j}$ ($n \geq 1$, $j \in \{1, \ldots, n\}$). Due to the continuity Theorem 2.4 and the multiplication rule for generating functions, we have to show

$$\lim_{n\to\infty} \prod_{j=1}^{n} g_{n,j}(s) = e^{\lambda(s-1)} \quad \text{for each } s \in [0, 1]. \tag{2.6}$$

Taking logarithms, this is equivalent to

$$\lim_{n\to\infty} \sum_{j=1}^{n} \log\left(1 - (1 - g_{n,j}(s))\right) = \lambda(s - 1) \quad \text{for each } s \in [0, 1]. \tag{2.7}$$

In view of Lemma 2.5 and the inequalities

$$1 - \frac{1}{t} \le \log t \le t - 1, \tag{2.8}$$

which hold for each $t > 0$, (2.7) in turn is the same as

$$\lim_{n\to\infty} \sum_{j=1}^{n} \left(1 - g_{n,j}(s)\right) = \lambda(1 - s) \quad \text{for each } s \in [0, 1]. \tag{2.9}$$

Self-question 2: Why do the inequalities (2.8) hold?

Putting

$$T_{n,1}(s) := (1-s) \sum_{j=1}^{n} \mathbb{P}(X_{n,j} > 0), \quad T_{n,2}(s) := \sum_{k=2}^{\infty} (s - s^k) \sum_{j=1}^{n} \mathbb{P}(X_{n,j} = k) \tag{2.10}$$

we obtain

$$\sum_{j=1}^{n} (1 - g_{n,j}(s)) = \sum_{j=1}^{n} \left[1 - \sum_{k=0}^{1} s^k \mathbb{P}(X_{n,j} = k)\right] - \sum_{j=1}^{n} \sum_{k=2}^{\infty} s^k \mathbb{P}(X_{n,j} = k)$$

$$= \sum_{j=1}^{n} \left[1 - \mathbb{P}(X_{n,j} = 0) - s\mathbb{P}(X_{n,j} = 1)\right]$$

$$+ \sum_{j=1}^{n} \sum_{k=2}^{\infty} (s - s^k)\mathbb{P}(X_{n,j} = k) - \sum_{j=1}^{n} s\mathbb{P}(X_{n,j} \ge 2)$$

$$= T_{n,1}(s) + T_{n,2}(s).$$

For each $k \geq 2$, we have $s(1 - s) \leq s(1 - s^{k-1}) = s - s^k \leq s$ and thus

$$s(1 - s) \sum_{j=1}^{n} \mathbb{P}(X_{n,j} > 1) \leq T_{n,2}(s) \leq s \sum_{j=1}^{n} \mathbb{P}(X_{n,j} > 1).$$

With condition (i), it follows that $\lim_{n\to\infty} T_{n,2}(s) = 0$. Moreover,

$$T_{n,1}(s) = (1 - s) \sum_{j=1}^{n} \mathbb{P}(X_{n,j} = 1) + (1 - s) \sum_{j=1}^{n} \mathbb{P}(X_{n,j} > 1),$$

so that (i) and (ii) yield the convergence $\lim_{n\to\infty} T_{n,1}(s) = (1 - s)\lambda$ and thus (2.9).

The proof of "$\Longrightarrow$" is fairly quick: First, $X_{n,1} + \ldots + X_{n,n} \xrightarrow{D} \mathrm{Po}(\lambda)$ implies (2.6) and thus, using (2.8), the convergence (2.9). Setting $s = 0$ in (2.9) yields

$$\lim_{n\to\infty} \sum_{j=1}^{n} \mathbb{P}(X_{n,j} > 0) = \lambda. \qquad (2.11)$$

By the part "$\Longleftarrow$" of the proof, we have

$$\sum_{j=1}^{n} \left(1 - g_{n,j}(s)\right) = (1 - s) \sum_{j=1}^{n} \mathbb{P}(X_{n,j} > 0) + T_{n,2}(s) = T_{n,1}(s) + T_{n,2}(s),$$

with $T_{n,2}(s)$ as in (2.10). Due to (2.9) and (2.11), $\lim_{n\to\infty} T_{n,2}(s) = 0$ for each $s \in [0, 1]$, and the inequalities

$$s(1 - s) \sum_{j=1}^{n} \mathbb{P}(X_{n,j} > 1) \leq T_{n,2}(s) \leq s \sum_{j=1}^{n} \mathbb{P}(X_{n,j} > 1)$$

then entail condition (i). Condition (ii) follows from

$$\sum_{j=1}^{n} \mathbb{P}(X_{n,j} = 1) = \sum_{j=1}^{n} \mathbb{P}(X_{n,j} > 0) - \sum_{j=1}^{n} \mathbb{P}(X_{n,j} > 1).$$

$\square$

The Poisson limit theorem suggests that the Poisson distribution may be a suitable model in a situation in which one deals with counts of how many of numerous possible, but individually relatively unlikely events occur. In addition to the decays of atoms, such as in the Rutherford–Geiger experiment (see, e.g., [HE1], p. 99), the number of registered photons or electrons at very low flow is approximately Poisson distributed. The same applies to the number of defective specimens in production series, the number of thunderstorms within a fixed period in a certain region or the number of accidents or suicides, related to a certain large population and a specified time period. The Poisson distribution is also the starting point of a fundamental class of *point processes*, the so-called *Poisson point processes* (see, e.g., [LAP]).

2.6 Poisson Convergence and Extreme Value Stochastics

In extreme value stochastics, among other things, one models probabilities for the occurrence of rare events. In the simplest case, such events are associated with maxima or minima of random variables. To clarify an essential idea, let $X_1, X_2, \ldots$ be i.i.d. random variables with distribution function F. If $1 - F(t) = \mathbb{P}(X_1 > t)$ is positive for each $t \in \mathbb{R}$, then the sequence (M_n) of random variables defined by $M_n := \max(X_1, \ldots, X_n)$ converges $\mathbb{P}$-a.s. to infinity as $n \to \infty$. Generally, $Y_n \xrightarrow{\text{a.s.}} \infty$ for a sequence (Y_n) means that there is a measurable set Ω_0 in the underlying probability space $(\Omega, \mathcal{A}, \mathbb{P})$, such that $\mathbb{P}(\Omega_0) = 1$ and for every $\omega \in \Omega_0$, the sequence $(Y_n(\omega))$ converges to infinity as $n \to \infty$, that is, for each $C > 0$ all but finitely many of the $Y_n(\omega)$ are greater than C.

Self-question 3: Why does $M_n \to \infty$ $\mathbb{P}$-a.s. hold?

Here, the immediate question arises whether one can find suitable *thresholds* $s_n(x)$, depending on a real parameter x, such that, as $n \to \infty$, the random number

$$U_n(x) := \sum_{j=1}^{n} \mathbf{1}\{X_j > s_n(x)\}$$

of exceedances over the threshold $s_n(x)$ among $X_1, \ldots, X_n$ converges in distribution to a Poisson distribution $\mathrm{Po}(\lambda(x))$ with $0 < \lambda(x) < \infty$. Since $U_n(x) = 0 \Longleftrightarrow M_n \leq s_n(x)$, it would then follow that

$$\lim_{n \to \infty} \mathbb{P}(U_n(x) = 0) = e^{-\lambda(x)} = \lim_{n \to \infty} \mathbb{P}(M_n \leq s_n(x)). \tag{2.12}$$

In a similar way,

$$V_n(x) := \sum_{j=1}^{n} \mathbf{1}\{X_j \leq b_n(x)\}$$

counts the number of X_j that do not exceed a threshold $b_n(x)$ among $X_1, \ldots, X_n$. Since $V_n(x) = 0 \Longleftrightarrow \min(X_1, \ldots, X_n) > b_n(x)$, a Poisson limit theorem for $V_n(x)$ would imply a result regarding the asymptotic behavior of the minimum of $X_1, \ldots, X_n$ as $n \to \infty$.

If we set $A_{n,j} := \{X_j > s_n(x)\}$, $j \in \{1, \ldots, n\}$, then $A_{n,1}, \ldots, A_{n,n}$ are independent events with the same probability $p_n(x) := 1 - F(s_n(x))$. If we could find a sequence $(s_n(x))$ with the property

$$\lim_{n \to \infty} n p_n(x) = \lim_{n \to \infty} n\big(1 - F(s_n(x))\big) = \lambda(x) \tag{2.13}$$

for some $\lambda(x) \in (0, \infty)$ then, by Theorem 2.3, we would obtain $U_n(x) \xrightarrow{\mathcal{D}} \mathrm{Po}(\lambda(x))$ and thus (2.12). The search for a sequence $(s_n(x))$ with (2.13) leads us to the upper tail of the distribution function F because, according to (2.13),

$$F(s_n(x)) \approx 1 - \frac{\lambda(x)}{n}.$$

If we restrict ourselves to special thresholds of the form $s_n(x) = a_n + b_n x$ with $b_n > 0$, then (2.12) turns into

$$\lim_{n \to \infty} \mathbb{P}\left(\frac{M_n - a_n}{b_n} \leq x\right) = H(x), \tag{2.14}$$

with $H(x) := e^{-\lambda(x)}$. In this context, the famous Fisher[2]–Tippett[3] Theorem (see, e.g., [EKM], p. 121) says: *If there are* sequences (a_n) and (b_n) with $b_n > 0$ for each n, such that the right-hand side of (2.14), when regarded as a function of x, is a non-degenerate distribution function, then, up to affine transformations of the argument of H, there are only three possible types of distributions, which are associated with the names Fréchet,[4] Weibull,[5] and Gumbel.[6]

---

[2] Sir Ronald Aylmer Fisher (1890–1962), British statistician and one of the pioneers of mathematical statistics (including experimental design, analysis of variance, maximum likelihood method, Fisher information, sufficiency).

[3] Leonard Henry Caleb Tippett (1902–1983), British statistician, one of the founders of stochastic extreme value theory.

[4] Maurice René Fréchet (1878–1973), French mathematician. Main field of work: Functional analysis. His name is associated with the terms Fréchet derivative, Fréchet filter, Fréchet metric, Fréchet distribution as well as with the information inequality of Fréchet–Cramér–Rao.

[5] Ernst Hjalmar Waloddi Weibull (1887–1979), Swedish engineer and mathematician.

[6] Emil Julius Gumbel (1891–1966), 1930 professor at the University of Heidelberg, 1932 emigration to France and later to the USA. Main fields of work: Probability theory and mathematical statistics.

Fig. 2.1 Density of the Gumbel extreme value distribution

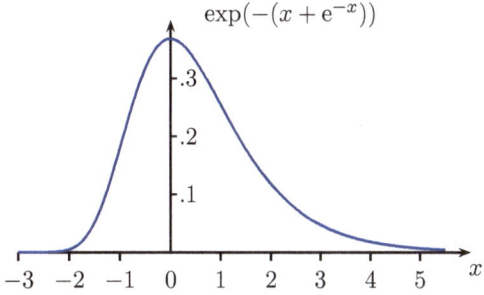
$$\exp(-(x + e^{-x}))$$

The most important of these three so-called *extreme value distributions for maxima* is given by the distribution function

$$G(x) := \exp\left(- e^{-x}\right), \quad x \in \mathbb{R},$$

and is named after Gumbel. Among other things, it occurs when the distribution of X_1 is an exponential distribution. Namely, suppose $F(t) = 1 - \exp(-t)$, $t > 0$, and fix any real number x. For each $n \geq 1$ with $x + \log n \geq 0$, we get

$$\mathbb{P}\left(M_n - \log n \leq x\right) = \left(\mathbb{P}(X_1 \leq x + \log n)\right)^n = \left(1 - e^{-(x+\log n)}\right)^n$$

$$= \left(1 - \frac{e^{-x}}{n}\right)^n$$

and thus $\lim_{n \to \infty} \mathbb{P}\left(M_n - \log n \leq x\right) = G(x)$. Figure 2.1 shows the density of the extreme value distribution of Gumbel. What stands out is a typical *right skewness* for limit distributions for maxima: The values of the density rise quickly and then fall off more slowly.

So far, we have considered the convergence in distribution of sums of *independent* non-negative integer random variables to a Poisson distribution using relatively elementary means. By Problem 2.5, a random variable X has the Poisson distribution $Po(\lambda)$ if and only if the equation

$$\mathbb{E}\left[Xf(X)\right] = \lambda \mathbb{E}\left[f(X + 1)\right]$$

holds for each bounded function $f : \mathbb{N}_0 \longrightarrow [0, \infty)$. This characterization of the Poisson distribution is the basis of a famous paper by L.H.Y. Chen[7] (see [CLH]), to obtain error

---

[7] Louis Hsiao Yun Chen (*1940), student of C. Stein, from 1981 professor at the then University of Singapore. Main field of work: probability theory, especially Poisson approximation.

estimates for the approximation of the distributions of a sum of indicator random variables by means of a Poisson distribution, in which the summands are allowed to be *dependent*. For further reading, see the monograph [BHJ]. In the literature, the techniques of C.M. Stein and L.H.Y. Chen, which represent a common approach to obtain error estimates for the approximation of distributions by other distributions (in particular the normal distribution and the Poisson distribution) are referred to as the *Stein–Chen method*. A more recent overview of results on Poisson approximations, obtained also with other methods, is provided by [NOV]. In the next chapter, we will see that convergence in distribution to a Poisson distribution can also be proved using the method of moments.

Answers to the Self-Questions

Answer 1 According to (2.5), $g_X(s) = \mathbb{E}(s^X)$. Since s^X and s^Y are independent, it follows that $g_{X+Y}(s) = \mathbb{E}(s^{X+Y}) = \mathbb{E}(s^X s^Y) = \mathbb{E}(s^X)\mathbb{E}(s^Y) = g_X(s)g_Y(s)$, $|s| \leq 1$.

Answer 2 Let $f(t) := \log t$, $t > 0$. The tangent to f at the point $t = 1$ is $y = t - 1$. Since $f''(t) = -\frac{1}{t^2} < 0$, the function f is concave, which yields $f(t) \leq t - 1$. The other inequality then follows if t is replaced with $\frac{1}{t}$.

Answer 3 For any $K \in (0, \infty)$, $\mathbb{P}(M_n > K) = 1 - (\mathbb{P}(X_1 \leq K))^n = 1 - F(K)^n \to 0$ as $n \to \infty$. Thus, $M_n \overset{\mathbb{P}}{\longrightarrow} \infty$, so $\mathbb{P}(M_n \geq C) \to 1$ for any $C > 0$. The $\mathbb{P}$-almost sure convergence follows because $M_n \leq M_{n+1}$, $n \geq 1$.

Problems

2.1 Problem Show that conditions (2.3) and (2.4) are equivalent.
Hint for " (2.4) $\Longrightarrow$ (2.3)" What does it mean that, for some $\varepsilon > 0$, the sequence (a_n) defined by

$$a_n := \max_{1 \leq j \leq n} \mathbb{P}(|X_{n,j}| > \varepsilon)$$

does not converge to zero?

2.2 Problem For each $n \geq 2$, suppose that $A_{n,1}, \ldots, A_{n,n}$ are independent events, and put $S_n := \sum_{j=1}^n \mathbf{1}\{A_{n,j}\}$. Prove the following assertions:

(a) If $\lim_{n \to \infty} \mathbb{E}(S_n) = \lim_{n \to \infty} \mathbb{V}(S_n) = \lambda$ for some $\lambda \in (0, \infty)$, then $S_n \overset{\mathcal{D}}{\longrightarrow} \mathrm{Po}(\lambda)$.

(b) If $\lim_{n\to\infty} \mathbb{V}(S_n) = \infty$, then

$$\frac{S_n - \mathbb{E}(S_n)}{\sqrt{\mathbb{V}(S_n)}} \xrightarrow{\mathcal{D}} N(0, 1).$$

Hint for (b) Use Theorem 1.20.

2.3 Problem For each $n \geq 1$, let $A_{n,1}, \ldots, A_{n,n}$ be independent events. Show using Theorem 2.3: If

$$\lim_{n\to\infty} \max_{1\leq j\leq n} \mathbb{P}(A_{n,j}) = 0$$

and

$$\lim_{n\to\infty} \sum_{j=1}^{n} \mathbb{P}(A_{n,j}) = \lambda$$

for some $\lambda \in (0, \infty)$, then $\mathbf{1}\{A_{n,1}\} + \ldots + \mathbf{1}\{A_{n,n}\} \xrightarrow{\mathcal{D}} \mathrm{Po}(\lambda)$.

2.4 Problem Prove the continuity Theorem 2.4 for generating functions.

2.5 Problem (Characterization of the Poisson distribution) Let X be a $\mathbb{N}_0$-valued random variable and $\lambda \in (0, \infty)$. Prove that the following statements are equivalent:

(a) X has a Poisson distribution with parameter λ.
(b) $\mathbb{E}[Xf(X)] = \lambda \mathbb{E}[f(X + 1)]$ for each bounded function $f : \mathbb{N}_0 \longrightarrow [0, \infty)$.

2.6 Problem The random variables $X_1, X_2, \ldots$ are i.i.d. with distribution function $F(t) := t^\vartheta, 0 \leq t \leq 1$, $F(t) := 1$ if $t > 1$, and $F(t) := 0$ if $t < 0$. Here, $\vartheta > 0$. Derive a Poisson limit theorem for $U_n(x) := \sum_{j=1}^{n} \mathbf{1}\{X_j > 1 - \frac{x}{n}\}$, $x > 0$, and thereby the limit distribution of $n(1 - \max(X_1, \ldots, X_n))$ as $n \to \infty$.

2.7 Problem The random variables $X_1, X_2, \ldots$ are i.i.d. with distribution function F, where $F(t) := 0$ if $t \leq 0$ and $F(t) = \lambda t^\alpha (1+o(1))$ as $t \downarrow 0$, that is, $\lim_{t\downarrow 0} F(t)/(\lambda t^\alpha) = 1$. Derive a Poisson limit theorem for $V_n(x) := \sum_{j=1}^{n} \mathbf{1}\{X_j \leq x/n^{1/\alpha}\}$, $x > 0$, and show that

$$\lim_{n\to\infty} \mathbb{P}\left(n^{1/\alpha} \min(X_1, \ldots, X_n) \leq x\right) = 1 - \exp\left(-\lambda x^\alpha\right), \quad x > 0. \tag{2.15}$$

The limit distribution of $n^{1/\alpha} \min(X_1, \ldots, X_n)$ is thus a Weibull distribution with parameters α and λ.

The Method of Moments

3

In this chapter, we meet a celebrated method for proving convergence in distribution, the so-called *method of moments*. Suppose $(X_n)_{n \geq 1}$ is a sequence of random variables with $\mathbb{E}|X_n|^k < \infty$ for each $k \geq 1$ and $n \geq 1$, that is, the k-th moment $\mathbb{E}(X_n^k)$ exists for all k and n. Suppose further X is a random variable with $\mathbb{E}|X|^k < \infty$ for each $k \geq 1$. Then the method of moments states that

$$\lim_{n \to \infty} \mathbb{E}(X_n^k) = \mathbb{E}(X^k) \quad \text{for each } k \geq 1 \tag{3.1}$$

implies $X_n \xrightarrow{\mathcal{D}} X$, *provided that the distribution* $\mathbb{P}^X$ *of* X *is uniquely determined by the sequence* $\left(\mathbb{E}(X^k)\right)_{k \geq 1}$ *of moments.* This result, which is stated as Theorem 3.6, is the main result of this chapter.

We start with a warning regarding convergence in distribution. Although, by Theorem 1.8, $X_n \xrightarrow{\mathcal{D}} X$ is equivalent to the convergence $\mathbb{E}h(X_n) \to \mathbb{E}h(X)$ for each continuous *bounded* function $h : \mathbb{R} \to \mathbb{R}$, it does not necessarily imply $\mathbb{E}(X_n) \to \mathbb{E}(X)$, that is, convergence of expectations, even if all those expectations exist. An example is as follows: By Slutsky's lemma (Lemma 1.23), $X + Y_n \xrightarrow{\mathcal{D}} X$ if the sequence (Y_n) converges to 0 in probability. Here, X has any distribution with $\mathbb{E}|X| < \infty$. The setting $\mathbb{P}(Y_n = n^2) := \frac{1}{n}$ and $\mathbb{P}(Y_n = 0) = 1 - \frac{1}{n}$ yields $Y_n \xrightarrow{\mathbb{P}} 0$ and thus $X + Y_n \xrightarrow{\mathcal{D}} X$, but $\mathbb{E}(X + Y_n) \to \infty$.

The following theorem provides a general inequality for expectations in the situation $X_n \xrightarrow{\mathcal{D}} X$.

N. Henze, *Asymptotic Stochastics*, Mathematics Study Resources 10, https://doi.org/10.1007/978-3-662-68923-3_3

3.1 Theorem (Fatou's Lemma for Convergence in Distribution)

Suppose $X, X_1, X_2, \ldots$ are random variables with $X_n \xrightarrow{\mathcal{D}} X$. Then, with the convention $\infty \leq \infty$,

$$\mathbb{E}|X| \leq \liminf_{n \to \infty} \mathbb{E}|X_n|.$$

Proof Let $(\widetilde{\Omega}, \widetilde{\mathcal{A}}, \widetilde{\mathbb{P}})$ and $Y, Y_1, Y_2, \ldots$ be as in Skorokhod's Theorem 1.24. Then $X \overset{\mathcal{D}}{=} Y$ and $0 \leq |Y_n| \to |Y|$ $\widetilde{\mathbb{P}}$-a.s. With Fatou's lemma (Theorem 1.29), we obtain

$$\mathbb{E}|X| = \mathbb{E}|Y| = \int |Y| \, d\widetilde{\mathbb{P}}$$

$$\leq \liminf_{n \to \infty} \int |Y_n| \, d\widetilde{\mathbb{P}} = \liminf_{n \to \infty} \mathbb{E}|Y_n| = \liminf_{n \to \infty} \mathbb{E}|X_n|,$$

which was to be shown.   □

The following notion allows to conclude $\mathbb{E}(X_n) \to \mathbb{E}(X)$ from $X_n \xrightarrow{\mathcal{D}} X$.

3.2 Definition (Uniform Integrability)

Let $X_1, X_2, \ldots$ be random variables on a probability space $(\Omega, \mathcal{A}, \mathbb{P})$. The sequence (X_n) is called *uniformly integrable* (in short: UI), if

$$\lim_{a \to \infty} \sup_{n \geq 1} \mathbb{E}\big[|X_n| \mathbf{1}\{|X_n| \geq a\}\big] = 0. \tag{3.2}$$

The term *uniformly* refers to the supremum over n. For each fixed n, the dominated convergence theorem (Theorem 1.31) gives $\mathbb{E}\big[|X_n| \mathbf{1}\{|X_n| \geq a\}\big] \to 0$ as $a \to \infty$, if the expectation of X_n exists.

Problem 3.2 shows that, under the assumption $X_n \xrightarrow{\mathbb{P}} X$, uniform integrability of the sequence (X_n) is equivalent to $\mathcal{L}^1$-convergence $\lim_{n \to \infty} \mathbb{E}|X_n - X| = 0$. The following properties are useful when dealing with uniform integrability.

3.3 Theorem (Conditions for Uniform Integrability)

(a) If $(X_n)_{n \geq 1}$ is UI then $\sup_{n \geq 1} \mathbb{E}|X_n| < \infty$.
(b) If $\sup_{n \geq 1} \mathbb{E}|X_n|^{1+\delta} < \infty$ for some $\delta > 0$ then (X_n) is UI.
(c) If $\sup_{n \geq 1} |X_n| \leq C < \infty$ for some C then (X_n) is UI.

Proof

(a) For any $a > 0$, we have

$$\mathbb{E}|X_n| = \mathbb{E}\big[|X_n|\mathbf{1}\{|X_n| < a\}\big] + \mathbb{E}\big[|X_n|\mathbf{1}\{|X_n| \geq a\}\big]$$
$$\leq a + \mathbb{E}\big[|X_n|\mathbf{1}\{|X_n| \geq a\}\big]$$
$$\leq a + \sup_{k \geq 1} \mathbb{E}\big[|X_k|\mathbf{1}\{|X_k| \geq a\}\big],$$

from which the claim follows.

(b) For any $a > 0$, we obtain

$$\mathbb{E}\big[|X_n|\mathbf{1}\{|X_n| \geq a\}\big] \leq \mathbb{E}\left[|X_n|\left(\frac{|X_n|}{a}\right)^{\delta}\right] = \frac{1}{a^{\delta}}\mathbb{E}|X_n|^{1+\delta}.$$

Since $\sup_{n \geq 1} \mathbb{E}|X_n|^{1+\delta} < \infty$, (3.2) is proved.

(c) is a consequence of (b).

$\square$

3.4 Theorem (Uniform Integrability and Convergence in Distribution)
If $X_n \xrightarrow{D} X$ and the sequence (X_n) is uniformly integrable, then

(a) $\mathbb{E}|X| < \infty$,
(b) $\lim_{n \to \infty} \mathbb{E}(X_n) = \mathbb{E}(X)$.

Proof Part (a) follows from Theorem 3.1 and Theorem 3.3 (a). To show (b), we use Theorem 1.24 (Skorokhod), according to which we may assume w.l.o.g. that $X_n \xrightarrow{\text{a.s.}} X$. For any $a > 0$, we define

$$\Delta_{n,1}(a) := \mathbb{E}\big[|X_n - X|\mathbf{1}\{|X_n - X| < a\}\big],$$
$$\Delta_{n,2}(a) := \mathbb{E}\big[|X_n - X|\mathbf{1}\{|X_n - X| \geq a\}\big]$$

and obtain

$$|\mathbb{E}X_n - \mathbb{E}X| \leq \mathbb{E}|X_n - X| = \Delta_{n,1}(a) + \Delta_{n,2}(a). \tag{3.3}$$

By the dominated convergence theorem (Theorem 1.31), $\lim_{n \to \infty} \Delta_{n,1}(a) = 0$. As for $\Delta_{n,2}(a)$, $|X_n - X| \geq a$ implies $\max(|X_n|, |X|) \geq a/2$.

Self-question 1: Why does $\max(|X_n|, |X|) \geq \frac{a}{2}$ follow from $|X_n - X| \geq a$?

As a consequence, we get

$$
\Delta_{n,2}(a) \leq 2\mathbb{E}\left[\max\left(|X_n|, |X|\right) \mathbf{1}\left\{\max\left(|X_n|, |X|\right) \geq \frac{a}{2}\right\}\right]
$$

$$
\leq 2\mathbb{E}\left[|X_n|\mathbf{1}\left\{|X_n| \geq \frac{a}{2}\right\}\right] + 2\mathbb{E}\left[|X|\mathbf{1}\left\{|X| \geq \frac{a}{2}\right\}\right] \tag{3.4}
$$

$$
\leq 2\sup_{k\geq 1}\mathbb{E}\left[|X_k|\mathbf{1}\left\{|X_k| \geq \frac{a}{2}\right\}\right] + 2\mathbb{E}\left[|X|\mathbf{1}\left\{|X| \geq \frac{a}{2}\right\}\right].
$$

Self-question 2: Why does the inequality (3.4) hold?

In this upper bound for $\Delta_{n,2}(a)$, the first term converges to 0 as $a \to \infty$ due to the uniform integrability of (X_n). The second term also converges to 0 as $a \to \infty$ in view of the dominated convergence theorem, since the expectation of X exists.

$\square$

3.5 Corollary *Let* $r \in \mathbb{N}$ *and* $\varepsilon > 0$. *If* $X_n \xrightarrow{\mathcal{D}} X$ *and* $\sup_{n\geq 1} \mathbb{E}|X_n|^{r+\varepsilon} < \infty$, *then*

(a) $\mathbb{E}|X|^r < \infty$,
(b) $\lim_{n\to\infty} \mathbb{E}(X_n^r) = \mathbb{E}(X^r)$.

Proof Because of $|X_n|^{r+\varepsilon} = |X_n^r|^{1+\varepsilon/r}$, the sequence $(X_n^r)_{n\geq 1}$ is uniformly integrable according to Theorem 3.3 (b). With the mapping theorem 1.22, we obtain $X_n^r \xrightarrow{\mathcal{D}} X^r$, and thus the claim follows from Theorem 3.4.

$\square$

We now come to the main result of this chapter.

3.6 Theorem (Method of Moments)
Let $X, X_1, X_2, \ldots$ *be random variables with* $\mathbb{E}|X|^k < \infty$ *and* $\mathbb{E}|X_n|^k < \infty$ *for all* $n, k \geq 1$. *Suppose the distribution of* X *is uniquely determined by the sequence* $(\mathbb{E}X^k)_{k\geq 1}$ *of moments. Then* $\lim_{n\to\infty} \mathbb{E}(X_n^k) = \mathbb{E}(X^k)$ *for each* $k \geq 1$ *implies* $X_n \xrightarrow{\mathcal{D}} X$.

Proof Let $a > 0$. Since $\mathbf{1}\{|X_n| > a\} \leq a^{-2}X_n^2$, it follows that

$$\mathbb{P}(|X_n| > a) \leq \frac{\mathbb{E}(X_n^2)}{a^2}.$$

By assumption, $\lim_{n \to \infty} \mathbb{E}(X_n^2) = \mathbb{E}(X^2)$, and thus for every $\varepsilon > 0$ there exists an a such that $\mathbb{P}(|X_n| \leq a) \geq 1 - \varepsilon$ for each $n \geq 1$. By Definition 1.9, the set $\{\mathbb{P}^{X_n} : n \geq 1\}$ is tight. In view of Theorem 1.10, there is a subsequence (X_{n_k}) and a random variable Y with $X_{n_k} \overset{\mathcal{D}}{\longrightarrow} Y$ as $k \to \infty$. By Corollary 3.5, $\lim_{k \to \infty} \mathbb{E}(X_{n_k}^r) = \mathbb{E}(Y^r), r \in \mathbb{N}$.

Self-question 3: Why can we conclude that $\lim_{k \to \infty} \mathbb{E}(X_{n_k}^r) = \mathbb{E}(Y^r), r \geq 1$?

Due to (3.1), it follows that $\mathbb{E}(X^k) = \mathbb{E}(Y^k)$ for each $k \geq 1$ and thus $X \overset{\mathcal{D}}{=} Y$, since $\mathbb{P}^X$ is uniquely determined by the sequence $\left(\mathbb{E}(X^k)\right)_{k \geq 1}$. The claim now follows from Corollary 1.11 (b).

$\square$

In order to apply the method of moments, we must know which distributions are uniquely determined by the sequence of their moments. The following criterion provides a corresponding sufficient condition. This condition holds in particular when the *moment generating function* of X, defined by

$$m_X(t) := \mathbb{E}\left[e^{tX}\right], \quad t \in \mathbb{R}, \tag{3.5}$$

takes finite values in a neighborhood of 0 (Problem 3.3). Note that the expectation in (3.5) may only be finite for $t = 0$. An example is a random variable X with the *Cauchy distribution* $C(0, 1)$, that is, with the density $f(x) = 1/(\pi(1 + x^2))$, $x \in \mathbb{R}$.

3.7 Theorem (Sufficient Condition for "$(\mathbb{E}X^k)_{k \geq 1}$ Determines $\mathbb{P}^X$")
Let X be a random variable with $\mathbb{E}|X|^k < \infty$ for each $k \geq 1$. If the power series

$$\sum_{k=1}^{\infty} \frac{\mathbb{E}(X^k)}{k!} t^k \tag{3.6}$$

has a non-vanishing radius of convergence, then $\mathbb{P}^X$ is uniquely determined by the sequence $(\mathbb{E}X^k)_{k \geq 1}$ of moments.

Proof Let $m_k := \mathbb{E}(X^k)$ and $b_k := \mathbb{E}(|X|^k)$, $k \in \mathbb{N}$. By assumption, there exists a $t_0 > 0$ with $\sum_{k=0}^{\infty} |m_k| t_0^k / k! < \infty$. Since $b_{2k} = m_{2k} = |m_{2k}|$, it follows that

$$\lim_{k \to \infty} \frac{b_{2k} h^{2k}}{(2k)!} = 0 \quad \text{for every } h \text{ with } |h| < t_0.$$

This statement remains valid if $2k$ is throughout replaced with $2k - 1$, because the inequality $|x|^{2k-1} \leq 1 + |x|^{2k}$, $x \in \mathbb{R}$, yields $b_{2k-1} \leq 1 + m_{2k}$. We thus obtain

$$\frac{b_{2k-1}|h|^{2k-1}}{(2k-1)!} \leq \frac{|h|^{2k-1}}{(2k-1)!} + \frac{m_{2k} t_0^{2k}}{(2k)!} \cdot \frac{2k|h|^{2k-1}}{t_0^{2k}}$$

and therefore

$$\lim_{n \to \infty} \frac{b_n h^n}{n!} = 0 \quad \text{for every } h \text{ with } |h| < t_0. \tag{3.7}$$

By induction over n, we get

$$\left| e^{itx} \left(e^{ihx} - \sum_{k=0}^{n} \frac{(ihx)^k}{k!} \right) \right| \leq \frac{|h|^{n+1}|x|^{n+1}}{(n+1)!}, \qquad t, h, x \in \mathbb{R}, \; n \in \mathbb{N}_0. \tag{3.8}$$

Denote φ the characteristic function of X. Since $\mathbb{E}|X|^k < \infty$ for each $k \geq 1$, φ is infinitely often continuously differentiable, and we have $\varphi^{(k)}(t) = \frac{d^k}{dt^k}\varphi(t) = \mathbb{E}\left[e^{itX}(iX)^k \right]$ for each $k \geq 1$. In view of (3.8), we obtain

$$\left| \varphi(t+h) - \sum_{k=0}^{n} \frac{h^k}{k!} \varphi^{(k)}(t) \right| \leq \frac{|h|^{n+1}}{(n+1)!} b_{n+1},$$

and (3.7) implies

$$\varphi(t+h) = \sum_{k=0}^{\infty} \frac{\varphi^{(k)}(t)}{k!} h^k, \qquad t \in \mathbb{R}, \; |h| < t_0. \tag{3.9}$$

If Y is another random variable with $\mathbb{E}(Y^k) = m_k$ for every $k \geq 1$ and characteristic function $\psi(t) = \mathbb{E}[e^{itY}]$, $t \in \mathbb{R}$, it follows as above that

$$\psi(t+h) = \sum_{k=0}^{\infty} \frac{\psi^{(k)}(t)}{k!} h^k, \qquad t \in \mathbb{R}, \; |h| < t_0. \tag{3.10}$$

Since $\psi^{(k)}(0) = \varphi^{(k)}(0) = i^k m_k$, $k \geq 1$, setting $t = 0$ in (3.9) and (3.10) yields $\psi(t) = \varphi(t)$ for every t with $|t| < t_0$. If one now sequentially chooses $t = \pm t_0/2$, $t = \pm t_0, \ldots$ in (3.9) and (3.10), it follows that $\psi(t) = \varphi(t)$ for every $t \in \mathbb{R}$ and thus $X \overset{\mathcal{D}}{=} Y$. $\square$

3.8 Examples (Normal Distribution)
If X has the standard normal distribution $N(0, 1)$, then due to reasons of symmetry $\mathbb{E}(X^{2k-1}) = 0$ for every $k \in \mathbb{N}$. Furthermore, we have

$$\mathbb{E}(X^{2k}) = 1 \cdot 3 \cdot \ldots \cdot (2k - 1) = \frac{(2k)!}{2^k k!}, \quad k \geq 1.$$

Hence, the power series in (3.6) has the form

$$\sum_{k=1}^{\infty} \frac{1}{2^k k!} t^{2k},$$

and thus the radius of convergence is ∞.

3.9 Examples (Poisson Distribution)
If X has the Poisson distribution $\mathrm{Po}(\lambda)$, then according to Problem 3.6 (b)

$$\mathbb{E}(X^{k+1}) \leq (k + \lambda)\mathbb{E}(X^k), \quad k \geq 1.$$

It follows that

$$\frac{\mathbb{E}(X^{k+1})}{(k + 1)!} \cdot \frac{k!}{\mathbb{E}(X^k)} \leq \frac{k + \lambda}{k + 1},$$

and therefore the power series in (3.6) has a positive radius of convergence. The Poisson distribution is thus uniquely determined by the sequence of its moments.

When using the method of moments in connection with the Poisson distribution, it is usually more convenient to consider the so-called *factorial moments* $\mathbb{E}[X(X - 1) \ldots (X - k + 1)]$, $k \geq 1$, rather than the moments $\mathbb{E}(X^k)$, $k \geq 1$. If X has the Poisson distribution $\mathrm{Po}(\lambda)$, some calculations yield

$$\mathbb{E}[X(X - 1) \ldots (X - k + 1)] = \lambda^k, \quad k \geq 1. \tag{3.11}$$

Self-question 4: Why does equation (3.11) hold?

The relationship between moments and factorial moments is generally given by

$$\mathbb{E}(X^{\ell}) = \sum_{k=1}^{\ell} \begin{Bmatrix} \ell \\ k \end{Bmatrix} \mathbb{E}[X(X-1)\dots(X-k+1)], \tag{3.12}$$

where $\begin{Bmatrix} \ell \\ k \end{Bmatrix}$ denotes the so-called *Stirling number of the second kind*. The Stirling number $\begin{Bmatrix} \ell \\ k \end{Bmatrix}$ is defined as the number of partitions of a set M that has ℓ elements into k disjoint subsets (see, e.g., [GKP], p. 258). Since $k!\begin{Bmatrix} \ell \\ k \end{Bmatrix}$ is the number of surjective functions from M onto a set having k elements (see Problem 3.8), we have

$$\begin{Bmatrix} \ell \\ k \end{Bmatrix} = \frac{1}{k!} \sum_{j=0}^{k} (-1)^j \binom{k}{j} (k-j)^{\ell}.$$

Due to the representation (3.12), one can show $X_n \xrightarrow{\mathcal{D}} \text{Po}(\lambda)$ by proving

$$\lim_{n\to\infty} \mathbb{E}\big[X_n(X_n-1)\dots(X_n-k+1)\big] = \lambda^k \quad \text{for each } k \in \mathbb{N}. \tag{3.13}$$

Problem 3.10 illustrates how to use the method of moments to prove Poisson convergence of sums of indicator random variables in certain cases. An application of this method of proof to the coupon collector's problem can be found in [SHE].

Problem 3.5 shows that any distribution concentrated on a compact interval is uniquely determined by the sequence of moments, but not the lognormal distribution.

Finally, we stress that there are also error estimates in connection with the method of moments analogous to (1.12) and (1.13), which are stated using so-called *cumulants*, see, e.g., [DJS].

Answers to the Self-Questions

Answer 1 By the triangle inequality, $|X_n - X| \le |X_n| + |X|$. If $\max(|X_n|, |X|) < \frac{a}{2}$, then $|X_n - X| < a$.

Answer 2 We distinguish between the cases $\max(|X_n|, |X|) = |X_n|$ and $\max(|X_n|, |X|) = |X|$. Since the event $\{|X_n| = |X|\}$ can have a positive probability, there is the sign "$\le$" at the beginning of (3.4).

Answer 3 We can draw this conclusion because for every $r \ge 1$ there is an $\varepsilon > 0$ with the property $\sup_{k\ge 1} \mathbb{E}|X_{n_k}|^{r+\varepsilon} < \infty$. If $r = 2\ell$ with $\ell \in \mathbb{N}$ is an even number, we can take

$\varepsilon = 2$, since $\lim_{k\to\infty} \mathbb{E}(X_{n_k}^{2\ell+2}) = \lim_{k\to\infty} \mathbb{E}(|X_{n_k}|^{2\ell+2}) = \mathbb{E}(|X|^{2\ell+2})$. If $r = 2\ell - 1$ with $\ell \in \mathbb{N}$ is odd, the inequality $\sup_{k\geq 1} \mathbb{E}|X_{n_k}|^{r+\varepsilon} < \infty$ holds for $\varepsilon = 1$.

Answer 4 The expectation in (3.11) is equal to

$$\sum_{j=k}^{\infty} j(j-1)\dots(j-k+1)e^{-\lambda}\frac{\lambda^j}{j!} = \lambda^k e^{-\lambda} \sum_{j=k}^{\infty} \frac{\lambda^{j-k}}{(j-k)!} = \lambda^k e^{-\lambda} e^{\lambda} = \lambda^k.$$

Problems

3.1 Problem Show: If the sequences $(X_n)_{n\geq 1}$ and $(Y_n)_{n\geq 1}$ are uniformly integrable, then the sequence $(X_n + Y_n)_{n\geq 1}$ is also uniformly integrable.

3.2 Problem Let $X, X_1, X_2, \dots$ be random variables on a probability space $(\Omega, \mathcal{A}, \mathbb{P})$ with $X_n \xrightarrow{\mathbb{P}} X$ as $n \to \infty$. Show that the following statements are equivalent:

(a) $X_n \xrightarrow{\mathcal{L}^1} X$.
(b) The sequence $(X_n)_{n\geq 1}$ is uniformly integrable.

Hint For "(a) $\Longrightarrow$ (b)" use the function $\Psi_C(x) := x$ if $0 \leq x \leq C - 1$, $\Psi_C(x) := (1 - C)(x - C)$ if $C - 1 \leq x \leq C$ and $\Psi_C(x) := 0$ if $x \geq C$, defined for $C \geq 1$, $\Psi_C : [0, \infty) \to [0, \infty)$ (sketch!).

3.3 Problem Let X be a random variable with $\mathbb{E}\left[e^{tX}\right] < \infty$ for every t with $|t| < \delta$, where $\delta > 0$. Show that the series

$$\sum_{k=0}^{\infty} \frac{\mathbb{E}(X^k)}{k!} t^k, \quad |t| < \delta,$$

converges.

Hint Use the inequality $e^{|tx|} \leq e^{tx} + e^{-tx}$, which holds for any real numbers x and t.

3.4 Problem Let $(X_n)_{n\geq 1}$ be a sequence of i.i.d. random variables with $\mathbb{E}(X_1) = 0$ and $\mathbb{V}(X_1) = 1$. Suppose there is a constant $M \in (0, \infty)$ with $\mathbb{P}(|X_1| \leq M) = 1$. Use the method of moments to show that for the sequence $(S_n)_{n\geq 1}$ of partial sums $S_n := \sum_{j=1}^{n} X_j$ we have

$$\frac{S_n}{\sqrt{n}} \xrightarrow{\mathcal{D}} N(0, 1).$$

3.5 Problem Let X be a random variable. Prove the following statements:

(a) If X is bounded, that is, if $\mathbb{P}(|X| \leq M) = 1$ for some $M \in (0, \infty)$, then the distribution of X is uniquely determined by the sequence $\mathbb{E}(X^k)$, $k \geq 1$, of moments.
(b) If X has a lognormal distribution with parameters $\mu = 0$ and $\sigma = 1$, that is, the density

$$f(t) = \frac{1}{t\sqrt{2\pi}} \cdot \exp\left(-\frac{1}{2}(\log t)^2\right), \quad t > 0,$$

and $f(t) := 0$, otherwise, then there exists a random variable Y with $\mathbb{P}^X \neq \mathbb{P}^Y$ and $\mathbb{E}(X^k) = \mathbb{E}(Y^k)$ for each $k \geq 1$.

Hint for (b) For each $k \in \mathbb{N}_0$ we have

$$\int_0^\infty t^k f(t) \sin(2\pi \log t)\, dt = 0. \tag{3.14}$$

3.6 Problem Let X be a random variable with the Poisson distribution $\mathrm{Po}(\lambda)$, $\lambda > 0$. Prove the following claims:

(a) $\mathbb{E}(X^{k+1}) = \lambda \sum_{\ell=0}^{k} \binom{k}{\ell} \mathbb{E}(X^\ell), \quad k \geq 1,$

(b) $\mathbb{E}(X^{k+1}) \leq (k+\lambda)\mathbb{E}(X^k), \quad k \geq 1.$

3.7 Problem Let $A_1, \ldots, A_n$ be events in a probability space $(\Omega, \mathcal{A}, \mathbb{P})$ and $S_n := \sum_{j=1}^{n} \mathbf{1}\{A_j\}$. Show:

$$\mathbb{E}\binom{S_n}{k} = \sum_{1 \leq i_1 < \ldots < i_k \leq n} \mathbb{P}(A_{i_1} \cap \ldots \cap A_{i_k}), \quad k \in \{1, \ldots, n\}.$$

3.8 Problem For $n \in \mathbb{N}$, put $\mathbb{N}_n := \{1, \ldots, n\}$. If ℓ and k are integers with $\ell \geq k$, show that there are

$$\sum_{j=0}^{k} (-1)^j \binom{k}{j} (k-j)^\ell$$

surjective functions from $\mathbb{N}_\ell$ onto $\mathbb{N}_k$.

Hint If M denotes the set of all functions $g : \mathbb{N}_\ell \to \mathbb{N}_k$ and A_i stands for those g in M with $g(m) \neq i$ for each $m \in \mathbb{N}_\ell$, then the number of surjective functions is $|M| - |\cup_{i=1}^{n} A_i|$.

3.9 Problem Suppose each of the $n!$ permutations of the numbers $1, \ldots, n$ is equally probable. Let F_n be the number of fixed points of such a purely random permutation. Using (3.13) and Problem 3.7, prove $F_n \overset{\mathcal{D}}{\longrightarrow} \mathrm{Po}(1)$ as $n \to \infty$.

3.10 Problem Let $\{A_{n,1}, \ldots, A_{n,n} : n \geq 1\}$ be a triangular array of *exchangeable* events in a probability space $(\Omega, \mathcal{A}, \mathbb{P})$, that is, $\mathbb{P}(A_{n,i_1} \cap \ldots \cap A_{n,i_k}) = \mathbb{P}(A_{n,1} \cap \ldots \cap A_{n,k})$ for each $n \geq 2$, each $k \in \{1, \ldots, n\}$ and any choice of $i_1, \ldots, i_k$ with $1 \leq i_1 < \ldots < i_k \leq n$. Show: If

$$\lim_{n \to \infty} n^k \mathbb{P}(A_{n,1} \cap \ldots \cap A_{n,k}) = \lambda^k \quad \text{for each } k \geq 1$$

for some $\lambda \in (0, \infty)$, then $\sum_{i=1}^{n} 1\{A_{n,i}\} \overset{\mathcal{D}}{\longrightarrow} \mathrm{Po}(\lambda)$.

Hint Consider Problem 3.7.

A Central Limit Theorem for Stationary m-Dependent Sequences

4

This chapter deals with sequences of random variables that may be dependent in a sense to be specified, as well as with a central limit theorem for such sequences. More precisely, let $Y_1, Y_2, \ldots$ be real-valued random variables that are defined on a common probability space $(\Omega, \mathcal{A}, \mathbb{P})$. For a non-empty set $T \subset \mathbb{N}$, let

$$\sigma\left(Y_t : t \in T\right) := \sigma\left(\bigcup_{t \in T} Y_t^{-1}(\mathcal{B})\right)$$

denote the sub-σ-field of $\mathcal{A}$ generated by $\{Y_t : t \in T\}$.

The following definition specifies the limited amount of dependence between the random variables $Y_1, Y_2, \ldots$ that is understood throughout this chapter.

4.1 Definition (m-Dependence)
Let $m \in \mathbb{N}_0$. The sequence $(Y_n)_{n \geq 1}$ is called *m-dependent* if the σ-fields $\sigma(Y_1, \ldots, Y_s)$ and $\sigma(Y_{s+m+j} : j \geq 1)$ are independent for each $s \geq 1$.

If $Y_1, Y_2, \ldots$ are independent, then, by definition, $\sigma(Y_1), \sigma(Y_2), \ldots$ are independent σ-fields. Consequently, for each $s \geq 1$, the σ-fields $\sigma(Y_1, \ldots, Y_s)$ and $\sigma(Y_{s+j} : j \geq 1)$ are also independent (see, e.g., [KLE], Theorem 2.13 (iv)). A sequence of independent random variables is therefore 0-dependent in the sense of the above definition. The converse of this statement also holds (Problem 4.1). A simple example of a 2-dependent sequence is $Y_n := \mathbb{1}\{A_n \cap A_{n+1} \cap A_{n+2}\}$, $n \geq 1$, where $A_1, A_2, \ldots$ are independent events. Here, $\sigma(Y_1, \ldots, Y_s) = \sigma(\{A_1, \ldots, A_{s+2}\})$ and $\sigma(Y_{s+2+j} : j \geq 1) = \sigma(\{A_{s+3}, A_{s+4}, \ldots\})$.

The following definition generalizes the notion of identical distributions.

© The Author(s), under exclusive license to Springer-Verlag GmbH, DE, part of Springer Nature 2024
N. Henze, *Asymptotic Stochastics*, Mathematics Study Resources 10, https://doi.org/10.1007/978-3-662-68923-3_4

4.2 Definition (Stationarity)

A sequence $(Y_n)_{n \geq 1}$ of random variables is called *stationary*, if for every $j \geq 1$ and any choice of $k \geq 0$, the distribution of the random vector $(Y_j, \ldots, Y_{j+k})$ does not depend on j.

Setting $k = 0$, it follows that the random variables in a stationary sequence are identically distributed. If we interpret the index n as *time*, then stationarity (more precisely: *strict* stationarity) means that every *finite-dimensional distribution*, that is, the distribution of a random vector of the form $(Y_j, \ldots, Y_{j+k})$, is invariant to time shifts. By forming marginal distributions, it follows that, for example, the pairs (Y_1, Y_4), (Y_2, Y_5), (Y_3, Y_6), \ldots also have the same distribution. Note that every sequence of i.i.d. random variables is stationary.

4.3 Functions of Blocks of an i.i.d. Sequence

Let $(X_n)_{n \geq 1}$ be an i.i.d. sequence of random variables. Furthermore, let ℓ be any integer and $f : \mathbb{R}^\ell \to \mathbb{R}$ be any measurable function. If we set

$$Y_j := f(X_j, X_{j+1}, \ldots, X_{j+\ell-1}), \quad j \geq 1,$$

then $(Y_j)_{j \geq 1}$ is a stationary sequence. For each $s \geq 1$, we have

$$\sigma(Y_1, \ldots, Y_s) \subset \sigma(X_1, \ldots, X_{s+\ell-1}),$$

$$\sigma(Y_{s+\ell-1+j} : j \geq 1) \subset \sigma(X_{s+\ell}, X_{s+\ell+1}, \ldots),$$

and thus $(Y_n)_{n \geq 1}$ is an $(\ell - 1)$-dependent sequence.

Functions of blocks of an i.i.d. sequence thus stand for a whole class of examples of stationary m-dependent sequences. We would like to emphasize two special cases separately, because they will be referred to later. In both cases, the numbering of the sequence members X_n starts with $n = 0$, which is irrelevant.

4.4 Examples

(a) For an i.i.d. sequence $X_0, X_1, X_2, \ldots$, let

$$Y_j := \mathbf{1}\{X_{j-1} > X_j < X_{j+1}\}, \quad j \geq 1.$$

If $X_0, X_1, X_2, \ldots$ is regarded as a *time series* that can be observed at the discrete time points $1, 2, \ldots$, then Y_j indicates whether there is a local minimum in the sequence (X_n) at time j or not. Obviously, the sequence $(Y_j)_{j \geq 1}$ is 2-dependent.

(b) In this example, let $X_0, X_1, \ldots$ be i.i.d. random variables with $\mathbb{P}(X_0 = 1) = p = 1 - \mathbb{P}(X_0 = 0)$ and $0 < p < 1$. Thus, there is a sequence of $\mathrm{Bin}(1, p)$-distributed random variables, often also referred to as a *Bernoulli sequence*. If the event $\{X_j = 1\}$ or $\{X_j = 0\}$ is interpreted as a *hit* or a *miss* at the j-th attempt, then the random variable Y_j, defined by

$$Y_j := (1 - X_{j-1}) X_j X_{j+1} \cdot \ldots \cdot X_{j+r-1} (1 - X_{j+r}), \quad j \geq 1,$$

indicates whether a so-called *lucky streak* of hits of exact length r starts on the j-th attempt. By construction, the sequence $(Y_j)_{j \geq 1}$ is $(r + 1)$-dependent.

For all further considerations, in addition to m-dependence and stationarity, we make the assumption $\mathbb{E}(Y_1^2) < \infty$ and set

$$\mu := \mathbb{E}(Y_1),$$

$$\sigma_{0,0} := \mathbb{V}(Y_1),$$

$$\sigma_{0,j} := \mathrm{Cov}(Y_1, Y_{1+j}).$$

Because of stationarity, $\mu = \mathbb{E}(Y_j)$ and $\sigma_{0,0} = \mathbb{V}(Y_j)$ for each $j \geq 1$. Moreover, $\sigma_{0,j} = \mathrm{Cov}(Y_i, Y_{i+j})$ for any choice of $i, j \geq 1$. Finally, due to m-dependence, $\sigma_{0,j} = 0$ if $j > m$.

We will now investigate whether a central limit theorem holds for the partial sums

$$S_n := Y_1 + \ldots + Y_n, \quad n \geq 1,$$

of $Y_1, Y_2, \ldots.$ In other words, we ask whether

$$\frac{S_n - \mathbb{E}(S_n)}{\sqrt{\mathbb{V}(S_n)}} \to N(0, 1) \quad \text{as } n \to \infty \tag{4.1}$$

holds. Such a limit theorem would generalize the Lindeberg–Lévy central limit theorem for an i.i.d. sequence to the present situation.

First, $\mathbb{E}(S_n) = n\mu$, and since taking covariances is a bilinear operator, m-dependence yields

$$\mathbb{V}(S_n) = \sum_{i=1}^{n} \sum_{j=1}^{n} \mathrm{Cov}(Y_i, Y_j)$$

$$= n\sigma_{0,0} + 2(n - 1)\sigma_{0,1} + \ldots + 2(n - m)\sigma_{0,m}.$$

Self-question 1: Can you derive this equation for $\mathbb{V}(S_n)$?

If we set

$$\sigma^2 := \sigma_{0,0} + 2 \sum_{j=1}^{m} \sigma_{0,j}, \tag{4.2}$$

then $\lim_{n \to \infty} \frac{1}{n} \mathbb{V}(S_n) = \sigma^2$. Problem 4.2 shows that the cases $\sigma^2 = 0$ and $\sigma_{0,0} > 0$ can occur.

We will see that (4.1) holds if $\sigma^2 > 0$. The proof of this result rests crucially on the following fact.

4.5 Proposition *Let $Z_{n,k}$, $X_{n,k}$ ($n, k \geq 1$) be random variables, and set*

$$T_n := Z_{n,k} + X_{n,k}, \quad n, k \geq 1.$$

We make the following assumptions:

(a) $\lim\limits_{k \to \infty} \sup\limits_{n \in \mathbb{N}} \mathbb{P}\left(|X_{n,k}| \geq \delta\right) = 0$ *for each $\delta > 0$.*

(b) *For each $k \geq 1$, there is a random variable Z_k with $Z_{n,k} \to Z_k$ as $n \to \infty$.*

(c) *There exists a random variable Z with $Z_k \to Z$ as $k \to \infty$.*

Then $T_n \xrightarrow{\mathcal{D}} Z$ as $n \to \infty$.

Proof Let $F, F_1, F_2, \ldots$ be the distribution functions of $Z, Z_1, Z_2, \ldots$, respectively, and denote C_G the set of continuity points of a function $G : \mathbb{R} \to \mathbb{R}$. Fix any $\varepsilon > 0$ and any $z \in C_F$. Because a distribution function has at most countably many discontinuity points, there is a $\delta > 0$ with

$$\mathbb{P}(|Z - z| \leq \delta) < \varepsilon \text{ and } \{z + \delta, z - \delta\} \subset C_F \cap \bigcap_{k=1}^{\infty} C_{F_k}. \tag{4.3}$$

According to (a), there exists a k_0 with

$$\mathbb{P}(|X_{n,k}| \geq \delta) < \varepsilon \quad \text{for every } n \geq 1 \text{ and every } k \geq k_0. \tag{4.4}$$

In view of (c), we find a $k_1 \geq k_0$ with

$$|\mathbb{P}(Z_k \leq z \pm \delta) - \mathbb{P}(Z \leq z \pm \delta)| < \varepsilon \quad \text{for every } k \geq k_1. \tag{4.5}$$

Due to (4.4), for every $k \geq k_1$

$$\mathbb{P}(T_n \leq z) = \mathbb{P}(Z_{n,k} + X_{n,k} \leq z)$$

$$= \mathbb{P}(Z_{n,k} + X_{n,k} \leq z, |X_{n,k}| < \delta) + \mathbb{P}(Z_{n,k} + X_{n,k} \leq z, |X_{n,k}| \geq \delta)$$

$$\leq \mathbb{P}(Z_{n,k} \leq z + \delta) + \mathbb{P}(|X_{n,k}| \geq \delta)$$

$$\leq \mathbb{P}(Z_{n,k} \leq z + \delta) + \varepsilon.$$

From assumption (b) and (4.3) as well as (4.5), we obtain

$$\limsup_{n \to \infty} \mathbb{P}(T_n \leq z) \leq \mathbb{P}(Z_k \leq z + \delta) + \varepsilon \leq \mathbb{P}(Z \leq z + \delta) + 2\varepsilon \leq \mathbb{P}(Z \leq z) + 3\varepsilon.$$

Here, (4.3) was used again in the last inequality. In the same way (with (4.5) and $z - \delta$), we get

$$\liminf_{n \to \infty} \mathbb{P}(T_n \leq z) \geq \mathbb{P}(Z \leq z) - 3\varepsilon,$$

and the claim follows, since ε was arbitrary.   □

4.6 Theorem (Central Limit Theorem for Stationary m-Dependent Sequences)
Suppose $(Y_j)_{j \geq 1}$ is a stationary m-dependent sequence with $\mathbb{E}(Y_1^2) < \infty$ and $\sigma^2 > 0$, where σ^2 is defined in (4.2). Let $S_n := Y_1 + \ldots + Y_n$, $n \geq 1$. Then

$$\frac{S_n - \mathbb{E}(S_n)}{\sqrt{\mathbb{V}(S_n)}} \xrightarrow{\mathcal{D}} \mathrm{N}(0,1) \quad \text{as } n \to \infty.$$

Proof We follow the proof given in [FER], p. 70. W.l.o.g. let $\mu = \mathbb{E}(Y_1) = 0$. The idea of the proof is to suitably decompose the sum S_n and apply the Lindeberg–Lévy central limit theorem and Slutsky's lemma. For this purpose, we choose an integer $k > m$, which is initially fixed. Then we define s and r by $n = s(k+m) + r$ and $0 \leq r < k+m$. Moreover, we set

$$V_{k,j} := \sum_{i=1}^{k} Y_{j(k+m)+i}, \qquad W_{k,j} := \sum_{i=k+1}^{k+m} Y_{j(k+m)+i}, \qquad j = 0, \ldots, s-1,$$

$$R_n := \sum_{i=1}^{r} Y_{s(m+k)+i}.$$

These random variables sum the Y_j over pairwise disjoint blocks, alternating between large blocks of length k and small blocks of length m, with a (possibly nonexistent) remainder of length r left over (see the following figure for the case $s = 6$, $m = 2$ and $r = 4$).

$$\underbrace{V_{k,0}} \quad \underbrace{W_{k,0}} \quad \underbrace{V_{k,1}} \quad \underbrace{W_{k,1}} \quad \underbrace{V_{k,2}} \quad \underbrace{W_{k,2}} \quad \cdots \quad \underbrace{V_{k,s-1}} \quad \underbrace{W_{k,s-1}} \quad \underbrace{R_n}$$

If we set

$$S_{n,1} := \sum_{j=0}^{s-1} V_{k,j}, \qquad S_{n,2} := \sum_{j=0}^{s-1} W_{k,j},$$

then

$$S_n = S_{n,1} + S_{n,2} + R_n. \tag{4.6}$$

Due to m-dependence and stationarity, $V_{k,0}, \ldots, V_{k,s-1}$ are i.i.d. random variables, to which we can apply the Lindeberg–Lévy central limit theorem. The components $S_{n,2}$ and R_n in (4.6) will turn out to be asymptotically negligible. We now apply Proposition 4.5, using

$$Z_{n,k} := \frac{S_{n,1} + R_n}{\sqrt{n}}, \qquad X_{n,k} := \frac{S_{n,2}}{\sqrt{n}}. \tag{4.7}$$

In doing so, the previously suppressed dependence of the random variables $S_{n,1}$, $S_{n,2}$ and R_n on k was highlighted. Thus

$$T_n = \frac{S_n}{\sqrt{n}} = Z_{n,k} + X_{n,k},$$

and we claim that assumptions (a), (b), and (c) of Proposition 4.5 are fulfilled. To show (a), we note that $\mathbb{E}(X_{n,k}) = 0$ and

$$\mathbb{V}(X_{n,k}) = \frac{1}{n}\mathbb{V}(S_{n,2}) = \frac{s}{n}\,\mathbb{V}(S_m).$$

The last equals sign holds due to the stationarity of the sequence (Y_j). Since $n = s(k + m) + r$ with $0 \leq r < k + m$, we have $\frac{s}{n} \leq \frac{1}{k+m}$. Chebyshev's inequality now yields

$$\sup_{n \in \mathbb{N}} \mathbb{P}(|X_{n,k}| \geq \delta) \leq \frac{\mathbb{V}(S_m)}{(k+m)\delta^2}$$

and thus (a). To prove (b), we highlight the dependence of s in the equation $n = s(k + m) + r$ on n by indexing with n. Thus

$$\frac{S_{n,1}}{\sqrt{n}} = \sqrt{\frac{s_n}{n}} \cdot \frac{1}{\sqrt{s_n}} \sum_{j=0}^{s_n-1} V_{k,j},$$

where $V_{k,0}, \ldots, V_{k,s_n-1}$ are i.i.d. random variables. By the Lindeberg–Lévy central limit theorem, and due to stationarity,

$$\frac{1}{\sqrt{s_n}} \sum_{j=0}^{s_n-1} V_{k,j} \xrightarrow{\mathcal{D}} \mathrm{N}(0, \mathbb{V}(S_k))$$

as $n \to \infty$. Since $\lim_{n\to\infty} \sqrt{s_n/n} = 1/\sqrt{k+m}$, it follows that

$$\frac{S_{n,1}}{\sqrt{n}} \xrightarrow{\mathcal{D}} Z_k, \quad \text{where } Z_k \sim \mathrm{N}\left(0, \frac{\mathbb{V}(S_k)}{k+m}\right).$$

Due to stationarity and the Cauchy–Schwarz inequality, the random variable R_n in (4.7) satisfies

$$\mathbb{E}\left[\frac{R_n}{\sqrt{n}}\right] = 0, \quad \mathbb{V}\left(\frac{R_n}{\sqrt{n}}\right) = \frac{\mathbb{V}(S_r)}{n} \leq \frac{(k+m)^2 \sigma_{0,0}}{n}.$$

Self-question 2: Can you derive the inequality $\mathbb{V}(S_r) \leq (k+m)^2 \sigma_{0,0}$?

Thus, $R_n/\sqrt{n} \xrightarrow{P} 0$, and Slutsky's lemma yields $Z_{n,k} \xrightarrow{\mathcal{D}} Z_k$. Hence, also assumption (b) of Proposition 4.5 is fulfilled. Since

$$\lim_{k\to\infty} \frac{\mathbb{V}(S_k)}{k+m} = \lim_{k\to\infty} \frac{\mathbb{V}(S_k)}{k} = \sigma^2 > 0,$$

it follows that

$$Z_k \xrightarrow{\mathcal{D}} Z \sim \mathrm{N}(0, \sigma^2) \text{ as } k \to \infty.$$

Consequently, assumption (c) of Proposition 4.5 also holds, and we obtain $T_n \xrightarrow{\mathcal{D}} Z$, that is, $S_n/\sqrt{n} \xrightarrow{\mathcal{D}} \mathrm{N}(0, \sigma^2)$. This proves the claim with Slutsky's lemma.

$\square$

Self-question 3: How does Slutsky's lemma come into play here?

4.7 Examples

(a) In Example 4.4 (a), that is, $Y_j = 1\{X_{j-1} > X_j < X_{j+1}\}$, suppose the distribution function F of X_1 is continuous. Then (Problem 4.3)

$$\sqrt{n}\left(\frac{S_n}{n} - \frac{1}{3}\right) \xrightarrow{\mathcal{D}} N\left(0, \frac{2}{45}\right) \quad \text{as } n \to \infty.$$

(b) In Example 4.4 (b), that is, $Y_j = (1 - X_{j-1})X_j \dots X_{j+r-1}(1 - X_{j+r})$, set $q := 1 - p$ and $S_n := Y_1 + \dots + Y_n$. Then (Problem 4.5)

$$\sqrt{n}\left(\frac{S_n}{n} - q^2 p^r\right) \xrightarrow{\mathcal{D}} N(0, \sigma^2),$$

where $\sigma^2 = q^2 p^r + 2q^3 p^{2r} - (2r+3)q^4 p^{2r}$.

    This chapter provided a small insight into the problem of establishing central limit theorems for sequences of dependent random variables. If $(Y_j)_{j\geq 1}$ is a sequence of random variables on a probability space $(\Omega, \mathcal{A}, \mathbb{P})$, one can try to weaken the property of m-dependence in such a way that—loosely speaking—events that relate to $Y_1, Y_2, \dots$ and are increasingly temporally separated should be less and less dependent. This requirement can be made more precise by introducing the σ-fields $\mathcal{F}_s := \sigma(Y_1, \dots, Y_s)$ and $\mathcal{F}_\ell^\infty := \sigma(Y_\ell, Y_{\ell+1}, \dots)$ for $s, \ell \geq 1$. If, for arbitrary sub-σ-fields $\mathcal{B}$ and $\mathcal{C}$ of $\mathcal{A}$, one defines

$$\alpha(\mathcal{B}, \mathcal{C}) := \sup_{B \in \mathcal{B}, C \in \mathcal{C}} |\mathbb{P}(B \cap C) - \mathbb{P}(B)\mathbb{P}(C)|,$$

the so-called *strong mixing property*, introduced in [ROS], states that the sequence $\beta(n)$ defined by

$$\beta(n) := \sup_{s \geq 1} \alpha(\mathcal{F}_s, \mathcal{F}_{s+n}^\infty)$$

converges to 0 as $n \to \infty$.

    If the sequence (Y_j) is m-dependent then, by Definition 4.1, the σ-fields $\mathcal{F}_s$ and $\mathcal{F}_{s+m+1}^\infty$ are independent for each $s \geq 1$, which results in $\beta(n) = 0$ for each $n \geq m + 1$. The strong mixing property is therefore weaker than the property of m-dependence, for each $m \geq 1$. In [ROS] a central limit theorem for not necessarily stationary sequences under a strong mixing property is proved. Numerous further central limit theorems for sequences of dependent random variables are contained in [HHE].

In Chap. 8 we will prove Theorems 8.9 and 8.24. These are two central limit theorems for sequences of random variables that have a certain dependence structure.

Answers to the Self-Questions

Answer 1 We have $\sum_{i,j=1}^{n} \mathrm{Cov}(Y_i, Y_j) = \sum_{i=1}^{n} \mathbb{V}(Y_i) + 2\sum_{i=1}^{n-1} \sum_{j=i+1}^{n} \mathrm{Cov}(Y_i, Y_j)$ and $\mathbb{V}(Y_i) = \sigma_{0,0}$ as well as $\mathrm{Cov}(Y_i, Y_j) = \sigma_{0,j-i}$. For each $k \in \{1, \ldots, m\}$, the term $\sigma_{0,k}$ appears exactly $(n-k)$ times in the last double sum. Since $\sigma_{0,k} = 0$ if $k > m$, the assertion follows.

Answer 2 Note that $\mathbb{V}(S_r) = \sum_{i=1}^{r} \sum_{j=1}^{r} \mathrm{Cov}(Y_i, Y_j)$. By the Cauchy–Schwarz inequality, $|\mathrm{Cov}(Y_i, Y_j)| \leq \sqrt{\mathbb{V}(Y_i)\mathbb{V}(Y_j)} = \sigma_{0,0}$. Because $r < m + k$, the assertion follows.

Answer 3 We have

$$\frac{S_n}{\sqrt{\mathbb{V}(S_n)}} = \frac{S_n}{\sigma\sqrt{n}} \cdot \frac{\sigma}{\sqrt{\mathbb{V}(S_n)/n}}.$$

The first factor on the right-hand side converges in distribution to $N(0, 1)$ according to the mapping theorem, and the second factor converges to 1. The assertion then follows from part (b) of Lemma 1.23.

Problems

4.1 Problem Show: If a sequence $(Y_j)_{j\geq 1}$ of random variables is 0-dependent, then $Y_1, Y_2, \ldots$ are independent.

4.2 Problem Provide an example that the case $\sigma_{0,0} > 0$ and $\sigma^2 = 0$ can occur.

Hint Use a function of blocks of length 2 of an i.i.d. sequence $X_1, X_2, \ldots$ with $X_1 \sim$ Bin$(1, \frac{1}{2})$.

4.3 Problem Let $X_0, X_1, \ldots$ be i.i.d. random variables with continuous distribution function F. The random variable

$$Y_j := \mathbf{1}\{X_{j-1} > X_j < X_{j+1}\}, \quad j = 1, 2, \ldots,$$

indicates whether a local minimum occurs at time j in the time series $(X_n)_{n\geq0}$. Let $S_n :=$ $Y_1 + \ldots + Y_n$ be the number of local minima in the sequence $X_0, X_1, \ldots$ up to time n. Show that

$$\sqrt{n}\left(\frac{S_n}{n} - \frac{1}{3}\right) \xrightarrow{D} N\left(0, \frac{2}{45}\right) \quad \text{as } n \to \infty.$$

4.4 Problem Let $X_0, X_1, \ldots$ be i.i.d. random variables with $X_0 \sim \text{Bin}(1, p)$, where $0 < p < 1$. Show that

$$\frac{1}{\sqrt{n}} \sum_{j=1}^{n} (X_{j-1}X_j - p^2) \xrightarrow{D} N(0, p^2(1 + 2p - 3p^2)) \quad \text{as } n \to \infty.$$

4.5 Problem In the setting of Example 4.4 (b), prove that

$$\sqrt{n}\left(\frac{S_n}{n} - q^2 p^r\right) \xrightarrow{D} N(0, \sigma^2),$$

where $\sigma^2 = q^2 p^r + 2q^3 p^{2r} - (2r+3)q^4 p^{2r}$.

4.6 Problem Suppose q is an integer and $(\varepsilon_j)_{j \geq 1-q}$ is a sequence of i.i.d. random variables with $\mathbb{E}(\varepsilon_0^2) < \infty$, $\mathbb{E}(\varepsilon_0) = 0$ and $\tau^2 := \mathbb{V}(\varepsilon_0) > 0$. Furthermore, let $\vartheta_0, \ldots, \vartheta_q$ be real numbers with $\vartheta_0 = 1$, $\vartheta_q \neq 0$ and $\vartheta_0 + \ldots + \vartheta_q \neq 0$. Then the sequence defined by

$$X_j := \sum_{\ell=0}^{q} \vartheta_\ell \varepsilon_{j-\ell}, \quad j = 1, 2, \ldots$$

is called a *moving average process of order q*.

Prove that there is a $\sigma^2 \in (0, \infty)$ such that

$$\frac{1}{\sqrt{n}} \sum_{j=1}^{n} X_j \xrightarrow{D} N(0, \sigma^2) \text{ as } n \to \infty.$$

Determine σ^2.

The Multivariate Normal Distribution

5

In this chapter, we will learn about the general d-variate (d-dimensional) normal distribution. Here, d denotes any integer. Elements of $\mathbb{R}^d$ and d-dimensional random vectors are written as row or column vectors, as appropriate. If $x \in \mathbb{R}^d$ is a column vector, the vector $x^\top$, denoted with a superscript transpose sign, is the row vector resulting from x. More generally, $A^\top$ stands for the matrix transposed to a matrix A. The zero vector in $\mathbb{R}^d$ is designated as 0_d. In this chapter, vectors are always column vectors. We need two terms in advance that might already be familiar from an introductory stochastics lecture.

5.1 Definition (Expectation of a Random Vector, Covariance Matrix)
Let $X = (X_1, \ldots, X_d)^\top$ be a d-dimensional random vector.

(a) If $\mathbb{E}|X_j| < \infty$ for each $j \in \{1, \ldots, d\}$, then

$$\mathbb{E}(X) := \left(\mathbb{E}X_1, \ldots, \mathbb{E}X_d\right)^\top$$

is called the *expectation* of X.

(b) If $\mathbb{E}(X_j^2) < \infty$ for each $j \in \{1, \ldots, d\}$, then the $d \times d$ matrix

$$\mathbb{C}\mathrm{ov}(X) := \left(\mathrm{Cov}(X_j, X_k)\right)_{1 \le j,k \le d}$$

is called the *covariance matrix* of X.

Thus, the expectation of a random vector is the vector of the expectations of its component random variables. If, more generally, $Y := \left(Y_{j,\ell}\right)_{1 \le j \le k, 1 \le \ell \le m}$ is written as a random $k \times m$ matrix ($\mathbb{R}^{k \cdot m}$-valued random vector) with $\mathbb{E}|Y_{j,\ell}| < \infty$ for each pair

N. Henze, *Asymptotic Stochastics*, Mathematics Study Resources 10, https://doi.org/10.1007/978-3-662-68923-3_5

(j, ℓ), then one sets

$$\mathbb{E}(Y) := \big(\mathbb{E}(Y_{j,\ell})\big)_{k \times m}.$$

With this definition, it follows that

$$\mathbb{Cov}(X) = \mathbb{E}\big[(X - \mathbb{E}X) \cdot (X - \mathbb{E}X)^\top\big]$$
$$= \mathbb{E}\big[XX^\top\big] - \mathbb{E}X \cdot (\mathbb{E}X)^\top,$$

which generalizes the equation $\mathbb{V}(X) = \mathbb{E}(X^2) - (\mathbb{E}X)^2$ for the variance of a real-valued random variable X.

We will several times encounter the behavior of expectations and covariance matrices under affine transformations, as stated below. These results follow from the linearity of the expectation operator, the bilinearity of taking covariances of pairs of random variables, and the definition of $\mathbb{E}(X)$ and $\mathbb{Cov}(X)$.

5.2 Remark (Affine Transformations)
If A is an $s \times d$ matrix and $b \in \mathbb{R}^s$, then

(a) $\mathbb{E}(AX + b) = A\mathbb{E}(X) + b$,
(b) $\mathbb{Cov}(AX + b) = A\,\mathbb{Cov}(X)\,A^\top$.

Self-question 1: What could a proof of 5.2 (b) look like?

5.3 Theorem (Properties of Covariance Matrices)
The covariance matrix $\mathbb{Cov}(X)$ of a random vector X has the following properties:

(a) $\mathbb{Cov}(X)$ is symmetric and positive-semidefinite.
(b) $\mathbb{Cov}(X)$ is singular if and only if there exists a vector $c \neq 0_d$ and a real number γ such that $\mathbb{P}(c^\top X = \gamma) = 1$.

Proof

(a) Since $\mathrm{Cov}(U, V) = \mathrm{Cov}(V, U)$ applies in general, $\mathbb{Cov}(X)$ is a symmetric matrix. Moreover, the bilinearity of taking covariances implies that for any vector $c := (c_1, \ldots, c_d)^\top \in \mathbb{R}^d$

$$0 \le \mathbb{V}(c^\top X) = \mathrm{Cov}\bigg(\sum_{j=1}^{d} c_j X_j, \sum_{k=1}^{d} c_k X_k\bigg) = \sum_{j=1}^{d}\sum_{k=1}^{d} c_j c_k\, \mathrm{Cov}(X_j, X_k)$$

$$= c^\top \mathbb{Cov}(X) c.$$

Thus, $\mathbb{C}\text{ov}(X)$ is a positive-semidefinite matrix.

(b) The covariance matrix $\mathbb{C}\text{ov}(X)$ is singular, that is, not invertible, if and only if there exists a $c = (c_1, \ldots, c_d)^\top \in \mathbb{R}^d$ with $c \neq 0_d$ such that $\mathbb{V}(c^\top X) = c^\top \mathbb{C}\text{ov}(X)c = 0$. The latter property is the same as $\mathbb{P}(c^\top X = \gamma) = 1$ for some real number γ.

$\square$

The singularity of a covariance matrix $\mathbb{C}\text{ov}(X)$ is thus characterized by the fact that, with probability 1, X takes on values in a suitable hyperplane $\mathcal{H} := \{x \in \mathbb{R}^d : c^\top x = \gamma\}$ of $\mathbb{R}^d$.

5.4 Example (Multinomial Distribution)

Consider n repeated independent trials, where each such trial can have one of s possible outcomes, called *hit of the j-th kind*. Here, $s \geq 2$ and $j \in \{1, \ldots, s\}$. If a hit of the j-th kind has probability p_j, and the random variable X_j models the number of achieved hits of j-th kind in the n trials ($j = 1, \ldots, s$), then the random vector $X := (X_1, \ldots, X_s)^\top$ has the multinomial distribution $\text{Mult}(n; p_1, \ldots, p_s)$. Because $\mathbb{P}(X_1 + \ldots + X_s = n) = 1$, the covariance matrix of X is singular. For this insight, one does not need the entries

$$\text{Cov}(X_i, X_j) = n(\delta_{i,j} p_i - p_i p_j), \quad i, j \in \{1, \ldots, s\},$$

of $\mathbb{C}\text{ov}(X)$. Here, $\delta_{i,j} := 1$, if $i = j$, and $\delta_{i,j} := 0$, otherwise, denotes the *Kronecker[1]-symbol*.

Usually, the normal distribution $\text{N}(\mu, \sigma^2)$ with expectation μ and variance σ^2 is introduced only for the case $\sigma^2 > 0$. To be able to define the general d-variate normal distribution, we extend the class of univariate (one-dimensional) normal distributions to include all *one-point distributions* (Dirac[2]-measures $\delta_\mu =: \text{N}(\mu, 0)$).

The following definition suggests how one could define the normal distribution also in infinite-dimensional Hilbert spaces (see Definition 17.24).

---

[1] Leopold Kronecker (1823–1891), his long-term management of a family-owned estate allowed him to live according to his mathematical inclinations from 1855 onwards. In 1883 he became a professor at the University of Berlin. Main areas of work: Algebra, number theory, complex analysis.

[2] Paul Adrien Maurice Dirac (1902–1984), theoretical physicist and mathematician, professor of mathematics at the universities of Cambridge (from 1940) and Oxford (from 1953), Nobel Prize 1933 for work on quantum mechanics. Among many other discoveries, Dirac derived an equation, named after him, which describes the behavior of fermions and predicted the existence of antimatter.

5.5 Definition (d-Variate Normal Distribution)

The random vector $X = (X_1, \ldots, X_d)^\top$ has a *d-variate normal distribution*, if for each $c = (c_1, \ldots, c_d)^\top \in \mathbb{R}^d$ the linear combination (the inner product)

$$c^\top X = \sum_{j=1}^{d} c_j X_j \tag{5.1}$$

has a univariate normal distribution.

Since in (5.1) any number of the components of c can be zero, we get the following result:

5.6 Corollary *If $X = (X_1, \ldots, X_d)^\top$ has a d-variate normal distribution, then:*

(a) *Let $s \in \{1, \ldots, d\}$ and $1 \le i_1 < \ldots < i_s \le d$. Then $(X_{i_1}, \ldots, X_{i_s})^\top$ has an s-dimensional normal distribution. In particular, each of the components $X_1, \ldots, X_d$ of X is normally distributed.*

(b) $\mathbb{E}(X_j^2) < \infty$ *for each $j \in \{1, \ldots, d\}$, and thus $\mathbb{E}(X)$ and $\mathbb{C}ov(X)$ exist.*

In the situation of Definition 5.5 we have

$$\mathbb{E}(c^\top X) = c^\top \mathbb{E}(X), \quad \mathbb{V}(c^\top X) = c^\top \mathbb{C}ov(X) c.$$

Because of Theorem 1.17, the distribution $\mathbb{P}^X$ of X is uniquely determined by $\mu := \mathbb{E}(X)$ and $\Sigma := \mathbb{C}ov(X)$. For this reason, one says that *X has a d-variate normal distribution with expectation μ and covariance matrix Σ* and writes briefly

$$X \sim N_d(\mu, \Sigma) \quad \text{or} \quad \mathbb{P}^X = N_d(\mu, \Sigma).$$

The following result states that normally distributed random vectors are transformed into normally distributed random vectors under affine mappings.

5.7 Corollary (Reproduction Theorem for the d-Variate Normal Distribution)

Suppose X has the normal distribution $N_d(\mu, \Sigma)$. If A is any $s \times d$ matrix and $b \in \mathbb{R}^s$, then

$$AX + b \sim N_s(A\mu + b, A\Sigma A^\top).$$

Proof For each $h \in \mathbb{R}^s$, the random variable $h^\top(AX + b) = (A^\top h)^\top X + h^\top b$ has a univariate normal distribution. $\qquad \square$

To ensure the existence of multivariate normal distributions, the following result from linear algebra is helpful.

5.8 Lemma *For each symmetric, positive-semidefinite $d \times d$ matrix Σ, there is a matrix A with $\Sigma = AA^{\top}$.*

Proof The matrix Σ has an orthonormal basis of eigenvectors $v_1, \ldots, v_d$ with associated non-negative eigenvalues $\lambda_1, \ldots, \lambda_d$, that is,

$$\Sigma v_j = \lambda_j v_j, \quad v_j^{\top} v_k = \delta_{j,k}, \quad j, k \in \{1, \ldots, d\}.$$

Let $V := (v_1 \cdots v_d)$ be the $d \times d$ matrix created by lining up the column vectors $v_1, \ldots, v_d$. Then $V^{\top} = V^{-1}$, and if we define Λ as the diagonal matrix $\Lambda := \operatorname{diag}(\lambda_1, \ldots, \lambda_d)$, we get $\Sigma V = V\Lambda$ and $\Sigma = V\Lambda V^{\top}$. Putting $A := V\Lambda^{1/2}$ with $\Lambda^{1/2} := \operatorname{diag}(\sqrt{\lambda_1}, \ldots, \sqrt{\lambda_d})$ then yields

$$\Sigma = V\Lambda^{1/2}\Lambda^{1/2}V^{\top} = AA^{\top}.$$

$\square$

5.9 Remark In Chap. 10 we will need that every symmetric positive-semidefinite matrix $\Sigma \in \mathbb{R}^{d \times d}$ has a symmetric positive-semidefinite square root, which is denoted by $\Sigma^{1/2}$. The matrix $\Sigma^{1/2}$ thus satisfies the equation $\Sigma = \Sigma^{1/2} \Sigma^{1/2}$. With the notations in the proof of Lemma 5.8 one only has to set $\Sigma^{1/2} := V\Lambda^{1/2}V^{\top}$. If Σ is positive definite then $\Sigma^{1/2}$ also shares this property.

5.10 Theorem (Existence of the Distribution $N_d(\mu, \Sigma)$)
For every $\mu \in \mathbb{R}^d$ and every symmetric positive-semidefinite $d \times d$ matrix Σ there exists a random vector X with $X \sim N_d(\mu, \Sigma)$.

Proof Let $Y_1, \ldots, Y_d$ be i.i.d. random variables with a standard normal distribution. By the addition theorem for the normal distribution, $Y := (Y_1, \ldots, Y_d)^{\top}$ has the normal distribution $N_d(0_d, I_d)$, where I_d denotes the $d \times d$ unit matrix. Let A be a $d \times d$ matrix with $\Sigma = AA^{\top}$. The reproduction theorem 5.7 yields

$$X := AY + \mu \sim N_d(\mu, \Sigma). \tag{5.2}$$

Each d-variate normal distribution thus arises as an affine transformation of the normal distribution $N_d(0_d, I_d)$.

$\square$

It is well-known that independent random variables X and Y with $\mathbb{E}(X^2) < \infty$ and $\mathbb{E}(Y^2) < \infty$ are uncorrelated, that is, $\operatorname{Cov}(X, Y) = 0$. The converse, however, does not generally hold. A notable exception is provided by the multivariate normal distribution (see also Problem 5.1).

5.11 Theorem ($\mathbf{N}_d(\mu, \Sigma)$ and Independence)
Let $X := (X_1, \ldots, X_k)^\top$ and $Y := (Y_1, \ldots, Y_\ell)^\top$ be k- and ℓ-dimensional random vectors, respectively, that are defined on the same probability space. If the random vector $(X^\top \, Y^\top)^\top$ has a $(k+\ell)$-variate normal distribution, then:

$$X, Y \text{ independent} \iff \mathrm{Cov}(X_i, Y_j) = 0 \quad \text{for every pair } (i, j) \text{ with } i \neq j.$$

Proof Since the implication "$\Longrightarrow$" is generally valid, only "$\Longleftarrow$" needs to be shown. Let $0_{r \times s}$ be the $r \times s$ zero matrix. By assumption, the covariance matrix Σ of $(X^\top \, Y^\top)^\top$ is given by

$$\Sigma = \begin{pmatrix} \mathbb{C}\mathrm{ov}(X) & 0_{k \times \ell} \\ 0_{\ell \times k} & \mathbb{C}\mathrm{ov}(Y) \end{pmatrix}.$$

By Lemma 5.8, there are $d \times d$ matrices A, B with $\mathbb{C}\mathrm{ov}(X) = AA^\top$ and $\mathbb{C}\mathrm{ov}(Y) = BB^\top$. Let $Z_1, \ldots, Z_{k+\ell}$ be i.i.d. random variables with a standard normal distribution. Then the random vector $(U^\top \, V^\top)^\top$, defined by

$$\begin{pmatrix} U_1 \\ \vdots \\ U_k \\ V_1 \\ \vdots \\ V_\ell \end{pmatrix} := \begin{pmatrix} A & 0_{k \times \ell} \\ 0_{\ell \times k} & B \end{pmatrix} \begin{pmatrix} Z_1 \\ \vdots \\ Z_k \\ Z_{k+1} \\ \vdots \\ Z_{k+\ell} \end{pmatrix} + \begin{pmatrix} \mathbb{E}X \\ \mathbb{E}Y \end{pmatrix}$$

with $U := (U_1, \ldots, U_k)^\top$ and $V := (V_1, \ldots, V_\ell)^\top$, has the $(k + \ell)$-variate normal distribution

$$\mathbf{N}_{k+\ell}\left(\begin{pmatrix} \mathbb{E}X \\ \mathbb{E}Y \end{pmatrix}, \begin{pmatrix} AA^\top & 0_{k \times \ell} \\ 0_{\ell \times k} & BB^\top \end{pmatrix} \right).$$

Since $U = A(Z_1 \cdots Z_k)^\top + \mathbb{E}(X)$ and $V = B(Z_{k+1} \cdots Z_{k+\ell})^\top + \mathbb{E}(Y)$ are functions of $Z_1, \ldots, Z_k$ and $Z_{k+1}, \ldots, Z_{k+\ell}$, respectively, they are independent. Furthermore, U and X as well as V and Y have the same distribution. Since $(X^\top Y^\top)^\top$ and $(U^\top V^\top)^\top$ also have the same distribution, the product measure notation "$\otimes$" yields

$$\mathbb{P}^{(X^\top Y^\top)^\top} = \mathbb{P}^{(U^\top V^\top)^\top} = \mathbb{P}^U \otimes \mathbb{P}^V = \mathbb{P}^X \otimes \mathbb{P}^Y$$

and thus the assertion.   □

By induction over d, we now obtain the following result:

5.12 Corollary *If $X = (X_1, \ldots, X_d)^\top$ has the normal distribution $N_d(\mu, \Sigma)$, then:*

$$X_1, \ldots, X_d \ \text{independent} \iff \text{Cov}(X) \ \text{is a diagonal matrix.}$$

As the next result shows, the addition theorem for the normal distribution also holds in the general multivariate case.

5.13 Theorem (Addition Theorem) *If X and Y are independent d-variate random vectors with the normal distributions $X \sim N_d(\mu, \Sigma)$ and $Y \sim N_d(\nu, T)$, then*

$$X + Y \sim N_d(\mu + \nu, \Sigma + T).$$

Proof The proof is left to you as Problem 5.2. In Problem 5.5 you are to show that covariance matrices generally add up when independent random vectors are added.

Among the normal distributions, the so-called *non-degenerate* ones play a prominent role. The normal distribution $N_d(\mu, \Sigma)$ is called *non-degenerate*, if the covariance matrix Σ is invertible, otherwise *degenerate*. If $d \geq 2$ then Σ is invertible if and only if $\det(\Sigma) > 0$. In the case $d = 1$, the degenerate normal distributions are the one-point distributions δ_μ with $\mu \in \mathbb{R}$.   □

5.14 Theorem (Density of a Non-degenerate Normal Distribution)
A random vector X with the non-degenerate normal distribution $N_d(\mu, \Sigma)$ has the Lebesgue density

$$f(x) = \frac{1}{(2\pi)^{d/2}\sqrt{\det(\Sigma)}} \exp\left(-\frac{1}{2}(x - \mu)^\top \Sigma^{-1}(x - \mu)\right), \qquad x \in \mathbb{R}^d. \qquad (5.3)$$

Proof Let $\Sigma = AA^\top$ and $Z := (Z_1, \ldots, Z_d)^\top \sim N_d(0_d, I_d)$. Due to the independence of $Z_1, \ldots, Z_d$ and $Z_j \sim N(0, 1)$ for each $j \in \{1, \ldots, d\}$, Z has the density

$$f_Z(z) = \prod_{j=1}^d \left(\frac{1}{\sqrt{2\pi}} \exp\left(-\frac{z_j^2}{2}\right)\right) = \frac{1}{(2\pi)^{d/2}} \exp\left(-\frac{z^\top z}{2}\right), \qquad z \in \mathbb{R}^d.$$

Here, we set $z := (z_1, \ldots, z_d)^\top$. Since X has the same distribution as $AZ + \mu$ and the density $f_{AZ+\mu}$ of $AZ + \mu$ is obtained from that of Z using the formula

$$f_{AZ+\mu}(x) = \frac{f_Z(A^{-1}(x - \mu))}{|\det(A)|}, \quad x \in \mathbb{R}^d$$

(see, e.g., [BI3], p. 261), the assertion follows by direct calculation.   □

Self-question 3: What could such a direct calculation look like?

According to (5.3), the density f of the normal distribution $N_d(\mu, \Sigma)$ is constant on the sets $\{x \in \mathbb{R}^d : (x - \mu)^\top \Sigma^{-1}(x - \mu) = c\}$ with $c > 0$, that is, on ellipsoids in $\mathbb{R}^d$ with center μ. Due to Lemma 5.8, $\Sigma = V \Lambda V^\top$ and thus $\Sigma^{-1} = V \Lambda^{-1} V^\top$. Here, $V = (v_1 \ldots v_d)$ is the matrix of normalized eigenvectors of Σ, and $\Lambda = \mathrm{diag}(\lambda_1, \ldots, \lambda_d)$ is the diagonal matrix consisting of the positive eigenvalues of Σ. If we set $\mu = 0_d$ for simplicity (the general case only causes a shift of the center of density), then

$$x^\top \Sigma^{-1} x = x^\top V \Lambda^{-1} V^\top x = y^\top \Lambda^{-1} y = \frac{y_1^2}{\lambda_1} + \ldots + \frac{y_d^2}{\lambda_d},$$

where $y := (y_1, \ldots, y_d)^\top = V^\top x$. After an orthogonal transformation, these ellipsoids are therefore axis-parallel, and the vector of the lengths of the major axes is a multiple of $(\sqrt{\lambda_1}, \ldots, \sqrt{\lambda_d})$. The orthogonal transformation $x \mapsto V^\top x$ is obsolete if $\Sigma = \mathrm{diag}(\sigma_1^2, \ldots, \sigma_d^2)$ is a diagonal matrix, that is, if the components of X are independent, see Fig. 5.1 for the case $d = 2$.

5.15 Principal Component Analysis

As in the proof of Lemma 5.8, let $\Sigma = V \Lambda V^\top$, where $\Lambda = \mathrm{diag}(\lambda_1, \ldots, \lambda_d)$, and $A := V \Lambda^{1/2}$. According to (5.2), if X has the normal distribution $N_d(\mu, \Sigma)$, then

$$X \overset{\mathcal{D}}{=} V \Lambda^{1/2} Y + \mu = \sqrt{\lambda_1} Y_1 v_1 + \ldots + \sqrt{\lambda_d} Y_d v_d + \mu. \tag{5.4}$$

Fig. 5.1 Contour lines of the density of a bivariate (two-dimensional) normal distribution in the case $\Sigma = \mathrm{diag}(\sigma_1^2, \sigma_2^2)$; the ratio of the major axes is σ_1/σ_2

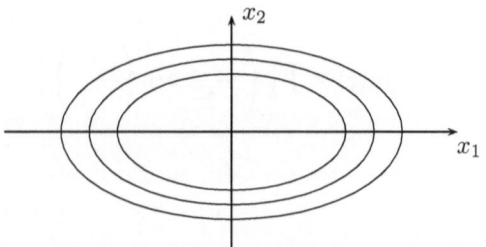

Fig. 5.2 Illustrating the
principal component
representation

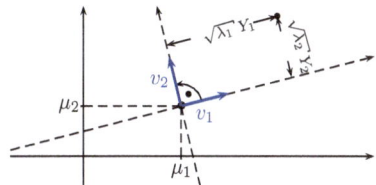

Here, $Y := (Y_1, \ldots, Y_d)^\top$, and $Y_1, \ldots, Y_d$ are independent and each $N(0, 1)$-distributed random variables. This method of generating the normal distribution $N_d(\mu, \Sigma)$ can be easily illustrated: At the point $\mu = (\mu_1, \ldots, \mu_d)^\top \in \mathbb{R}^d$ the (generally oblique) right-angled coordinate system of the orthonormal vectors $v_1, \ldots, v_d$ is plotted. After generating $Y_1, \ldots, Y_d$, $\sqrt{\lambda_j} Y_j$ is plotted in the direction of v_j ($j \in \{1, \ldots, d\}$) (see Fig. 5.2). If $\lambda_1 \geq \ldots \geq \lambda_d$, then $\sqrt{\lambda_j} Z_j v_j$ is called the *j-th principal component* of X, $j \in \{1, \ldots, d\}$, and (5.4) is the *principal component representation* or *principal component decomposition* of X.

The final result of this chapter is often applied. In this context, we recall the *Chi-square distribution with k degrees of freedom*, labeled χ_k^2. This arises as the distribution of $Y_1^2 + \ldots + Y_k^2$, where $Y_1, \ldots, Y_k$ are i.i.d. random variables with a standard normal distribution.

5.16 Theorem (Quadratic Forms and χ^2-Distribution)

If $X \sim N_d(\mu, \Sigma)$ with $det(\Sigma) > 0$, then

$$(X - \mu)^\top \Sigma^{-1} (X - \mu) \sim \chi_d^2.$$

Proof Recall $X \overset{\mathcal{D}}{=} AZ + \mu$, where $Z \sim N_d(0_d, I_d)$ and $\Sigma = AA^\top$. Putting $Z =: (Z_1, \ldots, Z_d)^\top$ yields

$$(X - \mu)^\top \Sigma^{-1} (X - \mu) \overset{\mathcal{D}}{=} (AZ)^\top \Sigma^{-1} (AZ) = Z^\top A^\top \left(AA^\top\right)^{-1} AZ$$

$$= Z^\top Z = \sum_{j=1}^{d} Z_j^2 \sim \chi_d^2.$$

$\square$

A generalization of the above statement can be found in Problem 12.3.

Answers to the Self-Questions

Answer 1 Let $X = (X_1, \ldots, X_d)^\top$, $A = (a_{mn})_{1 \le m \le s, 1 \le n \le d}$ and $b = (b_1, \ldots, b_s)^\top$. Further, let $Y = AX + b =: (Y_1, \ldots, Y_s)^\top$. Then $Y_i = \sum_{k=1}^d a_{ik} X_k + b_i$, $Y_j = \sum_{\ell=1}^d a_{j\ell} X_\ell + b_j$ $(1 \le i, j \le s)$, and it follows that $\mathrm{Cov}(Y_i, Y_j) = \sum_{k=1}^d \sum_{\ell=1}^d a_{ik} a_{j\ell} \mathrm{Cov}(X_k, X_\ell)$. This covariance matches the entry in the i-th row and j-th column of the matrix $A\mathbb{C}\mathrm{ov}(X)A^\top$.

Answer 2 Because two random vectors X and Y are independent if and only if the joint distribution of X and Y is the product of the distributions of X and of Y.

Answer 3 We have $|\det(A)| = \sqrt{\det(\Sigma)}$, and putting $y := x - \mu$ implies $(A^{-1}y)^\top A^{-1}y = y^\top (A^\top)^{-1} A^{-1}y = y^\top (AA^\top)^{-1}y = y^\top \Sigma^{-1}y$.

Problems

5.1 Problem Let f_ϱ be the density of a random vector $(X, Y)^\top$ with a bivariate normal distribution, where $\mathbb{E}(X) = \mathbb{E}(Y) = 0$, $\mathbb{V}(X) = \mathbb{V}(Y) = 1$, $\mathbb{E}(XY) = \varrho$ and $0 < |\varrho| < 1$. As a convex combination of two densities, $f := \alpha f_\varrho + (1 - \alpha) f_{-\varrho}$, $0 < \alpha < 1$, is a density (of a so-called *mixture distribution*). For a random vector $(U, V)^\top$ with density f, determine the marginal distributions of U and V and the correlation coefficient $\varrho(U, V)$. Conclude: There are uncorrelated normally distributed random variables, whose joint distribution is not a bivariate normal distribution.

Hint Represent the joint distribution of U and V using a suitable indicator variable.

5.2 Problem Prove the addition theorem 5.13 for the multivariate normal distribution.

5.3 Problem Let $d, n \in \mathbb{N}$ with $d \ge 2$, and let $x_1, \ldots, x_n \in \mathbb{R}^d$. Further, put

$$\overline{x}_n := \frac{1}{n} \sum_{j=1}^n x_j, \qquad s_n := \frac{1}{n} \sum_{j=1}^n (x_j - \overline{x}_n)(x_j - \overline{x}_n)^\top.$$

The $d \times d$ matrix s_n is called the *sample covariance matrix* of $x_1, \ldots, x_n$. Prove the following statements:

(a) If Y is a random vector with $\mathbb{P}(Y = x_j) = \frac{1}{n}$, $j \in \{1, \ldots, n\}$, then $\mathbb{E}(Y) = \overline{x}_n$ and $\mathbb{C}\mathrm{ov}(Y) = s_n$.
(b) If s_n is invertible, then $n \ge d + 1$.
(c) If, for every hyperplane $\mathcal{H} \subset \mathbb{R}^d$, there is a $j \in \{1, \ldots, n\}$ with $x_j \notin \mathcal{H}$, then s_n is invertible.

5.4 Problem Let $X_1, \ldots, X_n$, $n \geq 2$, be i.i.d. d-dimensional random vectors on a probability space $(\Omega, \mathcal{A}, \mathbb{P})$. Further, let $\overline{X}_n := n^{-1} \sum_{j=1}^{n} X_j$ and

$$S_n := \frac{1}{n} \sum_{j=1}^{n} (X_j - \overline{X}_n)(X_j - \overline{X}_n)^\top.$$

The random $d \times d$ matrix (matrix-valued mapping $S_n : \Omega \to \mathbb{R}^{d \times d}$) S_n is called *sample covariance matrix* of $X_1, \ldots, X_n$ (although, according to Problem 5.3, the same naming applies to each of the realizations $S_n(\omega)$, $\omega \in \Omega$).

Show: If $n \geq d + 1$, and X_1 has a density with respect to the Borel–Lebesgue measure on $\mathbb{R}^d$, then S_n is invertible with probability one.

5.5 Problem In this and the following problems, let $\| \cdot \|$ denote the Euclidean norm in $\mathbb{R}^d$. Let X and Y be independent d-dimensional random vectors with $\mathbb{E}\|X\|^2 < \infty$ and $\mathbb{E}\|Y\|^2 < \infty$ and covariance matrices Σ and T, respectively. Show that $X + Y$ has the covariance matrix $\Sigma + T$.

5.6 Problem Show that a random vector X with the normal distribution $N_d(\mu, \Sigma)$ has the characteristic function

$$\varphi_X(t) = \mathbb{E}\left[\exp\left(it^\top X \right) \right] = \exp\left(i\mu^\top t - \frac{t^\top \Sigma t}{2} \right), \quad t \in \mathbb{R}^d.$$

Note The characteristic function $\psi(u) = \mathbb{E}[\exp(iuY)]$, $u \in \mathbb{R}$, of a $N(0, 1)$-distributed random variable Y, is given by $\psi(u) = \exp(-u^2/2)$, $u \in \mathbb{R}$.

5.7 Problem Suppose X has a non-degenerate d-variate normal distribution with $\mathbb{E}(X) = 0_d$ and covariance matrix Σ. Let A be a symmetric $d \times d$ matrix. Show: The quadratic form $Q := X^\top A X$ has a χ_d^2-distribution if and only if $A = \Sigma^{-1}$.

Hint for "⟹" First consider the case $\Sigma = I_d$ and use the Cauchy–Schwarz inequality. Here, only the first two moments of Q are used.

5.8 Problem Let $X \sim N_d(0_d, I_d)$ and $a \in \mathbb{R}^d$ with $a \neq 0_d$. Prove the following claims:

(a) The distribution of $Y := \|X + a\|^2$ depends only on d and $\delta^2 := \|a\|^2$. It is called *non-central χ^2-distribution with d degrees of freedom and non-centrality parameter* δ^2 (in short: $Y \sim \chi_{d,\delta^2}^2$).

(b) $\mathbb{E}(Y) = d + \delta^2$, $\quad \mathbb{V}(Y) = 2d + 4\delta^2$.

(c) Let G_{d,δ^2} be the distribution function of the χ_{d,δ^2}^2-distribution. If $\delta^2 < \eta^2$, then $G_{d,\delta^2}(t) > G_{d,\eta^2}(t)$ for each $t > 0$.

Hint for (a) Employ a suitable orthogonal transformation.

Hint for (c) The density of X is a strictly decreasing function of $\|x\|$.

5.9 Problem Suppose the bivariate random vector $(X, Y)^\top$ has the normal distribution

$$\begin{pmatrix} X \\ Y \end{pmatrix} \sim N_2 \left(\begin{pmatrix} 0 \\ 0 \end{pmatrix}, \begin{pmatrix} \sigma^2 & \rho\sigma\tau \\ \rho\sigma\tau & \tau^2 \end{pmatrix} \right),$$

where $\sigma^2 > 0$, $\tau^2 > 0$ and $|\rho| < 1$. Prove the following statements:

(a) $\mathbb{E}(XY) = \rho\sigma\tau$,
(b) $\mathbb{E}(X^2 Y^2) = \sigma^2\tau^2 (1 + 2\rho^2)$,
(c) $\mathbb{E}(X^2 \cos(Y)) = \sigma^2(1 - \rho^2\tau^2)e^{-\frac{1}{2}\tau^2}$.

Hint Generate the distribution of $(X, Y)^\top$ using two independent standard normally distributed random variables.

5.10 Problem Let $\| \cdot \|$ be the Euclidean norm in $\mathbb{R}^d$ and $\mathcal{S}^d := \{x \in \mathbb{R}^d : \|x\| = 1\}$ the surface of the unit sphere in $\mathbb{R}^d$, where $d \geq 2$. The d-dimensional random vector Y has a *uniform distribution on $\mathcal{S}^d$* (in short: $Y \sim U(\mathcal{S}^d)$) if:

(i) $\mathbb{P}(Y \in \mathcal{S}^d) = 1$,   (ii) $\mathbb{P}^{HY} = \mathbb{P}^Y$ for every orthogonal $d \times d$ matrix H.

Let X be a random vector with $X \sim N_d(0_d, I_d)$. Show:

(a) $\frac{X}{\|X\|} \sim U(\mathcal{S}^d)$.
(b) If $Y \sim U(\mathcal{S}^d)$, then $\mathbb{E}(Y) = 0_d$ and $\mathbb{C}ov(Y) = \frac{1}{d} \cdot I_d$.

Hint for (b) Use symmetry arguments.

5.11 Problem Suppose $X_1, \ldots, X_n$, $n \geq 2$, are independent random variables with the same normal distribution $N(\mu, \sigma^2)$. Setting $\overline{X}_n := \frac{1}{n}\sum_{j=1}^{n} X_j$ and $S_n^2 := \frac{1}{n-1}\sum_{j=1}^{n}(X_j - \overline{X}_n)^2$, prove the following claims:

(a) $\overline{X}_n$ and S_n^2 are independent.
(b) $\frac{n-1}{\sigma^2}S_n^2 \sim \chi_{n-1}^2$.

Hint Consider $(Y_1, \ldots, Y_n)^\top = H(X_1, \ldots, X_n)^\top$, where H is any orthogonal $n \times n$ matrix with the property that the last row consists of all entries $1/\sqrt{n}$.

Convergence in Distribution and Central Limit Theorem in $\mathbb{R}^d$

6

Suppose $X, X_1, X_2, \ldots$ are *real-valued* random variables with distribution functions F, $F_1, F_2, \ldots$, respectively. The sequence (X_n) *converges in distribution* to X, and we write $X_n \xrightarrow{\mathcal{D}} X$, if

$$\lim_{n \to \infty} F_n(x) = F(x) \quad \text{for each continuity point } x \text{ of } F. \tag{6.1}$$

In particular when proving central limit theorems, it becomes apparent that we need handy conditions equivalent to (6.1) to prove convergence in distribution. Since $F(x) = \mathbb{P}^X\big((-\infty, x]\big) = \mathbb{E}\big[\mathbf{1}_{(-\infty,x]}(X)\big]$, (6.1) is the same as

$$\lim_{n \to \infty} \mathbb{E}\big[\mathbf{1}_{(-\infty,x]}(X_n)\big] = \mathbb{E}\big[\mathbf{1}_{(-\infty,x]}(X)\big] \quad \text{for each continuity point } x \text{ of } F,$$

that is, the convergence $\mathbb{E}[f(X_n)] \to \mathbb{E}[f(X)]$ for a certain collection of bounded, measurable functions f. In fact, (6.1) is equivalent to

$$\lim_{n \to \infty} \mathbb{E}\big[f(X_n)\big] = \mathbb{E}\big[f(X)\big] \quad \text{for each continuous bounded function } f : \mathbb{R} \to \mathbb{R} \tag{6.2}$$

(cf. Theorem 1.8) as well as to the convergence

$$\lim_{n \to \infty} \mathbb{E}\big[e^{it X_n}\big] = \mathbb{E}\big[e^{it X}\big], \quad t \in \mathbb{R},$$

of the associated characteristic functions (cf. Theorem 1.13).

In this chapter, we learn about the notion of convergence in distribution for d-dimensional random vectors. Also in this case, convergence in distribution can be defined using distribution functions, and so we first introduce the definition and the most

N. Henze, *Asymptotic Stochastics*, Mathematics Study Resources 10, https://doi.org/10.1007/978-3-662-68923-3_6

important properties of distribution functions for d-dimensional random vectors. However, property (6.2) will be crucial for the definition of convergence in distribution of random variables with values in general metric spaces (cf. Chap. 14). Up to Theorem 6.17, we write random vectors as row vectors. Only then does it make sense to switch to column vectors. For the entire chapter, let $\| \cdot \|$ denote the Euclidean norm in $\mathbb{R}^d$, and relations like $x \leq y$ or $x < y$ between vectors $x = (x_1, \ldots, x_d)$ and $y = (y_1, \ldots, y_d)$ are understood component-wise, that is, $x_j \leq y_j$ or $x_j < y_j$ for each $j \in \{1, \ldots, d\}$.

6.1 Definition (Distribution Function of a Random Vector)
If $X = (X_1, \ldots, X_d)$ is a d-dimensional random vector on a probability space $(\Omega, \mathcal{A}, \mathbb{P})$, then the function $F : \mathbb{R}^d \to [0, 1]$, defined by

$$F(x) := \mathbb{P}(X_1 \leq x_1, \ldots, X_d \leq x_d), \qquad x = (x_1, \ldots, x_d) \in \mathbb{R}^d,$$

is called the *distribution function of* X.

If $x = (x_1, \ldots, x_d) \in \mathbb{R}^d$ and $x^{(n)} = (x_{n1}, \ldots, x_{nd}) \in \mathbb{R}^d$, $n \geq 1$, we use the notation $x^{(n)} \downarrow x$ if for each $j \in \{1, \ldots, d\}$ the sequence (x_{nj}) converges monotonically decreasing to x_j.

6.2 Theorem (Properties of a Distribution Function F)

(a) For any $x, y \in \mathbb{R}^d$ with $x \leq y$, let

$$\Delta_x^y F := \sum_{(\varepsilon_1, \ldots, \varepsilon_d) \in \{0,1\}^d} (-1)^{d - \varepsilon_1 - \ldots - \varepsilon_d} F\left(y_1^{\varepsilon_1} x_1^{1 - \varepsilon_1}, \ldots, y_d^{\varepsilon_d} x_d^{1 - \varepsilon_d}\right).$$

Then $\Delta_x^y F \geq 0$ *(generalized monotonicity property of F).*
(b) F is continuous from above in the sense that $\lim_{n \to \infty} F(x^{(n)}) = F(x)$ if $x^{(n)} \downarrow x$.
(c) If $x_{nj} \to -\infty$ for at least one $j \in \{1, \ldots, d\}$ then $F(x^{(n)}) \to 0$.
    If $x_{nj} \to \infty$ for every $j \in \{1, \ldots, d\}$ then $F(x^{(n)}) \to 1$.

The proof is left to you as Problem 6.1. The generalized monotonicity property, which at first glance does not look very comprehensible, is easy to understand in the special case $d = 2$. In this case, it states that the alternating sum $F(y_1, y_2) - F(y_1, x_2) - F(x_1, y_2) + F(x_1, x_2)$ is non-negative. Since, generally speaking, $F(z_1, z_2) = \mathbb{P}(X_1 \leq z_1, X_2 \leq z_2)$ describes the probability mass "southwest of the point (z_1, z_2)", it becomes clear from Fig. 6.1 that $\Delta_x^y F$ equals the probability $\mathbb{P}(x_1 < X_1 \leq y_1, x_2 < X_2 \leq y_2)$ and is therefore non-negative.

If we set $(x, y] := \{z \in \mathbb{R}^d : x < z \leq y\}$, the class $\mathcal{H}^d := \{(x, y] : x, y \in \mathbb{R}^d, x \leq y\}$ of sets is a π-system, and it generates the Borel σ-field $\mathcal{B}^d$ in $\mathbb{R}^d$. Since $\Delta_x^y F = \mathbb{P}^X((x, y])$ (see Problem 6.1), it follows from the uniqueness theorem for measures (Theorem 1.28)

Fig. 6.1 On the generalized monotonicity property in the case $d = 2$

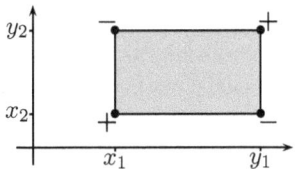

that the distribution $\mathbb{P}^X$ of a random vector X is uniquely determined by its distribution function. If a function $F : \mathbb{R}^d \rightarrow [0, 1]$ satisfies the above properties (a)–(c), there is exactly one probability measure Q on $\mathcal{B}^d$ with $Q((x, y]) = \Delta_x^y F$ for all $(x, y] \in \mathcal{H}^d$. This statement follows from the measure extension theorem (see, e.g., [BI3], p. 177). At this point, it should be noted that the distribution function of a random vector, whose components are all uniformly distributed on the unit interval, is also referred to as a *copula function*. Copula functions are used in various areas for modeling stochastic dependencies (see, e.g., [KUM], pp. 194–209). A mathematical reference on this topic is [DSE].

In the following, $\mathcal{O}^d$ and $\mathcal{A}^d$ denote the collections of open and closed subsets of $\mathbb{R}^d$, respectively. For a set $B \subset \mathbb{R}^d$, let $B° := \cup\{O \in \mathcal{O}^d : O \subset B\}$ be the *set of interior points* (the *interior*) of B, $\overline{B} := \cap\{A \in \mathcal{A}^d : A \supset B\}$ the *closed hull* (the *closure*) of B, and $\partial B := \overline{B} \setminus B°$ the *set of boundary points* (the *boundary*) of B. Furthermore, we write $\mathcal{C}_b(\mathbb{R}^d)$ for the *class of continuous and bounded functions* $f : \mathbb{R}^d \rightarrow \mathbb{R}$ and C_G for the set of all *continuity points* of a function $G : \mathbb{R}^d \rightarrow \mathbb{R}$. For a non-empty set $M \subset \mathbb{R}^d$ and $x \in \mathbb{R}^d$, let

$$\|x - M\| := \inf\{\|x - z\| : z \in M\}$$

be the Euclidean distance from x to M. Using the triangle inequality, it follows that the function $x \mapsto \|x - M\|$ is uniformly continuous. More precisely, we have

$$\big|\|x - M\| - \|y - M\|\big| \le \|x - y\|, \quad x, y \in \mathbb{R}^d. \tag{6.3}$$

Self-question 1: Can you prove inequality (6.3)?

Let $X, X_1, X_2, \ldots$ be d-dimensional random vectors on a probability space $(\Omega, \mathcal{A}, \mathbb{P})$ with associated distributions $Q := \mathbb{P}^X$, $Q_1 := \mathbb{P}^{X_1}$, $Q_2 := \mathbb{P}^{X_n}, \ldots$ and distribution functions $F(x) := \mathbb{P}(X \le x)$, $F_1(x) := \mathbb{P}(X_1 \le x)$, $F_2(x) := \mathbb{P}(X_2 \le x), \ldots$, respectively. Note that, in contrast to previous usage, X_j now does not stand for the j-th component of a random vector, but is itself a random vector.

The next theorem is due to A.D. Alexandrov[1] (see [ALA]); however, it is commonly known as the *Portmanteau theorem*. The reason for this is a naming that is probably due to the special humor of P. Billingsley.[2] In his book [BI1], Billingsley quotes the paper *Espoir pour l'ensemble vide?* (Hope for the empty set?) by Jean-Pierre Portmanteau (Annales de l'Université de Felletin, CXLI (1915), 322–325). However, neither this author nor the quoted university ever existed. Since the word *portmanteau* also means *suitcase*, one should always carry this theorem *in the suitcase* because of its importance. In France, the double arrows in the chain (a) $\Longleftrightarrow$ (b) $\Longleftrightarrow$ (c) $\Longleftrightarrow$ (d) $\Longleftrightarrow$ (e) are imagined as clothes racks on which coats can be hung[3] (French *porte-manteau* (coat rack)). In the following, each occurring integral extends over $\mathbb{R}^d$.

6.3 Theorem (Portmanteau Theorem)
The following statements are equivalent:

(a) $\lim\limits_{n\to\infty} \int f \, dQ_n = \int f \, dQ$ *for each function* $f \in C_b(\mathbb{R}^d)$,

(b) $\limsup\limits_{n\to\infty} Q_n(A) \le Q(A)$ *for each closed set* $A \subset \mathbb{R}^d$,

(c) $\liminf\limits_{n\to\infty} Q_n(O) \ge Q(O)$ *for each open set* $O \subset \mathbb{R}^d$,

(d) $\lim\limits_{n\to\infty} Q_n(B) = Q(B)$ *for each Borel set* $B \subset \mathbb{R}^d$ *with* $Q(\partial B) = 0$,

(e) $\lim\limits_{n\to\infty} F_n(x) = F(x)$ *for each continuity point* x *of* F.

Proof

"(a) $\Longrightarrow$ (b)":   The idea of the proof is to approximate the indicator function of a closed set by continuous and bounded functions. To this end, we define for each integer j a function $f_j : \mathbb{R}^d \to \mathbb{R}$ by

$$f_j(x) := \max\left(0, 1 - j\,\|x - A\|\right), \quad x \in \mathbb{R}^d. \tag{6.4}$$

Due to the continuity of the mapping $x \mapsto \|x - A\|$ and $0 \le f_j \le 1$, we have $f_j \in C_b(\mathbb{R}^d)$, and since A is closed, $\|x - A\| = 0$ is equivalent to $x \in A$. The sequence of functions f_j thus converges pointwise monotonically decreasing to the indicator function $\mathbf{1}_A$ of A. According to (a), $\lim_{n\to\infty} \int h_f \, dQ_n = \int f_j \, dQ$ for each $j \ge 1$.

---

[1] Alexander Danilovich Alexandrov (1912–1999), Russian mathematician. Main areas of work: Theoretical physics, geometry.

[2] Patrick Billingsley (1925–2011), American mathematician and actor, initially assistant professor from 1958 and professor from 1963 at the University of Chicago. Main area of work: Probability theory, especially application of limit theorems to number theory. Billingsley was also a theater and film actor, and he held a black belt in judo.

[3] personal communication from Nicolas Chenavier

Moreover, the inequality $\mathbf{1}_A \leq f_j$ entails $Q_n(A) = \int \mathbf{1}_A \, dQ_n \leq \int f_j \, dQ_n$ for each $n \geq 1$ and each $j \geq 1$. In view of condition (a), taking the limit $n \to \infty$ thus implies

$$\limsup_{n\to\infty} Q_n(A) \leq \int f_j \, dQ, \quad j \geq 1.$$

The equation $\lim_{j\to\infty} f_j = \mathbf{1}_A$ and the dominated convergence theorem now yield $\int f_j \, dQ \to \int \mathbf{1}_A \, dQ = Q(A)$ as $j \to \infty$ and thus (b).

"(b) $\Longleftrightarrow$ (c)":   Since the open sets are the complements of the closed sets and vice versa, this equivalence follows by taking complements.

"(b), (c) $\Longrightarrow$ (d)":   That property (d) follows from (b) and (c) can be seen from the chain of inequalities

$$Q(B^\circ) \leq \liminf_{n\to\infty} Q_n(B^\circ) \leq \liminf_{n\to\infty} Q_n(B) \leq \limsup_{n\to\infty} Q_n(B)$$

$$\leq \limsup_{n\to\infty} Q_n(\overline{B}) \leq Q(\overline{B})$$

$$= Q(B^\circ) + Q(\partial B),$$

which is valid for each Borel set B. Here, condition (c) was used for the first inequality and condition (b) for the fifth inequality. If $Q(\partial B) = 0$, then $\lim_{n\to\infty} Q_n(B) = Q(B)$, and (d) follows.

"(d) $\Longrightarrow$ (a)":   To prove this implication, we approximate an arbitrary function $f \in C_b(\mathbb{R}^d)$ by a function f_m of the form $f_m := \sum_{j=1}^m \alpha_j \mathbf{1}\{B_j\}$, where $Q(\partial B_j) = 0$ for each $j \in \{1, \ldots, m\}$. Fix any $\varepsilon > 0$ and denote $K := \|f\|_\infty := \sup_{x \in \mathbb{R}^d} |f(x)|$, which is a finite constant. Further, let $\alpha_0, \ldots, \alpha_m$ be real numbers with $\alpha_0 < \alpha_1 < \ldots < \alpha_m$ as well as $\alpha_0 < -K$, $\alpha_m > K$ and $\alpha_j - \alpha_{j-1} \leq \varepsilon$ for each $j \in \{1, \ldots, m\}$. If we choose B_j of the form $B_j := \{\alpha_{j-1} < f \leq \alpha_j\} \, (= \{x \in \mathbb{R}^d : \alpha_{j-1} < f(x) \leq \alpha_j\})$, then $\|f - f_m\|_\infty \leq \varepsilon$. Since for each boundary point x of B_j either $f(x) = \alpha_{j-1}$ or $f(x) = \alpha_j$, and since the random variable $f(X)$ can take on at most countably many values with positive probability, we can choose $\alpha_0, \ldots, \alpha_m$ such that $\mathbb{P}(f(X) \in \{\alpha_0, \ldots, \alpha_m\}) = 0$. This implies $Q(\partial B_j) = 0$ for each $j = 1, \ldots, m$. Due to the triangle inequality, we have

$$\left| \int f \, dQ_n - \int f \, dQ \right| \leq \left| \int (f - f_m) \, dQ_n \right| + \left| \int f_m \, dQ_n - \int f_m \, dQ \right| + \left| \int (f_m - f) \, dQ \right|$$

$$\leq \int |f - f_m| \, dQ_n + \left| \sum_{j=1}^m \alpha_j \left(Q_n(B_j) - Q(B_j) \right) \right| + \int |f_m - f| \, dQ$$

$$\leq 2\varepsilon + \left| \sum_{j=1}^m \alpha_j \left(Q_n(B_j) - Q(B_j) \right) \right|. \tag{6.5}$$

Here, the second term in (6.5) converges to 0 as $n \to \infty$ according to (d). It follows that $\limsup_{n\to\infty} |\int f \, dQ_n - \int f \, dQ| \le 2\varepsilon$ and thus (a), since $\varepsilon > 0$ is arbitrary.

"(d) $\Longrightarrow$ (e)":　If we set $B_x := (-\infty, x], x \in \mathbb{R}^d$, this implication follows, since $x \in C_F$ is the same as $Q(\partial B_x) = 0$.

"(e) $\Longrightarrow$ (c)":　Let D be a countable dense subset of $\mathbb{R}$ with the property $Q(\{(x_1, \ldots, x_d) \in \mathbb{R}^d : x_j = a\}) = 0$ for each $a \in D$ and each $j \in \{1, \ldots, d\}$. Then $D^d \subset C_F$. Let $\mathcal{M} := \{ \times_{j=1}^d (a_j, b_j] : a_j, b_j \in D \text{ and } a_j < b_j \text{ for } j \in \{1, \ldots, d\}\}$. From (e) we have

$$Q_n \left(\times_{j=1}^d (a_j, b_j] \right) = \Delta_a^b F_n \to \Delta_a^b F = Q \left(\times_{j=1}^d (a_j, b_j] \right),$$

that is, $Q_n(B) \to Q(B)$ for each $B \in \mathcal{M}$. Since the class $\mathcal{M} \cup \{\emptyset\}$ of sets is a π-system, the formula of inclusion and exclusion yields $Q_n(B) \to Q(B)$, if B is a *finite* union of sets from $\mathcal{M}$. Let O be any non-empty open set. Since the endpoints of the d-dimensional rectangles in $\mathcal{M}$ are dense in $\mathbb{R}^d$, there are $B_1, B_2, \ldots$ in $\mathcal{M}$ with $O = \cup_{j=1}^\infty B_j$. For each $k \ge 1$ we have $Q(\cup_{j=1}^k B_j) = \lim_{n\to\infty} Q_n(\cup_{j=1}^k B_j) \le \liminf_{n\to\infty} Q_n(O)$. Since Q is continuous from below, it follows that $Q(O) \le \liminf_{n\to\infty} Q_n(O)$.

$\square$

In the notation with random vectors $X, X_1, X_2, \ldots$, the statements of the Portmanteau theorem read as follows:

(a) $\lim_{n\to\infty} \mathbb{E}[f(X_n)] = \mathbb{E}[f(X)]$ for each function $f \in C_b(\mathbb{R}^d)$,

(b) $\limsup_{n\to\infty} \mathbb{P}(X_n \in A) \le \mathbb{P}(X \in A)$ for each closed set $A \subset \mathbb{R}^d$,

(c) $\liminf_{n\to\infty} \mathbb{P}(X_n \in O) \ge \mathbb{P}(X \in O)$ for each open set $O \subset \mathbb{R}^d$,

(d) $\lim_{n\to\infty} \mathbb{P}(X_n \in B) = \mathbb{P}(X \in B)$ for each Borel set $B \subset \mathbb{R}^d$ with $\mathbb{P}(X \in \partial B) = 0$,

(e) $\lim_{n\to\infty} F_n(x) = F(x)$ for each continuity point x of F.

6.4 Remark If $C_{j,F}$ denotes the set of continuity points of the marginal distribution function of the j-th component of X ($j \in \{1, \ldots, d\}$), then in the case $d > 1$ in (e) it is even sufficient to demand convergence only on the set, denoted C_F^*, of those continuity points $x = (x_1, \ldots, x_d)$ of F with $x_j \in C_{j,F}$ for each $j \in \{1, \ldots, d\}$ (see Theorem 5.58 in [WIM]). Since there are only countably many discontinuity points for the marginal distribution functions, there exist for any non-empty open set $O \subset \mathbb{R}^d$ and any $\varepsilon > 0$ finitely many pairwise disjoint subsets $M_1, \ldots, M_k$ of O of the form $(a_1, b_1] \times \ldots \times (a_d, b_d]$ with $\{a_j, b_j\} \subset C_{j,F}$ for each j and $\mathbb{P}(X \in O) \le \sum_{\ell=1}^k \mathbb{P}(X \in$

$M_\ell) + \varepsilon$. From $F_n(x) \to F(x)$ for each $x \in C_F^*$ it follows that $\mathbb{P}(X_n \in M_\ell) \to \mathbb{P}(X \in M_\ell)$ for each ℓ and thus, using the notation $\uplus$ for unions of (pairwise) disjoint sets,

$$\mathbb{P}(X \in O) \leq \lim_{n \to \infty} \mathbb{P}\left(X_n \in \biguplus_{\ell=1}^{k} M_\ell\right) + \varepsilon \leq \liminf_{n \to \infty} \mathbb{P}(X_n \in O) + \varepsilon.$$

Since $\varepsilon > 0$ is arbitrary, the assertion follows from part (c) of the Portmanteau theorem.

We choose statement (a) of the Portmanteau theorem to define the convergence in distribution of random vectors. Compared to (e), this choice (as well as (b)–(d)) has the advantage that it can be generalized in a straightforward manner to the case of random variables that take on values in general metric spaces (see Chap. 14).

6.5 Definition (Convergence in Distribution of Random Vectors, Weak Convergence)
If $X, X_1, X_2, \ldots$ are d-dimensional random vectors on a probability space $(\Omega, \mathcal{A}, \mathbb{P})$, we define

$$X_n \xrightarrow{\mathcal{D}} X :\Longleftrightarrow \lim_{n \to \infty} \mathbb{E}\big[f(X_n)\big] = \mathbb{E}\big[f(X)\big] \text{ for each function } f \in C_b(\mathbb{R}^d)$$

and say that X_n *converges in distribution to* X. Here, the hybrid notation $X_n \xrightarrow{\mathcal{D}} \mathbb{P}^X$ is also encountered. Equivalent to $X_n \xrightarrow{\mathcal{D}} X$ is the notation $Q_n \xrightarrow{\mathcal{D}} Q$. More generally (that is, detached from the meaning as distributions of random vectors), if $Q, Q_1, Q_2, \ldots$ are probability measures on $\mathcal{B}^d$, for which one of the statements of the Portmanteau theorem applies, we say that the sequence (Q_n) *converges weakly* to Q. The weak convergence $Q_n \xrightarrow{\mathcal{D}} Q$ of probability measures is thus defined by property (a) of the Portmanteau theorem.

The following frequently used fact, referred to as the *(continuous) mapping theorem*, states that, in generalization of Theorem 1.22, convergence in distribution is inherited under mappings that are almost everywhere continuous with respect to the limit distribution. It implicitly implies that the set C_h of continuity points of a function $h : \mathbb{R}^d \to \mathbb{R}^s$ is measurable (see Problem 14.12).

6.6 Theorem (Mapping Theorem)
Suppose $X_n \xrightarrow{\mathcal{D}} X$. *If* $s \geq 1$ *and* $h : \mathbb{R}^d \to \mathbb{R}^s$ *is a measurable mapping with the property* $\mathbb{P}\big(X \in C_h\big) = 1$, *then* $h(X_n) \xrightarrow{\mathcal{D}} h(X)$.

Proof We use criterion (b) of the Portmanteau theorem. Let $A \subset \mathbb{R}^s$ be any non-empty closed set. Since every point x of the closed hull of $h^{-1}(A)$ (for which there is a sequence

(x_n) with $x_n \in h^{-1}(A)$ and $\underline{x_n \to x})$ either belongs to $h^{-1}(A)$ or otherwise is a point of discontinuity of h, we have $\overline{h^{-1}(A)} \subset \mathbb{R}^d \setminus C_h \cup h^{-1}(A)$. It follows that

$$\limsup_{n \to \infty} \mathbb{P}\big(h(X_n) \in A\big) = \limsup_{n \to \infty} \mathbb{P}\big(X_n \in h^{-1}(A)\big)$$

$$\leq \limsup_{n \to \infty} \mathbb{P}\big(X_n \in \overline{h^{-1}(A)}\big)$$

$$\leq \mathbb{P}\big(X \in \overline{h^{-1}(A)}\big)$$

$$\leq \mathbb{P}\big(X \notin C_h\big) + \mathbb{P}\big(h(X) \in A\big)$$

$$= \mathbb{P}\big(h(X) \in A\big).$$

The second inequality is valid since $X_n \xrightarrow{\mathcal{D}} X$ and the set $\overline{h^{-1}(A)}$ is closed.   □

By the mapping theorem, for example, $\sin(U_n + V_n)e^{U_n} \xrightarrow{\mathcal{D}} \sin(U + V)e^U$ if the sequence of the bivariate random vectors (U_n, V_n) converges in distribution to (U, V). The following example shows, however, that convergence in distribution $(U_n, V_n) \xrightarrow{\mathcal{D}} (U, V)$ does not generally follow from $U_n \xrightarrow{\mathcal{D}} U$ and $V_n \xrightarrow{\mathcal{D}} V$ alone.

6.7 Example ($(U_n, V_n) \xrightarrow{\mathcal{D}} (U, V)$ does not follow from $U_n \xrightarrow{\mathcal{D}} U$ and $V_n \xrightarrow{\mathcal{D}} V$)
Let (U, V) be a random vector with a uniform distribution on the unit square $[0, 1]^2$ and $(U_n, V_n) := (U, V)$, $n \geq 1$. Then U_n and V_n are each uniformly distributed on $[0, 1]$, and we have $U_n \xrightarrow{\mathcal{D}} U$, $V_n \xrightarrow{\mathcal{D}} V$, and $(U_n, V_n) \xrightarrow{\mathcal{D}} (U, V)$. In this situation, in addition to $U_n \xrightarrow{\mathcal{D}} U$, we also have $V_n \xrightarrow{\mathcal{D}} U$. However, the sequence $(U_n, V_n)_{n \geq 1}$ does not converge in distribution to (U, U), because (U, U) has a uniform distribution on the diagonal $\{(x, x) : 0 \leq x \leq 1\}$ of the unit square.

However, the following applies (Problem 6.4), which is considerably generalized with Theorem 14.26 in Chap. 14.

6.8 Theorem (Convergence in Distribution and Independence)
On a common probability space, let $Y_1, Y_2, \ldots$ be k-dimensional and $Z_1, Z_2, \ldots$ be ℓ-dimensional random vectors with $Y_n \xrightarrow{\mathcal{D}} Y$ and $Z_n \xrightarrow{\mathcal{D}} Z$. If Y_n and Z_n are independent for each $n \geq 1$, then $(Y_n, Z_n) \xrightarrow{\mathcal{D}} (Y, Z)$, where Y and Z are independent.

Part (b) of Slutsky's lemma (Lemma 1.23) also applies more generally:

6.9 Theorem (Slutsky's Lemma)

If $X, X_1, X_2, \ldots$ and $Y_1, Y_2, \ldots$ are d-dimensional random vectors with $X_n \overset{D}{\longrightarrow} X$ and $Y_n \overset{\mathbb{P}}{\longrightarrow} 0_d$, then $X_n + Y_n \overset{D}{\longrightarrow} X$.

Proof The proof also uses criterion (b) of the Portmanteau theorem. For this purpose, let $A \subset \mathbb{R}^d$ be any non-empty closed set and $\varepsilon > 0$ be arbitrary. The set $A_\varepsilon := \{x \in \mathbb{R}^d : \|x - A\| \leq \varepsilon\}$ is closed, and we have

$$\{X_n + Y_n \in A\} \subset \{X_n \in A_\varepsilon\} \cup \{\|Y_n\| > \varepsilon\}. \tag{6.6}$$

Since $Y_n \overset{\mathbb{P}}{\longrightarrow} 0_d$, it follows that

$$\limsup_{n \to \infty} \mathbb{P}(X_n + Y_n \in A) \leq \limsup_{n \to \infty} \mathbb{P}(X_n \in A_\varepsilon) + 0 \leq \mathbb{P}(X \in A_\varepsilon).$$

The last inequality holds because $X_n \overset{D}{\longrightarrow} X$. Since A is closed, we get $A_\varepsilon \downarrow A$ as $\varepsilon \downarrow 0$, and the assertion follows, since $\mathbb{P}^X$ as a probability measure is continuous from above. $\square$

Self-question 2: Why does (6.6) apply?

Definition 1.9 can be generalized directly from the case $d = 1$ to arbitrary dimensions:

6.10 Definition (Tightness and Relative Compactness)
Let $\mathcal{Q}$ be a non-empty collection of probability measures on $\mathcal{B}^d$. $\mathcal{Q}$ is called

(a) *tight*, if for every $\varepsilon > 0$ there is a compact set $K \subset \mathbb{R}^d$ such that

$$Q(K) \geq 1 - \varepsilon \quad \text{for every } Q \in \mathcal{Q},$$

(b) *relatively compact*, if for every sequence (Q_n) in $\mathcal{Q}$ there is a subsequence (Q_{n_k}) and a probability measure Q with $Q_{n_k} \overset{D}{\longrightarrow} Q$ as $k \to \infty$.

It is important to note that the probability measure Q occurring in the definition of relative compactness does not necessarily have to belong to $\mathcal{Q}$. If $Q_n \overset{D}{\longrightarrow} Q$, then the set $\{Q_n : n \geq 1\}$ is necessarily relatively compact. The following result, which holds in a much more general setting (see Theorem 14.22), states that tightness and relative compactness are equivalent concepts.

6.11 Theorem (Prokhorov[4])

For every collection $\mathcal{Q} \neq \emptyset$ of probability measures on $\mathcal{B}^d$, we have:

$$\mathcal{Q} \text{ is tight} \iff \mathcal{Q} \text{ is relatively compact.}$$

Proof We only prove the implication "$\Longleftarrow$" and assume that $\mathcal{Q}$ is not tight. Then, putting $K_n := [-n, n]^d$ for $n \geq 1$, there exists an $\varepsilon > 0$ and a sequence (Q_n) in $\mathcal{Q}$ with $Q_n(K_n) < 1 - \varepsilon$ for each $n \geq 1$. Since $\mathcal{Q}$ is assumed to be relatively compact, there exist a subsequence (Q_{n_k}) and a probability measure Q with $Q_{n_k} \xrightarrow{\mathcal{D}} Q$. We set $K := [-M, M]^d$, where $M > 0$ is chosen such that $Q(K) \geq 1 - \frac{\varepsilon}{2}$ and $Q(\partial K) = 0$. Letting $k \to \infty$ then yields $Q_{n_k}(K) \to Q(K) \geq 1 - \frac{\varepsilon}{2}$. Since $K_{n_k} \supset K$ for sufficiently large k, we get $Q_{n_k}(K) \leq Q_{n_k}(K_{n_k}) < 1 - \varepsilon$ for each such k, which gives a contradiction. A proof of the implication "$\Longrightarrow$" can be found, e.g., in [BI3], p. 380 $\square$

The concept of relative compactness corresponds to that of boundedness of sequences in $\mathbb{R}^d$. A bounded sequence does not necessarily have to converge. However, for each subsequence, there is a sub-subsequence that does converge. Based on this analogy, the concept of *boundedness in probability*, which is equivalent to relative compactness (and thus to tightness), is found in particular in connection with random vectors. A sequence $(X_n)_{n \geq 1}$ of d-dimensional random vectors is called *bounded in probability*, if it is *tight* or—which is equivalent—*relatively compact*. Here, the sequence $(X_n)_{n \geq 1}$ is called *tight* or *relatively compact* if the set $\{\mathbb{P}^{X_n} : n \geq 1\}$ is tight or relatively compact.

Self-question 3: Why is every *finite* set of probability measures tight?

6.12 Example If $(X_n)_{n \geq 1}$ is a sequence of d-dimensional random vectors with the property $K := \sup_{n \geq 1} \mathbb{E}\|X_n\| < \infty$ then, for each C with $0 < C < \infty$ and each $n \geq 1$,

$$\mathbb{P}\big(\|X_n\| > C\big) \leq \frac{\mathbb{E}\|X_n\|}{C} \leq \frac{K}{C}.$$

Fix any $\varepsilon > 0$. If we choose C so large that $C \geq \frac{K}{\varepsilon}$, then $\mathbb{P}\big(\|X_n\| > C\big) \leq \varepsilon$ and thus $\mathbb{P}(\|X_n\| \leq C) \geq 1 - \varepsilon$ for each $n \geq 1$. Consequently, the sequence (X_n) is tight.

---

[4] Yuri Vasilyevich Prokhorov (1929–2013), Russian mathematician, PhD 1952, from 1960 head of the Department of Probability Theory of the Steklov Institute. Main field of work: Probability theory, in particular limit theorems.

We have seen that tightness is a necessary condition for convergence in distribution. If (a_n) is a bounded sequence of real numbers, it converges to a value a if and only if every subsequence contains a sub-subsequence that converges to a. The following analogue for random vectors will often play a role.

6.13 Remark Let (X_n) be a tight sequence of d-dimensional random vectors. If there is a probability measure Q such that every subsequence of (X_n) that converges in distribution at all converges to Q, then $X_n \xrightarrow{\mathcal{D}} X$, where the random vector X has the distribution Q.

In analogy to the O-notation and o-notation for sequences of real numbers introduced by E. Landau[5] to denote a bounded sequence and a sequence converging to 0, respectively, the following notations are often encountered.

6.14 Stochastic Landau Symbols
Let $X, X_1, X_2, \ldots$ be d-dimensional random vectors and (a_n) a sequence of real numbers with $a_n \neq 0$ for each $n \geq 1$. One defines

$$X_n = O_{\mathbb{P}}(1) :\Longleftrightarrow \text{ the sequence } (X_n)_{n \geq 1} \text{ is tight,}$$

$$X_n = O_{\mathbb{P}}(a_n) :\Longleftrightarrow \text{ the sequence } \left(\frac{X_n}{a_n}\right)_{n \geq 1} \text{ is tight,}$$

$$X_n = o_{\mathbb{P}}(1) :\Longleftrightarrow X_n \xrightarrow{\mathbb{P}} 0_d,$$

$$X_n = o_{\mathbb{P}}(a_n) :\Longleftrightarrow \frac{X_n}{a_n} \xrightarrow{\mathbb{P}} 0_d,$$

$$X_n = X + o_{\mathbb{P}}(1) :\Longleftrightarrow X_n - X \xrightarrow{\mathbb{P}} 0_d.$$

The most important rules for dealing with the stochastic Landau symbols are listed below.

6.15 Theorem (Properties of $O_{\mathbb{P}}$ and $o_{\mathbb{P}}$)
Let $X_n, Y_n, n \geq 1$, be d-dimensional random vectors and $(Z_n)_{n \geq 1}$ a sequence of real-valued random variables. Then the following hold:

(a) *If $X_n = O_{\mathbb{P}}(1)$ and $Y_n = O_{\mathbb{P}}(1)$ then $X_n + Y_n = O_{\mathbb{P}}(1)$,*
(b) *If $X_n = o_{\mathbb{P}}(1)$ and $Y_n = o_{\mathbb{P}}(1)$ then $X_n + Y_n = o_{\mathbb{P}}(1)$,*
(c) *If $X_n = O_{\mathbb{P}}(1)$ and $Z_n = O_{\mathbb{P}}(1)$ then $X_n \cdot Z_n = O_{\mathbb{P}}(1)$,*
(d) *If $X_n = O_{\mathbb{P}}(1)$ and $Z_n = o_{\mathbb{P}}(1)$ then $X_n \cdot Z_n = o_{\mathbb{P}}(1)$.*

---

[5] Edmund Landau (1877–1938), lecturer at the University of Berlin from 1899, professor in Göttingen from 1909, dismissed in 1933 after the Nazi seizure of power. Landau's main field of work was analytical number theory; he achieved significant results on the distribution of prime numbers.

(e) If $X_n = O_{\mathbb{P}}(1)$ and $h : \mathbb{R}^d \to \mathbb{R}^s$ is a continuous function then $h(X_n) = O_{\mathbb{P}}(1)$.

Proof (a) Fix any $\varepsilon > 0$. By assumption, there are positive constants K_1 and K_2 with $\mathbb{P}(\|X_n\| \leq K_1) \geq 1 - \frac{\varepsilon}{2})$ and $\mathbb{P}(\|Y_n\| \leq K_2) \geq 1 - \frac{\varepsilon}{2})$ for each $n \geq 1$. Using the triangle inequality, it follows that $\mathbb{P}(\|X_n + Y_n\| \leq K_1 + K_2) \geq 1 - \varepsilon$ for each $n \geq 1$. Thus, $(X_n + Y_n)$ is a tight sequence. The proof of (b)–(e) is the subject of Problem 6.7.   □

The next result is an addition to Slutsky's lemma (Theorem 6.9).

6.16 Corollary *Suppose $X, X_1, X_2, \ldots$ are d-dimensional random vectors with $X_n \xrightarrow{\mathcal{D}} X$, and (Z_n) is a sequence real-valued random variables with $Z_n \xrightarrow{\mathbb{P}} a$ for some $a \in \mathbb{R}$. Then $Z_n X_n \xrightarrow{\mathcal{D}} aX$.*

Proof We have $Z_n X_n = (Z_n - a)X_n + aX_n$. By the mapping theorem, $aX_n \xrightarrow{\mathcal{D}} aX$. Since $Z_n - a = o_{\mathbb{P}}(1)$ and $X_n = O_{\mathbb{P}}(1)$, Theorem 6.15 (d) yields $(Z_n - a)X_n = o_{\mathbb{P}}(1)$. The claim now follows from Slutsky's lemma (Theorem 6.9).   □

As announced, all vectors are noted as column vectors for the rest of this chapter. The next result generalizes Theorem 1.13 to the multivariate case.

6.17 Theorem (Continuity Theorem, Lévy–Cramér)
If $X, X_1, X_2, \ldots$ are d-dimensional random vectors with associated characteristic functions $\varphi, \varphi_1, \varphi_2, \ldots$, respectively, then:

$$X_n \xrightarrow{\mathcal{D}} X \iff \lim_{n \to \infty} \varphi_n(t) = \varphi(t) \quad \text{for each } t \in \mathbb{R}^d.$$

Proof "$\Longrightarrow$" follows by setting $h_1(x) := \cos(t^\top x)$ and $h_2(x) := \sin(t^\top x)$ for any $t \in \mathbb{R}^d$ and using the definition of convergence in distribution. To prove "$\Longleftarrow$", let $X_n =:$ $(X_{n1}, \ldots, X_{nd})^\top$, $X =: (Z_1, \ldots, Z_d)^\top$ and $e_j := (0, \ldots, 0, 1, 0, \ldots, 0)^\top$ be the j-th unit vector in $\mathbb{R}^d$, $j \in \{1, \ldots, d\}$. If we specifically set $t := \alpha e_j$, where $\alpha \in \mathbb{R}$, then

$$\varphi_{X_{nj}}(\alpha) = \mathbb{E}\left[\exp\left(i\alpha X_{nj}\right)\right] = \varphi_n(\alpha e_j),$$
$$\varphi_{Z_j}(\alpha) = \mathbb{E}\left[\exp\left(i\alpha Z_j\right)\right] = \varphi(\alpha e_j).$$

By assumption, $\varphi_n(\alpha e_j) \to \varphi(\alpha e_j)$ for each $\alpha \in \mathbb{R}$, and thus $X_{nj} \xrightarrow{\mathcal{D}} Z_j$ for every $j \in \{1, \ldots, d\}$ by Theorem 1.13. It follows that the sequence $(X_{nj})_{n \geq 1}$ is tight for each $j \in \{1, \ldots, d\}$. Problem 6.5 shows that this also makes $(X_n)_{n \geq 1}$ tight. In view of Theorem 6.11, there exist a subsequence (X_{n_k}) and a random variable Y with $X_{n_k} \xrightarrow{\mathcal{D}} Y$ as $k \to \infty$. If ψ denotes the characteristic function of Y, then part "$\Longrightarrow$" of the proof yields $\varphi = \psi$. With Corollary 1.16, we get $X \overset{D}{=} Y$ and thus $X_{n_k} \overset{D}{\longrightarrow} X$. The claim now follows from Remark 6.13.   □

As the following fundamental theorem, due to H. Cramér and H. Wold, shows, the Lévy–Cramér continuity theorem allows the proof of convergence in distribution of random vectors to be reduced to the one-dimensional case.

6.18 Theorem (Cramér–Wold Device)
If $X, X_1, X_2, \ldots$ are d-dimensional random vectors, then:

$$X_n \overset{D}{\longrightarrow} X \iff c^\top X_n \overset{D}{\longrightarrow} c^\top X \quad \text{for every } c \in \mathbb{R}^d.$$

Proof "$\Longrightarrow$" follows with $h(x) := c^\top x$ from the mapping theorem. To show the implication "$\Longleftarrow$", we use

$$\varphi_{X_n}(c) = \mathbb{E}\left[\exp\left(ic^\top X_n\right)\right] = \varphi_{c^\top X_n}(1),$$

$$\varphi_X(c) = \mathbb{E}\left[\exp\left(ic^\top X\right)\right] = \varphi_{c^\top X}(1).$$

Here, $\varphi_{c^\top X_n}$ and $\varphi_{c^\top X}$ denote the characteristic functions of $c^\top X_n$ and $c^\top X$, respectively. By the Lévy–Cramér continuity theorem (Theorem 1.13), $\varphi_{c^\top X_n}(1) \to \varphi_{c^\top X}(1)$. Thus, $\varphi_{X_n}(c) \to \varphi_X(c)$ for each $c \in \mathbb{R}^d$, and the claim follows from Theorem 6.17.   □

The next two theorems are the central results of this chapter.

6.19 Theorem (Multivariate Central Limit Theorem for i.i.d. Sequences)
Let $X_1, X_2, \ldots$ be i.i.d. d-dimensional random vectors with $\mathbb{E}\|X_1\|^2 < \infty$. Putting $\mu := \mathbb{E}(X_1)$ and $\Sigma := \mathrm{Cov}(X_1)$, we have

$$\frac{1}{\sqrt{n}}\left(\sum_{j=1}^n X_j - n\mu\right) \overset{D}{\longrightarrow} N_d(0_d, \Sigma) \quad \text{as } n \to \infty.$$

Proof The proof uses the Lindeberg–Lévy central limit theorem and the Cramér–Wold device (Theorem 6.18). To this end, let $Z_n := n^{-1/2}\left(\sum_{j=1}^n X_j - n\mu\right)$ and Y be a random vector with the distribution $N_d(0_d, \Sigma)$. We need to show $c^\top Z_n \overset{D}{\longrightarrow} c^\top Y$ for every $c \in \mathbb{R}^d$. Since

$$c^\top Z_n = \frac{1}{\sqrt{n}}\left(\sum_{j=1}^n c^\top X_j - nc^\top \mu\right)$$

and $\mathbb{E}(c^\top Z_n) = 0$, $\mathbb{V}(c^\top Z_n) = \mathbb{V}(c^\top X_1) = c^\top \Sigma c$, as well as $c^\top Y \sim N(0, c^\top \Sigma c)$, we may assume w.l.o.g. that $c^\top \Sigma c > 0$. With Theorem 1.22, applied to $(c^\top X_j)_{j\geq 1}$, it follows that

$$\frac{c^\top Z_n}{\sqrt{c^\top \Sigma c}} = \frac{\sum_{j=1}^n c^\top Z_n - nc^\top \mu}{\sqrt{nc^\top \Sigma c}} \xrightarrow{\mathcal{D}} N,$$

where $N \sim N(0, 1)$. The mapping theorem yields $c^\top Z_n \xrightarrow{\mathcal{D}} \sqrt{c^\top \Sigma c}\, N \sim N(0, c^\top \Sigma c)$. Because $c^\top Y$ has the normal distribution $N(0, c^\top \Sigma c)$, the claim follows. □

6.20 Example (Chi-Square Test)

The first application of Theorem 6.19 is the chi-square goodness-of-fit test for testing a simple hypothesis in a situation of n repeated independent trials, each of which has s possible outcomes $1, 2, \ldots, s$ (cf. Example 5.4). The test statistic of this test is

$$T_n := \sum_{k=1}^s \frac{(N_{n,j} - np_j)^2}{np_j}. \tag{6.7}$$

Here, $(N_{n,1}, \ldots, N_{n,s})$ has the multinomial distribution $\mathrm{Mult}(n; p_1, \ldots, p_s)$, and the hypothetical probabilities $p_1, \ldots, p_s$ of the respective outcomes are all positive. Moreover, we have $p_1 + \ldots + p_s = 1$. The connection between T_n and Theorem 6.19 is established by a sequence $X_1, X_2, \ldots$ of i.i.d. random vectors with $\mathbb{P}(X_1 = e_k) := p_k$, $k = 1, \ldots, s$. Here, e_k denotes the k-th unit vector in $\mathbb{R}^s$, $k \in \{1, \ldots, s\}$. We have $\sum_{j=1}^n X_j \overset{\mathcal{D}}{=} (N_{n,1}, \ldots, N_{n,s})^\top$ and

$$\mu := \mathbb{E}(X_1) = (p_1, \ldots, p_s)^\top, \qquad \Sigma := \mathrm{Cov}(X_1) = (p_k \delta_{k,\ell} - p_k p_\ell)_{1\leq k,\ell\leq s}. \tag{6.8}$$

By Theorem 6.19,

$$\frac{1}{\sqrt{n}} \left(\sum_{j=1}^n X_j - n\mu \right) \xrightarrow{\mathcal{D}} Z := (Z_1, \ldots, Z_s)^\top \sim N_s(0_s, \Sigma). \tag{6.9}$$

The matrix Σ is not invertible, but the matrix $A := (p_j \delta_{kj} - p_j p_k)_{1\leq j,k\leq s-1}$, formed by the first $s - 1$ rows and columns of Σ, is, and the inverse of A is given by $A^{-1} = (\delta_{jk} p_k^{-1} + p_s^{-1})_{1\leq j,k\leq s-1}$. If we now set $V_n := (N_{n,1}, \ldots, N_{n,s-1})^\top$, the mapping theorem yields

$$W_n := \frac{1}{\sqrt{n}} \left(V_n - n(p_1, \ldots, p_{s-1})^\top \right) \xrightarrow{\mathcal{D}} (Z_1, \ldots, Z_{s-1})^\top \sim N_{s-1}(0_{s-1}, A).$$

A further application of the mapping theorem implies

$$W_n^\top A^{-1} W_n \xrightarrow{\mathcal{D}} (Z_1, \ldots, Z_{s-1})^\top A^{-1} (Z_1, \ldots, Z_{s-1}) \overset{\mathcal{D}}{=} \chi^2_{s-1}.$$

The equality in distribution follows from Theorem 5.16. Using $N_{n,1} + \ldots + N_{n,s} = n$, a direct calculation yields $T_n = W_n^\top A^{-1} W_n$. Consequently, under the hypothesis that the unknown probabilities for the s possible outcomes of a multinomial experiment are $p_1, \ldots, p_s$, then the limit distribution of the chi-square test statistic T_n is a χ^2_{s-1}-distribution with $s - 1$ degrees of freedom.

Self-question 4: Why does $T_n = W_n^\top A^{-1} W_n$ apply?

6.21 Example (Number Lottery)

A prime example for the application of Theorem 6.19 is the draws in a general r/s-*lottery*. In such a lottery, you choose r numbers from among the numbers 1 to s, where $2 \leq r < s$. A drawing device contains s balls numbered from 1 to s, and a *draw* in such a r/s-lottery consists of drawing r balls one after the other purely at random as winning numbers, without putting them back into the drawing device.

We model the result of a k-th draw of an r/s-lottery by means of the s-dimensional random vector $X_k = (X_{k,1}, X_{k,2}, \ldots, X_{k,s})^\top$ with the components

$$X_{k,j} := \begin{cases} 1, & \text{if number } j \text{ is drawn on draw } k, \\ 0, & \text{otherwise.} \end{cases}$$

We assume that $X_1, X_2, \ldots$ are independent and each uniformly distributed over all $\binom{s}{r}$ possible s-tuples with r ones and $s - r$ zeros. These assumptions clarify the idea that the drawing machine has no memory, and that the balls are manufactured in the same way. The vector

$$H_n = (H_{n,1}, \ldots, H_{n,s})^\top := \sum_{k=1}^n X_k = \left(\sum_{k=1}^n X_{k,1}, \ldots, \sum_{k=1}^n X_{k,s} \right)^\top$$

then gives the frequencies of the individual winning numbers after n draws.

To investigate the asymptotic behavior of H_n as $n \to \infty$, we need the expectation and the covariance matrix of X_1. These are given by

$$\mathbb{E}(X_1) = \left(\frac{r}{s}, \frac{r}{s}, \cdots, \frac{r}{s} \right)^\top, \qquad \Sigma(X_1) = \frac{r}{s} \left(1 - \frac{r}{s} \right) \Sigma \tag{6.10}$$

with $\Sigma = (\sigma_{ij})_{1 \le i,j \le s}$ and $\sigma_{ij} = 1$ if $i = j$, and $\sigma_{ij} = -\frac{1}{s-1}$ for $i \ne j$ (Problem 6.11). By Theorem 6.19, it follows that

$$\frac{s}{\sqrt{r(s-r)}} \cdot \frac{1}{\sqrt{n}} (H_n - n\mathbb{E}(X_1)) \xrightarrow{\mathcal{D}} N_s(0_s, \Sigma) \quad \text{as } n \to \infty. \tag{6.11}$$

This convergence in distribution admits to study certain aspects of the winning number frequencies using the mapping theorem. If $T = (T_1, \ldots, T_s)^\top$ denotes a random vector with the distribution $N_s(0_s, \Sigma)$, an application of Theorem 6.6 to the function $h(x_1, \ldots, x_s) := \max_{1 \le j \le s} x_j$ yields

$$\frac{s}{\sqrt{r(s-r)}} \cdot \frac{1}{\sqrt{n}} \left(\max_{1 \le j \le s} H_{n,j} - n \cdot \frac{r}{s} \right) \xrightarrow{\mathcal{D}} \max_{1 \le j \le s} T_j. \tag{6.12}$$

In the same way, we have

$$\frac{s}{\sqrt{r(s-r)}} \cdot \frac{1}{\sqrt{n}} \left(\min_{1 \le j \le s} H_{n,j} - n \cdot \frac{r}{s} \right) \xrightarrow{\mathcal{D}} \min_{1 \le j \le s} T_j. \tag{6.13}$$

The message of (6.12) and (6.13) is that the deviation of the extreme winning frequencies $\max_{1 \le j \le s} H_{n,j}$ and $\min_{1 \le j \le s} H_{n,j}$ from the same expectation $n \cdot \frac{r}{s}$ for each winning frequency is *of the order of magnitude* $\sqrt{n}$. For the *range*

$$D_n := \max_{1 \le j \le s} H_{n,j} - \min_{1 \le j \le s} H_{n,j}$$

of the winning frequencies, (6.11) and the mapping theorem yield

$$\frac{s}{\sqrt{r(s-r)}} \cdot \frac{1}{\sqrt{n}} \cdot D_n \xrightarrow{\mathcal{D}} \max_{1 \le j \le s} T_j - \min_{1 \le j \le s} T_j$$

and thus in particular $D_n = O_{\mathbb{P}}(\sqrt{n})$. A χ^2-test for checking the equiprobability of lottery numbers is the subject of Problem 6.12.

Theorem 6.19 is a multivariate central limit theorem under relatively restrictive conditions. The following more general result (cf. [FH], Theorem 4.3.2) is a generalization analogous to the Lindeberg–Feller central limit theorem (see also Problems 6.9 and 6.10).

6.22 Theorem (Multivariate Central Limit Theorem for Triangular Arrays)
For each $n \geq 1$, let $X_{n,1}, \ldots, X_{n,r_n}$ be independent d-dimensional random vectors with $\mathbb{E}(X_{n,j}) = 0_d$ and $\mathbb{E}\|X_{n,j}\|^2 < \infty$ for each $n \geq 1$ and each $j \in \{1, \ldots, r_n\}$. Suppose there exists a $d \times d$ matrix Γ such that, for $S_n := X_{n,1} + \ldots + X_{n,r_n}$,

$$\lim_{n \to \infty} \mathbb{E}[S_n S_n^\top] = \Gamma. \tag{6.14}$$

Furthermore, let B be a subset of $\mathbb{R}^d$ such that each column of Γ is a linear combination of the elements of B. For any $c \in \mathbb{R}^d$ and any $\varepsilon > 0$, put

$$L_n(\varepsilon, c) := \sum_{j=1}^{r_n} \mathbb{E}\left[\left(c^\top X_{n,j}\right)^2 \mathbf{1}\{|c^\top X_{n,j}| > \varepsilon\}\right].$$

Under these assumptions, if

$$\lim_{n \to \infty} L_n(\varepsilon, b) = 0 \quad \text{for each } b \in B \text{ and each } \varepsilon > 0, \tag{6.15}$$

then

$$S_n \xrightarrow{\mathcal{D}} N_d(0_d, \Gamma) \quad \text{as } n \to \infty. \tag{6.16}$$

Proof By the Cramér–Wold device (Theorem 6.18), we have to show

$$c^\top S_n \xrightarrow{\mathcal{D}} N(0, c^\top \Gamma c) \tag{6.17}$$

for each $c \in \mathbb{R}^d$. If we set $\sigma_c^2 := c^\top \Gamma c$, $c \in \mathbb{R}^d$, and $N := \{c \in \mathbb{R}^d : \sigma_c^2 = 0\}$, then N is a subspace of $\mathbb{R}^d$, and (6.17) is valid for each $c \in N$.

Self-question 5: Why does $c^\top S_n \xrightarrow{\mathcal{D}} N(0, c^\top \Gamma c)$ apply for each $c \in N$?

Let $B^\perp := \{x \in \mathbb{R}^d : x^\top b = 0 \text{ for each } b \in B\}$ be the orthogonal complement of B, and fix any $x \in B^\perp$. By assumption, $x^\top y = 0$ for every column y of Γ, and it follows that $x^\top \Gamma = 0_d^\top$ and thus $\sigma_x^2 = 0$, so $B^\perp \subset N$. If $\text{Span}(B)$ denotes the linear hull of B, then every $c \in \mathbb{R}^d$ can be written in the form $c = c_1 + c_2$ with $c_1 \in \text{Span}(B)$ and $c_2 \in N$. Since statement (6.17) remains valid when c is replaced by αc with $\alpha \in \mathbb{R}$, we only need to show according to Slutsky's lemma (Theorem 6.9): If (6.17) holds for $c = v$ and $c = w$, where v and w are arbitrary vectors in B with $\sigma_v^2 > 0$, $\sigma_w^2 > 0$, and $\sigma_{v+w}^2 > 0$, then (6.17) also holds for $c = v + w$.

If v and w are two such vectors, we choose a $K > 0$ with $\sigma_v + \sigma_w < K\sigma_{v+w}$. Here, σ_c generally denotes the non-negative root of σ_c^2. Furthermore, we set $\sigma_n^2(c) := \mathbb{V}(c^\top S_n) = c^\top \mathbb{E}(S_n S_n^\top)c$, $c \in \mathbb{R}^d$, and $\sigma_n(c) := +\sqrt{\sigma_n^2(c)}$. Due to (6.14) and the assumption $\sigma_v^2 \sigma_w^2 \sigma_{v+w}^2 > 0$, $\sigma_n(v)\sigma_n(w)\sigma_n(v+w) > 0$ holds if n is sufficiently large. In addition, considering the inequality $\sigma_v + \sigma_w < K\sigma_{v+w}$, we can also assume $\sigma_n(v) + \sigma_n(w) \leq K\sigma_n(v+w)$ after possibly increasing n. For such n, we briefly set

$$Z_{n,j}(c) := \frac{|c^\top X_{n,j}|}{\sigma_n(c)}, \qquad c \in \{v, w, v+w\}.$$

For every $j \in \{1, \ldots, r_n\}$, we then get

$$Z_{n,j}(v+w) \leq Z_{n,j}(v) \cdot \frac{\sigma_n(v)}{\sigma_n(v+w)} + Z_{n,j}(w) \cdot \frac{\sigma_n(w)}{\sigma_n(v+w)}$$

$$\leq K\left(Z_{n,j}(v) \cdot \frac{\sigma_n(v)}{\sigma_n(v) + \sigma_n(w)} + Z_{n,j}(w) \cdot \frac{\sigma_n(w)}{\sigma_n(v) + \sigma_n(w)}\right)$$

$$\leq K \max\left(Z_{n,j}(v), Z_{n,j}(w)\right).$$

For the function $f : [0, \infty) \to [0, \infty)$, defined by $f(\xi) := \xi^2 \mathbf{1}_{(\varepsilon,\infty)}(\xi)$, we thus obtain

$$f\left(Z_{n,j}(v+w)\right) \leq f\left(K Z_{n,j}(v)\right) + f\left(K Z_{n,j}(w)\right).$$

Taking expectations on both sides of this inequality and summing over j from 1 to r_n, we obtain

$$L_n^*(\varepsilon, v+w) \leq K^2 L_n^*\left(\frac{\varepsilon}{K}, v\right) + K^2 L_n^*\left(\frac{\varepsilon}{K}, w\right), \tag{6.18}$$

where

$$L_n^*(\eta, c) := \frac{1}{\sigma_n^2(c)} \sum_{j=1}^{r_n} \mathbb{E}\left[(c^\top X_{n,j})^2 \mathbf{1}\{|c^\top X_{n,j}| > \eta\sigma_n(c)\}\right], \quad c \in \{v, w, v+w\}, \eta > 0.$$

Due to $\sigma_n^2(v) \to \sigma_v^2 > 0$ and $\sigma_n^2(w) \to \sigma_w^2 > 0$ as well as (6.15) for $b = v$ and $b = w$, we have $L_n^*(\eta, v) \to 0$ and $L_n^*(\eta, w) \to 0$ for every $\eta > 0$, and thus (6.18) yields

$$\lim_{n\to\infty} L_n^*(\varepsilon, v+w) = 0.$$

Since ε is arbitrary, the Lindeberg condition (1.9) holds for $(v+w)^\top X_{n,1}, \ldots, (v+w)^\top X_{n,r_n}$. From the Lindeberg–Feller central limit theorem (Theorem 1.19) it follows that (6.17) also holds for $c = v + w$.

$\square$

6.23 Remark A generalization of Theorem 6.22 to triangular arrays of random elements that take on values in a separable Hilbert space can be found in Chap. 17 (see Theorem 17.30). Condition (6.15), which formally differs from [FH], correlates with condition (17.40). Note that we can always choose the set B occurring in Theorem 6.22 as the set of canonical unit vectors of $\mathbb{R}^d$. It should also be noted that $\lim_{n\to\infty} L_n(\varepsilon, c) = 0$ for every $\varepsilon > 0$ and every $c \in \mathbb{R}^d$ follows from the frequently encountered condition

$$\lim_{n\to\infty} \sum_{j=1}^{r_n} \mathbb{E}\Big[\|X_{n,j}\|^2 \mathbf{1}\{\|X_{n,j}\| > \varepsilon\}\Big] = 0 \text{ for every } \varepsilon > 0.$$

Self-question 6: Why does the last claim hold?

6.24 Example (Chi-Square Test, Continuation of Example 6.20)
The setting underlying Example 6.20 was an i.i.d. sequence $X_1, X_2, \ldots$ with $\mathbb{P}(X_1 = e_k) = p_k$, $k \in \{1, \ldots, s\}$. Here, e_k denotes the k-th unit vector in $\mathbb{R}^s$, and $p_1, \ldots, p_s$ are positive numbers with $\sum_{j=1}^s p_j = 1$. Using Theorem 6.19, it turned out that, as $n \to \infty$, the chi-square test statistic T_n defined in (6.7) has an asymptotic χ^2_{k-1}-distribution. In a statistical context, this limit distribution is the limit distribution of T_n under the hypothesis H_0 that an unknown probability vector $q := (q_1, \ldots, q_s)$ in a multinomial experimental design (see Example 5.4) equals $p := (p_1, \ldots, p_s)$. In the following, we investigate what happens if this hypothesis does not hold. More precisely, we assume the setting of a triangular array $X_{n,1}, \ldots, X_{n,n}$, $n \geq 2$, of i.i.d. random vectors with

$$\mathbb{P}(X_{n,1} = e_k) := p_k^{(n)} := p_k + \frac{\delta_k}{\sqrt{n}}, \quad k \in \{1, \ldots, s\}. \tag{6.19}$$

Here, $\delta_1, \ldots, \delta_s$ are numbers with $\delta_1 + \ldots + \delta_s = 0$, which are not all equal to zero. Moreover, in view of asymptotic considerations, n is chosen so large that all $p_k^{(n)}$, $k \in \{1, \ldots, s\}$, are positive. Assumption (6.19) has the following consequence: As n increases, on the one hand, the data basis improves, because we have realizations of more and more random vectors available for carrying out the test. On the other hand, the test increasingly struggles to recognize the alternative (6.19), because the probability vectors $(p_1^{(n)}, \ldots, p_s^{(n)})$ converge to the hypothetical probability vector p. We will now see how to obtain the limit distribution of T_n under the sequence of so-called *local alternatives*, given by (6.19), to the hypothesis $H_0 : q = p$, using Theorem 6.22. To this end, we set

$$\widetilde{X}_{n,j} := \frac{1}{\sqrt{n}}\big(X_{n,j} - \mathbb{E}(X_{n,j})\big), \quad j \in \{1, \ldots, n\}.$$

Then $\widetilde{X}_{n,1}, \ldots, \widetilde{X}_{n,n}$ are i.i.d. s-dimensional random vectors with $\mathbb{E}(\widetilde{X}_{n,1}) = 0_s$ and $\mathbb{E}\|\widetilde{X}_{n,1}\|^2 < \infty$. Moreover, we have (cf. (6.8))

$$\mathbb{E}(X_{n,1}) = (p_1^{(n)}, \ldots, p_s^{(n)})^\top, \quad \Sigma_n := \mathrm{Cov}(X_{n,1}) = (p_k^{(n)}\delta_{k,\ell} - p_k^{(n)}p_\ell^{(n)})_{1 \leq k,\ell \leq s}.$$

If we set $S_n := \widetilde{X}_{n,1} + \ldots + \widetilde{X}_{n,n}$, then $\mathbb{E}(S_n S_n^\top) = n\mathbb{E}(\widetilde{X}_{n,1}\widetilde{X}_{n,1}^\top) = \Sigma_n$, and thus

$$\lim_{n \to \infty} \mathbb{E}(S_n S_n^\top) = \Sigma := (p_k\delta_{k,\ell} - p_k p_\ell)_{1 \leq k,\ell \leq s}.$$

With $X_{n,1}^* := X_{n,1} - \mathbb{E}(X_{n,1})$ and $\|X_{n,1}^*\|^2 \leq s$, as well as $c \in \mathbb{R}^s$, we further obtain

$$\sum_{j=1}^n \mathbb{E}\Big[(c^\top \widetilde{X}_{n,j})^2 \mathbf{1}\{|c^\top \widetilde{X}_{n,j}| > \varepsilon\}\Big] = n \cdot \mathbb{E}\Big[(c^\top \widetilde{X}_{n,1})^2 \mathbf{1}\{|c^\top \widetilde{X}_{n,1}| > \varepsilon\}\Big]$$

$$= \mathbb{E}\Big[(c^\top X_{n,j}^*)^2 \mathbf{1}\{|c^\top X_{n,1}^*| > \varepsilon\sqrt{n}\}\Big]$$

$$\leq \|c\|^2 \cdot s \cdot \mathbb{P}(|c^\top X_{n,1}^*| > \varepsilon\sqrt{n}).$$

Thus, for $\widetilde{X}_{n,1}, \ldots, \widetilde{X}_{n,n}$ with $d := s$, $r_n := n$, $\Gamma := \Sigma$ and $B := \mathbb{R}^s$, all prerequisites of Theorem 6.22 are fulfilled, and it follows that

$$S_n = \frac{1}{\sqrt{n}}\sum_{j=1}^n (X_{n,j} - \mathbb{E}(X_{n,j})) \xrightarrow{\mathcal{D}} N_s(0_s, \Sigma).$$

With $\Delta := (\delta_1, \ldots, \delta_s)^\top$ and $\mathbb{E}(X_{n,j}) = p + \Delta/\sqrt{n}$, the mapping theorem implies

$$\frac{1}{\sqrt{n}}\sum_{j=1}^n (X_{n,j} - p) \xrightarrow{\mathcal{D}} Z := (Z_1, \ldots, Z_s)^\top \sim N_s(\Delta, \Sigma).$$

In contrast to (6.9), the distribution of Z is a normal distribution with expectation Δ. Since the considerations made after display (6.9) remain valid, the limit distribution of the random variables T_n defined in (6.7) (with $(N_{n,1}, \ldots, N_{n,s}) \overset{\mathcal{D}}{=} \sum_{j=1}^n X_{n,j}$) as $n \to \infty$ is that of $(Z_1, \ldots, Z_{s-1})^\top A^{-1}(Z_1, \ldots, Z_{s-1})$, where $A^{-1} = (\delta_{jk}p_k^{-1} + p_s^{-1})_{1 \leq j,k \leq s-1}$. This limit distribution is a non-central χ_{s-1}^2 distribution with non-centrality parameter

$$\delta^2 = \sum_{j=1}^s \frac{\delta_j^2}{p_j}$$

(Problem 6.15).

It is often the case that transformations of asymptotically normally distributed random vectors play a role. As the following fact, referred to as the *delta method*, shows, the transformed random vectors are also asymptotically normally distributed under weak conditions.

6.25 Theorem (Delta Method)
Let (T_n) be a sequence of d-dimensional random vectors with

$$\sqrt{n}\,(T_n - \vartheta) \xrightarrow{\mathcal{D}} X \sim N_d(0_d, \Sigma) \tag{6.20}$$

for some $\vartheta \in \mathbb{R}^d$. Furthermore, let $g : \mathbb{R}^d \to \mathbb{R}^s$ be a measurable function, which is differentiable at the point ϑ. If $g'(\vartheta)$ denotes the $s \times d$ Jacobian matrix of g at ϑ, then

$$\sqrt{n}\big(g(T_n) - g(\vartheta)\big) \xrightarrow{\mathcal{D}} N_s \left(0_s, g'(\vartheta)\Sigma g'(\vartheta)^\top\right). \tag{6.21}$$

Proof From the definition of the differentiablility of g at ϑ (see, e.g., [ZR1], p. 432), it follows that

$$\sqrt{n}\,(g(T_n) - g(\vartheta)) = g'(\vartheta)\sqrt{n}(T_n - \vartheta) + \|\sqrt{n}(T_n - \vartheta)\|\, r(T_n - \vartheta), \tag{6.22}$$

pointwise on the underlying probability space. Here, $r(T_n - \vartheta) \to 0_s$ as $T_n \to \vartheta$. From (6.20), we have $\sqrt{n}(T_n - \vartheta) = O_{\mathbb{P}}(1)$ and thus $T_n - \vartheta \xrightarrow{\mathbb{P}} 0_d$. By the subsequence criterion 1.1 for convergence in probability, $r(T_n - \vartheta) \xrightarrow{\mathbb{P}} 0_s$. From (6.20) and the mapping theorem, it follows that $\|\sqrt{n}(T_n - \vartheta)\| \xrightarrow{\mathcal{D}} \|X\|$. Since stochastic convergence of random vectors is equivalent to component-wise stochastic convergence, Theorem 6.15 (d) yields $\|\sqrt{n}(T_n - \vartheta)\| \cdot r(T_n - \vartheta) \xrightarrow{\mathbb{P}} 0_s$. With (6.20) and the mapping theorem, we get $g'(\vartheta)\sqrt{n}(T_n - \vartheta) \xrightarrow{\mathcal{D}} g'(\vartheta)\,X$. By Theorem 5.7, $g'(\vartheta)\,X \sim N_s(0_s, g'(\vartheta)\Sigma g'(\vartheta)^\top)$, and so the claim follows from Slutsky's lemma. □

In (6.20), the letter ϑ was chosen deliberately because the delta method is of importance in statistics, and in this context ϑ stands for an unknown parameter (vector) that is to be estimated by T_n (cf. Chap. 9). The difference $T_n - \vartheta$ is then the random *estimation error*, and this converges according to (6.20) so quickly to 0_d that, after multiplication with $\sqrt{n}$, it has a limit distribution as $n \to \infty$. The *asymptotic covariance matrix* Σ generally depends on ϑ. The situation in (6.20) occurs in many methods of estimation, such as maximum likelihood estimation (cf. Chap. 10) or estimation by the method of moments (cf. Chap. 11). In principle, the covariance matrices occurring in (6.20) and (6.21) may not be positive-definite and thus not invertible.

With a view towards statistics, in the most important special case $d = s = 1$ the delta method reads as follows: If

$$\sqrt{n}(T_n - \vartheta) \xrightarrow{\mathcal{D}} N(0, \sigma^2(\vartheta)),$$

where $\sigma^2(\vartheta) > 0$, and if the measurable function g defined on the range of T_n is differentiable at the point ϑ and has a non-vanishing derivative there, then

$$\sqrt{n}(g(T_n) - g(\vartheta)) \xrightarrow{\mathcal{D}} N(0, \sigma^2(\vartheta)g'(\vartheta)^2). \qquad (6.23)$$

If $\vartheta \in \Theta$, where $\Theta \subset \mathbb{R}$ is an open interval, one can try to find a function g for which the limit distribution in (6.23) is the standard normal distribution $N(0, 1)$. Such a function g is called a *variance-stabilizing transformation*, and the condition for having this property is therefore

$$\sigma^2(\vartheta)g'(\vartheta)^2 = 1, \quad \vartheta \in \Theta. \qquad (6.24)$$

As an example, we consider the estimation of the unknown expectation ϑ of a Poisson distribution. If $X_1, X_2, \ldots$ are independent random variables with the same distribution $Po(\vartheta)$, where $\vartheta \in (0, \infty)$, and if $T_n := \overline{X}_n = n^{-1} \sum_{j=1}^{n} X_j$, then the Lindeberg–Lévy central limit theorem implies

$$\sqrt{n}(\overline{X}_n - \vartheta) \xrightarrow{\mathcal{D}} N(0, \vartheta).$$

In this case, equation (6.24) leads to $g'(\vartheta) = \frac{1}{\sqrt{\vartheta}}$ and thus to $g(\vartheta) = 2\sqrt{\vartheta}$.

6.26 On the Speed of Convergence
There are also studies on the speed of convergence with regard to the multivariate central limit theorem and the delta method. For example, [BRR] provides an introduction to multivariate analogues of the Berry–Essen bounds (1.12) and (1.13). More recent results can be found in [RAI]. The paper [PIM] contains error bounds for the delta method.

Answers to the Self-Questions

Answer 1 For any $z \in M$, $\|x - M\| \le \|x - z\| \le \|x - y\| + \|y - z\|$ and thus $\|x - M\| \le \|x - y\| + \|y - M\|$. Similarly, $\|y - M\| \le \|y - z\| \le \|y - x\| + \|x - z\|$ and thus $\|y - M\| \le \|x - y\| + \|x - M\|$, from which the assertion follows.

Answer 2 Fix any $\omega \in \Omega$ with $Z_n(\omega) := X_n(\omega) + Y_n(\omega) \in A$. Then either $\|Y_n(\omega)\| > \varepsilon$ or $\|Y_n(\omega)\| \le \varepsilon$. In the second case, $\|X_n(\omega) - Z_n(\omega)\| \le \varepsilon$, where $Z_n(\omega) \in A$, and thus $X_n(\omega) \in A_\varepsilon$.

Answer 3 If $Q_1, \ldots, Q_\ell$ are probability measures on $\mathcal{B}^d$, then for any $\varepsilon > 0$ there are compact sets $K_1, \ldots, K_\ell \subset \mathbb{R}^d$ with $\mathbb{P}(Q_j \in K_j) \ge 1 - \frac{\varepsilon}{\ell}, j \in \{1, \ldots, \ell\}$. The set $K := K_1 \cup \ldots \cup K_\ell$ is compact, and $\mathbb{P}(Q_j \in K) \ge 1 - \varepsilon$ for each $j \in \{1, \ldots, \ell\}$.

Answer 4 By definition of W_n and the representation of A^{-1}, putting $N_j := N_{n,j}$ gives

$$W_n^\top A^{-1} W_n = \frac{1}{n} \sum_{j,k=1}^{s-1} (N_j - np_j) \left(\frac{\delta_{j,k}}{p_k} + \frac{1}{p_s} \right) (N_k - np_k)$$

$$= \frac{1}{n} \sum_{j=1}^{s-1} \frac{(N_j - np_j)^2}{np_j} + \frac{1}{np_s} \left(\sum_{j=1}^{s-1} (N_j - np_j) \right)^2.$$

Since $\sum_{j=1}^{s-1} (N_j - np_j) = n - N_s - n(1 - p_s) = -(N_s - np_s)$, the assertion follows.

Answer 5 In this case, there is even convergence in probability, because $N(0, 0)$ is a point mass at 0, and $\mathbb{E}(c^\top S_n) = 0$ and $\mathbb{V}(c^\top S_n) = c^\top \mathbb{E}(S_n S_n^\top) c$. Due to (6.14), $\mathbb{V}(c^\top S_n) \to 0$ as $n \to \infty$.

Answer 6 Due to the Cauchy–Schwarz inequality $(c^\top X_{n,j})^2 \le \|c\|^2 \|X_{n,j}\|^2$, we have

$$(c^\top X_{n,j})^2 \mathbf{1}\{|c^\top X_{n,j}| > \varepsilon\} \le \|c\|^2 \|X_{n,j}\|^2 \mathbf{1}\left\{\|X_{n,j}\| > \frac{\varepsilon}{\|c\|}\right\}, \quad \varepsilon > 0, \ c \ne 0_d,$$

and thus the assertion (which is trivial if $c = 0_d$).

Problems

6.1 Problem Prove the properties of a distribution function stated in Theorem 6.2.

Hint With $x = (x_1, \ldots, x_d)$, $y = (y_1, \ldots, y_d)$ and $X = (X_1, \ldots, X_d)$ as well as $A_j := \{X \le y, X_j \le x_j\}$ for $j \in \{1, \ldots, d\}$, we have $\{X \le y\} = \{X \in (x, y]\} \uplus (\bigcup_{j=1}^{d} A_j)$.

6.2 Problem Make clear that a distribution function can have uncountably many points of discontinuity in the case $d \ge 2$.

6.3 Problem Let (U_n, V_n), $n \ge 1$, and (U, V) be two-dimensional random vectors with $(U_n, V_n) \xrightarrow{D} (U, V)$. Then the mapping theorem states that $U_n + V_n \xrightarrow{D} U + V$. Does this result also follow from $U_n \xrightarrow{D} U$ and $V_n \xrightarrow{D} V$ alone?

6.4 Problem Prove Theorem 6.8.

Hint Remark 6.4.

6.5 Problem Let $X_n = (X_{n1}, \ldots, X_{nd})$, $n \geq 1$, be d-dimensional random vectors. Show:

$$\{X_n : n \geq 1\} \text{ is tight } \Longleftrightarrow \{X_{nj} : n \geq 1\} \text{ is tight for each } j \in \{1, \ldots, d\}.$$

6.6 Problem Let $(\mu_n)_{n\geq 1}$ be a sequence in $\mathbb{R}^d$, and let $X_n \sim N_d(\mu_n, I_d)$, $n \geq 1$. Show that the sequence $(X_n)_{n\geq 1}$ is tight if and only if the sequence (μ_n) is bounded.

6.7 Problem Prove parts (b)–(e) of Theorem 6.15.

6.8 Problem Let $X, X_1, X_2, \ldots$ be d-dimensional random vectors with $X_n \xrightarrow{\mathcal{D}} X$. Furthermore, let $A, A_1, A_2, \ldots$ be $s \times d$ matrices with $\lim_{n\to\infty} A_n = A$. Show: $A_n X_n \xrightarrow{\mathcal{D}} AX$.

Hint For $x \in \mathbb{R}^d$ and an $s \times d$ matrix B, we have $\|Bx\| \leq \|B\|_{\mathrm{sp}}\|x\|$. Here, the *spectral norm* $\|B\|_{\mathrm{sp}}$ of B is defined by $\|B\|_{\mathrm{sp}} := \max_{\{x:\|x\|=1\}} \|Bx\|$.

6.9 Problem Show that Theorem 6.19 follows from the multivariate central limit theorem for triangular arrays (Theorem 6.22).

6.10 Problem Prove the following *multivariate central limit theorem for triangular arrays*: For each $n \geq 1$, let $X_{n,1}, \ldots, X_{n,n}$ independent random vectors in $\mathbb{R}^d$ with $\mathbb{E}\|X_{n,j}\|^2 < \infty$ $(j = 1, \ldots, n)$. The covariance matrix of $X_{n,j}$, assumed to be positive definite, is denoted $\Sigma_{n,j}$. Suppose

$$\lim_{n\to\infty} \left(\frac{1}{n} \sum_{j=1}^{n} \Sigma_{n,j} \right) = \Sigma$$

for a $d \times d$ matrix Σ and

$$\lim_{n\to\infty} \frac{1}{n} \sum_{j=1}^{n} \mathbb{E}\left(\|X_{n,j} - \mathbb{E}X_{n,j}\|^2 \, \mathbf{1}\{\|X_{n,j} - \mathbb{E}X_{n,j}\| > \varepsilon\sqrt{n}\} \right) = 0 \tag{6.25}$$

for every $\varepsilon > 0$. Then

$$\frac{1}{\sqrt{n}} \sum_{j=1}^{n} (X_{n,j} - \mathbb{E}X_{n,j}) \xrightarrow{\mathcal{D}} N_d(0_d, \Sigma).$$

Hint Cramér–Wold device.

6.11 Problem Assume the situation of Example 6.21 (number lottery). Show that $\mathbb{E}(X_1)$ and $\mathbb{C}ov(X_1)$ are given by (6.10).

6.12 Problem Assume the setting of Example 6.21 (number lottery). Prove the following statements:

(a) The matrix Σ occurring in (6.10) is singular.
(b) If A is the matrix formed from the first $s - 1$ rows and columns of Σ, then $B :=$ $(b_{ij})_{1 \leq i, j \leq s-1}$ with $b_{ii} := 2\left(1 - \frac{1}{s}\right)$ for $i = 1, \ldots, s$ and $b_{ij} := 1 - \frac{1}{s}$ for $i \neq j$ is the inverse of A.
(c) As $n \to \infty$, we have

$$T_n := \frac{s-1}{s-r} \sum_{j=1}^{s} \frac{\left(H_{n,j} - n\frac{r}{s}\right)^2}{n\frac{r}{s}} \xrightarrow{\mathcal{D}} \chi^2_{s-1}. \tag{6.26}$$

Note: T_n can be used as a test statistic for a test of equiprobability of the lottery numbers. (6.26) provides the asymptotic distribution of T_n under the hypothesis.
(d) What does the application of the χ^2-test described in (c) yield for the data in the following table, if a maximum value of 0.05 is assumed for the probability of a type I error (which occurs when the hypothesis is falsely rejected)? Here $r = 6$.

How often already drawn?
After 4510 drawings

1	2	3	4	5	6	7
563	554	571	556	555	613	556
8	9	10	11	12	13	14
509	555	550	584	530	495	524
15	16	17	18	19	20	21
527	541	559	554	556	529	515
22	23	24	25	26	27	28
582	532	549	568	579	559	515
29	30	31	32	33	34	35
554	527	584	589	582	532	543
36	37	38	39	40	41	42
564	545	589	549	544	566	571
43	44	45	46	47	48	49
569	536	500	530	555	558	593

6.13 Problem The random vector $X_n = (X_{n,1}, \ldots, X_{n,n})^\top$ has a uniform distribution on the set $M_n := \{x \in \mathbb{R}^n : \|x\| = \sqrt{n}\}$. Thus, $\mathbb{P}(X_n \in M_n) = 1$, and $X_n \overset{\mathcal{D}}{=} HX_n$ for every orthogonal $n \times n$ matrix, cf. Problem 5.10. Show: For fixed d, we have

$$(X_{n,1}, \ldots, X_{n,d})^\top \overset{\mathcal{D}}{\longrightarrow} N_d(0_d, I_d) \text{ as } n \to \infty.$$

6.14 Problem For each $n \geq 1$, let $W_{n,1}, \ldots, W_{n,r_n}$ be non-negative real-valued random variables with finite expectations. Suppose there exists a $K \in (0, \infty)$ with

$$\sum_{j=1}^{r_n} \mathbb{E}[W_{n,j}] \leq K, \qquad n \geq 1. \tag{6.27}$$

Prove that the following statements are equivalent:

(a) $\displaystyle\lim_{n\to\infty} \sum_{j=1}^{r_n} \mathbb{E}[W_{n,j} \min(1, W_{n,j})] = 0.$

(b) $\displaystyle\lim_{n\to\infty} \sum_{j=1}^{r_n} \mathbb{E}[W_{n,j} \min(1, W_{n,j}^\delta)] = 0$ for every $\delta > 0$.

(c) $\displaystyle\lim_{n\to\infty} \sum_{j=1}^{r_n} \mathbb{E}[W_{n,j} \mathbf{1}\{W_{n,j} > \varepsilon\}] = 0$ for every $\varepsilon > 0$.

Hint for "(c) $\Longrightarrow$ (b)": We have $x \min(1, x^\delta) \leq x\varepsilon^\delta + x\mathbf{1}_{[\varepsilon,\infty)}(x)$ for each $x \geq 0$ and each $\varepsilon > 0$.

6.15 Problem Let $Y := (Z_1, \ldots, Z_{s-1})^\top A^{-1}(Z_1, \ldots, Z_{s-1})$ with $(Z_1, \ldots, Z_{s-1})^\top \sim N_{s-1}(\Delta_{s-1}, A)$, $\Delta_{s-1} := (\delta_1, \ldots, \delta_{s-1})^\top$, and

$$A = (p_j \delta_{kj} - p_j p_k)_{1 \leq j,k \leq s-1}, \qquad A^{-1} = (\delta_{jk} p_k^{-1} + p_s^{-1})_{1 \leq j,k \leq s-1}.$$

Prove that Y has a non-central χ_{s-1}^2-distribution with non-centrality parameter

$$\delta^2 = \sum_{j=1}^{s} \frac{\delta_j^2}{p_j}.$$

Hint Consider Problem 5.8 and $\delta_1 + \ldots + \delta_{s-1} + \delta_s = 0$.

6.16 Problem The random variables $X_1, X_2, \ldots$ are independent with the same binomial distribution $\mathrm{Bin}(1, p)$, where $p \in (0, 1)$. Let $T_n := n^{-1} \sum_{j=1}^{n} X_j$. By the Lindeberg–Lévy central limit theorem, $\sqrt{n}(T_n - p) \overset{\mathcal{D}}{\longrightarrow} N(0, p(1 - p))$. Find a variance-stabilizing transformation $g : (0, 1) \to \mathbb{R}$ such that $\sqrt{n}(g(T_n) - g(p)) \overset{\mathcal{D}}{\longrightarrow} N(0, 1)$ for each p.

Empirical Distribution Function

7

The main result of this chapter is a theorem that was proved in 1933 by F. Cantelli and shortly before under more restrictive conditions by W. Glivenko.[1] Both proofs were published in the same issue of an Italian actuarial journal (see [GLI] and [CAN]). This result is purely probabilistic in nature. It, however, directly affects statistics, as can be seen from the sometimes used naming *Fundamental Theorem of Statistics*. In order to state the result of Glivenko and Cantelli, we need the notion of an *empirical distribution function*.

To this end, let $X_1, X_2, \ldots$ be a sequence of i.i.d. real-valued random variables defined on the same probability space $(\Omega, \mathcal{A}, \mathbb{P})$. The distribution function of X_1 is denoted by F; thus $F(x) = \mathbb{P}(X_1 \leq x)$, $x \in \mathbb{R}$.

7.1 Definition (Empirical Distribution Function)

The function $\widehat{F}_n : \Omega \times \mathbb{R} \to [0, 1]$, defined by

$$\widehat{F}_n^\omega(x) := \widehat{F}_n(\omega, x) := \frac{1}{n} \sum_{j=1}^{n} \mathbf{1}\{X_j(\omega) \leq x\}, \quad \omega \in \Omega, \ x \in \mathbb{R},$$

is called the *empirical distribution function of* $X_1, \ldots, X_n$.

For a fixed $\omega \in \Omega$, realizations $X_1(\omega), \ldots, X_n(\omega)$ of $X_1, \ldots, X_n$ are given data, and these data define the so-called *empirical distribution* of $X_1(\omega), \ldots, X_n(\omega)$. The empirical distribution is a discrete probability measure that assigns the same probability

---

[1] Valery Ivanovich Glivenko (1897–1940), Russian mathematician, from 1928 until his death professor at the Moscow Pedagogical Institute "Karl Liebknecht". Main areas of work: Probability theory and foundations of mathematics.

© The Author(s), under exclusive license to Springer-Verlag GmbH, DE, part of Springer Nature 2024
N. Henze, *Asymptotic Stochastics*, Mathematics Study Resources 10, https://doi.org/10.1007/978-3-662-68923-3_7

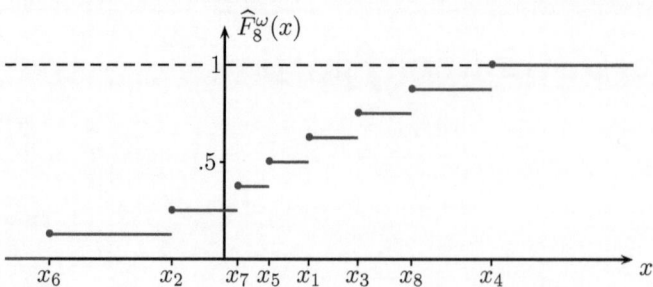

Fig. 7.1 Realization of an empirical distribution function for data $x_j = X_j(\omega)$, $j = 1, \ldots, 8$

mass n^{-1} to each $X_j(\omega)$. With δ_x denoting the Dirac measure at $x \in \mathbb{R}$, the empirical distribution is given by $n^{-1} \sum_{j=1}^{n} \delta_{X_j(\omega)}$. For a fixed ω, the empirical distribution function is the distribution function of this empirical distribution. It therefore has all properties of a distribution function, including right-continuity. Figure 7.1 shows the graph of an empirical distribution function. In this specific example, all realizations $x_j = X_j(\omega)$, $j = 1, \ldots, 8$, are different, so each jump height is $\frac{1}{8}$. Generally, if a realization occurs k times, the empirical distribution function has a jump of height $\frac{k}{n}$ at this point.

For any fixed $x \in \mathbb{R}$,

$$\widehat{F}_n(x) := \frac{1}{n} \sum_{j=1}^{n} \mathbf{1}\{X_j \le x\} = \frac{1}{n} \sum_{j=1}^{n} \mathbf{1}_{(-\infty, x]}(X_j)$$

is a random variable. Being a mean of i.i.d. random variables with a finite expectation, $\widehat{F}_n(x)$ converges $\mathbb{P}$-a.s. to $F(x)$ according to the strong law of large numbers. For each x, the sequence $(\widehat{F}_n(x))$ is thus a so-called *strongly consistent estimator* of the value $F(x) = \mathbb{P}(X_1 \le x)$ of the distribution function at the point x. There is therefore a set $A_x \in \mathcal{A}$ with $\mathbb{P}(A_x) = 1$ depending on x, with the property that for each $\omega \in A_x$ the sequence $\widehat{F}_n^\omega(x)$ of real numbers converges to $F(x)$. The following result goes far beyond this. It states that there exists a set $\Omega_0 \in \mathcal{A}$ with $\mathbb{P}(\Omega_0) = 1$, and that for each $\omega \in \Omega_0$ the sequence of functions $(\widehat{F}_n^\omega(\cdot))$ converges *uniformly on the real line* to $F(\cdot)$.

7.2 Theorem (Glivenko–Cantelli, Fundamental Theorem of Statistics)
If $X_1, X_2, \ldots$ are i.i.d. random variables with distribution function F, then

$$\lim_{n \to \infty} \sup_{x \in \mathbb{R}} \left| \widehat{F}_n(x) - F(x) \right| = 0 \quad \mathbb{P}\text{-a.s.}$$

Self-question 1: Why is the above supremum a random variable, i.e., measurable?

Proof Let

$$D_n := \sup_{x \in \mathbb{R}} \left| \widehat{F}_n(x) - F(x) \right|,$$

$$D_n^\omega := \sup_{x \in \mathbb{R}} \left| \widehat{F}_n^\omega(x) - F(x) \right|, \quad \omega \in \Omega.$$

We need to show that there is a set $\Omega_0 \in \mathcal{A}$ with $\mathbb{P}(\Omega_0) = 1$ and

$$\lim_{n \to \infty} D_n^\omega = 0 \quad \text{for every } \omega \in \Omega_0. \tag{7.1}$$

By the strong law of large numbers (Theorem 1.2), for each real number x there is a set $A_x \in \mathcal{A}$ with

$$\mathbb{P}(A_x) = 1 \quad \text{and} \quad \lim_{n \to \infty} \widehat{F}_n^\omega(x) = F(x) \text{ for each } \omega \in A_x. \tag{7.2}$$

Denote $H(x-) := \lim_{y \uparrow x, y < x} H(y)$ the left-hand limit at x of a non-decreasing function $H : \mathbb{R} \to \mathbb{R}$, which means that $H(a) \le H(b)$ whenever $a \le b$. An application of the strong law of large numbers to the sequence $(\mathbf{1}_{(-\infty, x)}(X_j))_{j \ge 1}$ shows that for each x there is a set $B_x \in \mathcal{A}$ with

$$\mathbb{P}(B_x) = 1 \quad \text{and} \quad \lim_{n \to \infty} \widehat{F}_n^\omega(x-) = F(x-) = \mathbb{P}(X_1 < x) \text{ for each } \omega \in B_x. \tag{7.3}$$

The idea of the proof now consists of using the sets A_x and B_x for countably many values of x in order to construct the set Ω_0. The fact that F and $\widehat{F}_n^\omega$ are non-decreasing functions is utilized significantly. These countably many values will be obtained using the *quantile function* F^{-1} to F. This quantile function $F^{-1} : (0, 1) \to \mathbb{R}$ is defined by

$$F^{-1}(p) := \inf\{x \in \mathbb{R} : F(x) \ge p\}, \quad 0 < p < 1 \tag{7.4}$$

(see Fig. 7.2).

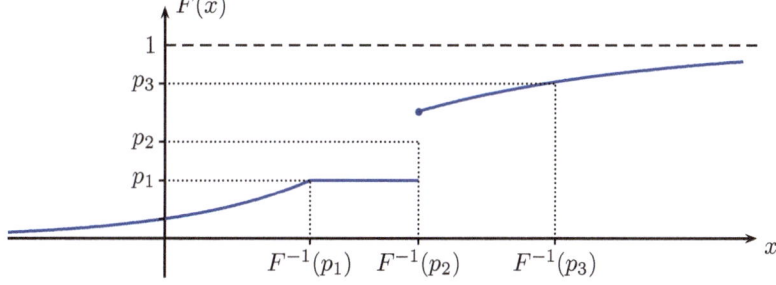

Fig. 7.2 Illustrating the definition of a quantile function

We will make decisive use of the inequalities

$$F\left(F^{-1}(p)-\right) \leq p \leq F\left(F^{-1}(p)\right) \tag{7.5}$$

that apply to each p in $(0, 1)$.

Self-question 2: Why do these inequalities hold?

The countably many values of x indicated above are chosen according to

$$x_{m,k} := F^{-1}\left(\frac{k}{m}\right),$$

where $m \geq 3$ and $k \in \{1, \ldots, m-1\}$. If we set $p = \frac{k}{m}$ once and $p = \frac{k-1}{m}$ once in (7.5), it follows that

$$F(x_{m,k}-) \leq \frac{k}{m} \leq F(x_{m,k}) \quad \text{and} \quad F(x_{m,k-1}-) \leq \frac{k-1}{m} \leq F(x_{m,k-1}).$$

By combining these inequalities, we obtain

$$F\left(x_{m,k}-\right) - F\left(x_{m,k-1}\right) \leq \frac{1}{m}, \qquad k \in \{2, \ldots, m-1\}. \tag{7.6}$$

Furthermore, we have

$$F\left(x_{m,1}-\right) \leq \frac{1}{m}, \qquad F\left(x_{m,m-1}\right) \geq 1 - \frac{1}{m}. \tag{7.7}$$

With $u \vee v := \max(u, v)$, let

$$D_{m,n}^{\omega} := \max_{1 \leq k \leq m-1} \left\{\left|\widehat{F}_n^{\omega}(x_{m,k}) - F(x_{m,k})\right| \vee \left|\widehat{F}_n^{\omega}(x_{m,k}-) - F(x_{m,k}-)\right|\right\}$$

be the maximum absolute deviation of the values and the left-hand limit values that are attained by the functions $\widehat{F}_n^{\omega}$ and F at the points $x_{m,1}, \ldots, x_{m,m-1}$. We claim that for every $m \geq 3, n \geq 1$ and $\omega \in \Omega$, the inequality

$$D_n^{\omega} \leq \frac{1}{m} + D_{m,n}^{\omega} \tag{7.8}$$

holds. For the proof, let x be any real number. We first consider the case where there is a $k \in \{2, \ldots, m-1\}$ with $x_{m,k-1} \leq x < x_{m,k}$. The monotonicity of $\widehat{F}_n^{\omega}$ and F as well as the definition of $D_{m,n}^{\omega}$ and (7.6) yield

$$\widehat{F}_n^{\omega}(x) \le \widehat{F}_n^{\omega}(x_{m,k}-) \le F(x_{m,k}-) + D_{m,n}^{\omega}$$

$$\le F(x_{m,k-1}) + \frac{1}{m} + D_{m,n}^{\omega} \le F(x) + \frac{1}{m} + D_{m,n}^{\omega}.$$

In the same way, one obtains $\widehat{F}_n^{\omega}(x) \ge F(x) - \frac{1}{m} - D_{m,n}^{\omega}$, so that together we get

$$\left| \widehat{F}_n^{\omega}(x) - F(x) \right| \le \frac{1}{m} + D_{m,n}^{\omega}. \tag{7.9}$$

The inequality (7.9) also holds in each of the remaining cases $x < x_{m,1}$ and $x \ge x_{m,m-1}$ and thus for every real number x. If $x < x_{m,1}$, then (7.7) yields

$$\widehat{F}_n^{\omega}(x) \le \widehat{F}_n^{\omega}(x_{m,1}-) \le F(x_{m,1}-) + D_{m,n}^{\omega} \le \frac{1}{m} + D_{m,n}^{\omega} \le F(x) + \frac{1}{m} + D_{m,n}^{\omega}$$

and

$$\widehat{F}_n^{\omega}(x) \ge 0 \ge F(x_{m,1}-) - \frac{1}{m} \ge F(x) - \frac{1}{m} \ge F(x) - \frac{1}{m} - D_{m,n},$$

which together also lead to (7.9). Similarly, the case $x \ge x_{m,m-1}$ also results in (7.9) (Problem 7.1). Now, we set

$$\Omega_0 := \bigcap_{m=3}^{\infty} \bigcap_{k=1}^{m-1} \left(A_{x_{m,k}} \cap B_{x_{m,k}} \right)$$

with A_x in (7.2) and B_x in (7.3), so $\Omega_0 \in \mathcal{A}$ and $\mathbb{P}(\Omega_0) = 1$.

Self-question 3: Why does $\mathbb{P}(\Omega_0) = 1$ apply?

If ω is in Ω_0 then, according to (7.2) and (7.3), as well as to the definition of $D_{m,n}^{\omega}$, we have $\lim_{n\to\infty} D_{m,n}^{\omega} = 0$ for every $m \ge 3$. In view of (7.8), it follows that $\limsup_{n\to\infty} D_n^{\omega} \le \frac{1}{m}$. Since this inequality is valid for every $m \ge 3$, we obtain (7.1). □

The Glivenko–Cantelli theorem raises very natural research questions: For a sequence of d-dimensional random vectors $X_1, X_2, \ldots$ on a probability space $(\Omega, \mathcal{A}, \mathbb{P})$ one can define a function $\widehat{P}_n : \Omega \times \mathcal{B}^d \to [0, 1]$ via

$$\widehat{P}_n^\omega(B) := \widehat{P}_n(\omega, B) := \frac{1}{n} \sum_{j=1}^{n} \mathbf{1}\{X_j(\omega) \in B\} = \frac{1}{n} \sum_{j=1}^{n} \mathbf{1}_B(X_j(\omega)).$$

For fixed $\omega \in \Omega$

$$\widehat{P}_n^\omega := \frac{1}{n} \sum_{j=1}^{n} \delta_{X_j(\omega)}$$

is called the *empirical distribution* of $X_1(\omega), \ldots, X_n(\omega)$. The probability measure $\widehat{P}_n^\omega$ assigns to each Borel set B the relative proportion of those "data" $X_1(\omega), \ldots, X_n(\omega)$ that take on values in the set B. In this more general view, the equation $\widehat{F}_n^\omega(x) = \widehat{P}_n^\omega((-\infty, x])$, $x \in \mathbb{R}$, applies to the empirical distribution function in the case $d = 1$.

If $B \in \mathcal{B}^d$ is any Borel set, then by the strong law of large numbers, applied to the indicator functions $\mathbf{1}_B(X_j)$, $j \geq 1$, there is a set $\Omega_B \in \mathcal{A}$ that depends on B, such that $\mathbb{P}(\Omega_B) = 1$ and

$$\lim_{n\to\infty} \widehat{P}_n^\omega(B) = \mathbb{P}(X_1 \in B) = \mathbb{P}^{X_1}(B) \quad \text{for each } \omega \in \Omega_B.$$

If $\mathcal{C} \subset \mathcal{B}^d$ is a non-empty collection of Borel sets, the question arises whether the sequence $(\widehat{P}_n)$ of empirical distributions converges with probability 1 *uniformly on the class $\mathcal{C}$* to the distribution $\mathbb{P}^{X_1}$. In other words: Is there a set $\Omega_0 \in \mathcal{A}$ with $\mathbb{P}(\Omega_0) = 1$ and

$$\lim_{n\to\infty} \sup_{B\in\mathcal{C}} \left| \widehat{P}_n^\omega(B) - \mathbb{P}^{X_1}(B) \right| = 0 \quad \text{for each } \omega \in \Omega_0? \tag{7.10}$$

The Glivenko–Cantelli theorem states that (7.10) holds in the case $d = 1$ and $\mathcal{C} := \{(-\infty, x] : x \in \mathbb{R}\}$. In fact, (7.10) holds for $\mathcal{C} = \{(-\infty, x] : x \in \mathbb{R}^d\}$, which means that the Glivenko–Cantelli theorem remains valid without any restriction regarding the distribution of X_1 in any dimension d. If we assume that the distribution of X_1 has a density with respect to the Borel–Lebesgue measure in $\mathbb{R}^d$, then (7.10) also holds for the class $\mathcal{C}$ of measurable convex sets in $\mathbb{R}^d$ (see, e.g., [GST], Chapter 3.).

We would like to conclude this chapter with a basic inequality that goes back to Dvoretsky,[2] Kiefer[3] and Wolfowitz[4] (see [DKW]), and was refined later (see ([MAS])). According to this inequality, the probability bound

$$\mathbb{P}\left(\sup_{x\in\mathbb{R}}\left|\widehat{F}_n(x) - F(x)\right| > t\right) \leq 2\exp\left(-2nt^2\right), \quad t > 0, \ n \in \mathbb{N}, \tag{7.11}$$

applies in the situation of the Glivenko–Cantelli theorem. It is sufficient to prove this inequality for the case that X_1 has a uniform distribution on the unit interval (Problem 7.2). Consider also that the Glivenko–Cantelli theorem follows from (7.11) (Problem 7.3).

Answers to the Self-Questions

Answer 1 Due to right-continuity of F and $\widehat{F}_n^\omega$ ($\omega \in \Omega$), it follows that

$$\sup_{x\in\mathbb{R}}\left|\widehat{F}_n(x) - F(x)\right| = \sup_{x\in\mathbb{Q}}\left|\widehat{F}_n(x) - F(x)\right|.$$

The supremum of countably many measurable functions is measurable.

Answer 2 The right inequality follows from the right-continuity of F. For the proof of the left inequality, we assume that $F(F^{-1}(p-)) > p$. Then there would be a t with $t < F^{-1}(p)$ and $F(t) > p$, thus $t > F^{-1}(p)$, which would be a contradiction.

Answer 3 The intersection of countably many sets, each having probability 1, also has probability 1. This follows when switching to complements and using the σ-subadditivity of $\mathbb{P}$.

---

[2] Aryeh Dvoretsky (1916–2006), Israeli mathematician, from 1951 professor at the Hebrew University of Jerusalem. Dvoretsky was, among other things, chief scientist of the Israeli Ministry of Defense. Main areas of work: functional analysis and stochastics.

[3] Jack Kiefer (1924–1981), American statistician, from 1953 professor at Cornell University; shortly before his death he had accepted a professorship at the University of California, Berkeley. Kiefer was instrumental in the development of optimal statistical experimental design.

[4] Jacob Wolfowitz (1910–1981), statistician who emigrated from Poland to the USA. Wolfowitz made significant contributions to nonparametric statistics (minimum distance estimation method), sequential procedures, optimal experimental design, and coding theory.

Problems

7.1 Problem In the proof of the Glivenko–Cantelli theorem, show that also the case $x \geq x_{m,m-1}$ leads to inequality (7.9).

7.2 Problem Prove the following claim: If inequality (7.11) holds when F is the distribution function of the uniform distribution on $[0, 1]$, then it holds for every distribution function F.

Hint Quantile transformation.

7.3 Problem Show that the Glivenko–Cantelli theorem follows from inequality (7.11).

7.4 Problem Let $X_1, X_2, \ldots$ be a sequence of i.i.d. random variables with distribution function F. Further, let $B_n(x) := \sqrt{n}\big(\widehat{F}_n(x) - F(x)\big)$, $x \in \mathbb{R}$. Show: For every $k \geq 1$ and every choice of $x_1, \ldots, x_k \in \mathbb{R}$, we have

$$\begin{pmatrix} B_n(x_1) \\ \vdots \\ B_n(x_k) \end{pmatrix} \xrightarrow{\mathcal{D}} N_k\left(0_k, \Sigma\right),$$

where $\Sigma = (\sigma_{ij})_{1 \leq i,j \leq k}$ and $\sigma_{ij} = F\left(\min(x_i, x_j)\right) - F(x_i)F(x_j)$, $1 \leq i, j \leq k$.

Limit Theorems for U-Statistics

8

This chapter deals with a class of random variables that have been referred to as U-*statistics* since a groundbreaking paper by W. Hoeffding[1] in 1948 (see [HOE]). The letter "U" stands for the word *unbiased*. U-statistics are indeed unbiased estimators, and we will soon see what exactly they estimate. There is an extensive literature on U-statistics, including the monographs [LEE] and [KOB]. We will focus on limit distributions for so-called *non-degenerate* U-statistics and for *simply-degenerate* U-statistics that have a degeneracy of first order. Moreover, we state and prove a central limit theorem for two-sample U-statistics. For further limit theorems or the Hoeffding decomposition, the reader is referred to the literature mentioned above.

To introduce the concept of a U-statistic, let $X_1, X_2, \ldots$ be i.i.d. d-dimensional random vectors defined on a common probability space $(\Omega, \mathcal{A}, \mathbb{P})$. The distribution function of X_1 is denoted F. Furthermore, let k be an integer and $h : (\mathbb{R}^d)^k \to \mathbb{R}$ be a measurable and *symmetric* function. Here, $(\mathbb{R}^d)^k$ stands for the k-fold Cartesian product of $\mathbb{R}^d$. The symmetry property means that the function h is invariant to permutations of its k arguments. If n is an integer, the inequality $n \geq k$ is tacitly assumed throughout this chapter.

---

[1] Wassily Hoeffding (1914–1991), received his doctorate at the University of Berlin in 1940. He emigrated to the U.S. in 1946, and from 1947 he was a professor at the University of North Carolina, Chapel Hill. Hoeffding was one of the founders of nonparametric statistics.

© The Author(s), under exclusive license to Springer-Verlag GmbH, DE,　　　　　103
part of Springer Nature 2024
N. Henze, *Asymptotic Stochastics*, Mathematics Study Resources 10,
https://doi.org/10.1007/978-3-662-68923-3_8

8.1 Definition (U-Statistic of Order k with Kernel h)

The random variable

$$U_n := U_n(X_1, \ldots, X_n) := \frac{1}{\binom{n}{k}} \sum_{1 \leq i_1 < \ldots < i_k \leq n} h\left(X_{i_1}, \ldots, X_{i_k}\right) \qquad (8.1)$$

is called (*one-sample-*) U-*statistic of order k with kernel h.*

Since the number of summands in (8.1) is equal to $\binom{n}{k}$, a U-statistic averages the values $h(X_{i_1}, \ldots, X_{i_k})$ over all selections of k of the random variables $X_1, \ldots, X_n$. We make the basic assumption

$$\mathbb{E}_F\left(h^2\right) = \mathbb{E}_F\left(h^2(X_1, \ldots, X_k)\right) < \infty$$

that the second moment of the kernel h exists. By indexing the expectation with F, we have emphasized its dependence on the distribution function, which is unknown in applications. We set $\mathcal{F} := \{F : F \text{ distribution function on } \mathbb{R}^d \text{ with } \mathbb{E}_F(h^2) < \infty\}$ and

$$\vartheta := \vartheta(F) := \mathbb{E}_F(h) = \mathbb{E}_F\left(h(X_1, \ldots, X_k)\right), \qquad F \in \mathcal{F}. \qquad (8.2)$$

For reasons of symmetry, $\vartheta = \mathbb{E}_F(U_n)$, and thus U_n is an unbiased estimator of ϑ. According to (8.2), $\vartheta : \mathcal{F} \to \mathbb{R}$ is a real-valued function defined on a certain set of distribution functions. Such a function is also called a *statistical functional*. Strictly speaking, a U-statistic estimates a statistical functional. Since $\vartheta(F)$ focuses on a specific aspect of interest (a *parameter*) of the underlying distribution of X_1, given by a real value, it is often casually said that the U-statistic estimates the parameter ϑ.

Since we have assumed $\mathbb{E}_F(h^2) < \infty$, the variance of U_n also exists, which we will discuss shortly. First, however, we introduce some examples of U-statistics.

8.2 Example

(a) If $k = 1$, then $U_n = \frac{1}{n} \sum_{j=1}^n h(X_j)$ is a mean of i.i.d. random variables, and we have $\vartheta = \mathbb{E}_F\left(h(X_1)\right)$. In this case, one can apply the Lindeberg–Lévy central limit theorem, so that the asymptotic behavior of U_n as $n \to \infty$ is known.

(b) Let $k = 2$, $d = 1$, and $h(x_1, x_2) = \frac{1}{2}(x_1 - x_2)^2$. In this case, we have

$$U_n = \frac{1}{\binom{n}{2}} \sum_{1 \leq i < j \leq n} \frac{1}{2}(X_i - X_j)^2 = \frac{1}{n-1} \sum_{j=1}^n (X_j - \overline{X}_n)^2, \qquad (8.3)$$

where $\overline{X}_n = n^{-1} \sum_{j=1}^n X_j$. This U-statistic is equal to the so-called *sample variance*, and the parameter $\vartheta(F) = \mathbb{V}_F(X_1)$ is the variance of the underlying distribution.

Self-question 1: Why does the second equal sign hold in (8.3)?

(c) In the case $k = 2$, $d = 1$, and $h(x_1, x_2) := |x_1 - x_2|$, we get *Gini's[2] mean absolute difference*

$$U_n := \frac{1}{\binom{n}{2}} \sum_{1 \le i < j \le n} |X_i - X_j|.$$

This U-statistic uses $\vartheta(F) = \mathbb{E}_F |X_1 - X_2|$ to estimate a measure of the dispersion of the underlying distribution. If $X_1 \sim \mathrm{N}(\mu, \sigma^2)$ then $\vartheta(F) = 2\sigma / \sqrt{\pi}$.

(d) In this example, $d = k = 2$. To avoid having to introduce double indices, we write the bivariate random vectors in the form (X_1, Y_1), (X_2, Y_2), ... The *sample covariance* of $(X_1, Y_1), \ldots, (X_n, Y_n)$ is defined by

$$SC_n := \frac{1}{n-1} \sum_{j=1}^{n} (X_j - \overline{X}_n)(Y_j - \overline{Y}_n),$$

where $\overline{X}_n := n^{-1} \sum_{j=1}^{n} X_j$ and $\overline{Y}_n := n^{-1} \sum_{j=1}^{n} Y_j$. If we set

$$h\big((x_1, y_1), (x_2, y_2)\big) := \frac{1}{2}(x_1 - x_2)(y_1 - y_2), \qquad (x_1, y_1), (x_2, y_2) \in \mathbb{R}^2,$$

it follows that

$$U_n = \frac{1}{\binom{n}{2}} \sum_{1 \le i < j \le n} \frac{1}{2}(X_i - X_j)(Y_i - Y_j) = SC_n. \tag{8.4}$$

Self-question 2: Why does the second equals sign hold in (8.4)?

In this case, the U-statistic is therefore equal to the sample covariance, and with the notation dF for $d\mathbb{P}^{X_1}$, the corresponding *covariance functional* is given by

$$\mathrm{Cov}_F(X_1, Y_1) = \int_{\mathbb{R}^2} \int_{\mathbb{R}^2} \frac{1}{2}(x_1 - x_2)(y_1 - y_2) \, dF(x_1, y_1) \, dF(x_2, y_2).$$

---

[2] Corrado Gini (1884–1965), Italian statistician. Among other things, Gini developed the Gini coefficient named after him, which can be used to represent the unequal distribution of income within an economy.

(e) For the next example, $d = 1$ and $k = 2$, and the kernel h is the indicator function
$h(x_1, x_2) := \mathbf{1}\{x_1 + x_2 > 0\}$. The corresponding U-statistic, often referred to as
Wilcoxon's[3] one-sample statistic, has the form

$$U_n = \frac{1}{\binom{n}{2}} \sum_{1 \le i < j \le n} \mathbf{1}\{X_i + X_j > 0\}.$$

The corresponding statistical functional is $\vartheta(F) = \mathbb{P}_F (X_1 + X_2 > 0)$. Here, the
dependence on F is indicated by indexing $\mathbb{P}$ with F.

(f) This example deals with the so-called *Kendall's τ-coefficient*.[4] Here, $d = k = 2$ and,
using the signum function, the kernel is defined by

$$h\big((x_1, y_1), (x_2, y_2)\big) := \mathrm{sgn}\big((x_1 - x_2)(y_1 - y_2)\big), \quad (x_1, y_1), (x_2, y_2) \in \mathbb{R}^2.$$

We have $h\big((x_1, y_1), (x_2, y_2)\big) > 0$ or $h\big((x_1, y_1), (x_2, y_2)\big) < 0$ according to whether
the line on which the points (x_1, y_1) and (x_2, y_2) lie has a positive or a negative
slope, respectively. In the first case, both points are called *concordant*, in the second
discordant. We assume $(X_1, Y_1), (X_2, Y_2), \ldots$ is a sequence of i.i.d. bivariate random
vectors ("random points") with the same distribution function F. Kendall's *measure
of discordance* is defined by $\tau(F) := \mathbb{E}_F[h\big((X_1, Y_1), (X_2, Y_2)\big)]$. From now on,
we make the additional assumption that F has a Lebesgue density, which implies
$\mathbb{P}_F(h\big((X_1, Y_1), (X_2, Y_2)\big) = 0) = 0$.

Self-question 3: Why does this equation hold if F has a density?

The U-statistic corresponding to the kernel h is the *empirical τ-coefficient*

$$\tau_n := \frac{1}{\binom{n}{2}} \sum_{1 \le i < j \le n} \mathrm{sgn}\big((X_i - X_j)(Y_i - Y_j)\big). \tag{8.5}$$

It uses the difference between the relative proportions of concordant and discordant pairs
of points among $(X_1, Y_1), \ldots, (X_n, Y_n)$ and thus only takes on values in the interval

---

[3] Frank Wilcoxon (1892–1965), US-American chemist and statistician, 1924 PhD in physical
chemistry at Cornell University. The signed-rank test named after him and the rank sum test also
bearing his name are among the most important nonparametric test procedures.

[4] Sir Maurice George Kendall (1907–1983), British statistician, known among other things for his
three-volume work *The Advanced Theory of Statistics* (with Alan Stuart), 1949–1961 Prof. at the
London School of Economics, from 1972 director of the *World Fertility Survey*, a project supported
by the United Nations, among others, to investigate fertility in industrial and developing countries.

$[-1, 1]$. Since the value of τ_n does not change if the components of the points involved are subjected to increasing continuous transformations in each coordinate, τ_n is a so-called *rank correlation coefficient*.

In the following, we often write $\mathbb{E} = \mathbb{E}_F$, $\mathbb{V} = \mathbb{V}_F$, $\mathbb{P} = \mathbb{P}_F$ etc.

To obtain the variance of U_n in (8.1), note that $\mathbb{V}(U_n) = \text{Cov}(U_n, U_n)$, and that $\text{Cov}(\cdot, \cdot)$ is a bilinear functional. In addition, due to the symmetry of h, the covariances between the individual summands in (8.1) only depend on how many random variables occur together as arguments of the kernel h. To this end, we define for each $c \in \{1, \ldots, k\}$

$$\sigma_c^2 := \text{Cov}\big(h(X_1, \ldots, X_c, X_{c+1}, \ldots, X_k), h(X_1, \ldots, X_c, X_{k+1}, \ldots, X_{2k-c})\big). \tag{8.6}$$

A direct calculation now provides the following result (Problem 8.1).

8.3 Theorem (Variance of a U-Statistic)
For the U-statistic U_n defined in (8.1), we have

$$\mathbb{V}(U_n) = \frac{1}{\binom{n}{k}} \sum_{c=1}^{k} \binom{k}{c}\binom{n-k}{k-c}\sigma_c^2. \tag{8.7}$$

If we employ the usual notation $a_n \sim b_n :\Longleftrightarrow \lim_{n\to\infty} \frac{a_n}{b_n} = 1$ for asymptotic equality of sequences of real numbers with $b_n \neq 0$ for each $n \geq 1$, then as $n \to \infty$

$$\binom{n}{k} \sim \frac{n^k}{k!}, \qquad \binom{n-k}{k-c} \sim \frac{n^{k-c}}{(k-c)!}.$$

Hence, the factor before σ_c^2 in (8.7) is of the order $O(n^{-c})$. In the case $\sigma_1^2 > 0$, $n\mathbb{V}(U_n)$ converges to $k^2\sigma_1^2$, whereas in the case $0 = \sigma_1^2 < \sigma_2^2$ we have $n^2\mathbb{V}(U_n) \to 2\binom{k}{2}^2\sigma_2^2$. If $\sigma_1^2 > 0$ then the U-statistic U_n is called *non-degenerate*. If $0 = \sigma_1^2 < \sigma_2^2$ applies, U_n is said to have a *degeneracy of first order*, and we use the naming that U_n is *simply-degenerate*. We will treat the asymptotics of U_n in each of these two cases. In doing so, the case of a simply-degenerate U-statistic proves to be significantly more difficult, but also much more interesting. Fundamental for further investigations are the functions $h_c : (\mathbb{R}^d)^c \to \mathbb{R}$, which are defined for each $c \in \{1, \ldots, k-1\}$ by

$$h_c(x_1, \ldots, x_c) := \mathbb{E}[h(x_1, \ldots, x_c, X_{c+1}, \ldots, X_k)] \tag{8.8}$$
$$= \mathbb{E}\big[h(X_1, \ldots, X_k)\big|X_1 = x_1, \ldots, X_c = x_c\big].$$

In addition, we define $h_k := h$.

Due to property 1.26 (a) of the conditional expectation, we have

$$\mathbb{E}(h_c) = \mathbb{E}[h_c(X_1, \ldots, X_c)] = \vartheta = \mathbb{E}(h), \quad c \in \{1, \ldots, k\}. \tag{8.9}$$

With the help of the function h_c, the variable σ_c^2 defined in (8.6) takes on a further meaning that also justifies the designation "σ^2", which usually stands for a variance.

8.4 Lemma *We have $\sigma_c^2 = \mathbb{V}\big(h_c(X_1, \ldots, X_c)\big)$.*

Proof In view of (8.9), the definition (8.6) of σ_c^2, and the equation $\mathrm{Cov}(U, V) = \mathbb{E}(UV) - \mathbb{E}(U)\mathbb{E}(V)$, valid for any random variables U and V, we obtain

$$\sigma_c^2 = \mathbb{E}\big[h(X_1, \ldots, X_k)h(X_1, \ldots, X_c, X_{k+1}, \ldots, X_{2k-c})\big] - \vartheta^2$$

$$= \mathbb{E}\Big[\mathbb{E}\big[h(X_1, \ldots, X_k)h(X_1, \ldots, X_c, X_{k+1}, \ldots, X_{2k-c})\big|X_1, \ldots, X_c\big]\Big] - \vartheta^2.$$

The second equals sign follows from the law of iterated expectation, cf. Theorem 1.26 (a). Conditionally on $X_1, \ldots, X_c$, $h(X_1, \ldots, X_k)$ and $h(X_1, \ldots, X_c, X_{k+1}, \ldots, X_{2k-c})$ are independent, and thus the above conditional expectation is equal to

$$\mathbb{E}\big[h(X_1, \ldots, X_k)\big|X_1, \ldots, X_c\big] \cdot \mathbb{E}\big[h(X_1, \ldots, X_c, X_{k+1}, \ldots, X_{2k-c})\big|X_1, \ldots, X_c\big]$$

$$= h_c(X_1, \ldots, X_c)h_c(X_1, \ldots, X_c).$$

It follows that $\sigma_c^2 = \mathbb{E}\big(h_c^2\big) - \big(\mathbb{E}(h_c)\big)^2 = \mathbb{V}(h_c)$. $\qquad\square$

8.5 Example (Continuation of Example 8.2 (b))
For the kernel $h(x_1, x_2) = \frac{1}{2}(x_1 - x_2)^2$, we put $\mu := \mathbb{E}(X_1)$ and $\mu_r := \mathbb{E}\big[(X_1 - \mu)^r\big]$, $r \in \{2, 4\}$. Then

$$h_1(x_1) = \mathbb{E}\left[\tfrac{1}{2}(x_1 - X_2)^2\right] = \tfrac{1}{2}\mathbb{E}\left[(X_2 - \mu + \mu - x_1)^2\right] = \tfrac{1}{2}\left(\mu_2 + (\mu - x_1)^2\right),$$

and thus

$$\sigma_1^2 = \mathbb{V}\big(\tfrac{1}{2}\big(\mu_2 + (X_1 - \mu)^2\big)\big) = \tfrac{1}{4}\big(\mu_4 - \mu_2^2\big),$$

$$\sigma_2^2 = \mathbb{V}\left(\tfrac{1}{2}(X_1 - X_2)^2\right) = \tfrac{1}{2}\big(\mu_4 + \mu_2^2\big). \qquad (8.10)$$

Self-question 4: Why does the second equal sign apply in (8.10)?

With Theorem 8.3 and (8.3) as well as direct calculation, it follows that the sample variance

$$S_n^2 = \frac{1}{n-1}\sum_{j=1}^{n}(X_j - \overline{X}_n)^2$$

has the variance

$$
\mathbb{V}(S_n^2) = \frac{2}{n(n-1)} \left[\binom{2}{1}\binom{n-2}{2-1}\sigma_1^2 + \binom{2}{2}\binom{n-2}{2-2}\sigma_2^2 \right]
$$
$$
= \frac{1}{n}\left(\mu_4 - \frac{n-3}{n-1}\mu_2^2 \right).
$$

The variance of the sample variance thus crucially depends on the fourth central moment μ_4 of the underlying distribution. Note that this moment exists due to the assumption $\mathbb{E}(h^2) < \infty$.

At this point, it is appropriate to briefly discuss laws of large numbers for U-statistics. Since $\mathbb{E}(U_n) = \vartheta$ and $\mathbb{V}(U_n) = O(n^{-c})$ with $c \geq 1$, $U_n \xrightarrow{\mathbb{P}} \vartheta$ as $n \to \infty$, and thus there is a weak law of large numbers for U-statistics. If $c \geq 2$, which is particularly true for the simply-degenerate case, the series criterion for almost sure convergence (Theorem 1.6) shows that even the strong law of large numbers holds, which means that $U_n \xrightarrow{a.s.} \vartheta$. With considerably more powerful mathematical techniques it can be shown that this strong law of large numbers remains valid under the weaker moment condition $\mathbb{E}[|h(X_1, \ldots, X_k)|] < \infty$. For a proof see, e.g., [LEE], p. 122 ff.

8.6 Theorem (Strong Law of Large Numbers for U-Statistics)
Let U_n be a U-statistic as in (8.1). If $\mathbb{E}[|h(X_1, \ldots, X_k)|] < \infty$ then

$$
U_n \xrightarrow{a.s.} \vartheta \text{ as } n \to \infty. \tag{8.11}
$$

We will now see that a non-degenerate U-statistic U_n satisfies a central limit theorem, that is, it has an asymptotic standard normal distribution as $n \to \infty$ after standardization. Since in the case $k \geq 2$ the summands $h(X_{i_1}, \ldots, X_{i_k})$ in (8.1) are identically distributed, but not independent in their entirety, the Lindeberg–Lévy central limit theorem cannot be applied here. The goal is achieved by the idea, traced back to W. Hoeffding (see [HOE]), to approximate U_n sufficiently well by a random variable for which a central limit theorem holds. The following naming is due to the fact that this idea was presented in a more general framework by J. Hájek[5] in [HAJ]. The so-called *projection lemma* of Hájek can be found as Problem 8.4.

---

[5] Jaroslav Hájek (1926–1974), Czech mathematician, from 1946 professor of probability theory and Statistics at Charles University in Prague. Main areas of work: Sampling theory as well as theoretical and nonparametric statistics. In 1967 his influential book *Theory of Rank Tests*, co-authored with Z. Šidák, was published.

8.7 Definition (Hájek Projection)

If U_n is a U-statistic as in (8.1), then

$$\widetilde{U}_n := \sum_{j=1}^{n} \mathbb{E}[U_n|X_j] - (n-1)\vartheta \tag{8.12}$$

is called the *Hájek projection of* U_n.

By the factorization theorem for conditional expectations (Theorem 1.27), the random variable $\mathbb{E}[U_n|X_j]$ is a measurable function of X_j. Apart from the term $(n-1)\vartheta$, the Hájek projection $\widetilde{U}_n$ is thus a sum of i.i.d. random variables. Since $\vartheta = \mathbb{E}(U_n) = \mathbb{E}\big(\mathbb{E}[U_n|X_j]\big)$, the subtrahend $(n-1)\vartheta$ shows up in (8.12) because it ensures the equality $\mathbb{E}(U_n) = \mathbb{E}(\widetilde{U}_n) = \vartheta$ of the expectations of U_n and $\widetilde{U}_n$. Moreover, since for each $j \in \{1, \ldots, n\}$ the conditional expectation $\mathbb{E}[U_n|X_j]$ provides the best possible approximation in quadratic mean of U_n by a measurable function of X_j (see, e.g., [KLE], Corollary 8.17), we can indeed hope that $\widetilde{U}_n$ provides a sufficiently good approximation of U_n for our purposes. Part (c) of the following auxiliary result shows that this hope is fulfilled.

8.8 Lemma *For the Hájek projection* $\widetilde{U}_n$ *of* U_n, *we have:*

(a) $\widetilde{U}_n = \dfrac{k}{n} \sum\limits_{j=1}^{n} \big(h_1(X_j) - \vartheta\big) + \vartheta,$

(b) $\mathbb{E}\big(U_n - \widetilde{U}_n\big)^2 = \sigma_1^2 \left\{ k \dfrac{\binom{n-k}{k-1}}{\binom{n}{k}} - \dfrac{k^2}{n} \right\} + \dfrac{1}{\binom{n}{k}} \sum\limits_{c=2}^{k} \binom{k}{c}\binom{n-k}{k-c} \sigma_c^2,$

(c) $\mathbb{E}\big(U_n - \widetilde{U}_n\big)^2 = O\big(\tfrac{1}{n^2}\big)$ *as* $n \to \infty$.

Proof

(a) For a k-element subset $A = \{i_1, \ldots, i_k\}$ of $\{1, \ldots, n\}$, we shortly put $h_A := h(X_{i_1}, \ldots, X_{i_k})$. By property 1.26 (c) of the conditional expectation,

$$\mathbb{E}[U_n|X_j] = \frac{1}{\binom{n}{k}} \sum_{A:|A|=k} \mathbb{E}\big[h_A|X_j\big].$$

Moreover, $\mathbb{E}[h_A|X_j] = \vartheta$ if $j \notin A$, and $\mathbb{E}[h_A|X_j] = h_1(X_j)$ if $j \in A$.

Self-question 5: Why do the last two equations apply?

If we count how often these two cases occur, the result is

$$\mathbb{E}[U_n | X_j] = \frac{1}{\binom{n}{k}}\left[\binom{n-1}{k-1}h_1(X_j) + \binom{n-1}{k}\vartheta\right] = \frac{k}{n}h_1(X_j) + \frac{n-k}{n}\vartheta.$$

(b) Since $\mathbb{E}(U_n) = \mathbb{E}(\tilde{U}_n)$ we may w.l.o.g. assume $\vartheta = 0$. It follows that

$$\mathbb{E}(U_n - \tilde{U}_n)^2 = \mathbb{V}(U_n) + \mathbb{V}(\tilde{U}_n) - 2\mathbb{E}(U_n\tilde{U}_n)$$

$$= \frac{1}{\binom{n}{k}}\sum_{c=1}^{k}\binom{k}{c}\binom{n-k}{k-c}\sigma_c^2 + \frac{k^2}{n^2}n\sigma_1^2$$

$$-2\frac{k}{n}\frac{1}{\binom{n}{k}}\sum_{A:|A|=k}\sum_{j=1}^{n}\mathbb{E}\big[h_A h_1(X_j)\big]. \tag{8.13}$$

Regarding the expectation occurring in (8.13), we distinguish between the two cases $j \notin A$ and $j \in A$. In the first case, h_A and $h_1(X_j)$ are independent, so this expectation vanishes due to the assumption $\vartheta = 0$. If $j \in A$, we can assume $j = 1$ and $A = \{1, \ldots, k\}$ for reasons of symmetry, which results in

$$\mathbb{E}\big[h_A h_1(X_j)\big] = \mathbb{E}\,[h(X_1, \ldots, X_k)h_1(X_1)]$$

$$= \mathbb{E}\big[\mathbb{E}\,[h(X_1, \ldots, X_k)h_1(X_1)|X_1]\big]$$

$$= \mathbb{E}\big[h_1(X_1)\mathbb{E}\,[h(X_1, \ldots, X_k)|X_1]\big]$$

$$= \mathbb{E}\,[h_1(X_1)h_1(X_1)]$$

$$= \mathbb{V}(h_1(X_1)) = \sigma_1^2.$$

For the third equals sign, we used property 1.26 (d) of conditional expectations. Thus, $\sum_{j=1}^{n}\mathbb{E}[h_A h_1(X_j)] = k\sigma_1^2$, and the expression in (8.13) becomes $-2k^2\sigma_1^2/n$. We get

$$\mathbb{E}(U_n - \tilde{U}_n)^2 = \frac{1}{\binom{n}{k}}\sum_{c=1}^{k}\binom{k}{c}\binom{n-k}{k-c}\sigma_c^2 - \frac{k^2}{n^2}n\sigma_1^2,$$

and (b) follows by separating the two terms containing σ_1^2 and factoring out σ_1^2.

(c) Since $c \geq 2$, the expression in (b) after the curly bracket is of the order $O(\frac{1}{n^2})$ as $n \to \infty$. The first summand on the right-hand side of (b) is equal to

$$\sigma_1^2\frac{k^2}{n}\left\{\frac{\binom{n-k}{k-1}}{\binom{n-1}{k-1}} - 1\right\},$$

and the curly bracket is of order $O(\frac{1}{n})$ as $n \to \infty$.

$\square$

Self-question 6: Why is the curly bracket of order $O\left(\frac{1}{n}\right)$?

The main result for non-degenerate U-statistics is as follows:

8.9 Theorem (Central Limit Theorem for Non-degenerate U-Statistics)
If U_n is a non-degenerate U-statistic, that is, if $\sigma_1^2 > 0$, then

$$\sqrt{n}(U_n - \vartheta) \xrightarrow{\mathcal{D}} N\left(0, k^2\sigma_1^2\right) \quad as \; n \to \infty.$$

Proof Putting $R_n := \sqrt{n}(U_n - \widetilde{U}_n)$, we get

$$\sqrt{n}\,(U_n - \vartheta) = \sqrt{n}(\widetilde{U}_n - \vartheta) + R_n.$$

By Lemma 8.8 (c), $\mathbb{E}(R_n^2) \to 0$ and thus $R_n \xrightarrow{\mathbb{P}} 0$. If we set $Y_j := k(h_1(X_j) - \vartheta)$, then $Y_1, \ldots, Y_n$ are i.i.d. random variables with $\mathbb{E}(Y_1) = 0$ and $\mathbb{V}(Y_1) = k^2\sigma_1^2$. The Lindeberg–Lévy central limit theorem yields

$$\sqrt{n}(\widetilde{U}_n - \vartheta) = \frac{1}{\sqrt{n}}\sum_{j=1}^{n} Y_j \xrightarrow{\mathcal{D}} N\left(0, k^2\sigma_1^2\right),$$

and the claim follows with Slutsky's lemma. $\square$

8.10 Example (Continuation of Example 8.2 (b))
Let $h(x_1, x_2) = \frac{1}{2}(x_1 - x_2)^2$ be the kernel associated with the sample variance $S_n^2 = \frac{1}{n-1}\sum_{j=1}^{n}(X_j - \overline{X}_n)^2$. In Example 8.5, we obtained

$$h_1(x_1) = \frac{1}{2}\left(\mu_2 + (\mu - x_1)^2\right), \quad \sigma_1^2 = \frac{1}{4}\left(\mu_4 - \mu_2^2\right),$$

where $\mu := \mathbb{E}(X_1)$ and $\mu_r := \mathbb{E}\left[(X_1 - \mu)^r\right], r \in \{2, 4\}$. To apply the central limit theorem, we have to check the condition $\sigma_1^2 > 0$. Since $\mu_4 - \mu_2^2 = \mathbb{V}\left[(X_1 - \mu)^2\right], \sigma_1^2$ is zero if and only if there is a $c \geq 0$ with $\mathbb{P}(X_1 \in \{\mu + \sqrt{c}, \mu - \sqrt{c}\}) = 1$. If we exclude this case, then, due to $\mu_2 = \sigma^2$, Theorem 8.9 yields

$$\sqrt{n}\left(S_n^2 - \sigma^2\right) \xrightarrow{\mathcal{D}} N\left(0, \mu_4 - \sigma^4\right) \quad as \; n \to \infty. \tag{8.14}$$

8.11 Example (Continuation of Example 8.2 (d))

For the kernel defined by $h(x_1, x_2) = \mathbf{1}\{x_1 + x_2 > 0\}$, we have

$$h_1(x_1) = \mathbb{E}\big(\mathbf{1}\{x_1 + X_2 > 0\}\big) = \mathbb{P}(X_2 > -x_1) = 1 - F(-x_1)$$

and thus $\sigma_1^2 = \mathbb{V}\big(1 - F(-X_1)\big) = \mathbb{V}\big(F(-X_1)\big)$. If the distribution function F is continuous, and if the distribution of X_1 is symmetric about 0, that is, if $X_1 \overset{\mathcal{D}}{=} -X_1$, then

$$F(-X_1) \overset{\mathcal{D}}{=} F(X_1) \overset{\mathcal{D}}{=} \mathrm{U}(0, 1). \tag{8.15}$$

Here, $\mathrm{U}(0, 1)$ denotes the uniform distribution on $(0, 1)$.

Self-question 7: Why does the second equality in distribution apply in (8.15)?

It follows that $\sigma_1^2 = \mathbb{V}\big(\mathrm{U}(0, 1)\big) = \frac{1}{12}$, and the central limit Theorem 8.9 thus yields

$$\sqrt{n}\left(\frac{1}{\binom{n}{2}} \sum_{1 \le i < j \le n} \mathbf{1}\{X_i + X_j > 0\} - \frac{1}{2}\right) \overset{\mathcal{D}}{\longrightarrow} \mathrm{N}\left(0, \frac{1}{3}\right).$$

We now examine the case that the U-statistic U_n in (8.1) is simply-degenerate, that is, we assume $0 = \sigma_1^2 < \sigma_2^2$. From (8.7), we then obtain

$$\mathbb{V}(U_n) = \frac{1}{\binom{n}{k}} \binom{k}{2}\binom{n-k}{k-2}\sigma_2^2 + O\left(\frac{1}{n^3}\right)$$

$$= \frac{2\binom{k}{2}^2}{n^2}\sigma_2^2 + O\left(\frac{1}{n^3}\right),$$

and thus

$$\lim_{n \to \infty} \mathbb{V}\big(n(U_n - \vartheta)\big) = 2\binom{k}{2}^2 \sigma_2^2.$$

In particular, it follows that the sequence $\big(n(U_n - \vartheta)\big)$ is tight. We can therefore conjecture that $n(U_n - \vartheta)$ has a non-degenerate limit distribution as $n \to \infty$. The following example shows what such a limit distribution could look like.

8.12 Example Let $s \geq 1$ and

$$h(x_1, x_2) := \sum_{\ell=1}^{s} \lambda_\ell \varphi_\ell(x_1) \varphi_\ell(x_2), \tag{8.16}$$

where $\varphi_1, \ldots, \varphi_s : \mathbb{R}^d \to \mathbb{R}$ are measurable functions and $\lambda_1, \ldots, \lambda_s$ are non-zero real numbers. Furthermore, let

$$\mathbb{E}[\varphi_\ell(X_1)] = 0, \quad \mathbb{E}[\varphi_\ell^2(X_1)] = 1, \tag{8.17}$$

$$\mathbb{E}[\varphi_\ell(X_1)\varphi_m(X_1)] = \delta_{\ell,m}, \quad \ell, m \in \{1, \ldots, s\}. \tag{8.18}$$

It follows that

$$U_n = \frac{1}{\binom{n}{2}} \sum_{1 \leq i < j \leq n} \sum_{\ell=1}^{s} \lambda_\ell \varphi_\ell(X_i) \varphi_\ell(X_j) = \sum_{\ell=1}^{s} \lambda_\ell \frac{1}{\binom{n}{2}} \frac{1}{2} \sum_{1 \leq i \neq j \leq n} \varphi_\ell(X_i) \varphi_\ell(X_j)$$

$$= \sum_{\ell=1}^{s} \lambda_\ell \frac{1}{\binom{n}{2}} \frac{1}{2} \left\{ \left(\sum_{j=1}^{n} \varphi_\ell(X_j) \right)^2 - \sum_{j=1}^{n} \varphi_\ell^2(X_j) \right\}$$

and thus

$$nU_n = \sum_{\ell=1}^{s} \lambda_\ell \frac{n}{n-1} \left\{ \left(\frac{1}{\sqrt{n}} \sum_{j=1}^{n} \varphi_\ell(X_j) \right)^2 - \frac{1}{n} \sum_{j=1}^{n} \varphi_\ell^2(X_j) \right\}.$$

By the strong law of large numbers (Theorem 1.2),

$$\frac{1}{n} \sum_{j=1}^{n} \varphi_\ell^2(X_j) \xrightarrow{\text{a.s.}} \mathbb{E}\left[\varphi_\ell^2(X_1)\right] = 1.$$

If $N_1, \ldots, N_s$ are independent random variables with a standard normal distribution, the multivariate central limit theorem (Theorem 6.19) implies

$$\frac{1}{\sqrt{n}} \sum_{j=1}^{n} \begin{pmatrix} \varphi_1(X_j) \\ \vdots \\ \varphi_s(X_j) \end{pmatrix} \xrightarrow{\mathcal{D}} N_s (0_s, I_s) \sim \begin{pmatrix} N_1 \\ \vdots \\ N_s \end{pmatrix}.$$

With the mapping theorem and Slutsky's lemma, we now obtain

$$nU_n \xrightarrow{\mathcal{D}} \sum_{\ell=1}^{s} \lambda_\ell \left(N_\ell^2 - 1\right) \quad \text{as } n \to \infty. \tag{8.19}$$

Thus, apart from the centering of the N_j^2, the limit distribution of nU_n is a weighted sum of independent χ_1^2-distributed random variables. The structure of this limit distribution is clearly different from that of the non-degenerate case, in which $\sqrt{n}(U_n - \vartheta)$ has an asymptotic normal distribution.

To obtain a limit distribution for $n(U_n - \vartheta)$, we proceed in two steps. First, we approximate a general kernel h, whose order k can be arbitrarily large, by a kernel $\widehat{h}$ of order 2. This approximation will prove to be so good that both $\mathbb{E}(\widehat{U}_n) = \vartheta$ and $n(U_n - \widehat{U}_n) = o_{\mathbb{P}}(1)$ hold for a random variable $\widehat{U}_n$ based on $\widehat{h}$, which is yet to be defined, and which will turn out to be a U-statistic after subtraction of ϑ. Consequently, $n(U_n - \vartheta)$ and $n(\widehat{U}_n - \vartheta)$ would have the same limit distribution as $n \to \infty$, if such a limit distribution exists at all. In a second step, and using tools from functional analysis, the kernel $\widehat{h}$ is then approximated by a kernel of the form (8.16).

The idea of the first approximation is not to take the conditional expectations $\mathbb{E}[U_n|X_j]$ with respect to *one* of the random variables $X_1, \ldots, X_n$ as in the Hájek projection, but to strive for the best approximation in quadratic mean with respect to functions of *two* of the random variables, that is, to use the conditional expectations $\mathbb{E}[U_n|X_i, X_j]$. Note that $\mathbb{E}[U_n|X_i, X_j]$ is a measurable function of X_i and X_j according to the factorization theorem for conditional expectations (Theorem 1.27).

The above approach leads to a random variable $\widehat{U}_n$, defined by

$$\widehat{U}_n := \sum_{1 \le i < j \le n} \mathbb{E}\left[U_n|X_i, X_j\right] - \binom{n}{2}\vartheta + \vartheta. \qquad (8.20)$$

Because of $\mathbb{E}\big[\mathbb{E}[U_n|X_i, X_j]\big] = \vartheta$ and the fact that the above sum has $\binom{n}{2}$ summands, it follows that $\mathbb{E}(\widehat{U}_n) = \vartheta$. The next result shows that the connection of U_n with $\widehat{U}_n$ is established via the function h_2 defined in (8.8). Furthermore, the quality of the approximation of U_n by $\widehat{U}_n$ is specified.

8.13 Lemma *For the random variable $\widehat{U}_n$ defined in (8.20), we have:*

(a) $\widehat{U}_n - \vartheta = \dfrac{1}{\binom{n}{2}}\binom{k}{2} \displaystyle\sum_{1 \le i < j \le n} \big(h_2(X_i, X_j) - \vartheta\big),$

(b) $\mathbb{E}\big(U_n - \widehat{U}_n\big)^2 = O\left(\dfrac{1}{n^3}\right)$ *as* $n \to \infty.$

Proof Putting $h_A := h(X_{i_1}, \ldots, X_{i_k})$ for a k-element subset $A = \{i_1, \ldots, i_k\}$ of $\{1, \ldots, n\}$, we get

$$\mathbb{E}\big[U_n|X_i, X_j\big] = \frac{1}{\binom{n}{k}} \sum_{A:|A|=k} \mathbb{E}\big[h_A|X_i, X_j\big].$$

For reasons of symmetry, and due to the definition of the functions h_1 and h_2,

$$
\mathbb{E}\left[h_A | X_i, X_j\right] = \begin{cases}
\vartheta, & \text{if } \{i, j\} \cap A = \emptyset, \\
h_1(X_j), & \text{if } i \notin A, \ j \in A, \\
h_1(X_i), & \text{if } i \in A, \ j \notin A, \\
h_2(X_i, X_j), & \text{if } \{i, j\} \subset A.
\end{cases}
$$

Since U_n is simply-degenerate, $0 = \sigma_1^2 = \mathbb{V}(h_1(X_1))$, and with $\vartheta = \mathbb{E}(h_1(X_1))$ it follows that

$$
\mathbb{E}\left[h_A | X_i, X_j\right] = \begin{cases}
\vartheta, & \text{if } |\{i, j\} \cap A| \leq 1, \\
h_2(X_i, X_j), & \text{otherwise.}
\end{cases}
$$

Now, (a) is obtained by direct calculation.

Self-question 8: What does $\mathbb{E}[U_n | X_i, X_j]$ look like?

b): Since $\mathbb{E}(U_n) = \mathbb{E}(\widehat{U}_n)$ we assume w.l.o.g. that $\vartheta = 0$, which implies $\mathbb{E}(U_n - \widehat{U}_n)^2 = \mathbb{V}(U_n) + \mathbb{V}(\widehat{U}_n) - 2\mathbb{E}(U_n \widehat{U}_n)$. The variance of U_n is given by (8.7), To determine $\mathbb{V}(\widehat{U}_n)$, we use Lemma 8.13 (a) and the formula $\mathbb{V}(\widehat{U}_n) = \mathrm{Cov}(\widehat{U}_n, \widehat{U}_n)$. After exploiting the bilinearity of the covariance functional $\mathrm{Cov}(\cdot, \cdot)$ and the symmetry of h, one has to tackle the expectation $\mathbb{E}[h_2(X_1, X_2)h_2(X_1, X_3)]$. This, however, vanishes since

$$
\begin{aligned}
\mathbb{E}\left[h_2(X_1, X_2)h_2(X_1, X_3)\right] &= \mathbb{E}\left[\mathbb{E}\left[h_2(X_1, X_2)h_2(X_1, X_3)|X_1\right]\right] \\
&= \mathbb{E}\left[\left(\mathbb{E}\left[h_2(X_1, X_2)|X_1\right]\right)^2\right] = \mathbb{E}\left[\left(h_1(X_1)\right)^2\right] \quad (8.21) \\
&= \mathbb{V}\left(h_1(X_1)\right) = \sigma_1^2 = 0.
\end{aligned}
$$

In view of $\mathbb{V}(h_2(X_1, X_2)) = \sigma_2^2$, we get

$$
\mathbb{V}(\widehat{U}_n) = \frac{\binom{k}{2}^2}{\binom{n}{2}} \sigma_2^2.
$$

Moreover,

$$
\mathbb{E}(2U_n \widehat{U}_n) = \frac{2\binom{k}{2}}{\binom{n}{k}\binom{n}{2}} \sum_{A:|A|=k} \sum_{1 \leq i < j \leq n} \mathbb{E}\left[h_A h_2(X_i, X_j)\right].
$$

Due to the assumption $\vartheta = 0$, the expectations on the right-hand side are equal to zero or equal to σ_2^2 depending on whether $|\{i, j\} \cap A| \leq 1$ or $|\{i, j\} \cap A| = 2$. Since there are $\binom{k}{2}$ ways to choose two-element subsets from each k-element set $A \subset \{1, \ldots, n\}$, we obtain

$$\mathbb{E}(2U_n \widehat{U}_n) = 2 \frac{\binom{k}{2}^2}{\binom{n}{k}\binom{n}{2}} \sigma_2^2,$$

and the assertion follows by direct calculation.

$\square$

By Lemma 8.13 (b), $\mathbb{E}\big[(n(U_n - \widehat{U}_n))^2\big] \to 0$ and thus $n(U_n - \widehat{U}_n) \xrightarrow{\mathrm{P}} 0$. Because of

$$n(U_n - \vartheta) = n(\widehat{U}_n - \vartheta) + n(U_n - \widehat{U}_n), \tag{8.22}$$

we therefore try to obtain a limit distribution for $n(\widehat{U}_n - \vartheta)$. For all further considerations, we set

$$\mathbb{K} := h_2 - \vartheta \tag{8.23}$$

for the *centered* kernel resulting from h_2, that is, we have

$$\mathbb{E}[\mathbb{K}(X_1, X_2)] = 0. \tag{8.24}$$

In view of Lemma 8.13 (a) and due to the symmetry of $\mathbb{K}$, the random variable $n(\widehat{U}_n - \vartheta)$ takes the form

$$n(\widehat{U}_n - \vartheta) = \binom{k}{2} \frac{1}{n-1} \sum_{1 \leq j \neq \ell \leq n} \mathbb{K}(X_j, X_\ell). \tag{8.25}$$

To proceed, we need some results from functional analysis. For background information, see, e.g., [HAA], [HEI], [MUS], or [WEI]. We again set $\mathrm{d}F := \mathrm{d}\mathbb{P}^{X_1}$ and consider the separable Hilbert space $L^2 := L^2(\mathbb{R}^d, \mathcal{B}^d, \mathrm{d}F)$ of $\mathrm{d}F$-square integrable real-valued functions on $\mathbb{R}^d$ with the inner product

$$\langle f, g \rangle_{L^2} := \int fg \, \mathrm{d}F := \int f(x)g(x) \, \mathrm{d}F(x)$$

and the norm $\|g\|_{L^2} := \big(\langle g, g \rangle_{L^2}\big)^{1/2}$. Each unspecified integral extends over $\mathbb{R}^d$. To have a Hilbert space, we must strictly speaking consider equivalence classes of functions that are almost everywhere equal with respect to $\mathrm{d}F$. As usual, we work with functions that represent their respective equivalence classes. Associated with the kernel $\mathbb{K}$ is the integral operator $A : L^2 \to L^2$ which, for each function g in L^2, is defined by

$$(Ag)(x) := \int \mathbb{K}(x, y)g(y)\,\mathrm{d}F(y), \qquad x \in \mathbb{R}^d.$$

Self-question 9: Why does $Ag \in L^2$ hold?

The operator A is linear, and the answer to the above self-question provides the inequality $\|Ag\|_{L^2} \le \|g\|_{L^2}\mathbb{E}[\mathbb{K}^2(X_1, X_2)]$. The operator A is thus continuous, and due to the symmetry of $\mathbb{K}$, it follows that A is self-adjoint, that is, we have $\langle Af, g\rangle_{L^2} = \langle f, Ag\rangle_{L^2}$ if $f, g \in L^2$ (see also Sect. 17.3).

Self-question 10: Which theorem of integration theory yields $\langle Af, g\rangle_{L^2} = \langle f, Ag\rangle_{L^2}$?

We now consider that A is a so-called *Hilbert-Schmidt integral operator*, named after D. Hilbert[6] and E. Schmidt.[7] To this end, we need to show that for any orthonormal basis (in short: ONB) $\{\varphi_1, \varphi_2, \ldots\}$ of L^2, we have

$$\sum_{j=1}^{\infty} \|A\varphi_j\|_{L^2}^2 = \iint \mathbb{K}^2(x, y)\,\mathrm{d}F(x)\mathrm{d}F(y) = \mathbb{E}[\mathbb{K}^2(X_1, X_2)] < \infty. \qquad (8.26)$$

---

[6] David Hilbert (1862–1942), professor at the universities of Königsberg (1893–1895) and Göttingen (1895–1930). Hilbert is considered one of the most significant mathematicians of modern times. Until the Nazi seizure of power, he built a leading center of mathematical research at the University of Göttingen. His speech with a list of 23 unsolved problems at the Mathematics Congress in Paris in 1900 had a lasting influence on mathematical research.

[7] Erhard Schmidt (1876–1959), received his doctorate in 1905 from David Hilbert with a thesis on integral equations. Schmidt was among other things 1917–1950 professor at the University of Berlin. He is considered one of the founders of functional analysis. The Gram-Schmidt orthogonalization process is a fixed part of every course on linear algebra.

To prove (8.26) (only the first equal sign is to be shown), let $\{\varphi_1, \varphi_2, \ldots\}$ be an arbitrary ONB of L^2. With $\mathbb{K}_x(y) := \mathbb{K}(x, y)$, it follows that

$$\sum_{j=1}^{\infty} \|A\varphi_j\|_{L^2}^2 = \sum_{j=1}^{\infty} \int \left(A\varphi_j(x)\right)^2 dF(x) = \sum_{j=1}^{\infty} \int \left(\int \mathbb{K}(x, y)\varphi_j(y) dF(y)\right)^2 dF(x)$$

$$= \sum_{j=1}^{\infty} \int \langle \mathbb{K}_x, \varphi_j \rangle_{L^2}^2 dF(x) = \int \sum_{j=1}^{\infty} \langle \mathbb{K}_x, \varphi_j \rangle_{L^2}^2 dF(x)$$

$$= \int \|\mathbb{K}_x\|_{L^2}^2 dF(x) = \int \mathbb{K}^2(x, y) dF(y) dF(x).$$

The fourth equals sign results from the monotone convergence theorem (Theorem 1.30), and the fifth equals sign is due to Parseval's equation (17.2).

A Hilbert–Schmidt integral operator A is *compact*, that is, for every bounded set $M \subset L^2$, its image $A(M)$ is *precompact* or—which is equivalent—*totally bounded* (see, e.g., [MUS], Theorem 15.31). In general, a subset M of a metric space with metric ϱ is called *precompact* if for every $\varepsilon > 0$ there are finitely many balls with radius ε with respect to the metric that cover the set M. In the case of the space L^2, the metric is given by $\varrho(f, g) := \|f - g\|_{L^2}$. The key for further considerations is the following fact (see, e.g., [HEI], Theorem 7.8.1):

8.14 Theorem (Expansion Theorem for Linear Self-adjoint Compact Operators)
If A is a compact self-adjoint linear operator on L^2, there exists a (possibly finite) orthonormal system $\{\varphi_1, \varphi_2, \ldots\}$ and a (possibly terminating) null sequence $\lambda_1, \lambda_2, \ldots$ of non-zero real numbers such that

$$Ag = \sum_{j \geq 1} \lambda_j \langle g, \varphi_j \rangle_{L^2} \varphi_j, \quad g \in L^2. \tag{8.27}$$

If $\{\psi_1, \psi_2, \ldots\}$ is an ONB of $\{g : Ag = 0\}$, then $\{\psi_1, \psi_2, \ldots\} \cup \{\varphi_1, \varphi_2, \ldots\}$ is an ONB of L^2.

The notation $\sum_{j \geq 1}$ deliberately includes the case of only finitely many summands. In the case that A is a symmetric $\ell \times \ell$ matrix with real entries, which defines a linear mapping of $\mathbb{R}^\ell$ onto itself via multiplication with column vectors x, display (8.27) is the spectral theorem for symmetric matrices known from linear algebra.

From (8.27) it follows that λ_j is an eigenvalue of A associated with the normalized eigenfunction φ_j, that is, we have $A\varphi_j = \lambda_j \varphi_j$, $j \geq 1$. Each λ_j occurs as often as its geometric multiplicity. Together with (8.26), we obtain

$$\sum_{j \geq 1} \lambda_j^2 = \sum_{j \geq 1} \|A\varphi_j\|_{L^2}^2 < \infty, \tag{8.28}$$

a result that is needed later.

In view of (8.27), we define for each integer s a function $\mathbb{K}_s : \mathbb{R}^d \times \mathbb{R}^d \to \mathbb{R}$ via

$$\mathbb{K}_s(x, y) := \sum_{\ell=1}^{s} \lambda_\ell \varphi_\ell(x)\varphi_\ell(y), \quad x, \, y \in \mathbb{R}^d. \tag{8.29}$$

This kernel function is of the same type as the one introduced in (8.16). Since the functions $\varphi_1, \ldots, \varphi_s$ are orthonormal with respect to dF, the equations in (8.18) also apply, which in particular includes the second equals sign in (8.17). To be able to use the result of Example 8.12, we therefore only need to prove the equations

$$\mathbb{E}\big[\varphi_j(X_1)\big] = 0, \quad j \geq 1. \tag{8.30}$$

In this regard, we anticipate a result that specifies in which form the kernel $\mathbb{K}_s$ approximates the kernel $\mathbb{K}$ given in (8.23). Here, we use the general notation $\iint L \, dF \otimes dF$ for $\iint L(x, y) dF(x) dF(y)$.

8.15 Lemma *We have*

$$\mathbb{E}\Big[\big(\mathbb{K}(X_1, X_2) - \mathbb{K}_s(X_1, X_2)\big)^2\Big] = \iint (\mathbb{K} - \mathbb{K}_s)^2 \, dF \otimes dF = \sum_{j=s+1}^{\infty} \lambda_j^2.$$

Proof By definition of the kernel $\mathbb{K}_s$,

$$\iint (\mathbb{K} - \mathbb{K}_s)^2 \, dF \otimes dF = \iint \mathbb{K}^2 dF \otimes dF$$

$$-2\sum_{j=1}^{s} \lambda_j \int \bigg[\int \mathbb{K}(x, y)\varphi_j(y)dF(y)\bigg] \varphi_j(x) \, dF(x)$$

$$+ \sum_{j,\ell=1}^{s} \lambda_j\lambda_\ell \int \varphi_j(x)\varphi_\ell(x)dF(x) \int \varphi_j(y)\varphi_\ell(y)dF(y).$$

Due to $\int \mathbb{K}(x, y)\varphi_j(y)dF(y) = \lambda_j\varphi_j(x)$, $\int \varphi_j^2(x)dF(x) = 1$, and $\int \varphi_j(x)\varphi_\ell(x)dF(x) = \delta_{j,\ell}$ we get $\iint (\mathbb{K} - \mathbb{K}_s)^2 \, dF \otimes dF = \iint \mathbb{K}^2 dF \otimes dF - \sum_{j=1}^{s} \lambda_j^2$. The assertion now follows from (8.26) and (8.28). □

In particular, we have $\lim_{s \to \infty} \iint (\mathbb{K} - \mathbb{K}_s)^2 \, dF \otimes dF = 0$.

To show (8.30), let $\tilde{h}_1 := h_1 - \vartheta$, where $h_1(x) = \mathbb{E}\big[h(x, X_2, \ldots, X_k)\big]$ (cf. (8.8)). Since $\tilde{h}_1(x) = \int \mathbb{K}(x, y)\,dF(y)$, the definition of $\mathbb{K}_s$ and the Cauchy–Schwarz inequality imply

$$\int \Big(\tilde{h}_1(x) - \sum_{j=1}^{s} \lambda_j \varphi_j(x)\int \varphi_j(y)\,dF(y)\Big)^2 dF(x)$$

$$= \int \Big(\int \big(\mathbb{K}(x, y) - \mathbb{K}_s(x, y)\big)\cdot 1\,dF(y)\Big)^2 dF(x)$$

$$\leq \iint (\mathbb{K} - \mathbb{K}_s)^2\,dF \otimes dF. \tag{8.31}$$

Since $0 = \sigma_1^2 = \mathbb{V}(h_1) = \mathbb{E}\big(\tilde{h}_1^2\big) = \int \tilde{h}_1^2(x)\,dF(x)$, we have $\tilde{h}_1 = 0$ dF-almost surely, and since the upper bound in (8.31) converges to 0 as $s \to \infty$, we obtain $\lim_{s\to\infty} I_s = 0$, where

$$I_s := \int \Big(\sum_{j=1}^{s} \lambda_j \varphi_j(x) \int \varphi_j\,dF\Big)^2 dF(x).$$

Now, $\int \varphi_i(x)\varphi_j(x)\,dF(x) = \delta_{i,j}$ yields

$$I_s = \sum_{i,j=1}^{s} \lambda_i \lambda_j \int \varphi_i\,dF \int \varphi_j\,dF \int \varphi_i(x)\varphi_j(x)\,dF(x) = \sum_{j=1}^{s} \lambda_j^2 \Big(\int \varphi_j\,dF\Big)^2.$$

Since $\lambda_j \neq 0$ for each $j \geq 1$, we obtain $\int \varphi_j\,dF = 0 = \mathbb{E}\big(\varphi_j(X_1)\big)$, $j \geq 1$ and thus (8.30).

If, after these technical preliminaries, we take a look at (8.22) and (8.25) and remember that we want to derive the limit distribution of $n(\widehat{U}_n - \vartheta)$, then, according to Slutsky's lemma, we can replace the factor $\frac{1}{n-1}$ before the double sum by $\frac{1}{n}$ without changing this limit distribution. Moreover, the binomial coefficient $\binom{k}{2}$ is only a constant factor, which we have to remember later. So we conveniently set

$$T_n := \frac{1}{n} \sum_{1\leq j\neq \ell\leq n} \mathbb{K}(X_j, X_\ell), \qquad T_{n,s} := \frac{1}{n} \sum_{1\leq j\neq \ell\leq n} \mathbb{K}_s(X_j, X_\ell), \quad s \geq 1, \tag{8.32}$$

and look for a limit distribution of T_n as $n \to \infty$. By Example 8.12, we have

$$T_{n,s} \xrightarrow{\mathcal{D}} \sum_{\ell=1}^{s} \lambda_\ell\big(N_\ell^2 - 1\big). \tag{8.33}$$

Here, $\lambda_1, \ldots, \lambda_s$ are given in (8.29), and $N_1, \ldots, N_s$ are independent standard normally distributed random variables. Since $\sum_{j=1}^{\infty} \lambda_j^2 < \infty$, Lemma 8.15 implies $\iint (\mathbb{K} - \mathbb{K}_s)^2 \mathrm{d}F \otimes \mathrm{d}F \to 0$ as $s \to \infty$. It may therefore be assumed that in the sum figuring in (8.33) we have to formally set $s = \infty$ to obtain the limit distribution of T_n. The following auxiliary result serves to prove this conjecture.

8.16 Lemma *We have* $\mathbb{E}(T_n - T_{n,s})^2 \le 2 \sum_{j=s+1}^{\infty} \lambda_j^2.$

Proof For $x, y \in \mathbb{R}^d$, let $G_s(x, y) := \mathbb{K}(x, y) - \mathbb{K}_s(x, y)$, and put

$$\Delta_n := \frac{1}{\binom{n}{2}} \sum_{1 \le i < j \le n} G_s(X_i, X_j).$$

Then $T_n - T_{n,s} = (n-1)\Delta_n$. Here, Δ_n is a U-statistic and, using (8.24) and (8.30), we get $\mathbb{E}[G_s(X_1, X_2)] = 0$. Due to Lemma 8.15, $\mathbb{E}[G_s^2(X_1, X_2)] = \sum_{j=s+1}^{\infty} \lambda_j^2$, and since $\sigma_1^2 = 0$ implies $\mathbb{E}[G_s(X_1, X_2)G_s(X_1, X_3)] = 0$, we obtain

$$\mathbb{E}(T_n - T_{n,s})^2 = (n-1)^2 \mathbb{V}(\Delta_n) = \frac{(n-1)^2}{\binom{n}{2}^2} \binom{n}{2} \mathbb{E}[G_s^2(X_1, X_2)] \le 2 \sum_{j=s+1}^{\infty} \lambda_j^2.$$

Self-question 11: Why does $\mathbb{E}[G_s(X_1, X_2)G_s(X_1, X_3)] = 0$ follow from $\sigma_1^2 = 0$?

The following result was independently discovered by G.G. Gregory ([GRG]) and R.J. Serfling ([SER], p. 194).

8.17 Theorem (Limit Distribution of Simply-Degenerate U-Statistics)
Let U_n be a U-statistic with $0 = \sigma_1^2 < \sigma_2^2$, and let $\mathbb{K} := h_2 - \vartheta$. Furthermore, let $\lambda_1, \lambda_2, \ldots$ be the non-zero eigenvalues of the integral operator associated with $\mathbb{K}$ on $L^2(\mathbb{R}^d, \mathcal{B}^d, \mathrm{d}F)$. Then

$$n(U_n - \vartheta) \xrightarrow{\mathcal{D}} \binom{k}{2} \sum_{j=1}^{\infty} \lambda_j (N_j^2 - 1) \quad as\ n \to \infty. \tag{8.34}$$

Here, $(N_j)_{j \ge 1}$ is a sequence of i.i.d. standard normally distributed random variables.

Proof We first show that the infinite series appearing in (8.34) is a well-defined random variable. The random variables $N_1, N_2, \ldots$ are all defined on a common probability space

$(\Omega_*, \mathcal{A}_*, \mathbb{P}_*)$, whereby $(\Omega, \mathcal{A}, \mathbb{P}) = (\Omega_*, \mathcal{A}_*, \mathbb{P}_*)$ does not necessarily apply. We also denote the expectation with respect to $\mathbb{P}_*$ by $\mathbb{E}$ and set

$$Y_s := \sum_{j=1}^{s} \lambda_j \left(N_j^2 - 1 \right), \quad s \geq 1.$$

The sequence $(Y_s)_{s \geq 1}$ is a Cauchy sequence in the space $\mathcal{L}_*^2 = \mathcal{L}_*^2(\Omega_*, \mathcal{A}_*, \mathbb{P}_*)$ of all random variables Z on Ω_* with $\mathbb{E}(Z^2) < \infty$ (Problem 8.10). Since $\mathcal{L}_*^2$ is complete (see, e.g., [BI3], p. 243), there is a random variable $Y \in \mathcal{L}_*^2$ with $\lim_{s \to \infty} \mathbb{E}(Y_s - Y)^2 = 0$, and we define

$$\sum_{j=1}^{\infty} \lambda_j \left(N_j^2 - 1 \right) := Y.$$

Since convergence in quadratic mean implies convergence in distribution, it follows that[8] $Y_s \xrightarrow{\mathcal{D}} Y$ as $s \to \infty$. In view of (8.25) and the definition of T_n in (8.32), it remains to prove that $T_n \xrightarrow{\mathcal{D}} Y$ as $n \to \infty$. We show

$$\lim_{n \to \infty} \mathbb{E}(e^{it T_n}) = \mathbb{E}(e^{it Y}), \quad t \in \mathbb{R},$$

from which the claim follows by the Lévy–Cramér continuity theorem (Theorem 1.13).

Fix any $t \in \mathbb{R}$ with $t \neq 0$. With $T_{n,s}$ as in (8.32), for fixed s we have

$$\left| \mathbb{E}e^{it T_n} - \mathbb{E}e^{it Y} \right| \leq \left| \mathbb{E}e^{it T_n} - \mathbb{E}e^{it T_{n,s}} \right| + \left| \mathbb{E}e^{it T_{n,s}} - \mathbb{E}e^{it Y_s} \right| + \left| \mathbb{E}e^{it Y_s} - \mathbb{E}e^{it Y} \right|$$

$$=: a_{n,s} + b_{n,s} + c_s.$$

Since $|e^{i\xi} - 1| \leq |\xi|$ for $\xi \in \mathbb{R}$, the Cauchy–Schwarz inequality and Lemma 8.16 yield

$$a_{n,s} \leq \mathbb{E}\left| e^{it T_n} - e^{it T_{n,s}} \right| = \mathbb{E}\left| \left(e^{it(T_n - T_{n,s})} - 1 \right) e^{it T_{n,s}} \right|$$

$$= \mathbb{E}\left| \left(e^{it(T_n - T_{n,s})} - 1 \right) \right| \leq |t| \, \mathbb{E}\left| T_n - T_{n,s} \right|$$

$$\leq |t| \left(\mathbb{E}(T_n - T_{n,s})^2 \right)^{1/2}$$

$$\leq |t| \left(2\sum_{j=s+1}^{\infty} \lambda_j^2 \right)^{1/2}.$$

---

[8] By Lévy's equivalence theorem (see, e.g., [DUD], Section 9.7), there is even almost sure convergence of the series in (8.34).

The convergence $\sum_{j=s+1}^{\infty} \lambda_j^2 \to 0$ as $s \to \infty$ ensures that for any given $\varepsilon > 0$ there exists an integer s_1 depending on ε and t such that $a_{n,s} \leq \varepsilon$ for each $n \geq 1$ and each s with $s \geq s_1$. Due to $Y_s \xrightarrow{D} Y$ as $s \to \infty$, there is an integer s_2 depending on ε and t such that $c_s \leq \varepsilon$, if $s \geq s_2$. Putting $s_0 := \max(s_1, s_2)$, we then obtain

$$\limsup_{n \to \infty} \left| \mathbb{E} e^{iT_n} - \mathbb{E} e^{iY} \right| \leq 2\varepsilon + \limsup_{n \to \infty} \left| \mathbb{E} e^{iT_{n,s_0}} - \mathbb{E} e^{iY_{s_0}} \right|.$$

By Example 8.12, $T_{n,s_0} \xrightarrow{D} Y_{s_0}$ as $n \to \infty$. Thus, the limit superior vanishes, and the assertion follows, since ε was arbitrary. $\square$

8.18 Example (Cramér–von Mises Statistic)

If $X_1, X_2, \ldots$ is a sequence of i.i.d. real-valued random variables with an unknown continuous distribution function F, and if F_0 is a given continuous distribution function, then a classical problem of mathematical statistics is to test the hypothesis $H_0 : F = F_0$, based on the sample $X_1, \ldots, X_n$. Suppose a test statistic $T_n = T_n(X_1, \ldots, X_n)$ has the property that its realizations do not change, if each of the X_j is subjected to the same continuous increasing transformation. Then, using the probability integral transform $x \mapsto F(x)$, it follows that the distribution of such a test statistic does not depend on F_0 if H_0 holds. Consequently, we may then w.l.o.g. assume that the distribution of X_1 is uniform on the unit interval. In the following, let $F(t) = t, 0 \leq t \leq 1$. Denote

$$\widehat{F}_n(t) := \frac{1}{n} \sum_{j=1}^{n} \mathbf{1}\{X_j \leq t\}, \quad 0 \leq t \leq 1,$$

the empirical distribution function of $X_1, \ldots, X_n$. Then a classical test statistic for the above testing problem is the *Cramér–von Mises*[9] statistic

$$\omega_n^2 := \int_0^1 \left(\sqrt{n}(\widehat{F}_n(t) - t) \right)^2 \, dt. \tag{8.35}$$

The random variable ω_n^2 is not a U-statistic; however, we have

$$\omega_n^2 = (n - 1) U_n + \frac{1}{6} + o_{\mathbb{P}}(1) \tag{8.36}$$

---

[9] Richard Edler von Mises (1883–1953), professor at the universities of Strasbourg (from 1909) and Dresden (from 1919). From 1920 he was professor and director of the newly founded Institute of Applied Mathematics in Berlin. 1933 emigration to Turkey and there professor at the University of Istanbul. From 1939 professor of aerodynamics and applied mathematics at Harvard University, Boston. Main areas of work: Numerical mathematics, mechanics, hydro- and aerodynamics, stochastics, theory of science.

with the U-statistic

$$U_n := \frac{1}{\binom{n}{2}} \sum_{1 \le i < j \le n} h(X_i, X_j),$$

where

$$h(x, y) := \frac{x^2}{2} + \frac{y^2}{2} - \max(x, y) + \frac{1}{3}, \quad x, y \in \mathbb{R} \tag{8.37}$$

(Problem 8.11). Since $\mathbb{E}[\max(X_1, X_2)] = \frac{2}{3}$ and $\mathbb{E}(X_1^2) = \frac{1}{3}$, it follows that $\mathbb{E}[h(X_1, X_2)] = 0 = \vartheta$. Furthermore, $\mathbb{E}[\max(x, X_2)] = \frac{1}{2}(1 + x^2)$ and thus

$$h_1(x) = \mathbb{E}[h(x, X_2)] = \frac{x^2}{2} + \frac{1}{6} - \frac{1 + x^2}{2} + \frac{1}{3} = 0, \quad 0 \le x \le 1.$$

Self-question 12: Why does $\mathbb{E}[\max(x, X_2)] = \frac{1}{2}(1 + x^2)$ apply?

It follows that $\sigma_1^2 = \mathbb{V}(h_1(X_1)) = 0$. With a little patience, we further obtain

$$\sigma_2^2 = \mathbb{V}(h(X_1, X_2)) = \mathbb{E}[h^2(X_1, X_2)] = \frac{1}{90}. \tag{8.38}$$

Thus, the U-statistic U_n is simply-degenerate. By Theorem 8.17,

$$nU_n \xrightarrow{\mathcal{D}} \sum_{j=1}^{\infty} \lambda_j (N_j^2 - 1),$$

where (N_j) is a sequence of i.i.d. standard normally distributed random variables, and $\lambda_1, \lambda_2, \dots$ are the non-zero eigenvalues of the integral operator defined by

$$Ag(x) = \int_0^1 h(x, y)g(y) \, dy \tag{8.39}$$

on $L^2 = L^2([0, 1], \mathcal{B} \cap [0, 1], U(0, 1))$. In view of $k = 2$ and $\vartheta = 0$, we have $h = h_2 = \mathbb{K}$.

How do we get to the solutions of equation (8.39)? Often, an integral equation can only be solved by numerical methods, but in this case all solutions are given in explicit form. We start by inserting the function $g_0 \equiv 1$, which is identically equal to one, into (8.39). In view of $0 = h_1(x) = \mathbb{E}[h(x, X_2)] = \int_0^1 h(x, y) dy$, g_0 is an eigenfunction of A with associated eigenvalue zero. Since we need all *non-zero* eigenvalues, this result seems of little use at first glance. However, it will be important that each eigenfunction is orthogonal to g_0 with respect to the inner product $\langle \cdot, \cdot \rangle_{L^2}$, because

$$\int_0^1 g(x)dx = \langle g, g_0 \rangle_{L^2} = \frac{1}{\lambda}\langle Ag, g_0 \rangle_{L^2} = \frac{1}{\lambda}\langle g, Ag_0 \rangle_{L^2} = \frac{1}{\lambda}\langle g, 0 \rangle_{L^2} = 0 \qquad (8.40)$$

for each $g \in L^2$ with $Ag = \lambda g$ and $\lambda \neq 0$. With $Ag = \lambda g$, equation (8.39) takes the form

$$\lambda g(x) = \frac{x^2}{2}\int_0^1 g(y)dy + \frac{1}{2}\int_0^1 y^2 g(y)dy - x\int_0^x g(y)dy - \int_x^1 yg(y)dy + \frac{1}{3}\int_0^1 g(y)dy.$$

Now, if we use (8.40) we obtain the simplified form

$$\lambda g(x) = \frac{1}{2}\int_0^1 y^2 g(y)dy - x\int_0^x g(y)dy - \int_x^1 yg(y)dy.$$

To proceed, a suitable method is to assume the solutions g of (8.39) to be differentiable sufficiently often and to differentiate both sides of this equation. Taking the first derivative yields $\lambda g'(x) = -\int_0^x g(y)dy$, and from a further differentiation we get the second-order differential equation

$$\lambda g''(x) = -g(x) \qquad (8.41)$$

for g. Setting $g(x) = \cos(ax)$ results in $g''(x) = -a^2 g(x)$, and a comparison with (8.41) shows that λ satisfies the equation $\lambda = \frac{1}{a^2}$. With (8.40) we get

$$0 = \int_0^1 g(x)dx = \frac{1}{a}\sin(ax)\Big|_0^1 = \frac{1}{a}\sin a$$

and thus $\sin a = 0$, which results in $a \in \{j\pi : j \in \mathbb{Z} \setminus \{0\}\}$. Therefore,

$$\lambda_j := \frac{1}{j^2\pi^2}, \quad j \geq 1, \qquad (8.42)$$

is an eigenvalue associated with the normalized eigenfunction $g_j(x) = 2^{-1/2}\cos(j\pi x)$, $0 \leq x \leq 1$.

The natural question arises whether we have obtained all non-zero eigenvalues of the equation (8.39) with (8.42). This question can be answered quickly, because according to (8.26) $\sum_{j=1}^{\infty}\|A\varphi_j\|_{L^2}^2 = \mathbb{E}[\mathbb{K}^2(X_1, X_2)]$ for any ONB $\{\varphi_1, \varphi_2, \ldots\}$ of L^2. The set $\{g_1, g_2, \ldots\}$ of the eigenfunctions corresponding to the non-zero eigenvalues in (8.42) is an orthonormal system, since eigenfunctions to different eigenvalues are orthogonal. Because of (8.38) and $\mathbb{K} = h$, we can in any case deduce the inequality

$$\sum_{j=1}^{\infty} \|Ag_j\|_{L^2} = \sum_{j=1}^{\infty} \lambda_j^2 \le \mathbb{E}[\mathbb{K}^2(X_1, X_2)] = \frac{1}{90}.$$

But now (see, e.g., [ABS], formula 23.2.24)

$$\sum_{j=1}^{\infty} \lambda_j^2 = \frac{1}{\pi^4} \sum_{j=1}^{\infty} \frac{1}{j^4} = \frac{1}{\pi^4} \frac{\pi^4}{90} = \frac{1}{90},$$

which means that we have really obtained all non-zero eigenvalues of the integral operator A. Theorem 8.17 thus implies

$$\omega_n^2 \xrightarrow{\mathcal{D}} \sum_{j=1}^{\infty} \frac{1}{\pi^2 j^2} \left(N_j^2 - 1\right) + \frac{1}{6} \qquad (8.43)$$

where $N_1, N_2, \ldots$ is a sequence of i.i.d. standard normally distributed random variables. The limit distribution is called *Cramér–von Mises distribution*. In Chap. 16 we will see that this distribution is related to the so-called *Brownian bridge*.

With regard to statistical applications, we have hitherto only dealt with the most important cases that a U-statistic is either non-degenerate or simply-degenerate, that is, that one of the cases $\sigma_1^2 > 0$ or $0 = \sigma_1^2 < \sigma_2^2$ applies. What can happen in cases of higher degeneracy—such as $0 = \sigma_1^2 = \sigma_2^2 < \sigma_3^2$ or $0 = \sigma_1^2 = \sigma_2^2 = \sigma_3^2 < \sigma_4^2$—is the topic of Problems 8.7 and 8.8. Further information can be found, e.g., in [LEE], p. 83 ff.

We now turn to so-called *two-sample U-statistics*. For the rest of this chapter we make the basic assumption that $X_1, X_2, \ldots$ and $Y_1, Y_2, \ldots$ are independent d-dimensional random vectors. Moreover, $X_1, X_2, \ldots$ are identically distributed with distribution function F, and $Y_1, Y_2, \ldots$ are identically distributed with distribution function G.

8.19 Definition (Two-Sample U-Statistic)
Let $h : (\mathbb{R}^d)^k \times (\mathbb{R}^d)^\ell \to \mathbb{R}$ be a measurable function, such that $h(x_1, \ldots, x_k, y_1, \ldots, y_\ell)$ is symmetric in $x_1, \ldots, x_k$ and in $y_1, \ldots, y_\ell$. Then

$$U_{m,n} := \frac{1}{\binom{m}{k}\binom{n}{\ell}} \sum_{1 \le i_1 < \ldots < i_k \le m} \sum_{1 \le j_1 < \ldots < j_\ell \le n} h(X_{i_1}, \ldots, X_{i_k}, Y_{j_1}, \ldots, Y_{j_\ell}) \qquad (8.44)$$

is called *two-sample U-statistic of order (k, ℓ) with kernel h*. Here and in the sequel, we tacitly assume $m \ge k$ and $n \ge \ell$.

Due to the assumptions about the symmetry of the kernel h,

$$\mathbb{E}_{F,G}(U_{m,n}) = \mathbb{E}_{F,G}\big[h(X_1, \ldots, X_k, Y_1, \ldots, Y_\ell)\big].$$

In analogy to the one-sample case, we have also emphasized the dependence of the expectation on the distribution functions F and G by indexing with F and G. Likewise, the assumption $\mathbb{E}_{F,G}\big[h^2(X_1, \ldots, X_k, Y_1, \ldots, Y_\ell)\big] < \infty$ which is made throughout is also analogous to the one-sample case. Thus, expectation and variance of $U_{m,n}$ exist, and the U-statistic $U_{m,n}$ is an unbiased estimator of

$$\vartheta = \vartheta(F, G) := \mathbb{E}_{F,G}\big[h(X_1, \ldots, X_k, Y_1, \ldots, Y_\ell)\big].$$

The above general setting of two independent samples with distribution functions F and G is a basic model in *nonparametric statistics*. Since F and G are unknown in applications, one of the classical problems, especially in the case $d = 1$ of real-valued random variables, is to test the hypothesis $H_0 : F = G$, based on (realizations of) $X_1, \ldots, X_m$ and $Y_1, \ldots, Y_n$. This problem is referred to as the *two-sample problem*. The following example concerns a widely used test of H_0, proposed by H.B. Mann[10] and his student D.R. Whitney[11], which is equivalent to the *Wilcoxon rank sum test*.

8.20 Example (Mann–Whitney U-Statistic)
The Mann–Whitney U-statistic arises in the case $d = k = \ell = 1$ with the kernel $h(x, y) = 1\{x \le y\}$ and is thus given by

$$U_{m,n} = \frac{1}{mn} \sum_{i=1}^{m} \sum_{j=1}^{n} 1\{X_i \le Y_j\}.$$

It is an estimator of the probability

$$\vartheta(F, G) = \mathbb{E}_{F,G}[1\{X_1 \le Y_1\}] = \mathbb{P}_{F,G}(X_1 \le Y_1)$$

that X_1 is less than or equal to Y_1. Under the hypothesis $H_0 : F = G$, we have $\vartheta(F, F) = \frac{1}{2}$ if the distribution function F is continuous.

---

[10] Henry Berthold (Heinrich) Mann (1905–2000), professor at Ohio State University (1946–1964). After receiving his doctorate from the University of Vienna, Mann, who was Jewish, emigrated to the U.S. in 1938. He proved the famous Schnirelmann–Landau conjecture of additive number theory.
[11] Donald Ransom Whitney (1915–2001), American statistician, received his doctorate degree in 1947 from H. Mann at Ohio State University.

We briefly set $h_{A,B}$ for $h(X_{i_1}, \ldots, X_{i_k}, Y_{j_1}, \ldots, Y_{j_\ell})$, where $A = \{i_1, \ldots, i_k\}$ is a k-element subset of $\{1, \ldots, m\}$ and B is an ℓ-element subset of $\{1, \ldots, n\}$. Then $U_{m,n}$ takes the compact form

$$U_{m,n} = \frac{1}{\binom{m}{k}\binom{n}{\ell}} \sum_{A \subset \{1,\ldots,m\}:|A|=k} \sum_{B \subset \{1,\ldots,n\}:|B|=\ell} h_{A,B}.$$

Being the covariance $\mathrm{Cov}(U_{m,n}, U_{m,n})$, the variance of $U_{m,n}$ is given by

$$\mathbb{V}(U_{m,n}) = \frac{1}{\binom{m}{k}^2 \binom{n}{\ell}^2} \sum_{A_1:|A_1|=k} \sum_{A_1:|A_1|=k} \sum_{B_1:|B_1|=\ell} \sum_{B_2:|B_2|=\ell} \mathrm{Cov}\left(h_{A_1,B_1}, h_{A_2,B_2}\right).$$

Due to the symmetry of h, the covariances occurring here depend only on the numbers $b := |A_1 \cap A_2|$ and $c := |B_1 \cap B_2|$ with $b \in \{0, \ldots, k\}$ and $c \in \{0, \ldots, \ell\}$. To this end, we define $\sigma_{00}^2 := 0$ and for b, c with $b + c \geq 1$

$$\sigma_{b,c}^2 := \mathrm{Cov}\left(h_{A_1,B_1}, h_{A_2,B_2}\right),$$

if $|A_1 \cap A_2| = b$ and $|B_1 \cap B_2| = c$. Then some combinatorial reasoning yields the following result.

8.21 Theorem (Variance of a Two-Sample U-Statistic)
We have

$$\mathbb{V}(U_{m,n}) = \frac{1}{\binom{m}{k}\binom{n}{\ell}} \sum_{b=0}^{k} \sum_{c=0}^{\ell} \binom{k}{b}\binom{m-k}{k-b}\binom{\ell}{c}\binom{n-\ell}{\ell-c} \sigma_{b,c}^2.$$

Self-question 13: What does the combinatorial reasoning look like?

We will soon see that a two-sample U-statistic is asymptotically normally distributed under certain conditions, provided that m and n tend to infinity in a suitable way. Analogous to the one-sample case, the functions $h_{1,0} : \mathbb{R}^d \to \mathbb{R}$ and $h_{0,1} : \mathbb{R}^d \to \mathbb{R}$, defined by

$$h_{1,0}(x_1) := \mathbb{E}\left[h(x_1, X_2, \ldots, X_k, Y_1, \ldots, Y_\ell)\right]$$
$$= \mathbb{E}\left[h(X_1, X_2, \ldots, X_k, Y_1, \ldots, Y_\ell)\big|X_1 = x_1\right],$$
$$h_{0,1}(y_1) := \mathbb{E}\left[h(X_1, X_2, \ldots, X_k, y_1, Y_2, \ldots, Y_\ell)\right]$$
$$= \mathbb{E}\left[h(X_1, X_2, \ldots, X_k, Y_1, \ldots, Y_\ell)\big|Y_1 = y_1\right],$$

will play a role. By the law of iterated expectation (Theorem 1.26 (a)), we get $\mathbb{E}[h_{1,0}(X_1)] = \vartheta = \mathbb{E}[h_{0,1}(Y_1)]$, and conditioning on X_1 yields

$$
\begin{aligned}
\sigma_{1,0}^2 &= \mathrm{Cov}\big(h(X_1, \ldots, X_k, Y_1, \ldots, Y_\ell), h(X_1, X_{k+1}, \ldots, X_{2k-1}, Y_{\ell+1}, \ldots, Y_{2\ell})\big) \\
&= \mathbb{E}\Big[\mathbb{E}\big[(h(X_1, \ldots, X_k, Y_1, \ldots, Y_\ell) \\
&\quad\quad \times h(X_1, X_{k+1}, \ldots, X_{2k-1}, Y_{\ell+1}, \ldots, Y_{2\ell})\big|X_1\big]\Big] - \vartheta^2 \\
&= \mathbb{E}\big[h_{1,0}(X_1)h_{1,0}(X_1)\big] - \vartheta^2 \\
&= \mathbb{V}\big(h_{1,0}(X_1)\big).
\end{aligned}
$$

Self-question 14: Why does the third equals sign hold?

Analogously, $\sigma_{0,1}^2 = \mathbb{V}\big(h_{0,1}(Y_1)\big)$. In addition to $h_{1,0}$ and $h_{0,1}$, for b with $1 \le b \le k$ and c with $1 \le c \le \ell$ we set

$$
\begin{aligned}
h_{b,c}(x_1, &\ldots, x_b, y_1, \ldots, y_c) \\
&:= \mathbb{E}\,[h(x_1, \ldots, x_b, X_{b+1}, \ldots, X_k, y_1, \ldots, y_c, Y_{c+1}, \ldots, Y_\ell)]\,.
\end{aligned}
$$

By suitable conditioning, we then get $\sigma_{b,c}^2 = \mathbb{V}\big(h_{b,c}(X_1, \ldots, X_c, Y_1, \ldots, Y_d)\big)$.

In order to establish a central limit theorem for $U_{m,n}$, one will try to approximate $U_{m,n}$ by a sum of independent random variables. This goal is served by a suitable adaptation of the Hájek projection from (8.12).

8.22 Definition (Hájek Projection, Two-Sample Case)
For a two-sample U-statistic as in (8.44),

$$
\tilde{U}_{m,n} := \sum_{i=1}^{m} \mathbb{E}[U_{m,n}|X_i] + \sum_{j=1}^{n} \mathbb{E}[U_{m,n}|Y_j] - (m+n-1)\vartheta
$$

is called the *Hájek projection of* $U_{m,n}$.

In contrast to $U_{m,n}$, $\tilde{U}_{m,n}$ is a sum of independent random variables, to which, at least in principle, the Lindeberg–Feller central limit theorem can be applied. Note that the subtraction of $(m+n-1)\vartheta$ ensures that $U_{m,n}$ and $\tilde{U}_{m,n}$ have the same expectation. The next result provides an explicit formula for $\tilde{U}_{m,n}$.

8.23 Lemma *For the Hájek projection* $\tilde{U}_{m,n}$, *we have:*

(a) $\tilde{U}_{m,n} = \dfrac{k}{m} \displaystyle\sum_{i=1}^{m} \left(h_{1,0}(X_i) - \vartheta \right) + \dfrac{\ell}{n} \displaystyle\sum_{j=1}^{n} \left(h_{0,1}(Y_j) - \vartheta \right) + \vartheta,$

(b) *Putting* $(a)_j := a(a-1) \cdot \ldots \cdot (a - j + 1)$, *it follows that*

$$
\mathbb{E}\left(U_{m,n} - \tilde{U}_{m,n}\right)^2 = \frac{k^2}{m} \left\{ \frac{(m-k)_{k-1}}{(m-1)_{k-1}} \frac{(n-\ell)_\ell}{(n)_\ell} - 1 \right\} \sigma_{1,0}^2
$$

$$
+ \frac{\ell^2}{n} \left\{ \frac{(m-k)_k}{(m)_k} \frac{(n-\ell)_{\ell-1}}{(n-1)_{\ell-1}} - 1 \right\} \sigma_{0,1}^2
$$

$$
+ \frac{1}{\binom{m}{k}\binom{n}{\ell}} \sum_{b=0}^{k} \sum_{c=0}^{\ell} \binom{k}{b}\binom{m-k}{k-b}\binom{\ell}{c}\binom{n-\ell}{\ell-c} \sigma_{b,c}^2.
$$
$$
{\scriptstyle b+c\geq 2}
$$

Proof The proof is completely analogous to the proof of Lemma 8.8. $\qquad\qquad\square$

Before we state and prove a central limit theorem for $U_{m,n}$, we clarify what it means that m and n tend to infinity in a suitable way. A frequently occurring scenario in two-sample problems is to let $m = m_s \to \infty$ and $n = n_s \to \infty$ as $s \to \infty$ in a linked manner so that

$$
\lim_{s\to\infty} \frac{m_s}{m_s + n_s} = \tau \tag{8.45}
$$

for some τ with $0 < \tau < 1$. Condition (8.45) means in particulat that none of the two samples "disappears asymptotically" if the sample sizes increase across all limits. So, for example, the case $m_s = s$ and $n_s = \lfloor \sqrt{s} \rfloor$ is excluded.

8.24 Theorem (Central Limit Theorem for Two-Sample U-Statistics)
For the U-statistic $U_{m,n}$ *in (8.44), let* $\sigma_{1,0}^2 > 0$ *and* $\sigma_{0,1}^2 > 0$. *If* $m \to \infty$ *and* $n \to \infty$
subject to (8.45), then

$$
\sqrt{m+n}\left(U_{m,n} - \vartheta\right) \xrightarrow{\mathcal{D}} N\left(0, \sigma^2\right),
$$

where

$$
\sigma^2 := \frac{k^2 \sigma_{1,0}^2}{\tau} + \frac{\ell^2 \sigma_{0,1}^2}{1 - \tau}. \tag{8.46}
$$

Proof We have

$$\sqrt{m+n}\,(U_{m,n}-\vartheta) = \sqrt{m+n}(\tilde{U}_{m,n}-\vartheta) + R_{m,n},$$

where $R_{m,n} = \sqrt{m+n}(U_{m,n}-\tilde{U}_{m,n})$. If $\sigma_{1,0}^2 > 0$ and $\sigma_{0,1}^2 > 0$, then the first and the second curly bracket in Lemma 8.23 (b) is of the order $O(\frac{1}{m})$ and $O(\frac{1}{n})$, respectively. Under the condition (8.45), we have $\mathbb{E}(R_{m,n}^2) \to 0$ and thus $R_{m,n} \xrightarrow{\mathbb{P}} 0$. It therefore remains to prove

$$\sqrt{m+n}\,(\tilde{U}_{m,n}-\vartheta) \xrightarrow{\mathcal{D}} \mathrm{N}(0,\sigma^2),$$

where σ^2 is given in (8.46). For the sake of clarity, we will write m_s instead of m and n_s instead of n in the following. With Lemma 8.23 (a), we obtain

$$\sqrt{m_s + n_s}\,(\tilde{U}_{m_s,n_s}-\vartheta) = \sum_{i=1}^{m_s+n_s} Z_{s,i},$$

where

$$Z_{s,i} := \begin{cases} \sqrt{m_s+n_s}\,\dfrac{k}{m_s}\,\big(h_{1,0}(X_i)-\vartheta\big), & \text{if } i \in \{1,\dots,m_s\}, \\[3mm] \sqrt{m_s+n_s}\,\dfrac{\ell}{n_s}\,\big(h_{0,1}(Y_{i-m_s})-\vartheta\big), & \text{if } i \in \{m_s+1,\dots,m_s+n_s\}. \end{cases}$$

$(Z_{s,1},\dots,Z_{s,m_s+n_s})_{s\geq 1}$ is a triangular array of row-wise independent random variables. Moreover, we have $\mathbb{E}(Z_{s,i}) = 0$ for each i and

$$\mathbb{V}(Z_{s,i}) = \begin{cases} k^2\,\dfrac{m_s+n_s}{m_s^2}\,\sigma_{1,0}^2, & \text{if } i \leq m_s, \\[3mm] \ell^2\,\dfrac{m_s+n_s}{n_s^2}\,\sigma_{0,1}^2, & \text{if } i > m_s. \end{cases}$$

If we define

$$\sigma_s^2 := \sum_{i=1}^{m_s+n_s} \mathbb{V}(Z_{s,i}),$$

then $\lim_{s \to \infty} \sigma_s^2 = \sigma^2$. We now check whether the Lindeberg condition (1.9) holds. For any $\varepsilon > 0$ we get

$$L_s(\varepsilon) := \frac{1}{\sigma_s^2} \sum_{i=1}^{m_s+n_s} \mathbb{E}\big[Z_{s,i}^2 \mathbf{1}\{|Z_{s,i}| > \varepsilon \sigma_s\}\big]$$

$$= \frac{m_s}{\sigma_s^2} \mathbb{E}\big[Z_{s,1}^2 \mathbf{1}\{|Z_{s,1}| > \varepsilon \sigma_s\}\big] + \frac{n_s}{\sigma_s^2} \mathbb{E}\big[Z_{s,m_s+1}^2 \mathbf{1}\{|Z_{s,m_s+1}| > \varepsilon \sigma_s\}\big].$$

By definition of $Z_{s,1}$, we conclude

$$m_s \mathbb{E}\left[Z_{s,1}^2 \mathbf{1}\{|Z_{s,1}| > \varepsilon \sigma_s\}\right]$$

$$= k^2 \frac{m_s + n_s}{m_s} \mathbb{E}\left[(h_{1,0}(X_1) - \vartheta)^2 \mathbf{1}\left\{|h_{1,0}(X_1) - \vartheta| > \frac{\varepsilon \sigma_s m_s}{k\sqrt{m_s + n_s}}\right\}\right]. \quad (8.47)$$

Since

$$\lim_{s \to \infty} \frac{m_s + n_s}{m_s} = \frac{1}{\tau}, \qquad \lim_{s \to \infty} \frac{\varepsilon \sigma_s m_s}{k\sqrt{m_s + n_s}} = \infty$$

it follows that

$$\lim_{s \to \infty} \frac{m_s}{\sigma_s^2} \mathbb{E}\left[Z_{s,1}^2 \mathbf{1}\{|Z_{s,1}| > \varepsilon \sigma_s\}\right] = 0.$$

$\square$

Self-question 15: Which theorem justifies this conclusion?

In the same way, we obtain

$$\lim_{s \to \infty} \frac{n_s}{\sigma_s^2} \mathbb{E}\left[Z_{s,m_s+1}^2 \mathbf{1}\{|Z_{s,m_s+1}| > \varepsilon \sigma_s\}\right] = 0,$$

which shows that the Lindeberg condition is fulfilled. The Lindeberg–Feller central limit theorem (Theorem 1.19) yields $\frac{1}{\sigma_s} \sum_{i=1}^{m_s+n_s} Z_{s,i} \xrightarrow{D} N(0, 1)$, and due to $\sigma_s^2 \to \sigma^2$ the assertion follows from Slutsky's lemma and the mapping theorem.

8.25 Example (Mann–Whitney U-Statistic, Continued)
In the case of the Mann–Whitney U-statistic of Example 8.20 with $\vartheta = \mathbb{P}(X_1 \le Y_1)$ and
the kernel $h(x, y) = \mathbf{1}\{x \le y\}$, we have

$$\sigma_{1,0}^2 = \mathrm{Cov}\big(\mathbf{1}\{X_1 \le Y_1\}, \mathbf{1}\{X_1 \le Y_2\}\big) = \mathbb{P}(X_1 \le Y_1, X_1 \le Y_2) - \vartheta^2,$$

$$\sigma_{0,1}^2 = \mathrm{Cov}\big(\mathbf{1}\{X_1 \le Y_1\}, \mathbf{1}\{X_2 \le Y_1\}\big) = \mathbb{P}(X_1 \le Y_1, X_2 \le Y_1) - \vartheta^2.$$

If $\sigma_{1,0}^2 > 0$ and $\sigma_{0,1}^2 > 0$, then by Theorem 8.24

$$\sqrt{m+n}\left(\frac{1}{mn}\sum_{i=1}^{m}\sum_{j=1}^{n}\mathbf{1}\{X_i \le Y_j\} - \vartheta\right) \xrightarrow{\mathcal{D}} \mathrm{N}\left(0, \frac{\sigma_{1,0}^2}{\tau} + \frac{\sigma_{0,1}^2}{1-\tau}\right)$$

as $m, n \to \infty$ subject to (8.45).

The Mann–Whitney U-statistic $U_{m,n}$, along with the Kolmogorov–Smirnov statistic
(see Sect. 16.19), is a widely used statistic to test the hypothesis $H_0 : F = G$ in the
two-sample problem. If we assume that F and G are continuous, then $\vartheta = \frac{1}{2}$ and

$$\sigma_{1,0}^2 = \sigma_{0,1}^2 = \frac{1}{3} - \frac{1}{4} = \frac{1}{12},$$

provided that the hypothesis H_0 holds.

Self-question 16: Why do these equations hold?

In this case, we thus have

$$\sqrt{m+n}\left(U_{m,n} - \frac{1}{2}\right) \xrightarrow{\mathcal{D}} \mathrm{N}\left(0, \frac{1}{12\tau(1-\tau)}\right).$$

The statistic $U_{m,n}$ degenerates if $\mathbb{P}(X_1 < Y_1) = 1$ or $\mathbb{P}(Y_1 < X_1) = 1$. In each of these
cases, $\sigma_{1,0}^2 = \sigma_{0,1}^2 = 0$, and $U_{m,n}$ takes the value 1 or the value 0 with probability 1.

The U-statistic $U_{m,n}$ by Mann and Whitney is equivalent to the so-called *Wilcoxon's
rank sum statistic*. To understand this, we further assume that F and G are continuous and
therefore all X_i and Y_j are pairwise distinct with probability one. The random variable

$$r(X_i) := \sum_{j=1}^{m}\mathbf{1}\{X_j \le X_i\} + \sum_{\ell=1}^{n}\mathbf{1}\{Y_\ell \le X_i\}, \quad i \in \{1, \dots, m\},$$

counts the number of all $X_1, \ldots, X_m, Y_1, \ldots, Y_n$ that are less than or equal to X_i. By definition, it is the *rank* of X_i in the joint sample of all X_i and Y_j. Wilcoxon's rank sum statistic is defined by

$$W_{m,n} := \sum_{i=1}^{m} r(X_i)$$

Since

$$\mathbb{P}\left(\sum_{i=1}^{m}\sum_{j=1}^{m} \mathbf{1}\{X_j \le X_i\} = \frac{m(m+1)}{2}\right) = 1$$

and

$$\sum_{i=1}^{m}\sum_{\ell=1}^{n} \mathbf{1}\{Y_\ell \le X_i\} = mn - \sum_{i=1}^{m}\sum_{\ell=1}^{n} \mathbf{1}\{X_i < Y_\ell\},$$

it follows that $W_{m,n}$ is $\mathbb{P}$-almost surely equal to an affine transformation of the Mann–Whitney U statistic, namely to $\frac{m(m+1)}{2} - mn(1 - U_{m,n})$.

8.26 V-Statistics
Closely related to U-statistics are the V-*statistics*, named after R. von Mises. The V-*statistic with kernel h of order k associated with* a U-statistic

$$U_n = \frac{1}{\binom{n}{k}} \sum_{1 \le i_1 < \ldots < i_k \le n} h(X_{i_1}, \ldots, X_{i_k})$$

as in (8.1) is defined by

$$V_n := \frac{1}{n^k} \sum_{i_1=1}^{n} \cdots \sum_{i_k=1}^{n} h(X_{i_1}, \ldots, X_{i_k}).$$

As the solution to Problem 8.11 shows, the Cramér–von Mises statistic ω_n^2 in (8.35) is a V-statistic of order 2 with the kernel given in (8.37). The essential difference between U_n and V_n is that in V_n the arguments of h can also be random variables with the same indices. For V-statistics, one requires

$$\mathbb{E}\big[h^2(X_{i_1}, \ldots, X_{i_k})\big] < \infty \quad \text{for all } i_1, \ldots, i_k \text{ with } i_1, \ldots, i_k \in \{1, \ldots, k\}.$$

Thus, in the case $k = 2$, there is the assumption $\mathbb{E}\big[h^2(X_1, X_1)\big] < \infty$ in addition to the familiar condition $\mathbb{E}\big[h^2(X_1, X_2)\big] < \infty$. The example $h(x_1, x_2) = x_1 x_2$ and a distribution with $\mathbb{E}(X_1^2) < \infty$ and $\mathbb{E}(X_1^4) = \infty$ show that one cannot generally infer the additional condition from the familiar one.

If V_n is a V-statistic of order 2 with kernel h, then

$$V_n = \frac{1}{n^2} \sum_{i,j=1}^{n} h(X_i, X_j) = \frac{1}{n^2} \sum_{i=1}^{n} h(X_i, X_i) + \frac{2}{n^2} \sum_{1 \le i < j \le n} h(X_i, X_j)$$

$$= \frac{1}{n} U_n^{(1)} + \frac{n-1}{n} U_n^{(2)}, \tag{8.48}$$

with the U-statistics

$$U_n^{(1)} := \frac{1}{n} \sum_{j=1}^{n} h(X_j, X_j), \qquad U_n^{(2)} := \frac{1}{\binom{n}{2}} \sum_{1 \le i < j \le n} h(X_i, X_j)$$

of orders one and two, respectively. Putting $\mu := \mathbb{E}[h(X_1, X_1)]$ and $\vartheta := \mathbb{E}[h(X_1, X_2)]$, and retaining the meaning of $\lambda_1, \lambda_2, \ldots$ and $N_1, N_2, \ldots$ as in Theorem 8.17, it follows that (Problem 8.13)

$$\sqrt{n}(V_n - \vartheta) \xrightarrow{D} \mathrm{N}(0, 4\sigma_1^2), \quad \text{if } \sigma_1^2 > 0, \tag{8.49}$$

$$n(V_n - \vartheta) \xrightarrow{D} \sum_{j=1}^{\infty} \lambda_j (N_j^2 - 1) + \mu - \vartheta, \quad \text{if } 0 = \sigma_1^2 < \sigma_2^2. \tag{8.50}$$

A detailed treatment of V-statistics can be found, e.g., in [SER], Chapter 6.

8.27 Concluding Remarks Results on the speed of convergence in the central limit Theorem 8.9 are given in [BJZ]. Comparable results in the situation of Theorem 8.17— interestingly with the rate of convergence n^{-1} instead of $n^{-1/2}$—can be found in [BGO]. Finally, it should be emphasized that there is also a theory of Poisson limit theorems for U-statistics, see, e.g., [LEE], Section 3.2.4.

Answers to the Self-Questions

Answer 1 We have $\sum_{i<j}(X_i - X_j)^2 = \frac{1}{2} \sum_{i,j=1}^{n} ((X_i - \overline{X}_n) - (X_j - \overline{X}_n))^2 = n \sum_{i=1}^{n}(X_i - \overline{X}_n)^2$, since $\sum_{i=1}^{n}(X_i - \overline{X}_n) = 0$.

Answer 2 We put $D_i := X_i - \overline{X}_n$ and $E_i := Y_i - \overline{Y}_n$, $i = 1, \ldots, n$. Since $\sum_{i=1}^{n} D_i = 0 = \sum_{i=1}^{n} E_i$, we have $U_n = \frac{2}{n(n-1)} \cdot \frac{1}{2} \cdot \frac{1}{2} \sum_{i,j=1}^{n}(D_i - D_j)(E_i - E_j) = \frac{1}{n(n-1)} \frac{1}{2} \left(2n \sum_{i=1}^{n} D_i E_i + 0\right) = SC_n$.

Answer 3 We have $h\big((X_1, Y_1), (X_2, Y_2)\big) = 0$ if and only if either $(X_1, Y_1) = (X_2, Y_2)$ or if the line connecting (X_1, Y_1) and (X_2, Y_2) is parallel to one of the coordinate axes. The probability of this event is 0 if F has a density.

Answer 4 With $X_j^* := X_j - \mu$, $j = 1, 2$, we have $\mathbb{V}(\frac{1}{2}(X_1 - X_2)^2) = \frac{1}{4}\big(\mathbb{E}(X_1^* - X_2^*)^4 - \big(\mathbb{E}(X_1^* - X_2^*)^2\big)^2\big)$. Since $\mathbb{E}(X_1^*) = 0 = \mathbb{E}(X_2^*)$, it follows that $\mathbb{E}(X_1^* - X_2^*)^4 = 2\mu_4 + 6\mu_2^2$ and $\mathbb{E}(X_1^* - X_2^*)^2 = 2\mu_2$, and the claim is proved.

Answer 5 The first equation holds because in the case $j \notin A$ the random variables h_A and X_j are independent (see Theorem 1.26 g)). For the second equals sign, we let $A = \{1, \ldots, k\}$ and $j = 1$ for reasons of symmetry. By definition, $h_1(X_1) = \mathbb{E}[h(X_1, \ldots, X_k)|X_1]$.

Answer 6 Let $a_n = \prod_{j=k}^{2k-2}\left(1 - \frac{j}{n}\right)$, $b_n = \prod_{j=1}^{k-1}\left(1 - \frac{j}{n}\right)$. After writing out the binomial coefficients using factorials, and by cancelling out, the curly bracket equals $\frac{a_n}{b_n} - 1$, from which the assertion follows.

Answer 7 Generally, if X is a random variable with a *continuous* distribution function F, the so-called *probability integral transform* $X \mapsto U := F(X)$ yields a random variable U with the distribution $U(0, 1)$. With the quantile function F^{-1} defined in (7.4) and the fact that, due to the continuity of F, we have $F(F^{-1}(p)) = p$ for each $p \in (0, 1)$ (cf. (7.5)), it follows that $\mathbb{P}(U < p) = \mathbb{P}(F(X) < p) = \mathbb{P}(X < F^{-1}(p)) = \mathbb{P}(X \le F^{-1}(p)) = F(F^{-1}(p)) = p$ and thus also $\mathbb{P}(U \le p) = \lim_{n \to \infty} \mathbb{P}(U_n < p + n^{-1}) = p$, $0 < p < 1$.

Answer 8 Since in the case $\{i, j\} \subset A$ the remaining elements of A can be chosen in $\binom{n-2}{k-2}$ ways, it follows that $\mathbb{E}[U_n|X_i, X_j] = \binom{n}{k}^{-1}\left\{\binom{n-2}{k-2}h_2(X_i, X_j) + \left(\binom{n}{k} - \binom{n-2}{k-2}\right)\vartheta\right\}$.

Answer 9 With the Cauchy–Schwarz inequality, it follows that

$$\int (Ag)^2 \, dF = \int (Ag(x))^2 \, dF(x) = \int \left(\int \mathbb{K}(x, y)g(y)dF(y)\right)^2 dF(x)$$
$$\le \int \left(\int \mathbb{K}^2(x, y)dF(y)\int g^2(y)dF(y)\right) dF(x)$$
$$= \|g\|_{L^2}^2 \int\int \mathbb{K}^2(x, y)dF(x)dF(y) = \|g\|_{L^2}^2 \, \mathbb{E}(\mathbb{K}^2(X_1, X_2)) < \infty.$$

Answer 10 The third equals sign follows from Fubini's theorem, which implies

$$\langle Af, g\rangle_{L^2} = \int (Af)(x)g(x) \, dF(x) = \int \left(\int \mathbb{K}(x, y)f(y)dF(y)\right) g(x) \, dF(x)$$
$$= \int f(y) \left(\int \mathbb{K}(y, x)g(x)dF(x)\right) dF(y) = \int f(y)(Ag)(y)dF(y)$$
$$= \langle f, Ag\rangle_{L^2}.$$

Answer 11 Since $G_s = \mathbb{K} - \mathbb{K}_s = h_2 - \vartheta - \mathbb{K}_s$, we have to show

$$0 = \mathbb{E}\big[\big(h_2(X_1, X_2) - \vartheta - \mathbb{K}_s(X_1, X_2)\big)\big(h_2(X_1, X_3) - \vartheta - \mathbb{K}_s(X_1, X_3)\big)\big] = 0.$$

After expanding, the assertion follows by writing out $\mathbb{K}_s$ and using (8.21), (8.30), the multiplication formula for expectations, and $\mathbb{E}[h_2(X_1, X_2)] = \vartheta$.

Answer 12 If $x \in (0, 1)$ then

$$\mathbb{E}[\max(x, X_2)] = x\,\mathbb{E}[\max(x, X_2)|X_2 \leq x] + (1-x)\,\mathbb{E}[\max(x, X_2)|X_2 > x]$$

and $\mathbb{E}[\max(x, X_2)|X_2 \leq x] = x$. Under the condition $X_2 > x$, X_2 is uniformly distributed on $(x, 1)$, and it follows that $\mathbb{E}[\max(x, X_2)|X_2 > x] = \frac{1}{2}(1 + x)$.

Answer 13 The set $A_1 \subset \{1, \ldots, m\}$ can be formed in $\binom{m}{k}$ ways. Then one decides which b elements of A_1 should also belong to A_2. There are $\binom{k}{b}$ ways for such a choice. Afterwards, A_2 can be supplemented in $\binom{m-k}{k-b}$ ways to form an m-element set. The product $\binom{n}{\ell}\binom{\ell}{c}\binom{n-\ell}{\ell-c}$ originates in the same way.

Answer 14 Under the condition X_1, the two factors to the left of the condition are independent, so the multiplication rule for expectations applies.

Answer 15 It is the dominated convergence theorem (Theorem 1.31). The expectation in (8.47) is of the type $\int_\Omega Z^2 \mathbf{1}\{|Z| > a_s\}\,d\mathbb{P}$, where $a_s \to \infty$. The integrand converges to zero pointwise on Ω, and for each a the function $Z^2\mathbf{1}\{|Z| > a_s\}$ is dominated by the integrable function Z^2.

Answer 16 X_1, Y_1, and Y_2 are identically distributed, and they are pairwise distinct with probability one, since F is continuous. For reasons of symmetry, the probability that $X_1 = \min(X_1, Y_1, Y_2)$ is equal to $\frac{1}{3}$. Similarly, $\mathbb{P}(X_1 \leq Y_1, X_2 \leq Y_1) = \mathbb{P}(Y_1 = \max(X_1, X_2, Y_1)) = \frac{1}{3}$.

Problems

8.1 Problem Prove Theorem 8.3.

8.2 Problem Let $X_1, X_2, \ldots$ be an i.i.d. sequence of random variables with $\mathbb{E}(X_1^2) < \infty$ and $\sigma^2 := \mathbb{V}(X_1) > 0$. Further, let $\mu := \mathbb{E}(X_1)$ and $h : \mathbb{R}^2 \to \mathbb{R}$ be defined by

$h(x_1, x_2) := x_1 x_2$. Moreover, let U_n be the U-statistic U_n with kernel h. Show: If $\mu \neq 0$ then

$$\sqrt{n}(U_n - \mu^2) \xrightarrow{D} \mathrm{N}(0, 4\mu^2\sigma^2) \quad \text{as } n \to \infty.$$

What happens in the case $\mu = 0$?

8.3 Problem Let τ_n be the empirical τ-coefficient of Kendall defined in (8.5) from Example 8.2 (e). Assume X_1 and Y_1 are independent and have the continuous distribution functions G and H, respectively. Prove the following claims:

(a) $\mathbb{V}(\tau_n) = \dfrac{2(2n+5)}{9n(n-1)}$,

(b) $\sqrt{n}\tau_n \xrightarrow{D} \mathrm{N}\left(0, \dfrac{4}{9}\right)$.

8.4 Problem (Projection Lemma, Hájek) Let $X_1, \ldots, X_n$ be independent d-dimensional random vectors and $s : (\mathbb{R}^d)^n \to \mathbb{R}$ a measurable function such that $\mathbb{E}(S^2) < \infty$, where $S = s(X_1, \ldots, X_n)$. Furthermore, let $L := \sum_{i=1}^{n} \ell_i(X_i)$, where $\ell_i : \mathbb{R}^d \to \mathbb{R}$, $i \in \{1, \ldots, n\}$, are measurable functions with $\mathbb{E}(\ell_i(X_i)^2) < \infty$ for each $i \in \{1, \ldots, n\}$. Show that the random variable

$$\widehat{S} := \sum_{j=1}^{n} \mathbb{E}[S|X_j] - (n-1)\mathbb{E}(S)$$

satisfies the equation $\mathbb{E}(S - L)^2 = \mathbb{E}(S - \widehat{S})^2 + \mathbb{E}(\widehat{S} - L)^2$.

Among all linear combinations of functions of $X_1, \ldots, X_n$, $\widehat{S}$ is thus the best approximation in quadratic mean of S.

Hint Since $\mathbb{E}(S) = \mathbb{E}(\widehat{S})$ (!), we may assume w.l.o.g. that $\mathbb{E}(S) = 0$. Then it needs to be shown that $\mathbb{E}[(S - \widehat{S})(\widehat{S} - L)] = 0$. What is $\mathbb{E}[S - \widehat{S}|X_i]$?

8.5 Problem Let $U_n = U_n(X_1, \ldots, X_n)$ and $V_n = V_n(X_1, \ldots, X_n)$ be U-statistics with kernels f and g of order k and ℓ, respectively, and $\vartheta = \mathbb{E}(U_n)$, $\eta = \mathbb{E}(V_n)$. Assume $\mathbb{E}[f^2(X_1, \ldots, X_k)] < \infty$ and $\mathbb{E}[g^2(X_1, \ldots, X_k)] < \infty$. Provide conditions under which the bivariate random vector $\sqrt{n}(U_n - \vartheta, V_n - \eta)^\top$ has an asymptotic bivariate normal distribution as $n \to \infty$, and specify the parameters of this distribution.

8.6 Problem Let $X_1, X_2, \ldots$ be a sequence of i.i.d. random variables with $m_4 := \mathbb{E}(X_1^4) < \infty$ and $0 < \sigma^2 := \mathbb{V}(X_1)$. Additionally, let $X_1 \stackrel{\mathcal{D}}{=} -X_1$. Show: For the U-statistic

$$U_n := \frac{1}{\binom{n}{2}} \sum_{1 \le i < j \le n} h(X_i, X_j)$$

with kernel $h(x_1, x_2) := x_1 x_2 + (x_1^2 - \sigma^2)(x_2^2 - \sigma^2)$, we have

$$nU_n \stackrel{\mathcal{D}}{\longrightarrow} \sigma^2(N_1^2 - 1) + (m_4 - \sigma^4)(N_2^2 - 1) \quad \text{as } n \to \infty.$$

Here, N_1 and N_2 are independent standard normally distributed random variables.

8.7 Problem Let $X_1, X_2, \ldots$ be a sequence of i.i.d. random variables with $\mathbb{E}|X_1|^3 < \infty$, $\mathbb{E}(X_1) = 0$, and $\mathbb{V}(X_1) = 1$. Show: If U_n is the U-statistic associated with the kernel $h : \mathbb{R}^3 \to \mathbb{R}$ defined by $h(x_1, x_2, x_3) := x_1 x_2 x_3$, then

$$n^{3/2} U_n \stackrel{\mathcal{D}}{\longrightarrow} Z^3 - 3Z, \quad \text{where } Z \sim \mathrm{N}(0, 1).$$

8.8 Problem Let $X_1, X_2, \ldots$ be a sequence of i.i.d. random variables with $\mathbb{E}(X_1^4) < \infty$, $\mathbb{E}(X_1) = 0$, and $\mathbb{V}(X_1) = 1$. Show: For the U-statistic U_n associated with the kernel $h : \mathbb{R}^4 \to \mathbb{R}$ defined by $h(x_1, x_2, x_3, x_4) := x_1 x_2 x_3 x_4$, we have

$$n^2 U_n \to Z^4 - 6Z^2 + 3, \quad \text{where } Z \sim \mathrm{N}(0, 1).$$

8.9 Problem Let $X, X_1, X_2, \ldots$ be a sequence of i.i.d. random variables with distribution function F, where $X \stackrel{\mathcal{D}}{=} -X$, $\mathbb{E}(X^6) < \infty$ and $\mathbb{V}(X) > 0$. Further, let $h(x, y) := xy(1 + x^2 y^2)$, $x, y \in \mathbb{R}$, and

$$U_n := \frac{1}{\binom{n}{2}} \sum_{1 \le i < j \le n} h(X_i, X_j).$$

Prove the following statements:

(a) $\mathbb{E}(U_n) = 0$, and the U-statistic U_n is simply-degenerate.
(b) Each eigenfunction associated with the integral operator A, defined by

$$(Ag)(x) := \int_{\mathbb{R}} h(x, y) g(y) \, dF(y), \quad x \in \mathbb{R},$$

has the form $g(x) = ax + bx^3$, $x \in \mathbb{R}$. Here, a and b are suitable non-zero real numbers.

(c) With $m_j := \mathbb{E}(X^{2j})$, $j \in \{1, 2, 3\}$, and

$$\lambda_1 = \frac{m_6 + m_2}{2} + \frac{1}{2}\sqrt{(m_6 - m_2)^2 + 4m_4^2}, \tag{8.51}$$

$$\lambda_2 = \frac{m_6 + m_2}{2} - \frac{1}{2}\sqrt{(m_6 - m_2)^2 + 4m_4^2}, \tag{8.52}$$

we have $nU_n \to \lambda_1(N_1^2 - 1) + \lambda_2(N_2^2 - 1)$ as $n \to \infty$. Here, N_1 and N_2 are independent and each standard normally distributed random variables. What results in the special case $X \sim N(0, 1)$?

8.10 Problem Show that the random variables $Y_s = \sum_{j=1}^{s} \lambda_j(N_j^2 - 1)$, $s \geq 1$, in the proof of Theorem 8.17 are a Cauchy sequence.

8.11 Problem Derive the representation (8.36) for the Cramér–von Mises statistic defined in (8.35).

8.12 Problem Let $X_1, \ldots, X_m, Y_1, \ldots, Y_n$ be independent random variables, where $X_1, \ldots, X_m$ have the same distribution function F and $Y_1, \ldots, Y_n$ have the same distribution function G. Further, let $\vartheta := \mathbb{P}(X_1 < Y_1, X_2 < Y_1)$.

(a) Provide a two-sample U-statistic $U_{m,n}$ that estimates ϑ in an unbiased way.
(b) What is the asymptotic distribution of $U_{m,n}$ as m, $n \to \infty$ under the condition (8.45) and the specific assumption $X_1 \sim U(0, 1)$ and $Y_1 \sim U(-1, 1)$?

8.13 Problem Prove the statements (8.49) and (8.50).

Basic Concepts of Estimation Theory

9

In this chapter, we set the general framework for some topics of asymptotic statistics that are covered in this and in the following chapters. This framework also includes a specific notation. In addition, terms such as *parametric model, statistical space*, and *canonical model* are introduced. We also learn the notion of an *estimator* and desirable properties of such an estimator. Basic knowledge of mathematical statistics is helpful, but not necessarily required. Comprehensive introductions to mathematical statistics are provided, among others, in [SHA], [SPD], [BD1], and [BD2]. The book [LEC] is exclusively dedicated to methods of estimation.

Unless stated otherwise, we assume a sequence $X_1, X_2, \ldots$ of i.i.d. random variables that are all defined on a common probability space $(\Omega, \mathcal{A}, \mathbb{P})$, which is ultimately not of interest. These random variables take on values in a set $\mathcal{X}_0$, which is called the *sample space*, and which is equipped with a σ-field $\mathcal{B}_0$. The term *random variable*, which includes the $(\mathcal{A}, \mathcal{B}_0)$-measurability of the mappings $X_j : \Omega \to \mathcal{X}_0$, is therefore to be understood in general terms. In most cases, we have $\mathcal{X}_0 \subset \mathbb{R}^d$ so that we deal with d-dimensional random vectors. In this case, $\mathcal{X}_0$ is a Borel set, and $\mathcal{B}_0 = \{\mathcal{X}_0 \cap B : B \in \mathcal{B}^d\}$ is the trace of $\mathcal{B}^d$ in $\mathcal{X}_0$. Here, the most important special case is that of real-valued random variables, that is, the case $d = 1$.

In the following, let

$$\mathcal{M}^1 := \{P : P \text{ is a probability measure on } \mathcal{B}_0\}$$

denote the set of all probability measures on the σ-field $\mathcal{B}_0$. In statistics, it is assumed that the distribution $\mathbb{P}^{X_1}$ of X_1 is not fully known. A so-called *parametric model* makes a relatively restrictive fundamental assumption on this distribution.

© The Author(s), under exclusive license to Springer-Verlag GmbH, DE, part of Springer Nature 2024
N. Henze, *Asymptotic Stochastics*, Mathematics Study Resources 10,
https://doi.org/10.1007/978-3-662-68923-3_9

9.1 Definition (Parametric Model)

A *parametric model* for $\mathbb{P}^{X_1}$ is a subset $\mathcal{P}$ of $\mathcal{M}^1$ with the following property: There is an integer k and a non-empty Borel set $\Theta \subset \mathbb{R}^k$ as well as a bijective mapping $Q : \Theta \to \mathcal{P}$ from Θ onto $\mathcal{P}$. Here, one usually sets $Q_\vartheta := Q(\vartheta)$, $\vartheta \in \Theta$. We thus have

$$\mathcal{P} = \{Q_\vartheta : \vartheta \in \Theta\}.$$

The set Θ is called the *parameter space*.

9.2 Example (Normal Distribution)

Suppose a physical or technical quantity is measured n times under the same, mutually unaffected conditions, and that the results of these measurements are realizations of i.i.d. random variables $X_1, \ldots, X_n$. In this situation, it is often assumed that X_1 has a normal distribution $N(\mu, \sigma^2)$, whereby μ and σ^2 are not known. Here, μ is the "true value" of the quantity to be measured, and the variance σ^2 models the inaccuracy of the measurement method. In this situation,

$$\Theta = \mathbb{R} \times \mathbb{R}_{>0} = \{\vartheta = (\mu, \sigma^2) : \mu \in \mathbb{R}, \ \sigma^2 > 0\}.$$

The parameter ϑ is thus a vector with two components, which is why we also speak of a *two-parametric model*. If, as in this example, $k \geq 2$, it can happen that only one component of this vector is of interest. In the general case, there is a function $\gamma : \Theta \to \mathbb{R}$, and there may be only a focus on $\gamma(\vartheta)$. In this specific example, we usually have $\gamma(\vartheta) = \mu$. So, one is interested in the expectation of the normal distribution, but not in its variance. Since the latter is then only an annoying, necessary accessory, σ^2 is called a *nuisance parameter*.

9.3 Example (Exponential Distribution)

If the random variables $X_1, X_2, \ldots$ model random lifetimes, that is, if X_1 is positive with probability one, then the simplest parametric model is the assumption of an *exponential distribution*, that is, one sets

$$\Theta = (0, \infty), \quad Q_\vartheta = \text{Exp}(\vartheta).$$

In this context, the parameter ϑ is usually referred to as λ.

9.4 Example (Bernoulli Sequence)

In the case of a so-called *dichotomous feature*, which can only take two different values (which are then conveniently chosen as 1 and 0), X_1 has the binomial distribution $\text{Bin}(1, \vartheta)$ with $\vartheta \in \Theta := (0, 1)$. In this case, ϑ can be considered as a probability of *success*, if the realization 1 of the X_j is regarded as a success, whatever such success may be in a particular situation. Note that by choosing $\Theta = (0, 1)$, the extreme probabilities of success are excluded. In this context, it is common to deviate from the notation ϑ and to write p for the probability of success.

9.5 Example (Bivariate Normal Distribution)

A five-parametric statistical model results when $X_1, X_2, \ldots$ are two-dimensional random vectors and some unspecified non-degenerate normal distribution is assumed for the distribution of X_1. The model assumption is therefore

$$X_1 \sim N_2 \left(\begin{pmatrix} \mu \\ \nu \end{pmatrix}, \begin{pmatrix} \sigma^2 & \rho\sigma\tau \\ \rho\sigma\tau & \tau^2 \end{pmatrix} \right),$$

where $\vartheta := (\mu, \nu, \sigma^2, \tau^2, \rho)$ and

$$\Theta = \left\{ \vartheta = (\mu, \nu, \sigma^2, \tau^2, \rho) : \mu, \ \nu \in \mathbb{R}, \ \sigma^2 > 0, \ \tau^2 > 0, \ -1 < \rho < 1 \right\}.$$

9.6 The Canonical Model

At the beginning of this chapter, it was mentioned that the probability space on which the random variables $X_1, X_2, \ldots$ are defined has no relevance, a fact that will now be specified. For this purpose, we define the set

$$\mathcal{X} := \mathcal{X}_0^{\mathbb{N}} := \left\{ \mathbf{x} = (x_1, x_2, \ldots) : x_j \in \mathcal{X}_0 \text{ for every } j \geq 1 \right\}$$

as the countably infinite Cartesian product of $\mathcal{X}_0$ with itself and endow this set with the *product σ-field* $\mathcal{B} := \mathcal{B}_0^{\mathbb{N}}$, that is, with the smallest σ-field on $\mathcal{X}$ with respect to which all *coordinate projections* $\mathcal{X} \ni \mathbf{x} = (x_1, x_2, \ldots) \mapsto x_j$, $j \geq 1$, are $(\mathcal{B}, \mathcal{B}_0)$-measurable. Finally, for Q_ϑ with $\vartheta \in \Theta$, the infinite product probability measure on $\mathcal{B}$ is denoted by $Q_\vartheta^{\mathbb{N}}$. With $\mathbb{P}_\vartheta := Q_\vartheta^{\mathbb{N}}$, $(\mathcal{X}, \mathcal{B}, \mathbb{P}_\vartheta)_{\vartheta \in \Theta}$ is a collection of probability spaces. This collection is often called a *statistical space*. If we define $X := \mathrm{id}_{\mathcal{X}}$ as the identity map on $\mathcal{X}$, that is, if we set $X_j(\mathbf{x}) := x_j$, $j \geq 1$, then $X = (X_1, X_2, \ldots)$, and $X_1, X_2, \ldots$ are independent random variables with the same distribution Q_ϑ. This construction is called the *canonical model*. Unless otherwise agreed, we will always base our discussions on the canonical model. With this, for example, we have

$$\mathbb{P}_\vartheta \left(A \times (\times_{j=n+1}^{\infty} \mathcal{X}_0) \right) = \mathbb{P}_\vartheta \left((X_1, \ldots, X_n) \in A \right)$$

for each $n \geq 1$ and each choice of $A \in \mathcal{B}_0 \otimes \ldots \otimes \mathcal{B}_0$ (n factors). Here, we have emphasized by indexing with ϑ that the probability measure $\mathbb{P}_\vartheta$ was used as the model. In the same way, we make this fact clear by the notations $\mathbb{E}_\vartheta$ and $\mathbb{V}_\vartheta$ for expectations and variances as well as by $\xrightarrow{\mathbb{P}_\vartheta}$ and $\xrightarrow{\mathcal{D}_\vartheta}$ for convergence in probability and convergence in distribution, respectively.

A few general considerations are appropriate at this point. A parametric model is tempting because, in particular, given $\vartheta \in \Theta$, one can generate pseudorandom numbers according to the distribution Q_ϑ and thus perform simulations. In addition, it is sometimes possible to arrive at optimal estimators or tests *within a parametric model*. Of course, it must be specified how exactly the pleasantly vague word "optimal" is to be understood.

On the other hand, every model is only a more or less good approximation of reality. If one wants to be more cautious compared to the assumption of a parametric model, there is no other choice but to considerably enlarge the class of distributions considered possible for X_1. If $X_1, X_2, \ldots$ is an i.i.d. sequence of d-dimensional random vectors, that is, if $\mathcal{X}_0 \subset \mathbb{R}^d$, then the distribution of X_1 and thus also the distribution of the sequence $\mathrm{X} = (X_j)_{j \geq 1}$ is uniquely determined by the distribution function of X_1, denoted by F. This distribution function can also be seen as a (not finite-dimensional) "parameter", and as a general setting, the model assumption can be that F belongs to a given set $\mathcal{F}$ of distribution functions. Such an assumption is referred to as a *nonparametric model*. We considered such models in Chap. 8, where in the case of one-sample U-statistics $\mathcal{F}$ denoted the set of all distribution functions F with $\mathbb{E}_F[h^2(X_1, \ldots, X_k)] < \infty$. There, we have written the parameter F as an index of expectations, variances, covariances, and probabilities in order to highlight their dependence on F and to emphasize that the stochastic model is only determined by the specification of F. Ultimately, however, only a real-valued aspect of F was of interest, namely $\vartheta(F) = \mathbb{E}_F[h(X_1, \ldots, X_k)]$.

In the following, let $(\mathcal{X}, \mathcal{B}, \mathbb{P}_\vartheta)_{\vartheta \in \Theta}$ be a statistical space with $\Theta \subset \mathbb{R}^k$, s an integer with $s \in \{1, \ldots, k\}$, and $\gamma : \mathbb{R}^k \to \mathbb{R}^s$. The Euclidean norm in $\mathbb{R}^s$ is denoted by $\| \cdot \|$. The goal is to estimate the value of $\gamma(\vartheta)$ based on $\mathbf{x} = (x_1, x_2, \ldots)$. However, we do not have the entire sequence $(x_1, x_2, \ldots)$ available as a data basis for estimation, but only an initial section $x_1, x_2, \ldots, x_n$ of length n. The number n is called the *sample size*.

9.7 Definition (Estimator)

An *estimator* of $\gamma(\vartheta)$ (based on the sample size n) is a measurable mapping $T_n : \mathcal{X} \to \gamma(\Theta)$ with the property that, for each $\mathbf{x} = (x_1, x_2, \ldots)$ in $\mathcal{X}$, the estimate $T_n(\mathbf{x})$ depends only on $x_1, \ldots, x_n$. To emphasize this fact, we also set $T_n(x_1, \ldots, x_n) := T_n(\mathbf{x})$, and we write briefly $T_n := T_n(X_1, \ldots, X_n)$, when we are dealing with the random variable $T_n((X_1, X_2, \ldots))$ in the canonical model. The measurability of T_n implies that $\gamma(\Theta)$ is equipped with the trace σ-field $\gamma(\Theta) \cap \mathcal{B}^s$.

For mathematical reasons, it is sometimes necessary that an estimator as above can also take on values in a proper superset of $\gamma(\Theta)$. This may seem surprising, but quickly becomes clear when considering the problem of estimating an unknown probability of success in Bernoulli trials. Even if one chooses the parameter space $\Theta := (0, 1)$ for good reason and thus excludes the extreme probabilities 0 and 1, it can happen that no single success has occurred in a sample of size n, which at least for the relative frequency of successes as an estimator yields the estimate 0.

The estimator T_n is a *random variable* defined on the sample space $\mathcal{X}$, whose distribution depends on the unknown parameter ϑ. Clearly, we would like this distribution to be strongly concentrated around the value $\gamma(\vartheta)$ when $\mathbb{P}^{X_1} = Q_\vartheta$, and this should of course apply to every $\vartheta \in \Theta$. If the expectation $\mathbb{E}_\vartheta(T_n)$ exists for every $\vartheta \in \Theta$, T_n is called *unbiased for* $\gamma(\vartheta)$, if

$$\mathbb{E}_\vartheta(T_n) = \gamma(\vartheta) \quad \text{for every } \vartheta \in \Theta.$$

In the case $s > 1$, this expectation is by definition the vector of the expectations of the individual components. Unbiased estimators are "impartial" with respect to ϑ in the sense that the physical center of gravity of the distribution of T_n is equal to the value to be estimated, regardless of which ϑ in Θ is assumed as the distribution of X_1 via Q_ϑ. However, there may be specific situations in which no unbiased estimator exists (see Problem 9.1).

9.8 Definition (Sequences of Estimators and Their Properties)

A *sequence of estimators of* $\gamma(\vartheta)$ is a sequence (T_n) of measurable mappings $T_n : \mathcal{X} \to \gamma(\Theta)$ such that, for each n, the mapping T_n is an estimator of $\gamma(\vartheta)$ for the sample size n. The sequence (T_n) is called

(a) *asymptotically unbiased for* $\gamma(\vartheta)$ if

$$\lim_{n \to \infty} \mathbb{E}_\vartheta(T_n) = \gamma(\vartheta) \quad \text{for every } \vartheta \in \Theta,$$

(b) *(weakly) consistent for* $\gamma(\vartheta)$ if

$$\lim_{n \to \infty} \mathbb{P}_\vartheta\left(\|T_n - \gamma(\vartheta)\| > \varepsilon\right) = 0 \quad \text{for every } \varepsilon > 0 \text{ and every } \vartheta \in \Theta,$$

(c) *strongly consistent* for $\gamma(\vartheta)$ if $\lim_{n \to \infty} T_n = \gamma(\vartheta)$ $\mathbb{P}_\vartheta$-a.s. for every $\vartheta \in \Theta$,

(d) $\sqrt{n}$-*consistent* for $\gamma(\vartheta)$ if $\sqrt{n}\left(T_n - \gamma(\vartheta)\right) = O_{\mathbb{P}_\vartheta}(1)$ for every $\vartheta \in \Theta$.

Part (a) of the above definition assumes that the expectation exists. In the following, we omit the attribute "weakly" if we mean weakly consistent. The property of consistency states that, regardless of which value of ϑ in Θ is assumed for the underlying distribution Q_ϑ of X_1, the sequence (T_n) of estimators converges in $\mathbb{P}_\vartheta$-probability to $\gamma(\vartheta)$. This property is in a certain way indispensable because, as the sample size n increases, the distribution of the estimator T_n should concentrate more and more around the true value ϑ to be estimated, regardless of that true value which specifies the distribution of X_1, because ϑ is unknown. Figure 9.1 shows this desirable effect using the example of estimating a probability of success in Bernoulli trials by the random relative frequency of successes denoted with T_n, for the sample sizes $n = 20$ and $n = 50$. For this figure, the probability of success was set to $\vartheta = 0.7$.

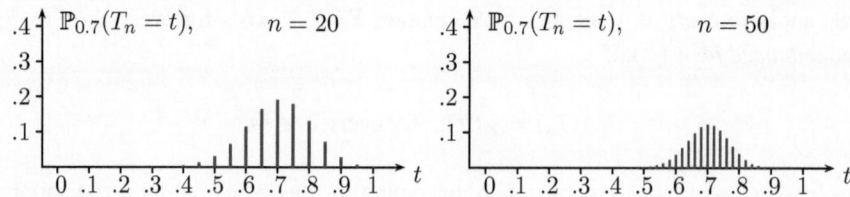

Fig. 9.1 Bar charts of the distribution of the random relative frequency of successes ($n = 20$ and $n = 50$, $\vartheta = 0.7$)

The property of $\sqrt{n}$-consistency holds in particular when, for every $\vartheta \in \Theta$, the *estimation error* $T_n - \gamma(\vartheta)$, multiplied by $\sqrt{n}$, has a limit distribution as $n \to \infty$, because convergence in distribution implies tightness.

Self-question 1: Why does consistency follow from $\sqrt{n}$-consistency?

If we set $T_n =: (T_{n,1}, \ldots, T_{n,k})$, then (T_n) is consistent for $\gamma(\vartheta)$, if (T_n) is asymptotically unbiased for $\gamma(\vartheta)$, and if $\lim_{n \to \infty} \mathbb{V}_\vartheta(T_{n,j}) = 0$ for each $j \in \{1, \ldots, k\}$ and each $\vartheta \in \Theta$ (Problem 9.2). Here, we assume $\mathbb{E}_\vartheta(T_{n,j}^2) < \infty$ for each $j \in \{1, \ldots, k\}$ and each $\vartheta \in \Theta$.

9.9 Example (Estimation of a Variance)

Let $X_1, X_2, \ldots$ be a sequence of i.i.d. random variables with unknown normal distribution $N(\mu, \sigma^2)$, so $\vartheta := (\mu, \sigma^2)$ and $\Theta = \mathbb{R} \times \mathbb{R}_{>0}$. We want to estimate the variance $\gamma(\vartheta) := \sigma^2$. By Example 8.2, the sample variance

$$S_n^2 = \frac{1}{n-1} \sum_{j=1}^{n} (X_j - \overline{X}_n)^2 \tag{9.1}$$

is an unbiased estimator of σ^2, not only in the context of a normal distribution, but more generally in a nonparametric setting. With Example 8.5, and since $\mu_4 = 3\sigma^4$ and $\mu_2 = \sigma^2$, we have

$$\mathbb{V}_\vartheta(S_n^2) = \frac{\sigma^4}{n}\left(3 - \frac{n-3}{n-1}\right) \to 0 \quad \text{as } n \to \infty.$$

Therefore, $S_n^2 \xrightarrow{\mathbb{P}_\vartheta} \gamma(\vartheta)$, $\vartheta \in \Theta$, that is, the sequence (S_n^2) is consistent for σ^2. By (8.14), $\sqrt{n}(S_n^2 - \sigma^2) \xrightarrow{\mathcal{D}_\vartheta} N(0, 2\sigma^4)$, and thus the sequence (S_n^2) is $\sqrt{n}$-consistent for σ^2. Note that the result (8.14) is much more general.

Self-question 2: Is $(S_n^2)_{n \geq 2}$ a *strongly* consistent sequence of estimators of σ^2?

9.10 Example (Estimation of the Right Endpoint of a Uniform Distribution)

Let $X_1, X_2, \ldots$ be an i.i.d. sequence of random variables with the uniform distribution $X_1 \sim \mathrm{U}(0, \vartheta)$, where $\vartheta \in \Theta$ and $\Theta := (0, \infty)$. The goal is to estimate ϑ. Since $\mathbb{E}_\vartheta(X_1) = \frac{\vartheta}{2}$, it seems obvious to estimate ϑ using $X_1, \ldots, X_n$ by

$$T_n := 2\overline{X}_n, \qquad \text{where} \quad \overline{X}_n := \frac{1}{n} \sum_{j=1}^{n} X_j.$$

The word *obvious* here refers to a general method of estimation, which is called the *method of moments* (see Chap. 11). That method thus has the same naming as the method of inferring convergence in distribution from the convergence of all moments discussed in Chap. 3.

We have $\mathbb{E}_\vartheta(T_n) = \vartheta$, $\vartheta \in \Theta$, that is, the estimator T_n is unbiased for ϑ. Furthermore,

$$\mathbb{V}(T_n) = \frac{4}{n} \mathbb{V}_\vartheta(X_1) = \frac{4\vartheta^2}{n} \cdot \frac{1}{12} = \frac{\vartheta^2}{3n}, \tag{9.2}$$

and thus Chebyshev's inequality shows that the sequence (T_n) is consistent for ϑ. Note that $X_j \sim \vartheta U_j$, $j \geq 1$, where $U_1, U_2, \ldots$ is an i.i.d. sequence of random variables with the uniform distribution on $(0, 1)$, which simplifies calculations. With the Lindeberg–Lévy central limit theorem, it follows that

$$\sqrt{n}(T_n - \vartheta) \xrightarrow{\mathcal{D}_\vartheta} \mathrm{N}\left(0, \frac{\vartheta^2}{3}\right), \qquad \vartheta \in \Theta.$$

Therefore, the sequence (T_n) is $\sqrt{n}$-consistent for ϑ. Since, by the strong law of large numbers, T_n converges $\mathbb{P}_\vartheta$-a.s. to ϑ, (T_n) is also strongly consistent for ϑ.

Although T_n as an estimator of ϑ has many desirable properties, there is a better estimator, namely

$$S_n := \frac{n+1}{n} \max(X_1, \ldots, X_n).$$

This estimator is unbiased for ϑ, and it has the variance

$$\mathbb{V}_\vartheta(S_n) = \frac{n\vartheta^2}{(n+2)(n+1)^2},$$

which is uniformly smaller as a function of ϑ compared to (9.2) (Problem 9.4). Compared to that of T_n, it is also of the smaller order of magnitude $O(n^{-2})$. In particular, the sequence (S_n) is consistent for ϑ. It is even strongly consistent.

Self-question 3: Why is the sequence (S_n) strongly consistent for ϑ?

Since $n(S_n - \vartheta)$ has a non-degenerate limit distribution as $n \to \infty$ (Problem 9.4), $\sqrt{n}(S_n - \vartheta) = o_{\mathbb{P}}(1)$. Thus, (S_n) is $\sqrt{n}$-consistent, but it has the stronger property that the estimation error $S_n - \vartheta$ is so small that it must be multiplied by n and not by $\sqrt{n}$ to obtain a non-degenerate limit distribution.

An estimator T_n as in Definition 9.7 is often also called a *point estimator* because its realizations are individual values (or "points") in the set $\gamma(\Theta)$. In contrast, a *confidence set* assigns subsets of $\gamma(\Theta)$ to the data. Loosely speaking, these subsets should be as small as possible, but one should be practically certain that they contain $\gamma(\vartheta)$, regardless of which value of ϑ characterizes the distribution of X_1. To clarify these ideas, let $\alpha \in (0, 1)$ be a given small number. Common values here are $\alpha = 0.05$, $\alpha = 0.1$ or $\alpha = 0.01$. The following definition is based on the setting described before Definition 9.7.

9.11 Definition (Confidence Set)
A *confidence set with level of significance* $1 - \alpha$ for $\gamma(\vartheta)$ (for the sample size n) is a mapping $C_n : \mathcal{X} \to \mathcal{P}(\gamma(\Theta))$ with the following properties:

(a) $C_n(\mathbf{x})$ depends on $\mathbf{x} = (x_1, x_2, \ldots) \in \mathcal{X}$ only via $x_1, \ldots, x_n$.
(b) For each $\vartheta \in \Theta$, $\{\mathbf{x} \in \mathcal{X} : C_n(\mathbf{x}) \ni \gamma(\vartheta)\} \in \mathcal{B}$.
(c) For each $\vartheta \in \Theta$, $\mathbb{P}_\vartheta\big(\{\mathbf{x} \in \mathcal{X} : C_n(\mathbf{x}) \ni \gamma(\vartheta)\}\big) \geq 1 - \alpha$.

The crucial part of this definition is property (c), as the technical condition (b) only ensures that the probability in (c) is well-defined. A confidence set is thus an *estimator* that assigns subsets of $\gamma(\Theta)$ to the elements of the sample space. Because of condition (a) we also write $C_n(x_1, \ldots, x_n) := C_n(\mathbf{x})$. Since we are working with the canonical model, due to condition (a) we can write property (c) in the form

$$\mathbb{P}_\vartheta\big(C_n(X_1, \ldots, X_n) \ni \gamma(\vartheta)\big) \geq 1 - \alpha \quad \text{for every } \vartheta \in \Theta.$$

The estimator is thus designed so that, no matter which $\vartheta \in \Theta$ is assumed, the *random subset* $C_n(X_1, \ldots, X_n)$ of $\gamma(\Theta)$ contains or—which is also a frequently used expression—*covers* the value $\gamma(\vartheta)$ with the minimum probability $1 - \alpha$. The notation $C_n(X_1, \ldots, X_n) \ni \gamma(\vartheta)$ was deliberately preferred to $\gamma(\vartheta) \in C_n(X_1, \ldots, X_n)$ to prevent a common misinterpretation. If a realization $C_n(x_1, \ldots, x_n)$ of the random set

$C_n(X_1, \ldots, X_n)$ is present, there is no longer any randomness, because ϑ (and thus also $\gamma(\vartheta)$) is unknown, but not random. Therefore, one cannot speak of the probability that "$\gamma(\vartheta)$ lies in the set $C_n(x_1, \ldots, x_n)$". In any method to generate confidence sets, the sets $C_n(x_1, \ldots, x_n)$ should, of course, be "as small as possible". In this respect, note that the set $C_n(x_1, \ldots, x_n) = \gamma(\Theta)$ always contains $\gamma(\vartheta)$.

Synonymous with confidence set is the term *confidence region*. If $s = 1$ and thus $\gamma(\vartheta)$ is real-valued, and if $C_n(\mathbf{x})$ is an interval for each $\mathbf{x}$ in $\mathcal{X}$, then C_n is called a *confidence interval*. It is also common to refer to the level of significance $1 - \alpha$ as *confidence level* or simply as *level*.

General principles for constructing confidence regions for $\gamma(\vartheta)$ can be found in Chapter 7 of [SHA]. At this point, we only want to give a historically and practically important example that is directly related to Problem 5.11.

9.12 Example (Confidence Interval for μ With Unknown σ^2 Under Normality)

Suppose in the situation of Example 9.2 ($X_1, \ldots, X_n$ independent with the same normal distribution $N(\mu, \sigma^2)$) that we want a confidence interval for $\mu := \gamma(\vartheta)$, where $\vartheta = (\mu, \sigma^2)$. Thus, the variance σ^2 is a nuisance parameter. Let $\overline{X}_n := n^{-1} \sum_{j=1}^{n} X_j$ and S_n^2 be as in (9.1). With the addition theorem and properties of the normal distribution, we get

$$\frac{\sqrt{n}}{\sigma}(\overline{X}_n - \mu) \sim N(0, 1). \tag{9.3}$$

According to Problem 5.11,

$$\frac{n-1}{\sigma^2} S_n^2 \sim \chi_{n-1}^2, \tag{9.4}$$

where, again by Problem 5.11, the random variables in (9.3) and (9.4) are independent. If we divide the random variable in (9.3) by the positive root of the random variable in (9.4), after having divided the latter by the factor $n - 1$, the nuisance parameter σ^2 cancels out, and it follows that

$$\frac{\sqrt{n}(\overline{X}_n - \mu)}{S_n} \sim \frac{N(0, 1)}{\sqrt{\frac{\chi_{n-1}^2}{n-1}}}. \tag{9.5}$$

The distribution of the quotient on the right-hand side of the tilde is called *Student's t-distribution with $n - 1$ degrees of freedom*. In general, the *Student's t-distribution with k degrees of freedom* (short: t_k-distribution, where k is an integer) is the distribution of the quotient

$$\frac{N_0}{\sqrt{\frac{1}{k}(N_1^2 + \ldots + N_k^2)}}. \tag{9.6}$$

Here, $N_0, N_1, \ldots, N_k$ are i.i.d. standard normal random variables. The naming of this distribution goes back to the British statistician W. Gosset.[1] Gosset discovered the t-distribution in 1908, but his contract with the Guinness Brewery did not allow publication under his name, which is why Gosset published under the pseudonym *Student*. However, the t-distribution already appeared in a work by J. Lüroth[2] published in 1876 (see, e.g., [PSH]). Since the t_k-distribution is the distribution of the quotient in (9.6), it is symmetric around 0, and it converges to the standard normal distribution as $k \to \infty$.

Self-question 4: Why does the t_k-distribution converge to N(0, 1) as $k \to \infty$?

If $t_{n-1;1-\alpha/2}$ denotes the $(1 - \frac{\alpha}{2})$-quantile of the t_{n-1}-distribution, it follows from (9.5) that

$$\mathbb{P}_\vartheta \big(C_n(X_1, \ldots, X_n) \ni \mu \big) = 1 - \alpha, \quad \vartheta = (\mu, \sigma^2),$$

where

$$C_n(X_1, \ldots, X_n) := \left[\overline{X}_n - \frac{S_n}{\sqrt{n}} t_{n-1;1-\frac{\alpha}{2}}, \overline{X}_n + \frac{S_n}{\sqrt{n}} t_{n-1;1-\frac{\alpha}{2}} \right]. \tag{9.7}$$

The shape of this confidence interval for μ immediately shows the influence of the sample size n and the level $1 - \alpha$ on the length of this interval. Note that the random variable S_n converges to σ with probability one under $\mathbb{P}_\vartheta$ as $n \to \infty$.

The following definition is also based on the setting described before Definition 9.7.

9.13 Definition (Asymptotic Confidence Set)
Let $\alpha \in (0, 1)$. An *asymptotic confidence set for* $\gamma(\vartheta)$ *with level* $1 - \alpha$ is a sequence (C_n), where for each n the mapping C_n fulfills properties (a) and (b) of Definition 9.11 as well as the condition

$$\liminf_{n \to \infty} \mathbb{P}_\vartheta \big(C_n(X_1, \ldots, X_n) \ni \gamma(\vartheta) \big) \geq 1 - \alpha \quad \text{for each } \vartheta \in \Theta.$$

---

[1] William Sealy Gosset (1876–1937), British statistician, employee of the Guinness Brewery in Dublin and from 1935 in London.

[2] Jacob Lüroth (1844–1910), German mathematician, professor at the TH Karlsruhe (from 1869), at the TH Munich (from 1880 as successor of Felix Klein) and at the Albert-Ludwigs-University Freiburg im Breisgau (from 1883). Main field of research: Geometry.

9.14 Example (Poisson Distribution)

Let $X_1, X_2, \ldots$ be an i.i.d. sequence of random variables with the Poisson distribution $\mathrm{Po}(\vartheta)$, where $\vartheta \in \Theta := (0, \infty)$ is unknown. With $\overline{X}_n := n^{-1} \sum_{j=1}^{n} X_j$ and $\mathbb{E}_\vartheta(X_1) = \mathbb{V}_\vartheta(X_1) = \vartheta$, the Lindeberg–Lévy central limit theorem yields

$$\frac{\sqrt{n}(\overline{X}_n - \vartheta)}{\sqrt{\vartheta}} \xrightarrow{\mathcal{D}_\vartheta} \mathrm{N}(0, 1) \quad \text{for every } \vartheta \in \Theta.$$

If we replace ϑ in the denominator by $\overline{X}_n$, Slutsky's lemma gives

$$\frac{\sqrt{n}(\overline{X}_n - \vartheta)}{\sqrt{\overline{X}_n}} \xrightarrow{\mathcal{D}_\vartheta} \mathrm{N}(0, 1) \quad \text{for every } \vartheta \in \Theta.$$

If $z_\alpha := \Phi^{-1}(1 - \frac{\alpha}{2})$ denotes the $(1 - \frac{\alpha}{2})$-quantile of the distribution $\mathrm{N}(0, 1)$, it follows that

$$\lim_{n \to \infty} \mathbb{P}_\vartheta\left(\left|\frac{\sqrt{n}(\overline{X}_n - \vartheta)}{\sqrt{\overline{X}_n}}\right| \le z_\alpha\right) = 1 - \alpha \quad \text{for every } \vartheta \in \Theta. \tag{9.8}$$

Equivalent to this is $\lim_{n \to \infty} \mathbb{P}_\vartheta\left(\mathcal{C}_n(X_1, \ldots, X_n) \ni \vartheta\right) = 1 - \alpha$, $\vartheta \in \Theta$, where

$$\mathcal{C}_n(X_1, \ldots, X_n) := \left[\overline{X}_n - \frac{z_\alpha}{\sqrt{n}}\sqrt{\overline{X}_n}, \, \overline{X}_n + \frac{z_\alpha}{\sqrt{n}}\sqrt{\overline{X}_n}\right] \cap (0, \infty). \tag{9.9}$$

Thus, the sequence $(\mathcal{C}_n)$ is an asymptotic confidence interval for ϑ with level $1 - \alpha$.

Self-question 5: Why is there an intersection symbol in (9.9)?

Alternatively, one can use the delta method (cf. Theorem 6.25) and the variance-stabilizing transformation $g(\vartheta) = 2\sqrt{\vartheta}$ (see the explanations after (6.23)). From $\sqrt{n}(\overline{X}_n - \vartheta) \xrightarrow{\mathcal{D}_\vartheta} \mathrm{N}(0, \vartheta)$, it follows that

$$\sqrt{n}\left(2\sqrt{\overline{X}_n} - 2\sqrt{\vartheta}\right) \xrightarrow{\mathcal{D}_\vartheta} \mathrm{N}(0, 1) \quad \text{for every } \vartheta \in \Theta,$$

and we get

$$\lim_{n \to \infty} \mathbb{P}_\vartheta\left(\left|\sqrt{\overline{X}_n} - \sqrt{\vartheta}\right| \le \frac{z_\alpha}{2\sqrt{n}}\right) = 1 - \alpha \quad \text{for every } \vartheta \in \Theta.$$

Solving the above inequality for ϑ yields $\lim_{n\to\infty} \mathbb{P}_\vartheta\left(\widetilde{C}_n(X_1,\ldots,X_n) \ni \vartheta\right) = 1 - \alpha$ for every $\vartheta \in \Theta$, where

$$\widetilde{C}_n(X_1,\ldots,X_n) := \left[\left(\sqrt{\overline{X}_n} - \frac{z_\alpha}{2\sqrt{n}}\right)^2, \left(\sqrt{\overline{X}_n} + \frac{z_\alpha}{2\sqrt{n}}\right)^2\right]. \qquad (9.10)$$

Hence, the sequence $(\widetilde{C}_n)$ is also an asymptotic confidence interval for ϑ with level $1 - \alpha$. Note that the intervals in (9.9) and (9.10) are of equal length; the second interval is only shifted to the right by $z_\alpha^2/(4n)$ compared to the first.

9.15 Example (Confidence Interval for an Expectation)

In this example, we assume a *nonparametric* model. Let $X_1, X_2, \ldots$ be a sequence of i.i.d. random variables with unknown distribution function F, where

$$F \in \mathcal{F} := \left\{F : \mathbb{E}_F\left(X_1^2\right) = \int_{\mathbb{R}} x^2 \, dF(x) < \infty,\ 0 < \sigma^2(F) := \mathbb{V}_F\left(X_1\right) > 0\right\}.$$

The expectation $\mu_F(X_1) := \mathbb{E}_F(X_1)$ is of interest. Putting $\overline{X}_n := \frac{1}{n}\sum_{j=1}^n X_j$, the Lindeberg–Lévy central limit theorem implies

$$\frac{\sqrt{n}\left(\overline{X}_n - \mu(F)\right)}{\sigma(F)} \xrightarrow{\mathcal{D}_F} N(0,1) \quad \text{as } n \to \infty \text{ for each } F \in \mathcal{F}.$$

Here, $\sigma(F)$ stands for the positive root of $\sigma^2(F)$, that is, for the standard deviation of the underlying distribution. Note that we have also indexed the symbol $\mathcal{D}$ for convergence in distribution with F to emphasize that the distribution function F is used to specify the model. In view of the assumption $\mathbb{E}_F(X_1^2) < \infty$ the sequence defined by

$$S_n := \sqrt{\frac{1}{n-1} \sum_{j=1}^n \left(X_j - \overline{X}_n\right)^2}$$

converges almost surely to $\sigma(F)$. With the help of Slutsky's lemma it follows that

$$\frac{\sqrt{n}\left(\overline{X}_n - \mu(F)\right)}{S_n} \xrightarrow{\mathcal{D}_F} N(0,1) \quad \text{for each } F \in \mathcal{F}.$$

Analogous to (9.8) we now obtain

$$\lim_{n\to\infty} \mathbb{P}_F\left(\left|\frac{\sqrt{n}(\overline{X}_n - \mu(F))}{S_n}\right| \le z_\alpha\right) = 1 - \alpha \quad \text{for each } F \in \mathcal{F}, \qquad (9.11)$$

and thus the sequence $(\mathcal{C}_n)_{n \geq 2}$, defined by

$$\mathcal{C}_n(X_1, \ldots, X_n) := \left[\overline{X}_n - \frac{z_\alpha S_n}{\sqrt{n}}, \overline{X}_n + \frac{z_\alpha S_n}{\sqrt{n}}\right], \tag{9.12}$$

is an *asymptotic* confidence interval with level $1 - \alpha$ for the expectation $\mu(F)$ of the underlying distribution. Note the similarity with the exact confidence interval for the expectation *of an assumed normal distribution* given in (9.7).

Finally, it should be emphasized that the convergence in (9.11) is not uniform in $F \in \mathcal{F}$. In order to still obtain property (c) in Definition 9.11 for a suitable α at finite n, one must suitably reduce $\mathcal{F}$ or modify the interval (9.12), see [GUT] and Chapter 4 of [MES].

Answers to the Self-Questions

Answer 1 Fix any $\delta > 0$ and $\vartheta \in \Theta$. Due to the $\sqrt{n}$-consistency, there is a $K > 0$ with $\mathbb{P}_\vartheta(\|\sqrt{n}(S_n - \gamma(\vartheta))\| > K) \leq \delta, n \geq 1$. If $\varepsilon > 0$ is arbitrary, then for sufficiently large n $\{\|S_n - \gamma(\vartheta)\| > \varepsilon\} \subset \{\sqrt{n}\|S_n - \gamma(\vartheta)\| > K\}$, and thus $\limsup_{n \to \infty} \mathbb{P}_\vartheta(\|S_n - \gamma(\vartheta)\| > \varepsilon) \leq \delta$. Since δ was arbitrary, the consistency of (S_n) follows.

Answer 2 Yes, because $\sum_{j=1}^n (X_j - \overline{X}_n)^2 = \sum_{j=1}^n X_j^2 - n\overline{X}_n^2$. It thus follows that

$$S_n^2 = \frac{n}{n-1} \cdot \frac{1}{n} \sum_{j=1}^n X_j^2 - \frac{n}{n-1} \cdot \overline{X}_n^2.$$

By the strong law of large numbers, for each $\vartheta \in \Theta$ the right-hand side converges $\mathbb{P}_\vartheta$-a.s. to $\mathbb{E}_\vartheta(X_1^2) - (\mathbb{E}_\vartheta(X_1))^2 = \mathbb{V}_\vartheta(X_1) = \gamma(\vartheta) = \sigma^2$.

Answer 3 Since $M_n := \max(X_1, \ldots, X_n)$ is monotonically increasing and $M_n \xrightarrow{\mathbb{P}_\vartheta} \vartheta$, M_n converges $\mathbb{P}_\vartheta$-a.s. to ϑ, and thus also $\frac{n+1}{n} M_n$.

Answer 4 In (9.6), $k^{-1}(N_1^2 + \ldots + N_k^2)$, and thus also the root of this expression converges in probability to 1 as $k \to \infty$. With Slutsky's lemma, it then follows that a random variable with the t_k-distribution converges in distribution to the standard normal distribution as $k \to \infty$.

Answer 5 By definition, $\mathcal{C}_n$ takes on values in $\gamma(\Theta)$. Without performing the intersection with the set $(0, \infty)$, the left interval limit can be less than or equal to zero with positive $\mathbb{P}_\vartheta$-probability. Performing such an intersection with the parameter range is usually implicitly assumed.

Problems

9.1 Problem The random variable X has the binomial distribution $\text{Bin}(\ell, \vartheta)$, where $\vartheta \in \Theta := (0, 1)$ is unknown. The task is to estimate $\gamma(\vartheta) := \vartheta^{-1}$. Show: There is no unbiased estimator of $\gamma(\vartheta)$, that is, there is no function $T : \{0, 1, \ldots, \ell\} \to \mathbb{R}$ with

$$\mathbb{E}_\vartheta\big(T(X)\big) = \gamma(\vartheta) \quad \text{for every } \vartheta \in \Theta.$$

9.2 Problem In the setting of Definition 9.8, let (T_n) with $T_n =: (T_{n,1}, \ldots, T_{n,k})$ be a sequence of estimators of $\gamma(\vartheta)$. Assume $\mathbb{E}_\vartheta(T_{n,j}^2) < \infty$ $(j \in \{1, \ldots, k\}, \vartheta \in \Theta)$. Show: If (T_n) is asymptotically unbiased for $\gamma(\vartheta)$, and $\lim_{n \to \infty} \mathbb{V}_\vartheta(T_{n,j}) = 0$ for every $j \in \{1, \ldots, k\}$ and every $\vartheta \in \Theta$, then the sequence (T_n) is consistent for $\gamma(\vartheta)$.

9.3 Problem In the setting of Definition 9.7 with $k = 1$, let T_n be an unbiased estimator of ϑ, that is, we have $\mathbb{E}_\vartheta(T_n) = \vartheta, \vartheta \in \Theta$. Does it follow that T_n^2 is an unbiased estimator of ϑ^2?

9.4 Problem Assume the setting of Example 9.10, and put $S_n := \frac{n+1}{n} \max(X_1, \ldots, X_n)$. Prove the following assertions:

(a) $\mathbb{E}_\vartheta(S_n) = \vartheta, \quad \vartheta \in \Theta,$

(b) $\mathbb{V}_\vartheta(S_n) = \dfrac{n\vartheta^2}{(n+2)(n+1)^2}, \quad \vartheta \in \Theta,$

(c) $\displaystyle\lim_{n \to \infty} \mathbb{P}_\vartheta\big(n(S_n - \vartheta) \le t\big) = \begin{cases} 1, & \text{if } t \ge \vartheta, \\ \exp\left(\frac{t}{\vartheta} - 1\right), & \text{if } t < \vartheta. \end{cases}$

9.5 Problem In the situation of Example 9.10, let $M_n := \max(X_1, \ldots, X_n)$. Show that for $\alpha \in (0, 1)$

$$\mathbb{P}_\vartheta\big(\big[M_n, M_n\alpha^{-1/n}\big] \ni \vartheta\big) = 1 - \alpha \quad \text{for every } \vartheta > 0,$$

that is, $\big[M_n, M_n\alpha^{-1/n}\big]$ is a confidence interval for ϑ with level $1 - \alpha$.

9.6 Problem Let $X_1, X_2, \ldots$ be a sequence of i.i.d. random variables with the binomial distribution $\text{Bin}(1, \vartheta)$, where $\vartheta \in \Theta := (0, 1)$ and $\overline{X}_n := n^{-1} \sum_{j=1}^n X_j$. With the help of Problem 6.16, show that the sequence, defined by

$$C_n(X_1, \ldots, X_n) := \left[\sin^2\left(\arcsin\sqrt{\overline{X}_n} - \frac{z_\alpha}{2\sqrt{n}}\right), \sin^2\left(\arcsin\sqrt{\overline{X}_n} + \frac{z_\alpha}{2\sqrt{n}}\right)\right],$$

is an asymptotic confidence interval for ϑ with level $1 - \alpha$. Here, z_α denotes the $(1 - \frac{\alpha}{2})$-quantile of the standard normal distribution, and we put $\sin^2(t) := (\sin t)^2$.

Remark With the addition theorem for the sine and Taylor expansions of the sine and cosine functions at the point 0, it follows that the endpoints of the above interval are equal to $\overline{X}_n \pm \frac{z_\alpha}{\sqrt{n}} \sqrt{\overline{X}_n(1 - \overline{X}_n)}$ up to terms of order $O_{\mathbb{P}_\vartheta}(n^{-1})$.

9.7 Problem Let $X_1, X_2, \ldots$ be a sequence of i.i.d. d-dimensional random vectors on a probability space $(\Omega, \mathcal{A}, \mathbb{P})$ with $\mathbb{E}\|X_1\|^2 < \infty$. The problem is to estimate the unknown covariance matrix $\Sigma := \mathbb{Cov}(X_1)$. Let

$$S_n := \frac{1}{n} \sum_{j=1}^n (X_j - \overline{X}_n)(X_j - \overline{X}_n)^\top$$

be the sample covariance matrix of $X_1, \ldots, X_n$, where $\overline{X}_n := n^{-1} \sum_{j=1}^n X_j$ (cf. Problem 5.4). Show that

(a) $\mathbb{E}(S_n) = \dfrac{n-1}{n} \Sigma, \ \ n \geq 2,$

(b) $S_n \xrightarrow{\text{a.s.}} \Sigma$ as $n \to \infty$.

Maximum Likelihood Estimation

10

In this chapter, we will learn a fundamental method to construct estimators of unknown parameters in parametric models. This method is primarily associated with the name R.A. Fisher, who introduced the term *likelihood* in 1920 in connection with estimation problems (see, e.g., [EDW]). Although in terms of philosophy, the words *likelihood* and *probability* have the same meaning, we must differentiate well in view of the *method of maximum likelihood estimation* propagated by Fisher, who studied this method extensively from a mathematical point of view. However, the underlying idea of the maximum likelihood method was already known to J. Lambert,[1] D. Bernoulli[2] and J. Lagrange[3] (see, e.g., [HAL]).

Since we will be dealing with asymptotic properties of (yet to be defined) maximum likelihood estimators, we assume, as in Chap. 9, the situation of a sequence $X_1, X_2, \ldots$ of random variables, each taking on values in a set $\mathcal{X}_0$ equipped with a σ-field $\mathcal{B}_0$ and having the distribution Q_ϑ. Here, ϑ in Θ with $\Theta \subset \mathbb{R}^k$ is an unknown parameter, and the mapping $\Theta \ni \vartheta \mapsto Q_\vartheta$ is injective. We also work again with the canonical model $(\mathcal{X}, \mathcal{B}, \mathbb{P}_\vartheta)_{\vartheta \in \Theta}$, that is, the set $\mathcal{X} = \mathcal{X}_0^{\mathbb{N}}$, equipped with the infinite product σ-field $\mathcal{B} = \mathcal{B}_0^{\mathbb{N}}$ of all sequences

---

[1] Johann Heinrich Lambert (1728–1777), mathematician, natural scientist and philosopher; his work on the parallel postulate paved the way for non-Euclidean geometry. His work on the number π is famous; Lambert provided the first flawless proof that π is an irrational number.

[2] Daniel Bernoulli (1700–1782), among others physicist, mathematician and physician. Member of the St. Petersburg Academy of Sciences, professor at the University of Basel (initially for botany and anatomy, from 1750 for physics).

[3] Joseph Louis Lagrange (1736–1813), became a professor at the University of Turin at the age of 19, member of the Berlin and Paris Academy of Sciences. Lagrange contributed to algebra, number theory, probability theory, mechanics and astronomy, as well as to differential and difference equations and infinite series.

© The Author(s), under exclusive license to Springer-Verlag GmbH, DE,
part of Springer Nature 2024
N. Henze, *Asymptotic Stochastics*, Mathematics Study Resources 10,
https://doi.org/10.1007/978-3-662-68923-3_10

$\mathbf{x} = (x_1, x_2, \ldots)$ with $x_j \in \mathcal{X}_0$ for each $j \geq 1$, and with the infinite product probability measure $\mathbb{P}_\vartheta := Q_\vartheta^{\mathbb{N}}$. In this model, $X_j(\mathbf{x}) := x_j$ for each $j \geq 1$ and each $\mathbf{x} = (x_1, x_2, \ldots)$ in $\mathcal{X}$.

Compared to Chap. 9, a new aspect comes into play, leading to the central concept of *likelihood function*. Namely, there is a σ-finite measure ν on $\mathcal{B}_0$, which dominates all probability measures Q_ϑ with $\vartheta \in \Theta$ in the sense that every ν-null set B in $\mathcal{B}_0$ satisfies $Q_\vartheta(B) = 0$ for each $\vartheta \in \Theta$. Then, by the Radon–Nikodým theorem (Theorem 1.32), there are non-negative measurable functions $f(\cdot, \vartheta) : \mathcal{X}_0 \to \mathbb{R}$, $\vartheta \in \Theta$, such that $Q_\vartheta = f(\cdot, \vartheta)\, \nu$, that is, we have

$$Q_\vartheta(B) = \int_B f(x, \vartheta)\, \nu(\mathrm{d}x), \quad B \in \mathcal{B}_0. \tag{10.1}$$

The probability measure Q_ϑ thus has a *density* $f(\cdot, \vartheta)$ with respect to ν. This is equivalent to the fact that each of the random variables $X_1, X_2, \ldots$ has the density

$$f(\cdot, \vartheta) =: \frac{\mathrm{d}Q_\vartheta}{\mathrm{d}\nu}(\cdot)$$

with respect to ν.

If $X_1, X_2, \ldots$ are d-dimensional random vectors and $\mathcal{X}_0$ is a subset of $\mathbb{R}^d$, then the measure ν is either the restriction of the Borel–Lebesgue measure λ^d to $\mathcal{X}_0$, denoted by $\lambda^d_{|\mathcal{X}_0}$, or the counting measure on a countable subset D of $\mathcal{X}_0$. In the latter case, ν can be written in the form $\nu = \sum_{t \in D} \delta_t$ as a sum of Dirac measures. In the first case, the distribution of X_1 is absolutely continuous with respect to $\lambda^d_{|\mathcal{X}_0}$, and the probability $\mathbb{P}_\vartheta(X_1 \in B) = Q_\vartheta(B)$ given in (10.1) is obtained as a Lebesgue integral or, depending on a particular set B and the density $f(\cdot, \vartheta)$, even as a Riemann integral. If ν is the counting measure on a countable set D, then X_1 has a discrete distribution, and we have $f(t, \vartheta) = Q_\vartheta(\{t\}) = \mathbb{P}_\vartheta(X_1 = t)$ for each $t \in D$.

Self-question 1: Why do $Q_\vartheta(B) = \mathbb{P}_\vartheta(X_1 \in B)$ and $Q_\vartheta(\{t\}) = \mathbb{P}_\vartheta(X_1 = t)$ hold?

10.1 Example (Continuation of Examples 9.2–9.5)

(a) In Example 9.2 (normal distribution), $\nu = \lambda^1$, and with $\vartheta = (\mu, \sigma^2)$ we have

$$f(x, \vartheta) = \frac{1}{\sigma \sqrt{2\pi}} \exp\left(-\frac{(x - \mu)^2}{2\sigma^2}\right), \quad x \in \mathbb{R}.$$

(b) In Example 9.3 (exponential distribution), $\nu = \lambda^1_{|[0,\infty)}$ and

$$f(x, \vartheta) = \vartheta e^{-\vartheta x}, \quad x \geq 0,$$

with $f(x, \vartheta) = 0$, otherwise.
(c) In Example 9.4 (Bernoulli sequence), $\nu = \delta_0 + \delta_1$ is the counting measure on $D :=$ $\{0, 1\}$, and

$$f(x, \vartheta) = \vartheta^x (1 - \vartheta)^{1-x}, \quad x \in D,$$

with $f(x, \vartheta) := 0$, otherwise.
(d) In Example 9.5 (bivariate normal distribution), $\nu = \lambda^2$, and with $\vartheta = (a, b, \sigma^2, \tau^2, \rho)$ we have

$$f(x, y, \vartheta) = \frac{1}{2\pi\sigma\tau\sqrt{1-\rho^2}} \exp\left(-\frac{\tau^2(x-a)^2 - 2\rho\sigma\tau(x-a)(y-b) + \sigma^2(y-b)^2}{2\sigma^2\tau^2(1-\rho^2)}\right),$$

$(x, y) \in \mathbb{R}^2$.

We will now define a maximum likelihood estimator of ϑ for the sample size n. From now on, *maximum likelihood* is abbreviated to ML. Since $X_1, \ldots, X_n$ are independent random variables, the function $f_n(\cdot, \vartheta)$, defined by

$$f_n(x_1, \ldots, x_n, \vartheta) := \prod_{j=1}^n f(x_j, \vartheta), \quad (x_1, \ldots, x_n) \in \mathcal{X}_0^n,$$

is the density of $(X_1, \ldots, X_n)$ with respect to the n-fold product measure $\nu_n := \nu \otimes \nu \otimes \ldots \otimes \nu$ (n factors). For asymptotics, we regard $x_1, \ldots, x_n$ as the initial segment of length n of a sequence $\mathbf{x} = (x_1, x_2, \ldots) \in \mathcal{X}$ and write

$$f_n(\mathbf{x}, \vartheta) := f_n(x_1, \ldots, x_n, \vartheta) = \prod_{j=1}^n f(x_j, \vartheta).$$

Thus, f_n is formally a function defined on $\mathcal{X} \times \Theta$.

To arrive at the concept of the likelihood function, we change our perspective. For a fixed value ϑ, $f_n(x_1, \ldots, x_n, \vartheta)$, regarded as a function of $x_1, \ldots, x_n$, is a *probability density* with respect to ν_n. This density receives the label *likelihood function* if it is regarded as a function of ϑ at fixed $x_1, \ldots, x_n$.

10.2 Definition (Likelihood Function, ML Estimator)

(a) For any $\mathbf{x} \in \mathcal{X}$, the function defined by

$$L_{n,\mathbf{x}}(\vartheta) := f_n(\mathbf{x}, \vartheta) = f_n(x_1, \ldots, x_n, \vartheta) = \prod_{j=1}^{n} f(x_j, \vartheta), \quad \vartheta \in \Theta,$$

is called the *likelihood function of* $x_1, \ldots, x_n$.
(b) Any value $\widehat{\vartheta}_n(\mathbf{x}) \in \Theta$ with

$$f_n\big(\mathbf{x}, \widehat{\vartheta}_n(\mathbf{x})\big) = \sup_{\vartheta \in \Theta} f_n(\mathbf{x}, \vartheta) \tag{10.2}$$

is called a ML *estimate of* ϑ *given* $x_1, \ldots, x_n$.
(c) A measurable mapping $\widehat{\vartheta}_n : \mathcal{X} \to \Theta$ with (10.2) for every $\mathbf{x} \in \mathcal{X}$ is called a ML *estimator of* ϑ (for the sample size n).

In what follows, MLE will be shorthand for *maximum likelihood estimator*. The inherent idea of ML estimation is thus to consider, within a parametric model with given data $x_1, \ldots, x_n$, the value of ϑ as the "most credible" that, in the case of a counting measure ν on $\mathcal{B}^d$, assigns the greatest probability of occurrence to these data. In the general case, the value of ϑ is sought for which the probability *density*, evaluated at the point $(x_1, \ldots, x_n)$, attains its maximum value. The above definition assumes that the supremum in (10.2) is attained, which is not necessarily the case. If, putting $\vartheta =: (\vartheta_1, \ldots, \vartheta_k)$, the function $f_n(x_1, \ldots, x_n, \vartheta)$ is partially differentiable with respect to $\vartheta_1, \ldots, \vartheta_k$, one will try to obtain a solution $\widehat{\vartheta}_n(\mathbf{x})$ of (10.2) by differentiation. Since $f_n(x_1, \ldots, x_n, \vartheta)$ is a product of functions, it is usually more convenient to consider the so-called *log-likelihood function* defined by

$$\log L_{n,\mathbf{x}}(\vartheta) = \sum_{j=1}^{n} \log f(x_j, \vartheta).$$

Because of the strict monotonicity of the logarithm function, the functions $L_{n,\mathbf{x}}$ and $\log L_{n,\mathbf{x}}$ attain their maxima at the same value of ϑ. Therefore, we solve the so-called *log-likelihood equations* (or *likelihood equations* for simplicity)

$$\sum_{j=1}^{n} \frac{\partial}{\partial \vartheta_\ell} \log f(x_j, \vartheta) = 0, \quad \ell = 1, \ldots, k. \tag{10.3}$$

These equations are generally nonlinear, and they usually cannot be solved in closed form (see, e.g., Problem 10.3), but only with the help of iterative numerical methods.

Moreover, (10.3) can have different solutions, not all of which must lead to relative maxima. Note also that solutions of (10.2) can also lie on the boundary of the set Θ and thus are not captured by (10.3).

It should be emphasized that we will use the notation $\widehat{\vartheta}_n$ exclusively for the MLE and, as already in Chap. 9 (with $\widehat{\vartheta}_n$ instead of T_n), we briefly write $\widehat{\vartheta}_n = \widehat{\vartheta}_n(X_1, \ldots, X_n)$.

10.3 Example (Bernoulli Sequence)

The random variables $X_1, X_2, \ldots$ are independent with the same binomial distribution $\mathrm{Bin}(1, \vartheta)$, where $\vartheta \in \Theta := [0, 1]$. This case fits into our general setting with $\mathcal{X}_0 := \{0, 1\}$, $\mathcal{X} := \mathcal{X}_0^{\mathbb{N}}$, $\nu := \delta_0 + \delta_1$, and $f(x, \vartheta) = \vartheta^x(1 - \vartheta)^{1-x}$. Setting $t := x_1 + \ldots + x_n$, we have

$$L_{n,\mathbf{x}}(\vartheta) = \prod_{j=1}^{n} \left(\vartheta^{x_j}(1 - \vartheta)^{1-x_j}\right) = \vartheta^t(1 - \vartheta)^{n-t}.$$

In the case $0 < t < n$, the maximum of $L_{n,\mathbf{x}}(\vartheta)$ is attained in the open interval $(0, 1)$. Since $\log L_{n,\mathbf{x}}(\vartheta) = t \log \vartheta + (n - t) \log(1 - \vartheta)$, $0 < \vartheta < 1$, the likelihood equation leads to

$$\frac{t}{\vartheta} - \frac{n - t}{1 - \vartheta} = 0$$

and thus to the ML estimate $\widehat{\vartheta}_n(x_1, \ldots, x_n) = \frac{t}{n}$, which can be interpreted as the relative frequency of hits.

Self-question 2: Why is there a *maximum* of the likelihood function at $\frac{t}{n}$?

In the case $t = 0$, we have $L_{n,\mathbf{x}}(\vartheta) = (1 - \vartheta)^n$ and thus $\widehat{\vartheta}_n(x_1, \ldots, x_n) = 0 = \frac{t}{n}$. For $t = n$, we get $L_{n,\mathbf{x}}(\vartheta) = \vartheta^n$ and thus $\widehat{\vartheta}_n(x_1, \ldots, x_n) = 1 = \frac{t}{n}$. Consequently, in the cases $t = 0$ and $t = 1$, no ML estimate would exist if we had defined the *open* unit interval as the parameter space.

The MLE $\widehat{\vartheta}_n = n^{-1} \sum_{j=1}^{n} X_j$ is unbiased for ϑ, and the sequence of estimators $(\widehat{\vartheta}_n)$ is strongly consistent for ϑ according to the strong law of large numbers. In addition, $\widehat{\vartheta}_n$ has the (uniformly in ϑ) smallest variance among all unbiased estimators of ϑ (see Problem 11.2). Finally, by the Lindeberg–Lévy central limit theorem,

$$\sqrt{n}\left(\widehat{\vartheta}_n - \vartheta\right) \xrightarrow{\mathcal{D}_\vartheta} \mathrm{N}\left(0, \vartheta(1 - \vartheta)\right), \quad 0 < \vartheta < 1. \tag{10.4}$$

10.4 Example (Uniform Distribution, Continuation of Example 9.10)

In the setting of Example 9.10, $X_1, X_2, \ldots$ are independent random variables with a uniform distribution on $[0, \vartheta]$, where ϑ in $\Theta := (0, \infty)$ is unknown. In this case, for $\mathbf{x} = (x_1, x_2, \ldots)$ in $\mathcal{X} := \mathbb{R}_{>0}^{\mathbb{N}}$ the likelihood function is given by

$$L_{n,\mathbf{x}}(\vartheta) = \prod_{j=1}^{n} \left(\frac{1}{\vartheta} \mathbf{1}_{[0,\vartheta]}(x_j) \right) = \frac{1}{\vartheta^n} \mathbf{1}_{[0,\vartheta]}\left(\max(x_1, \ldots, x_n) \right).$$

We thus conclude that $\widehat{\vartheta}_n(x_1, \ldots, x_n) := \max(x_1, \ldots, x_n)$ is the ML estimate of ϑ. According to the results of Example 9.10, $\mathbb{E}_\vartheta (\widehat{\vartheta}_n(X_1, \ldots, X_n)) = \frac{n}{n+1}\vartheta$, and thus the sequence $(\widehat{\vartheta}_n)$ of MLE's is asymptotically unbiased for ϑ. The sequence $(\widehat{\vartheta}_n)$ is also strongly consistent for ϑ (cf. Self-question 3 in Chap. 9), and the sequence $n(\widehat{\vartheta}_n - \vartheta)$ has a non-degenerate limit distribution as $n \to \infty$ (Problem 10.4).

10.5 Example (Normal Distribution)
Continuing Examples 9.2 and 10.1 (a), let $X_1, X_2, \ldots$ be independent with the same normal distribution $N(\mu, \sigma^2)$, where $\vartheta := (\mu, \sigma^2) \in \Theta := \mathbb{R} \times \mathbb{R}_{>0}$. For $\mathbf{x} = (x_1, x_2, \ldots)$ in $\mathcal{X} := \mathbb{R}^{\mathbb{N}}$, we have

$$L_{n,\mathbf{x}}(\vartheta) = \prod_{j=1}^{n} \left(\frac{1}{\sigma \sqrt{2\pi}} \exp \left(-\frac{(x_j - \mu)^2}{2\sigma^2} \right) \right)$$

$$= \frac{1}{\left(\sigma \sqrt{2\pi} \right)^n} \exp \left(-\frac{1}{2\sigma^2} \sum_{j=1}^{n} (x_j - \mu)^2 \right).$$

Here, maximization of the likelihood function can be done in two steps. For a fixed σ^2, the sum of squares $\sum_{j=1}^{n} (x_j - \mu)^2$ attains its minimum value it we set $\mu = \bar{x}_n$, where $\bar{x}_n = n^{-1} \sum_{j=1}^{n} x_j$. In a second step, we have to minimize the function

$$\sigma^2 \mapsto \frac{1}{\sigma^n} \exp \left(-\frac{1}{2\sigma^2} \sum_{j=1}^{n} (x_j - \bar{x}_n)^2 \right)$$

with respect to σ^2. Taking the logarithm and differentiating shows that this function attains its minimum value if we set $\sigma^2 := n^{-1} \sum_{j=1}^{n} (x_j - \bar{x}_n)^2$. In the case $x_1 = \ldots = x_n$, we obtain the value zero, which is not admitted as an estimate of σ^2. However, for each $\vartheta \in \Theta$ this case only occurs with $\mathbb{P}_\vartheta$-probability 0. A possible remedy could be to extend the set of normal distributions by the one-point distributions δ_μ with $\mu \in \mathbb{R}$. The thus enlarged class, however, is no longer dominated by the Borel–Lebesgue measure λ^1. Despite these drawbacks, it is common to denominate the estimate $\widehat{\vartheta}_n := (\widehat{\mu}_n, \widehat{\sigma_n^2})$ with

$$\widehat{\mu}_n(X_1, \ldots, X_n) := \bar{X}_n = \frac{1}{n} \sum_{j=1}^{n} X_j, \quad \widehat{\sigma_n^2}(X_1, \ldots, X_n) := \frac{1}{n} \sum_{j=1}^{n} (X_j - \bar{X}_n)^2 \quad (10.5)$$

the MLE of the parameters of a normal distribution. We also adopt the common manner of speaking "the MLE's of μ and σ^2 of a normal distribution are $\widehat{\mu}_n = \overline{X}_n$ and $\widehat{\sigma^2_n} = n^{-1}\sum_{j=1}^{n}(X_j - \overline{X}_n)^2$", even though in the case of a vector-valued parameter we have not made an ML estimate for a real-valued aspect $\gamma(\vartheta)$ such as $\gamma(\vartheta) = \mu$, but only identified $\widehat{\mu}_n$ and $\widehat{\sigma^2_n}$ as the *components of the MLE* $\widehat{\vartheta}_n$ of $\vartheta = (\mu, \sigma^2)$. Of course, the estimator $\widehat{\gamma(\vartheta)}_n := \gamma(\widehat{\vartheta}_n)$, derived from an MLE $\widehat{\vartheta}_n : \mathcal{X} \to \Theta$ of ϑ, is generally suitable for estimating $\gamma(\vartheta)$ when a statistical model $(\mathcal{X}, \mathcal{B}, (\mathbb{P}_\vartheta)_{\vartheta \in \Theta})$ with $\Theta \subset \mathbb{R}^k$ is given and $\gamma(\vartheta)$ is to be estimated, where $\gamma : \Theta \to \mathbb{R}$.

With the Lindeberg–Lévy central limit theorem it follows that $\sqrt{n}(\widehat{\mu}_n - \mu) \xrightarrow{\mathcal{D}_\vartheta} N(0, \sigma^2)$, and according to Example 9.9 we have $\sqrt{n}(S_n^2 - \sigma^2) \xrightarrow{\mathcal{D}_\vartheta} N(0, 2\sigma^4)$, where $S_n^2 = \frac{n}{n-1}\widehat{\sigma^2_n}$. With Slutsky's lemma we now get $\sqrt{n}(\widehat{\sigma^2_n} - \sigma^2) \xrightarrow{\mathcal{D}_\vartheta} N(0, 2\sigma^4)$. Since, by Problem 5.11, $\widehat{\mu}_n$ and $\widehat{\sigma^2_n}$ are independent for each $n \geq 2$, Theorem 6.8 yields

$$\sqrt{n}(\widehat{\vartheta}_n - \vartheta) = \sqrt{n}\begin{pmatrix} \widehat{\mu}_n - \mu \\ \widehat{\sigma^2_n} - \sigma^2 \end{pmatrix} \xrightarrow{\mathcal{D}_\vartheta} N_2\left(\begin{pmatrix} 0 \\ 0 \end{pmatrix}, \begin{pmatrix} \sigma^2 & 0 \\ 0 & 2\sigma^4 \end{pmatrix} \right). \tag{10.6}$$

10.6 Example (Multivariate Normal Distribution)

Let $d \geq 2$, and let $X_1, X_2, \ldots$ be an i.i.d. sequence of d-dimensional random vectors with the normal distribution $N_d(\mu, \Sigma)$. Here, μ and Σ are unknown, but we assume that the covariance matrix Σ is positive-definite and thus invertible. If we write $\mathbb{R}^{d \times d}_{\text{p.d.}}$ for the set of all $d \times d$ symmetric positive-definite matrices with real entries, then the unknown parameter in this parametric statistical model is given by $\vartheta := (\mu, \Sigma)$, and the parameter space Θ is $\Theta = \mathbb{R}^d \times \mathbb{R}^{d \times d}_{\text{p.d.}}$. According to (5.3), the likelihood function for $\mathbf{x} = (x_1, x_2, \ldots)$ in $\mathcal{X} := (\mathbb{R}^d)^{\mathbb{N}}$ is

$$L_{n,\mathbf{x}}(\vartheta) = \frac{1}{(2\pi)^{nd/2}|\Sigma|^{n/2}} \exp\left(-\frac{1}{2}\sum_{j=1}^{n}(x_j - \mu)^{\top}\Sigma^{-1}(x_j - \mu) \right),$$

where we set $|\Sigma| := \det(\Sigma)$.

As in Example 10.5, maximization of the likelihood function can be done in two steps. In view of Problem 10.6, with fixed Σ, the sum $\sum_{j=1}^{n}(x_j - \mu)^{\top}\Sigma^{-1}(x_j - \mu)$ is minimal when we set $\mu = \overline{x}_n$, where $\overline{x}_n = n^{-1}\sum_{j=1}^{n}x_j$. In a second step, the task remains to maximize the function

$$\Sigma \mapsto \frac{1}{|\Sigma|^{n/2}} \exp\left(-\frac{1}{2}\sum_{j=1}^{n}(x_j - \overline{x}_n)^{\top}\Sigma^{-1}(x_j - \overline{x}_n) \right)$$

with respect to $\Sigma \in \mathbb{R}_{\text{p.d.}}^{d \times d}$. After taking the logarithm and multiplying by $2/n$, this task is equivalent to having to solve the following problem: Find a matrix $\Sigma \in \mathbb{R}_{\text{p.d.}}^{d \times d}$ for which the function

$$\Sigma \mapsto h(\Sigma) := -\log|\Sigma| - \frac{1}{n}\sum_{j=1}^{n}(x_j - \overline{x}_n)^\top \Sigma^{-1}(x_j - \overline{x}_n) \tag{10.7}$$

attains its maximum value.

We would like to anticipate the answer: If $n \geq d + 1$, and if the sample covariance matrix of $x_1, \ldots, x_n$ (cf. Problems 5.3 and 5.4), denoted by

$$S_n := \frac{1}{n}\sum_{j=1}^{n}(x_j - \overline{x}_n)(x_j - \overline{x}_n)^\top,$$

is positive-definite, then $h(\Sigma)$ attains its maximum value if we put $\Sigma := s_n$. To justify this claim, we need some properties of the *trace* of a square matrix $A = (a_{i,j})_{1 \leq i,j, \leq k}$, $k \in \mathbb{N}$, defined by $\text{tr}(A) := \sum_{j=1}^{k} a_{j,j}$. On the set $\mathbb{R}^{k \times k}$ of $k \times k$ matrices (which includes the case $k = 1$), the trace is a linear functional, and—as one can calculate directly—$\text{tr}(CD) = \text{tr}(DC)$ for not necessarily square matrices $C \in \mathbb{R}^{k \times \ell}$ and $D \in \mathbb{R}^{\ell \times k}$, where $k, \ell \in \mathbb{N}$. Furthermore, $\text{tr}(A(B + C)) = \text{tr}(AB) + \text{tr}(AC)$ if $A, B, C \in \mathbb{R}^{k \times k}$, $k \in \mathbb{N}$. This gives

$$\frac{1}{n}\sum_{j=1}^{n}(x_j - \overline{x}_n)^\top \Sigma^{-1}(x_j - \overline{x}_n) = \text{tr}\left(\frac{1}{n}\sum_{j=1}^{n}(x_j - \overline{x}_n)^\top \Sigma^{-1}(x_j - \overline{x}_n)\right)$$

$$= \frac{1}{n}\sum_{j=1}^{n}\text{tr}\left((x_j - \overline{x}_n)^\top \Sigma^{-1}(x_j - \overline{x}_n)\right)$$

$$= \text{tr}\left(\Sigma^{-1}s_n\right),$$

and thus the function h defined in (10.7) takes the form

$$h(\Sigma) = -\log|\Sigma| - \text{tr}\left(\Sigma^{-1}s_n\right). \tag{10.8}$$

To proceed, we need an inequality between the determinant and the trace of a positive-definite symmetric matrix $A \in \mathbb{R}^{k \times k}$. Such a matrix can be written in the form $A = B\text{diag}(\lambda_1, \ldots, \lambda_k)B^\top$ with an orthogonal matrix B, and $\text{diag}(\lambda_1, \ldots, \lambda_k)$ is a diagonal matrix consisting of the positive eigenvalues of A. Thus, we have $|A| =$

$|\mathrm{diag}(\lambda_1, \ldots, \lambda_k)| = \prod_{j=1}^{k} \lambda_j$ and $\mathrm{tr}(A) = \mathrm{tr}(\mathrm{diag}(\lambda_1, \ldots, \lambda_k)) = \sum_{j=1}^{k} \lambda_j$. The inequality $\log t \leq t - 1, t > 0$, now yields

$$\log |A| = \sum_{j=1}^{k} \log \lambda_j \leq \sum_{j=1}^{k} (\lambda_j - 1) = \mathrm{tr}(A) - k.$$

Since the matrix s_n is assumed to be positive-definite, $\Sigma^{-1/2} s_n \Sigma^{-1/2}$ is also positive-definite (and symmetric). Here, $\Sigma^{-1/2}$ denotes the symmetric positive-definite square root of Σ^{-1}, which exists by Remark 5.9. From the above inequality, we get

$$\log |\Sigma^{-1} s_n| = \log \left| \Sigma^{-1/2} s_n \Sigma^{-1/2} \right| \leq \mathrm{tr}\left(\Sigma^{-1/2} s_n \Sigma^{-1/2} \right) - d = \mathrm{tr}\left(\Sigma^{-1} s_n \right) - d.$$

In view of (10.8) and $|\Sigma^{-1}| = 1/|\Sigma|$, it follows that

$$
\begin{aligned}
h(\Sigma) &\leq -\log |\Sigma| - \log |\Sigma^{-1} s_n| + d \\
&= \log |\Sigma| - \log |\Sigma^{-1}| - \log |s_n| + d \\
&= -\log |s_n| + \mathrm{tr}(s_n^{-1} s_n) \\
&= h(s_n).
\end{aligned}
$$

Thus, the likelihood function $L_{n,\mathbf{x}}(\vartheta)$ is maximized at the point $\vartheta = (\overline{x}_n, s_n)$ if the sample covariance matrix s_n of $x_1, \ldots, x_n$ is positive-definite. Under the condition $n \geq d+1$, this case occurs with probability one (see Problem 5.4). Putting

$$\overline{X}_n := \frac{1}{n} \sum_{j=1}^{n} X_j, \qquad S_n := \frac{1}{n} \sum_{j=1}^{n} (X_j - \overline{X}_n)(X_j - \overline{X}_n)^{\top}$$

we can therefore generalize from (10.5) that $\widehat{\vartheta}_n := (\overline{X}_n, S_n)$ is the MLE of $\vartheta = (\mu, \Sigma)$.

Based on the Examples 10.3–10.5 and Problem 9.7, one can hope that MLE's have good properties under general conditions. We first turn to the question under which conditions a sequence $(\widehat{\vartheta}_n)$ of MLE's is strongly consistent. To avoid indices, let X be a random variable with the same distribution as X_1, depending on $\vartheta \in \Theta$, that is, the distribution Q_ϑ with the density $f(\cdot, \vartheta)$ with respect to ν. An important role in connection with ML estimation is played by the following notion.

10.7 Definition (Kullback[4]–Leibler[5]-Information)
For $\vartheta \in \Theta$ and $\vartheta' \in \Theta$,

$$I_{KL}(\vartheta : \vartheta') := \mathbb{E}_\vartheta \left(\log \frac{f(X, \vartheta)}{f(X, \vartheta')} \right) = \int_{X_0} \log \frac{f(x, \vartheta)}{f(x, \vartheta')} f(x, \vartheta) \, v(\mathrm{d}x) \qquad (10.9)$$

is called the *Kullback–Leibler information of* $f(\cdot, \vartheta)$ *with respect to* $f(\cdot, \vartheta')$ *(or of* Q_ϑ *with respect to* $Q_{\vartheta'}$*).*

In (10.9), we set $0 \log 0 := 0$ and

$$\log \frac{f(x, \vartheta)}{f(x, \vartheta')} := +\infty, \quad \text{if } f(x, \vartheta) > 0 \text{ and } f(x, \vartheta') = 0. \qquad (10.10)$$

Generally, if Q and Q' are probability measures on $\mathcal{B}_0$ with the property $Q \ll Q'$, then

$$I_{KL}(Q : Q') := \int_{X_0} \log \frac{\mathrm{d}Q}{\mathrm{d}Q'}(x) \, Q(\mathrm{d}x) \qquad (10.11)$$

is the *Kullback–Leibler information of* Q *with respect to* Q'. Here, $\frac{\mathrm{d}Q}{\mathrm{d}Q'}$ is a Radon–Nikodým density of Q with respect to Q'. If Q is not absolutely continuous with respect to Q', one sets $I_{KL}(Q : Q') := \infty$.

10.8 Example (Uniform Distribution)
Let $\Theta := (0, \infty)$ and $Q_\vartheta = U(0, \vartheta)$. If $\vartheta \le \vartheta'$ then

$$I_{KL}(\vartheta : \vartheta') = \int_{-\infty}^{\infty} \log \frac{\frac{1}{\vartheta} \mathbf{1}_{(0,\vartheta)}(x)}{\frac{1}{\vartheta'} \mathbf{1}_{(0,\vartheta')}(x)} \frac{1}{\vartheta} \mathbf{1}_{(0,\vartheta)}(x) \, \mathrm{d}x = \log \frac{\vartheta'}{\vartheta}.$$

If $\vartheta' < \vartheta$ then $I_{KL}(\vartheta : \vartheta') = +\infty$.

10.9 Example (Normal Distribution)
Let $\Theta = \mathbb{R}^d$ and

$$f(x, \vartheta) = \frac{1}{(2\pi)^{d/2}} \exp\left(-\frac{\|x - \vartheta\|^2}{2} \right), \quad x \in \mathbb{R}^d,$$

where $\| \cdot \|$ denotes the Euclidean norm. Since

$$\log \frac{f(x, \vartheta)}{f(x, \vartheta')} = -\frac{1}{2} \left(\|x - \vartheta\|^2 - \|x - \vartheta'\|^2 \right) = (\vartheta - \vartheta')^\top x + \frac{1}{2} \left(\|\vartheta'\|^2 - \|\vartheta\|^2 \right)$$

---

[4] Solomon Kullback (1907–1994), American mathematician and cryptologist.
[5] Richard Arthur Leibler (1914–2003), American mathematician and cryptologist.

with $X \sim \mathrm{N}_d(\vartheta, \mathrm{I}_d)$, we obtain

$$I_{KL}(\vartheta : \vartheta') = \mathbb{E}_\vartheta\left(\log \frac{f(X, \vartheta)}{f(X, \vartheta')}\right) = \mathbb{E}_\vartheta\left((\vartheta - \vartheta')^\top X + \frac{1}{2}\left(\|\vartheta'\|^2 - \|\vartheta\|^2\right)\right)$$

$$= (\vartheta - \vartheta')^\top \vartheta + \frac{1}{2}\left(\|\vartheta'\|^2 - \|\vartheta\|^2\right)$$

$$= \frac{1}{2}\|\vartheta - \vartheta'\|^2.$$

The following fact is important in connection with the Kullback–Leibler information.

10.10 Lemma
For the Kullback–Leibler information, we have

$$I_{KL}(\vartheta : \vartheta') = \mathbb{E}_\vartheta\left(\log \frac{f(X, \vartheta)}{f(X, \vartheta')}\right) \geq 0.$$

The equals sign holds if and only if $f(x, \vartheta) = f(x, \vartheta')$ for ν-almost all $x \in \mathcal{X}_0$, that is, if $Q_\vartheta = Q_{\vartheta'}$ and thus $\vartheta = \vartheta'$.

Proof W.l.o.g. let $I_{KL}(\vartheta : \vartheta') < \infty$. Thus, the case in (10.10) only occurs for values of x in a ν-null set. Since $\log t \geq 1 - \frac{1}{t}, t > 0$, with equality only if $t = 1$, it follows that

$$I_{KL}(\vartheta : \vartheta') \geq \mathbb{E}_\vartheta\left(1 - \frac{f(X, \vartheta')}{f(X, \vartheta)}\right) = 1 - \int_{\mathcal{X}_0} \frac{f(x, \vartheta')}{f(x, \vartheta)} f(x, \vartheta)\nu(\mathrm{d}x) = 1 - 1 = 0.$$

Here, the equals sign occurs if and only if $f(\cdot, \vartheta) = f(\cdot, \vartheta')$ ν-a.e. and thus if $\vartheta = \vartheta'$. □

The same reasoning remains valid in the more general case of the Kullback–Leibler information $I_{KL}(Q : Q')$ in (10.11). We have $I_{KL}(Q : Q') \geq 0$, with equality if only if $Q = Q'$.

In addition to the *identifiability condition* that the mapping $\Theta \ni \vartheta \mapsto Q_\vartheta$ is injective, that is, different parameter values correspond to different distributions, we now assume that the distributions Q_ϑ with $\vartheta \in \Theta$ have a common support. Specifically, we require:

The set $\{x \in \mathcal{X}_0 : f(x, \vartheta) > 0\}$ does not depend on $\vartheta \in \Theta$. $\qquad$ (10.12)

Furthermore, we make the assumption

$$\mathbb{E}_\vartheta|\log f(X, \vartheta')| < \infty, \quad \vartheta, \vartheta' \in \Theta. \qquad (10.13)$$

Here, X is a random variable with the distribution Q_ϑ, that is, with the density $f(\cdot, \vartheta)$ with respect to ν.

What role do the Kullback–Leibler information and Lemma 10.10 play in ML estimation? To this end, we fix any value ϑ_0 in Θ. ML estimation requires to maximize $\prod_{j=1}^n f(X_j, \vartheta)$ or, which is equivalent,

$$\frac{1}{n} \sum_{j=1}^n \left(\log(f(X_j, \vartheta) - \log f(X_j, \vartheta_0)) \right) = \frac{1}{n} \sum_{j=1}^n \log \frac{f(X_j, \vartheta)}{f(X_j, \vartheta_0)}$$

with respect to ϑ. By the strong law of large numbers, these means converge for each $\vartheta \in \Theta$ $\mathbb{P}_{\vartheta_0}$-almost surely to

$$\mathbb{E}_{\vartheta_0} \left(\log \frac{f(X, \vartheta)}{f(X, \vartheta_0)} \right) = -I_{KL}(\vartheta_0 : \vartheta). \tag{10.14}$$

Self-question 3: Why is the strong law of large numbers applicable here?

By Lemma 10.10, this value is less than 0 for each ϑ different from ϑ_0. If the parameter space Θ is a *finite* set, no matter how large, then the strong consistency of the sequence of MLE's follows from the above considerations, that is, we have

$$\mathbb{P}_{\vartheta_0} \left(\{ \mathbf{x} \in \mathcal{X} : \lim_{n \to \infty} \widehat{\vartheta}_n(\mathbf{x}) = \vartheta_0 \} \right) = 1, \quad \vartheta_0 \in \Theta. \tag{10.15}$$

To provide the short proof, for each $\vartheta \in \Theta$ with $\vartheta \neq \vartheta_0$ we set

$$B_\vartheta := \left\{ \mathbf{x} = (x_1, x_2, \dots) \in \mathcal{X} : \lim_{n \to \infty} \frac{1}{n} \sum_{j=1}^n \log \frac{f(x_j, \vartheta)}{f(x_j, \vartheta_0)} = -I_{KL}(\vartheta_0 : \vartheta) \right\}. \tag{10.16}$$

In view of the above reasoning, $\mathbb{P}_{\vartheta_0}(B_\vartheta) = 1$. Putting $B := \cap_{\vartheta \in \Theta \setminus \{\vartheta_0\}} B_\vartheta$ it follows that $\mathbb{P}_{\vartheta_0}(B) = 1$, and $\mathbf{x} \in B$ implies $\lim_{n \to \infty} \widehat{\vartheta}_n(\mathbf{x}) = \vartheta_0$ (Problem 10.9).

There are various results regarding the strong consistency (10.15) of the sequence $(\widehat{\vartheta}_n)$ (see, e.g., [WAL], [FER], Theorem 17, as well as [PF1], Section 6.5, and [LIM], Section 7.5). In specific cases, (10.15) can often be verified directly.

In reference to [FER], Theorem 17, and [WAL], we state and prove a fairly general result, which requires a weak condition on the density $f : \mathcal{X}_0 \times \Theta \to \mathbb{R}$, but presupposes that the parameter space Θ is a *compact* subset of $\mathbb{R}^k$. We assume that for each x in $\mathcal{X}_0$ the function $\Theta \ni \vartheta \mapsto f(x, \vartheta)$ is *upper semi-continuous*. This property is weaker than continuity, as it only requires that for each $\vartheta \in \Theta$ and each sequence (ϑ_n) in Θ converging to ϑ, we have $\limsup_{n \to \infty} f(x, \vartheta_n) \leq f(x, \vartheta)$. In other words, the values $f(x, \vartheta')$ of the function $f(x, \cdot)$ for arguments ϑ' close to ϑ are not allowed to "jump upwards".

Equivalently, for each $x \in \mathcal{X}_0$ and each $t \in \mathbb{R}$ the set $\{\vartheta \in \Theta : f(x, \vartheta) < t\}$ is *open* in Θ (that is, with respect to the Euclidean relative topology on Θ). Using this characterization of upper semi-continuity—which by the way serves as the *definition* of upper semi-continuity for functions on general topological spaces—one quickly recognizes that for each $x \in \mathcal{X}_0$ the maximum of the values $f(x, \vartheta)$, $\vartheta \in \Theta$, is attained when the set Θ is compact, that is, when every open cover of Θ has a finite subcover. In order to understand this assertion, set for fixed $x \in \mathcal{X}_0$

$$t_0 := \sup\{f(x, \vartheta) : \vartheta \in \Theta\}.$$

We will show that the assumption that $f(x, \vartheta) < t_0$ for each $\vartheta \in \Theta$, which implies that the above supremum is not attained, will lead to a contradiction. To this end, let (t_n) be any sequence of real numbers with $t_n < t_{n+1}$ for each $n \geq 1$ and $\lim_{n \to \infty} t_n = t_0$. Then $\{O_n : n \geq 1\}$ with $O_n := \{\vartheta \in \Theta : f(x, \vartheta) < t_n\}$ is an open cover of Θ consisting of increasing sets. Since Θ is compact, there is an n with $\Theta = \{\vartheta \in \Theta : f(x, \vartheta) < t_n\}$. The inequality $t_n < t_0$, however, contradicts the definition of t_0, since t_0 is the supremum of the set $\{f(x, \vartheta) : \vartheta \in \Theta\}$.

In the following, we fix any ϑ_0 in Θ and write briefly

$$V(x, \vartheta) := \log \frac{f(x, \vartheta)}{f(x, \vartheta_0)}, \qquad x \in \mathcal{X}_0, \; \vartheta \in \Theta. \tag{10.17}$$

Furthermore, we set

$$v(\vartheta) := \mathbb{E}_{\vartheta_0}[V(X, \vartheta)] = \int_{\mathcal{X}_0} \log \frac{f(x, \vartheta)}{f(x, \vartheta_0)} f(x, \vartheta_0)\, v(dx), \qquad \vartheta \in \Theta. \tag{10.18}$$

Thus, $v(\vartheta) = -I_{KL}(\vartheta_0 : \vartheta)$, cf. (10.9). In particular, it follows that $v(\vartheta_0) = 0$ and $v(\vartheta) < 0$ for every $\vartheta \in \Theta \setminus \{\vartheta_0\}$.

10.11 Theorem (Strong Consistency of ML Estimators)
Let $X_1, X_2, \ldots$ be a sequence of i.i.d. $\mathcal{X}_0$-valued random variables with density $f(\cdot, \vartheta)$, $\vartheta \in \Theta$, with respect to a σ-finite measure ν on $\mathcal{B}_0$. Suppose that

(a) the parameter space Θ is a compact subset of $\mathbb{R}^k$,
(b) for every $x \in \mathcal{X}_0$, the function $\Theta \ni \vartheta \mapsto f(x, \vartheta)$ is upper semi-continuous,
(c) there exists a measurable function $K : \mathcal{X}_0 \to \mathbb{R}$ with $\mathbb{E}_{\vartheta_0}|K(X)| < \infty$ and $V(x, \vartheta) \leq K(x)$, $x \in \mathcal{X}_0$, $\vartheta \in \Theta$, where $V(x, \vartheta)$ is defined in (10.17),
(d) for every $\vartheta \in \Theta$ and every sufficiently small $r > 0$, the function $\overline{f}_{\vartheta,r} : \mathcal{X}_0 \to [0, \infty]$, defined by

$$\overline{f}_{\vartheta,r}(x) := \sup \{f(x, \vartheta') : \vartheta' \in \Theta, \; \|\vartheta - \vartheta'\| < r\}, \tag{10.19}$$

is measurable and integrable with respect to $f(\cdot, \vartheta_0)\nu$.

Then for every sequence $(\widehat{\vartheta}_n)$ *of MLE's of* ϑ *we have* (10.15), *that is,*

$$\lim_{n\to\infty} \widehat{\vartheta}_n = \vartheta_0 \quad \mathbb{P}_{\vartheta_0}\text{-almost surely.}$$

Proof For arbitrary $\vartheta_0 \in \Theta$ and $r > 0$, let $S := \{\vartheta \in \Theta : \|\vartheta - \vartheta_0\| \geq r\}$. As a closed subset of the compact set Θ, S is compact. From condition (c) and Problem 1.8, it follows that the function v defined in (10.18) is upper semi-continuous. If we could show

$$\mathbb{P}_{\vartheta_0}\left(\limsup_{n\to\infty} \sup_{\vartheta\in S} \frac{1}{n} \sum_{j=1}^n V(X_j, \vartheta) \leq \sup_{\vartheta\in S} v(\vartheta) \right) = 1, \tag{10.20}$$

the proof would be quickly completed: As a semi-continuous function, v attains a (strictly) negative maximum on S, denoted by δ, and so we could replace $\sup_{\vartheta\in S} v(\vartheta)$ with δ in (10.20). But then

$$\mathbb{P}_{\vartheta_0}\left(\left\{ \mathbf{x} = (x_1, x_2, \dots) \in \mathcal{X} : \exists n_0 \text{ with } \sup_{\vartheta\in S} \frac{1}{n} \sum_{j=1}^n V(x_j, \vartheta) \leq \frac{\delta}{2} < 0 \,\forall n \geq n_0 \right\} \right) = 1.$$

On the other hand, by definition of the MLE it follows that

$$\frac{1}{n} \sum_{j=1}^n V\big(x_j, \widehat{\vartheta}_n(\mathbf{x})\big) = \sup_{\vartheta\in\Theta} \frac{1}{n} \sum_{j=1}^n V(x_j, \vartheta) \geq 0,$$

since the second sum attains the value 0 if $\vartheta = \vartheta_0$. For each $n \geq n_0$, we then have $\widehat{\vartheta}_n(\mathbf{x}) \notin S$ and thus $\|\widehat{\vartheta}_n(\mathbf{x}) - \vartheta_0\| < r$. Since $r > 0$ was arbitrary, the claim follows if the proof of (10.20) is finished.

To provide this proof, set

$$h(x, \vartheta, r) := \sup \big\{ V(x, \vartheta') : \vartheta' \in \Theta, \ \|\vartheta - \vartheta'\| < r \big\},$$

where $x \in \mathcal{X}_0$, $\vartheta \in S$, and $r > 0$. Due to condition (d) and $h(x, \vartheta, r) = \log \overline{f}_{\vartheta, r}(x) - \log f(x, \vartheta_0)$, $h(\cdot, \vartheta, r)$ is a measurable function for each $\vartheta \in S$ and each sufficiently small r. According to (c), $h(x, \vartheta, r) \leq K(x)$ ($x \in \mathcal{X}_0$, $\vartheta \in S$, $r > 0$) with a function K that is integrable with respect to Q_{ϑ_0} ($= f(\cdot, \vartheta_0)\nu$), and due to condition (b), $h(x, \vartheta, r) \downarrow V(x, \vartheta)$ as $r \downarrow 0$. The monotone convergence theorem (Theorem 1.30), applied to $K(\cdot) - h(\cdot, \vartheta, r)$ as $r \downarrow 0$, yields

$$\lim_{r\downarrow 0} \int_{\mathcal{X}_0} h(x, \vartheta, r)\, f(x, \vartheta_0)\, \nu(\mathrm{d}x) = \int_{\mathcal{X}_0} V(x, \vartheta)\, f(x, \vartheta_0)\, \nu(\mathrm{d}x) = v(\vartheta).$$

We now use condition (a) and thus again the compactness of S and fix any $\varepsilon > 0$. For each $\vartheta \in S$ there is an $r(\vartheta) > 0$ with

$$\mathbb{E}_{\vartheta_0}\big[h(X, \vartheta, r(\vartheta))\big] = \int_{\mathcal{X}_0} h(x, \vartheta, r(\vartheta)) f(x, \vartheta_0)\, v(dx) < v(\vartheta) + \varepsilon. \qquad (10.21)$$

The sets $S(\vartheta, r(\vartheta)) := \{\vartheta' \in \Theta : \|\vartheta - \vartheta'\| < r(\vartheta)\}$, $\vartheta \in S$, are an open cover of S. Since S is compact, there is an integer m and $\vartheta_1, \dots, \vartheta_m \in S$ as well as positive numbers $r(\vartheta_1), \dots, r(\vartheta_m)$ with $S \subset \bigcup_{i=1}^m S(\vartheta_i, r(\vartheta_i))$. By definition of h, $\vartheta \in S(\vartheta_i, r(\vartheta_i))$ implies $V(x, \vartheta) \le h(x, \vartheta_i, r(\vartheta_i))$, $x \in \mathcal{X}_0$, and it follows that

$$\sup_{\vartheta \in S} \frac{1}{n} \sum_{j=1}^n V(X_j, \vartheta) \le \max_{i=1,\dots,m} \frac{1}{n} \sum_{j=1}^n h\big(X_j, \vartheta_i, r(\vartheta_i)\big).$$

By the strong law of large numbers (Theorem 1.2), for every $i \in \{1, \dots, m\}$

$$\lim_{n \to \infty} \frac{1}{n} \sum_{j=1}^n h(X_j, \vartheta_i, r(\vartheta_i)) = \mathbb{E}_{\vartheta_0}\big[h(X, \vartheta_i, r(\vartheta_i))\big] \quad \mathbb{P}_{\vartheta_0}\text{-almost surely.}$$

Since the intersection of finitely many events of probability 1 has probability 1, (10.21) yields

$$\mathbb{P}_{\vartheta_0}\left(\limsup_{n \to \infty} \sup_{\vartheta \in S} \frac{1}{n} \sum_{j=1}^n V(X_j, \vartheta) \le \max_{i=1,\dots,m} v(\vartheta_i) + \varepsilon \right) = 1.$$

Since $\max_{i=1,\dots,m} v(\vartheta_i) \le \sup_{\vartheta \in S} v(\vartheta)$, (10.20) now follows, because $\varepsilon > 0$ was arbitrary. $\qquad \square$

We now turn to the so-called main theorem of ML estimation. This main result states, loosely speaking, that *under certain conditions* the sequence $(\sqrt{n}(\widehat{\vartheta}_n - \vartheta))$ is asymptotically normally distributed. We assume that the parameter space $\Theta \subset \mathbb{R}^k$ contains interior points, that is, $\Theta^\circ \ne \emptyset$. The so-called *regularity conditions* refer in particular to smoothness properties of the v-densities $f(x, \vartheta)$ for fixed x with respect to $\vartheta = (\vartheta_1, \dots, \vartheta_k)$. Note that, for the rest of this chapter and in contrast to the notation used in the above proof, $\vartheta_1, \dots, \vartheta_k$ are not elements of Θ, but components of $\vartheta \in \Theta$.

10.12 Regularity Conditions

(a) For each $x \in \mathcal{X}_0$ the partial derivatives $\frac{\partial^2}{\partial \vartheta_i \partial \vartheta_j} f(x, \vartheta)$, $i, j \in \{1, \dots, k\}$, exist, and they are continuous functions on Θ°.

(b) For each $\vartheta \in \Theta^\circ$ and each $i \in \{1, \dots, k\}$, we have

$$0 = \mathbb{E}_\vartheta \left[\frac{\partial}{\partial \vartheta_i} \log f(X, \vartheta) \right] \qquad \left(= \mathbb{E}_\vartheta \left[\frac{\frac{\partial}{\partial \vartheta_i} f(X, \vartheta)}{f(X, \vartheta)} \right] \right).$$

(c) For each $\vartheta \in \Theta^\circ$ and any choice of $i, j \in \{1, \dots, k\}$, we have

$$0 = \mathbb{E}_\vartheta \left[\frac{1}{f(X, \vartheta)} \cdot \frac{\partial^2 f(X, \vartheta)}{\partial \vartheta_i \partial \vartheta_j} \right].$$

(d) For each $\vartheta \in \Theta^\circ$ there exist a $\delta_\vartheta > 0$ with $U(\vartheta, \delta_\vartheta) := \{y \in \mathbb{R}^k : \|y - \vartheta\| < \delta_\vartheta\} \subset \Theta^\circ$ and a non-negative measurable function $M(\cdot, \vartheta)$ on $\mathcal{X}_0$ with $\mathbb{E}_\vartheta M(X, \vartheta) < \infty$ and

$$\left| \frac{\partial^2}{\partial \vartheta_i \partial \vartheta_j} \log f(\cdot, \vartheta') \right| \leq M(\cdot, \vartheta) \; \forall \vartheta' \in U(\vartheta, \delta_\vartheta) \; \forall i, j \in \{1, \dots, k\}.$$

(e) For each $\vartheta \in \Theta^\circ$ the so-called *Fisher information matrix*

$$I_1(\vartheta) := \left(\mathbb{E}_\vartheta \left[\frac{\partial}{\partial \vartheta_i} \log f(X, \vartheta) \frac{\partial}{\partial \vartheta_j} \log f(X, \vartheta) \right] \right)_{1 \leq i, j \leq k}$$

is invertible.

By assumption (b), we may differentiate under the integral, that is, we have

$$\frac{\partial}{\partial \vartheta_i} \int_{\mathcal{X}_0} f(x, \vartheta) \, \nu(dx) = \frac{\partial}{\partial \vartheta_i} 1 = 0$$

and

$$\int_{\mathcal{X}_0} \frac{\partial}{\partial \vartheta_i} f(x, \vartheta) \, \nu(dx) = \int_{\mathcal{X}_0} \frac{\frac{\partial}{\partial \vartheta_i} f(x, \vartheta)}{f(x, \vartheta)} f(x, \vartheta) \, \nu(dx)$$

$$= \int_{\mathcal{X}_0} \frac{\partial}{\partial \vartheta_i} \log f(x, \vartheta) f(x, \vartheta) \, \nu(dx) = \mathbb{E}_\vartheta \left[\frac{\partial}{\partial \vartheta_i} \log f(X, \vartheta) \right].$$

Self-question 4: Why does condition (c) also allow differentiation under the integral?

In the case $k = 1$, the Fisher information matrix $I_1(\vartheta)$ is the scalar quantity

$$I_1(\vartheta) = \mathbb{E}_\vartheta \left(\frac{\partial}{\partial \vartheta} \log f(X, \vartheta) \right)^2,$$

which is briefly referred to as *Fisher information*. Since $\frac{\partial}{\partial \vartheta} \log f(x, \vartheta)$ represents the rate of change of the logarithm of $f(x, \cdot)$ with respect to ϑ, $I_1(\vartheta)$ describes the integrated square rate of change of the logarithmic density at the point ϑ. The larger $I_1(\vartheta)$ is, the faster the underlying distribution changes depending on ϑ, and one can hope to obtain better estimates of the unknown parameter ϑ.

Generally, we write $\frac{d}{d\vartheta} := \left(\frac{\partial}{\partial \vartheta_1}, \ldots, \frac{\partial}{\partial \vartheta_k} \right)^\top$ for the column vector of partial derivatives with respect to the components of ϑ. The random vector

$$U_1(\vartheta) := \frac{d}{d\vartheta} \log f(X, \vartheta)$$

is called the *score vector of X*. Due to condition 10.12 (b), the score vector is centered, that is, we have $\mathbb{E}_\vartheta[U_1(\vartheta)] = 0_k$. Thus, the Fisher information matrix is the covariance matrix of the score vector. It follows that

$$\Sigma_\vartheta\big(U_1(\vartheta)\big) := \mathbb{E}_\vartheta\big[U_1(\vartheta)U_1(\vartheta)^\top\big] = I_1(\vartheta). \tag{10.22}$$

If $X_1, \ldots, X_n$ are independent with the same density $f(\cdot, \vartheta)$ with respect to ν, then

$$U_n(\vartheta) := \frac{d}{d\vartheta} \log f_n(X_1, \ldots, X_n, \vartheta) \tag{10.23}$$

is the *score vector of $(X_1, \ldots, X_n)$*. This random vector is also centered, and its covariance matrix is given by $nI_1(\vartheta)$ (*additivity of Fisher information*, see Problem 10.11).

10.13 Example (Normal Distribution, Continuation of Example 10.5)
Let $\vartheta = (\vartheta_1, \vartheta_2) =: (\mu, \sigma^2)$, $\Theta = \mathbb{R} \times \mathbb{R}_{>0}$, and

$$f(x, \vartheta) = \frac{1}{\sigma\sqrt{2\pi}} \exp\left(-\frac{(x - \mu)^2}{2\sigma^2} \right), \quad x \in \mathbb{R}.$$

Because of

$$\log f(x, \vartheta) = -\frac{1}{2}\log(2\pi) - \frac{1}{2}\log\sigma^2 - \frac{(x - \mu)^2}{2\sigma^2},$$

putting $Y := \frac{X-\mu}{\sigma}$ yields

$$\frac{\partial}{\partial \mu} \log f(X, \vartheta) = \frac{Y}{\sigma}, \qquad \frac{\partial}{\partial \sigma^2} \log f(X, \vartheta) = \frac{1}{2\sigma^2} \left(Y^2 - 1 \right).$$

Since Y has a standard normal distribution, it follows that $\mathbb{E}(Y^2) = 1$, $\mathbb{E}(Y^3) = 0$, and $\mathbb{E}(Y^4) = 3$, and we obtain

$$\mathbb{E}_\vartheta \left(\frac{\partial}{\partial \mu} \log f(X, \vartheta) \right)^2 = \frac{1}{\sigma^2}, \qquad \mathbb{E}_\vartheta \left(\frac{\partial}{\partial \sigma^2} \log f(X, \vartheta) \right)^2 = \frac{1}{2\sigma^4},$$

as well as $\mathbb{E}_\vartheta \left[\frac{\partial}{\partial \mu} \log f(X, \vartheta) \cdot \frac{\partial}{\partial \sigma^2} \log f(X, \vartheta) \right] = 0$. The Fisher information matrix thus takes the form

$$I_1(\vartheta) = \begin{pmatrix} \frac{1}{\sigma^2} & 0 \\ 0 & \frac{1}{2\sigma^4} \end{pmatrix}.$$

As we will see shortly, it is no coincidence that this matrix is the inverse of the covariance matrix of the limit distribution of $\sqrt{n}(\widehat{\vartheta}_n - \vartheta)$ in (10.6).

Before we state the main theorem of ML estimation, we emphasize a technical difficulty. As already pointed out after Definition 10.2, the supremum in (10.2) need not be attained. On the other hand, as the sample size n increases, the (not necessarily measurable!) subsets of all $\mathbf{x} = (x_1, x_2, \ldots)$ in the sample space $\mathcal{X}$, for which the supremum in (10.2) is attained, typically are supersets of measurable sets whose probabilities converge to 1 as $n \to \infty$ for any $\vartheta \in \Theta$. Thus, the following modification of the definition of a MLE suggests itself with regard to asymptotic investigations (cf. [WIM], p. 177).

10.14 Definition (Asymptotic Maximum Likelihood Estimator)

Let $X_1, X_2, \ldots$ be a sequence of i.i.d. random variables with ν-density $f(\cdot, \vartheta)$, where $\vartheta \in \Theta \subset \mathbb{R}^k$. Moreover, let

$$M_n := \bigcup_{\vartheta \in \Theta} \left\{ \mathbf{x} \in \mathcal{X} : f_n(\mathbf{x}, \vartheta) = \sup_{t \in \Theta} f_n(\mathbf{x}, t) \right\}$$

be the set of those $\mathbf{x} = (x_1, x_2, \ldots)$ in $\mathcal{X}$ for which an ML estimate exists for the sample size n. If there are sets $M_n' \subset M_n$ with $M_n' \in \mathcal{B}$ and $\mathbb{P}_\vartheta(M_n') \to 1$ for each $\vartheta \in \Theta$, we call any sequence $(\widehat{\vartheta}_n)$ of measurable mappings $\widehat{\vartheta}_n : \mathcal{X} \to \Theta$ with

$$f_n\big(\mathbf{x}, \widehat{\vartheta}_n(\mathbf{x})\big) = \sup_{t \in \Theta} f_n(\mathbf{x}, t) \qquad \text{for each } \mathbf{x} \in M_n' \tag{10.24}$$

an *asymptotic ML estimator*.

10.15 Theorem (Main Theorem on Maximum Likelihood Estimation)

Let $(\widehat{\vartheta}_n)$ be an asymptotic MLE of ϑ, and suppose that the regularity conditions 10.12 (a)–(e) hold. If the sequence $(\widehat{\vartheta}_n)$ is consistent for ϑ, then

$$\sqrt{n}(\widehat{\vartheta}_n - \vartheta) \xrightarrow{\mathcal{D}_\vartheta} N_k\left(0, I_1(\vartheta)^{-1}\right) \qquad \text{for every } \vartheta \in \Theta^\circ.$$

Proof The reasoning follows [WIM], pp. 203–205, without going into every technical detail. Let $\vartheta \in \Theta^\circ$ be arbitrary and $M'_n \subset \mathcal{X}$ as in Definition 10.14. Furthermore, let $U(\vartheta, \delta_\vartheta) \subset \Theta^\circ$ be as in 10.12 (d). We set

$$V_n := \left\{\mathbf{x} \in \mathcal{X} : \widehat{\vartheta}_n(\mathbf{x}) \in U(\vartheta, \delta_\vartheta)\right\}.$$

Since the sequence $(\widehat{\vartheta}_n)$ is assumed to be consistent, $\lim_{n\to\infty} \mathbb{P}_\vartheta(V_n) = 1$. Let

$$\overline{\vartheta}_n(\mathbf{x}) := \widehat{\vartheta}_n(\mathbf{x})\mathbf{1}_{\{M'_n \cap V_n\}}(\mathbf{x}) + \vartheta \mathbf{1}_{\{(M'_n \cap V_n)^c\}}(\mathbf{x}), \quad \mathbf{x} \in \mathcal{X}.$$

Thus, $\overline{\vartheta}_n \xrightarrow{\mathbb{P}_\vartheta} \vartheta$, and because $\mathbb{P}_\vartheta(\overline{\vartheta}_n \neq \widehat{\vartheta}_n) \to 0$ it follows that $\sqrt{n}(\overline{\vartheta}_n - \widehat{\vartheta}_n) = o_{\mathbb{P}_\vartheta}(1)$.

Self-question 5: Why do $\mathbb{P}_\vartheta(\overline{\vartheta}_n \neq \widehat{\vartheta}_n) \to 0$ and $\sqrt{n}(\overline{\vartheta}_n - \widehat{\vartheta}_n) = o_{\mathbb{P}_\vartheta}(1)$ hold?

By Slutsky's lemma, we thus have to show

$$\sqrt{n}(\overline{\vartheta}_n - \vartheta) \xrightarrow{\mathcal{D}_\vartheta} N_k\left(0, I_1(\vartheta)^{-1}\right) \qquad \text{for every } \vartheta \in \Theta^\circ.$$

Let

$$U_n(t) := \sum_{j=1}^{n} \frac{\mathrm{d}}{\mathrm{d}\vartheta} \log f(X_j, \vartheta)\Big|_{\vartheta = t}, \qquad t \in \Theta^\circ. \tag{10.25}$$

For further reasoning, it is crucial to note that, on the intersection $M'_n \cap V_n$, $\overline{\vartheta}_n$ satisfies the likelihood equation $0_k = U_n(\overline{\vartheta}_n)$. The idea now is to perform a Taylor expansion of $U_n(t)$ around $t = \vartheta$. For this purpose, we set

$$H_n(t) := \frac{\mathrm{d}}{\mathrm{d}\vartheta^\top} U_n(\vartheta)\Big|_{\vartheta = t} = \left(\sum_{\ell=1}^{n} \frac{\partial^2}{\partial\vartheta_i \partial\vartheta_j} \log f(X_\ell, \vartheta)\Big|_{\vartheta = t}\right)_{1 \le i, j \le k}, \tag{10.26}$$

so

$$0_k = U_n(\overline{\vartheta}_n) = U_n(\vartheta) + H_n(\vartheta)(\overline{\vartheta}_n - \vartheta) + R_n(\vartheta, \overline{\vartheta}_n - \vartheta)$$

with a remainder term $R_n(\vartheta, \overline{\vartheta}_n - \vartheta)$ that is not specified further. If we divide both sides by $\sqrt{n}$ we get

$$0_k = \frac{1}{\sqrt{n}} U_n(\vartheta) + \frac{1}{n} H_n(\vartheta) \cdot \sqrt{n}(\overline{\vartheta}_n - \vartheta) + \frac{1}{\sqrt{n}} R_n(\vartheta, \overline{\vartheta}_n - \vartheta).$$

The regularity conditions (in particular 10.12 (d)) show that the last term on the right-hand side converges to 0 in probability, and it follows that

$$\frac{1}{n} H_n(\vartheta) \cdot \sqrt{n}(\overline{\vartheta}_n - \vartheta) = -\frac{1}{\sqrt{n}} U_n(\vartheta) + o_{\mathbb{P}_\vartheta}(1). \tag{10.27}$$

Since the score vector $U_n(\vartheta)$ of $(X_1, \ldots, X_n)$ (cf. (10.23)) is a sum of i.i.d. centered random vectors with covariance matrix $I_1(\vartheta)$ (cf. (10.22)), the multivariate central limit theorem, the mapping theorem and Slutsky's lemma yield

$$-\frac{1}{\sqrt{n}} U_n(\vartheta) + o_{\mathbb{P}_\vartheta}(1) \xrightarrow{\mathcal{D}_\vartheta} N_k(0, I_1(\vartheta)). \tag{10.28}$$

Self-question 6: Why is the mapping theorem needed?

We now claim that

$$\lim_{n \to \infty} \frac{1}{n} H_n(\vartheta) = -I_1(\vartheta) \quad \mathbb{P}_\vartheta\text{-almost surely}. \tag{10.29}$$

According to the quotient rule, for each $\ell \in \{1, \ldots, n\}$ and any choice of $i, j \in \{1, \ldots, k\}$

$$\frac{\partial}{\partial \vartheta_j}\left[\frac{\partial}{\partial \vartheta_i} \log f(X_\ell, \vartheta)\right] = \frac{\partial}{\partial \vartheta_j} \frac{\frac{\partial}{\partial \vartheta_i} f(X_\ell, \vartheta)}{f(X_\ell, \vartheta)}$$

$$= \frac{1}{f(X_\ell, \vartheta)} \frac{\partial^2 f(X_\ell, \vartheta)}{\partial \vartheta_i \partial \vartheta_j} - \frac{\partial}{\partial \vartheta_i} \log f(X_\ell, \vartheta) \frac{\partial}{\partial \vartheta_j} \log f(X_\ell, \vartheta).$$

Due to regularity condition 10.12 (c), the expectation of the first term on the right-hand side vanishes under $\mathbb{P}_\vartheta$, and the corresponding expectation of the subtrahend is the entry in the i-th row and j-th column of the Fisher information matrix $I_1(\vartheta)$. Thus, $\mathbb{E}_\vartheta[H_n(\vartheta)] = -nI_1(\vartheta)$, and (10.29) follows from the strong law of large numbers, since $n^{-1} H_n(\vartheta)$ is a mean of a sequence of random matrices with the same expectation $-I_1(\vartheta)$. Since $I_1(\vartheta)$ is invertible, $n^{-1}H_n(\vartheta)$ is also invertible with probability 1 for sufficiently large n. For such n, (10.27) implies

$$\sqrt{n}(\widehat{\vartheta}_n - \vartheta) = \left(\frac{1}{n}H_n(\vartheta)\right)^{-1}\left(-\frac{1}{\sqrt{n}}U_n(\vartheta) + o_{\mathbb{P}_\vartheta}(1)\right).$$

Since $(n^{-1}H_n(\vartheta))^{-1}$ converges $\mathbb{P}_\vartheta$-almost surely to $I_1(\vartheta)^{-1}$, (10.28) together with the mapping theorem and Slutsky's lemma entail

$$\sqrt{n}(\widehat{\vartheta}_n - \vartheta) \xrightarrow{\mathcal{D}_\vartheta} I_1(\vartheta)^{-1}N_k\big((0_k, I_1(\vartheta))\big) \overset{\mathcal{D}}{=} N_k\big(0_k, I_1(\vartheta)^{-1}\big),$$

which was to be shown.   □

10.16 Corollary (Representation of the Estimation Error)
Under the standing assumption, we have

$$\sqrt{n}(\widehat{\vartheta}_n - \vartheta) = \frac{1}{\sqrt{n}}\sum_{j=1}^{n}\ell(X_j, \vartheta) + o_{\mathbb{P}_\vartheta}(1) \quad \text{as } n \to \infty, \tag{10.30}$$

where

$$\ell(X_j, \vartheta) = I_1(\vartheta)^{-1}\frac{d}{d\vartheta}\log f(X_j, \vartheta).$$

Proof From the proof of Theorem 10.15, it follows that

$$\sqrt{n}(\widehat{\vartheta}_n - \vartheta) = I_1(\vartheta)^{-1}\frac{1}{\sqrt{n}}U_n(\vartheta) + o_{\mathbb{P}_\vartheta}(1).$$

□

The function $\ell(\cdot, \cdot)$ in (10.30) satisfies the equations $\mathbb{E}_\vartheta(\ell(X_1, \vartheta)) = 0$, $\vartheta \in \Theta$. A representation of the form (10.30) for the estimation error $\widehat{\vartheta}_n - \vartheta$ is valid not only for the MLE $\widehat{\vartheta}_n$, but also for other methods of estimation, such as the method of moments, as we will see in Chap. 11 (cf. Problem 11.7). The function $\ell(\cdot, \vartheta) : \mathcal{X}_0 \to \mathbb{R}$ appearing in (10.30) is called the *influence function*.

Note that the proof of the main theorem only made use of the consistency of the sequence $(\widehat{\vartheta}_n)$ and the fact that the likelihood equations (10.3) are satisfied. The message of the main theorem is that *under certain regularity conditions* the estimation error $\widehat{\vartheta}_n - \vartheta$, multiplied by $\sqrt{n}$, is asymptotically normally distributed. In the next chapter we will see that the asymptotic covariance matrix $I_1(\vartheta)^{-1}$ is in a certain sense optimal, again under certain conditions.

From the multitude of refinements of Theorem 10.15 and Corollary 10.16, we mention a version with weaker conditions or local uniformity in the parameter in [PF1], Theorem 7.5.5, an error bound of the Berry–Esseen type in [PIN] as well as a refinement of the normal approximation by an asymptotic expansion in [PFW], Proposition 10.3.1.

Problems 10.4 and 10.5 show that, when dealing with ML estimation, there are situations in which the estimation error $\widehat{\vartheta}_n - \vartheta$, *after multiplication with n* (instead of with $\sqrt{n}$), has a non-degenerate limit distribution. These and other cases are not covered by the main theorem because at least one of the regularity conditions 10.12 (a)–(e) is violated.

10.17 Example Let $\Theta := \mathbb{R}_{>0} \times \mathbb{R}_{>0}$, and for $\vartheta =: (\vartheta_1, \vartheta_2)$ let $X_1, X_2, \ldots$ be an i.i.d. sequence of random variables with the same Lebesgue density

$$f(x, \vartheta) := \frac{1}{\vartheta_1 + \vartheta_2} \exp\left(-\frac{x}{\vartheta_1} \mathbf{1}_{[0,\infty)}(x) + \frac{x}{\vartheta_2} \mathbf{1}_{(-\infty,0)}(x)\right), \quad x \in \mathbb{R}.$$

Direct calculation shows that the regularity conditions 10.12 (a)–(d) hold for f, and the Fisher information matrix is given by

$$I_1(\vartheta) = \frac{1}{(\vartheta_1 + \vartheta_2)^2} \begin{pmatrix} \frac{\vartheta_1 + 2\vartheta_2}{\vartheta_1} & -1 \\ -1 & \frac{2\vartheta_1 + \vartheta_2}{\vartheta_2} \end{pmatrix} \tag{10.31}$$

(Problem 10.13 (a)) and is thus invertible. With

$$s_n(\mathbf{x}) := \sum_{j=1}^{n} x_j \mathbf{1}_{[0,\infty)}(x_j), \qquad t_n(\mathbf{x}) := -\sum_{j=1}^{n} x_j \mathbf{1}_{(-\infty,0)}(x_j)$$

for $\mathbf{x} = (x_1, x_2, \ldots) \in \mathcal{X} := \mathbb{R}^{\mathbb{N}}$, the log-likelihood function is

$$\log L_{n,\mathbf{x}}(\vartheta) = -n \log(\vartheta_1 + \vartheta_2) - \frac{1}{\vartheta_1} s_n(\mathbf{x}) - \frac{1}{\vartheta_2} t_n(\mathbf{x}).$$

For the case $s_n := s_n(\mathbf{x}) > 0$ and $t_n := t_n(\mathbf{x}) > 0$, $\widehat{\vartheta}_n(\mathbf{x}) =: (\widehat{\vartheta}_{n,1}(\mathbf{x}), \widehat{\vartheta}_{n,2}(\mathbf{x}))$ with

$$\widehat{\vartheta}_{n,1}(\mathbf{x}) := \frac{\sqrt{s_n}}{n}\left(\sqrt{s_n} + \sqrt{t_n}\right), \qquad \widehat{\vartheta}_{n,2}(\mathbf{x}) := \frac{\sqrt{t_n}}{n}\left(\sqrt{s_n} + \sqrt{t_n}\right) \tag{10.32}$$

is the only solution of the likelihood equations (Problem 10.13 (b)). If $s_n(\mathbf{x}) > 0$ and $t_n(\mathbf{x}) = 0$ or $s_n(\mathbf{x}) = 0$ and $t_n(\mathbf{x}) > 0$ (the case $s_n(\mathbf{x}) = t_n(\mathbf{x}) = 0$ is not possible), then $\log L_{n,\mathbf{x}}(\vartheta)$ is also maximized for $\widehat{\vartheta}_n(\mathbf{x})$ in (10.32).

Self-question 7: Why does the last statement hold?

Put $S_n := \sum_{j=1}^n X_j \mathbf{1}_{[0,\infty)}(X_j)$ and $T_n := -\sum_{j=1}^n X_j \mathbf{1}_{(-\infty,0)}(X_j)$. Since $\lim_{n\to\infty} \mathbb{P}_\vartheta(S_n = 0$ or $T_n = 0) = 0$ for each $\vartheta \in \Theta$, and since the sequence of estimators $(\widehat{\vartheta}_n) =: ((\widehat{\vartheta}_{n,1}, \widehat{\vartheta}_{n,2}))$ with

$$\widehat{\vartheta}_{n,1} := \frac{\sqrt{S_n}}{n}\left(\sqrt{S_n}+\sqrt{T_n}\right), \quad \widehat{\vartheta}_{n,2} := \frac{\sqrt{T_n}}{n}\left(\sqrt{S_n}+\sqrt{T_n}\right) \tag{10.33}$$

is consistent for ϑ (Problem 10.13 (c)), the main Theorem 10.15 yields

$$\sqrt{n}(\widehat{\vartheta}_n - \vartheta) \xrightarrow{\mathcal{D}_\vartheta} N_2(0_2, I_1(\vartheta)^{-1}).$$

Here, $I_1(\vartheta)^{-1}$ is given by

$$I_1(\vartheta)^{-1} = \frac{(\vartheta_1 + \vartheta_2)^2 \vartheta_1 \vartheta_2}{4\vartheta_1\vartheta_2 + 2\vartheta_1 + 2\vartheta_2}\begin{pmatrix} (2\vartheta_1 + \vartheta_2)/\vartheta_2 & 1 \\ 1 & (\vartheta_1 + 2\vartheta_2)/\vartheta_1 \end{pmatrix}.$$

Answers to Self-Questions

Answer 1 Since $\{X_1 \in B\} = \{(x_1, x_2, \ldots) \in \mathcal{X} : x_1 \in B\}$, the first equation follows from $\mathbb{P}_\vartheta = Q_\vartheta^{\mathbb{N}}$. Putting $B = \{t\}$, the second equation is a special case of the first.

Answer 2 The function $\vartheta \mapsto \frac{t}{\vartheta} - \frac{n-t}{1-\vartheta}$ is positive for $0 < \vartheta < \frac{t}{n}$ and negative for $\frac{t}{n} < \vartheta < 1$.

Answer 3 Because of the condition (10.13), the expectation occurring in (10.14) exists.

Answer 4 We have $\frac{\partial^2}{\partial\vartheta_i\partial\vartheta_j} \int_{\mathcal{X}_0} f(x, \vartheta)\, \nu(\mathrm{d}x) = \frac{\partial^2}{\partial\vartheta_i\partial\vartheta_j} 1 = 0$ and

$$\mathbb{E}_\vartheta\left[\frac{1}{f(X, \vartheta)} \cdot \frac{\partial^2 f(X, \vartheta)}{\partial\vartheta_i\partial\vartheta_j}\right] = \int_{\mathcal{X}_0} \frac{1}{f(x, \vartheta)} \cdot \frac{\partial^2 f(x, \vartheta)}{\partial\vartheta_i\partial\vartheta_j} f(x, \vartheta)\, \nu(\mathrm{d}x)$$

$$= \int_{\mathcal{X}_0} \frac{\partial^2 f(x, \vartheta)}{\partial\vartheta_i\partial\vartheta_j}\nu(\mathrm{d}x).$$

Answer 5 Note that $\mathbb{P}_\vartheta(\overline{\vartheta}_n \neq \widehat{\vartheta}_n) \leq \mathbb{P}_\vartheta\big(M_n' \cap V_n\big)^c\big) \leq \mathbb{P}_\vartheta(M_n'^c) + \mathbb{P}_\vartheta(V_n^c) \to 0$. The second statement holds because, for each $\varepsilon > 0$, $\sqrt{n}|\overline{\vartheta}_n - \widehat{\vartheta}_n| > \varepsilon$ implies $\overline{\vartheta}_n \neq \widehat{\vartheta}_n$.

Answer 6 Because of the minus sign.

Answer 7 If $s_n(\mathbf{x}) > 0$ and $t_n(\mathbf{x}) = 0$ then the log-likelihood function is $\log L_{n,\mathbf{x}}(\vartheta) = -n \log(\vartheta_1 + \vartheta_2) - s_n(\mathbf{x})/\vartheta_1$. For a fixed ϑ_1 the maximum over ϑ_2 is attained at the boundary point $\vartheta_2 = 0$ of the interval $(0, \infty)$, and the function $(0, \infty) \ni \vartheta_1 \mapsto -n \log(\vartheta_1) - s_n(\mathbf{x})/\vartheta_1$ attains its maximum value for $\widehat{\vartheta}_{n,1}(\mathbf{x}) = \frac{1}{n} s_n(\mathbf{x})$, which agrees with (10.32).

Problems

10.1 Problem Show that ML estimates are independent of the specific choice of the dominating measure ν. More precisely, let $Q_\vartheta \ll \nu$ for each $\vartheta \in \Theta$. For a second σ-finite measure $\widetilde{\nu}$ on $\mathcal{B}_0$, let $\nu \ll \widetilde{\nu}$ and thus also $Q_\vartheta \ll \widetilde{\nu}$ for each $\vartheta \in \Theta$. Let $f(\cdot, \vartheta)$ and $\widetilde{f}(\cdot, \vartheta)$ be the densities of Q_ϑ with respect to ν and $\widetilde{\nu}$, respectively. Show that the ML estimates obtained using f coincide $\nu \otimes \ldots \otimes \nu$-almost everywhere with those obtained from $\widetilde{f}$.

10.2 Problem The random variables $X_1, \ldots, X_n$ are independent and have the same Lebesgue density

$$f(x, \vartheta) = \frac{\vartheta^\alpha}{\Gamma(\alpha)} x^{\alpha-1} e^{-\vartheta x}, \quad x > 0,$$

and $f(x, \vartheta) = 0$, otherwise. Here, $\alpha > 0$ is known and $\vartheta \in \Theta := (0, \infty)$ is unknown.

What is the MLE $\widehat{\vartheta}_n = \widehat{\vartheta}_n(X_1, \ldots, X_n)$ of ϑ?

10.3 Problem Let $X_1, X_2, \ldots$ be a sequence of i.i.d. random variables with the same Gamma distribution $\Gamma(\alpha, \lambda)$, that is, with the same density

$$f(x, \vartheta) = \frac{\lambda^\alpha}{\Gamma(\alpha)} x^{\alpha-1} e^{-\lambda x}, \quad x > 0,$$

and $f(x, \vartheta) = 0$, otherwise. Putting $\Theta := \{\vartheta := (\alpha, \lambda) : \alpha > 0, \lambda > 0\}$, show that the likelihood equations (10.3) are equivalent to

$$\frac{1}{n} \sum_{j=1}^{n} \log x_j = \Psi(\alpha) - \log \lambda, \qquad \frac{1}{n} \sum_{j=1}^{n} x_j = \frac{\alpha}{\lambda}.$$

Here, $\Psi(t) := \frac{d}{dt} \log \Gamma(t), t > 0$, is the so-called *Digamma function* or *Psi function*.

10.4 Problem In the setting of Example 10.4, prove that

$$\lim_{n\to\infty} \mathbb{P}_\vartheta\left(n(\widehat{\vartheta}_n - \vartheta) \le u\right) = \begin{cases} 1, & \text{if } u \ge 0, \\ \exp\left(\frac{u}{\vartheta}\right), & \text{if } u < 0. \end{cases}$$

Hint Use Problem 9.4.

10.5 Problem Let $\Theta := \{\vartheta := (\vartheta_1, \vartheta_2) : -\infty < \vartheta_1 < \vartheta_2 < \infty\}$. The random variables $X_1, X_2, \ldots$ are independent and identically distributed, with $X_1 \sim U([\vartheta_1, \vartheta_2])$. Prove the following claims:

(a) The MLE of ϑ for the sample size n is given by $\widehat{\vartheta}_n = (U_n, V_n)$ with $U_n := \min(X_1, \ldots, X_n)$ and $V_n := \max(X_1, \ldots, X_n)$.

(b) Putting $\Delta := \vartheta_2 - \vartheta_1$, we have

$$\lim_{n\to\infty} \mathbb{P}_\vartheta\left(n(U_n - \vartheta_1) \le s, n(\vartheta_2 - V_n) \le t\right) = \left(1 - \exp\left(-\frac{s}{\Delta}\right)\right)\left(1 - \exp\left(-\frac{t}{\Delta}\right)\right)$$

for all positive real numbers s and t.

Hint for (b) Note that $\{n(U_n - \vartheta_1) \le s\} = \cup_{j=1}^n \{X_j \le \vartheta_1 + \frac{s}{n}\}$.

10.6 Problem Let $X_1, \ldots, X_n$ be independent d-dimensional random vectors with the same normal distribution $N_d(\vartheta, \Sigma)$. Here, Σ is a known positive-definite matrix, and $\vartheta \in \Theta := \mathbb{R}^d$ is unknown. Show that the MLE of ϑ is given by $\overline{X}_n := n^{-1}\sum_{j=1}^n X_j$. What is the distribution of $\sqrt{n}(\overline{X}_n - \vartheta)$?

10.7 Problem Let $\Theta = (0, \infty)$ and $Q_\vartheta = \mathrm{Exp}(\vartheta)$, so $f(x, \vartheta) = \vartheta e^{-\vartheta x}$ if $x \ge 0$ and $f(x, \vartheta) = 0$, otherwise. Show that

$$I_{KL}(\vartheta : \vartheta') = \log\frac{\vartheta}{\vartheta'} + \frac{\vartheta'}{\vartheta} - 1.$$

10.8 Problem In the situation of Definition 10.7, let Θ be an open interval, and suppose the conditions (10.12) and (10.13) hold. Show that

$$\lim_{\varepsilon\downarrow 0} \frac{2}{\varepsilon^2} I_{KL}(\vartheta + \varepsilon : \vartheta) = I_1(\vartheta), \quad \vartheta \in \Theta.$$

Assume that $f(x, \vartheta)$ is differentiable with respect to ϑ for fixed x, and that terms of the order $o(\varepsilon^2)$ can throughout be neglected.

Hint Note that $t \log t = t - 1 + \frac{1}{2}(t-1)^2 + o((t-1)^2)$ as $t \to 1$.

10.9 Problem Let B_ϑ be as in (10.16), and set $B := \bigcap_{\vartheta \in \Theta \setminus \{\vartheta_0\}} B_\vartheta$, where Θ is a finite set. Show that $\mathbf{x} \in B$ implies $\lim_{n \to \infty} \widehat{\vartheta}_n(\mathbf{x}) = \vartheta_0$.

10.10 Problem Let X be a random variable with the uniform distribution $U(0, \vartheta)$, where $\vartheta \in \Theta := [1, 2]$. Prove the following claim: With the definition

$$f(x, \vartheta) := \frac{1}{\vartheta} \mathbf{1}_{[0,\vartheta]}(x), \quad x \in \mathbb{R},$$

the conditions of Theorem 10.11 hold, but not if $f(x, \vartheta)$ is set to $\frac{1}{\vartheta}$ on the *open* interval $(0, 1)$ and to 0 outside that interval.

10.11 Problem Show that the random vector $U_n(\vartheta)$ defined in (10.23) is centered and has the covariance matrix $n I_1(\vartheta)$.

10.12 Problem Show by calculating the Fisher information that the convergence in distribution in (10.4) matches the result of the main Theorem 10.15.

10.13 Problem Assume the setting of Example 10.17. Prove the following claims:

(a) The Fisher information matrix $I_1(\vartheta)$ is given by (10.31).
(b) In the case $s_n > 0$ and $t_n > 0$, (10.32) is the only solution of the likelihood equations.
(c) The sequence of estimators $(\widehat{\vartheta}_n) = ((\widehat{\vartheta}_{n,1}, \widehat{\vartheta}_{n,2}))$ with $\widehat{\vartheta}_{n,1}$ and $\widehat{\vartheta}_{n,2}$ as in (10.33) is consistent for ϑ.

10.14 Problem Assume the setting of the main Theorem 10.15 with $k = 1$, where Θ is an open interval. Show: If the function $\Theta \ni \vartheta \mapsto I_1(\vartheta)$ is continuous, then for every $\alpha \in (0, 1)$:

$$\lim_{n \to \infty} \mathbb{P}_\vartheta \left(\widehat{\vartheta}_n - \frac{\Phi^{-1}(1 - \alpha/2)}{\sqrt{n}\sqrt{I_1(\widehat{\vartheta}_n)}} \leq \vartheta \leq \widehat{\vartheta}_n + \frac{\Phi^{-1}(1 - \alpha/2)}{\sqrt{n}\sqrt{I_1(\widehat{\vartheta}_n)}} \right) = 1 - \alpha \quad \text{for every } \vartheta \in \Theta.$$

Here, Φ and Φ^{-1} denote the distribution function and the quantile function of the standard normal distribution, respectively.

What does an asymptotic confidence interval for ϑ look like for the special case that $X_1 \sim \mathrm{Bin}(1, \vartheta)$?

Asymptotic (Relative) Efficiency of Estimators

11

This chapter deals with the quality of estimators of unknown parameters in the presence of large sample sizes. Our setting is that stated at the beginning of Chap. 10, that is, a sequence $X_1, X_2, \ldots$ of independent $\mathcal{X}_0$-valued random variables with the same distribution Q_ϑ, $\vartheta \in \Theta$, and $\Theta \subset \mathbb{R}^k$. The probability measure Q_ϑ has a density $f(\cdot, \vartheta)$ with respect to a σ-finite dominating measure ν on $\mathcal{B}_0$, and this density satisfies the regularity conditions stated in 10.12.

To get started, we ask ourselves, just as M. Fréchet, H. Cramér and C.R. Rao[1] did independently of each other, whether there is a lower bound for the variance of an estimator $T_n = T_n(X_1, \ldots, X_n)$ of of a real-valued parameter ϑ for a fixed sample size n. In what follows, we assume $\mathbb{E}_\vartheta\left(T_n^2\right) < \infty$ for each $\vartheta \in \Theta$, where Θ is an open interval.

Since the score vector $U_n(\vartheta) = \frac{\partial}{\partial \vartheta} \log f_n(X_1, \ldots, X_n, \vartheta)$ defined in (10.23) has expectation zero, we have for each $\vartheta \in \Theta$

$$
\begin{aligned}
\mathrm{Cov}_\vartheta\left(U_n(\vartheta), T_n\right) &= \mathbb{E}_\vartheta\left[U_n(\vartheta)\, T_n\right] = \mathbb{E}_\vartheta\left[\frac{\frac{\partial}{\partial\vartheta} f_n(X_1, \ldots, X_n, \vartheta)}{f_n(X_1, \ldots, X_n, \vartheta)}\, T_n\right] \\
&= \int_{\mathcal{X}_0^n} \frac{\partial}{\partial\vartheta} f_n(x_1, \ldots, x_n, \vartheta)\, T_n(x_1, \ldots, x_n)\, \nu(\mathrm{d}x_1) \ldots \nu(\mathrm{d}x_n)
\end{aligned}
$$

---

[1] Calyampudi Radhakrishna Rao (1920–2023), one of the internationally most renowned mathematicians and statisticians. Rao was Professor Emeritus of Pennsylvania State University and Research Professor at the University of Buffalo. He held various other professorships and had been honored with numerous honorary doctorates and other awards. His fundamental contributions include, among others, estimation theory, linear statistical models, multivariate statistics and experimental design.

© The Author(s), under exclusive license to Springer-Verlag GmbH, DE, part of Springer Nature 2024
N. Henze, *Asymptotic Stochastics*, Mathematics Study Resources 10,
https://doi.org/10.1007/978-3-662-68923-3_11

$$= \frac{\partial}{\partial \vartheta} \int_{\mathcal{X}_0^n} T_n(x_1, \ldots, x_n) \, f_n(x_1, \ldots, x_n, \vartheta) \, \nu(dx_1) \ldots \nu(dx_n)$$

$$= \frac{\partial}{\partial \vartheta} \mathbb{E}_\vartheta (T_n). \tag{11.1}$$

Here, at the fourth equals sign, we assumed that differentiation with respect to ϑ and integration with respect to ν can be interchanged. Due to the Cauchy–Schwarz inequality, $\mathrm{Cov}_\vartheta^2\big(U_n(\vartheta), T_n\big) \leq \mathbb{V}_\vartheta(U_n(\vartheta)) \mathbb{V}_\vartheta(T_n)$. Since $\mathbb{V}_\vartheta(U_n(\vartheta)) = n I_1(\vartheta)$ (see Problem 10.11), we then obtain the famous *information inequality*

$$\mathbb{V}_\vartheta(T_n) \geq \frac{\left(\frac{\partial}{\partial \vartheta} \mathbb{E}_\vartheta(T_n)\right)^2}{n I_1(\vartheta)}, \quad \vartheta \in \Theta, \tag{11.2}$$

of Fréchet–Cramér–Rao.[2] It follows in particular that the variance of an unbiased estimator of ϑ is bounded from below by

$$\mathbb{V}_\vartheta(T_n) \geq \frac{1}{n I_1(\vartheta)}, \quad \vartheta \in \Theta. \tag{11.3}$$

The larger the Fisher information is, the smaller the variance of an unbiased estimator of ϑ can be. Although the lower bound in the information inequality can only be attained if the underlying distribution belongs to a so-called *one-parameter exponential family* (see, e.g., [SHA], p. 172, or [BD1], p. 182), (11.3) can sometimes be used to show that an estimator T_n of ϑ has the uniformly smallest variance within the class of all unbiased estimators (see Problem 11.1). If one defines as a quality criterion for an estimator S_n of ϑ a smallest possible *mean squared deviation* $\mathbb{E}_\vartheta\big[(S_n - \vartheta)^2\big]$, it may happen that an unbiased estimator T_n of ϑ exists, which is *Cramér–Rao-efficient* in the sense that for each $\vartheta \in \Theta$ equality in (11.3) holds. However, such an estimator can be inferior to another (not unbiased) estimator in terms of mean squared deviation (see Problem 11.3). In view of

$$\mathbb{E}_\vartheta\big[(S_n - \vartheta)^2\big] = \big(\mathbb{E}_\vartheta(S_n) - \vartheta\big)^2 + \mathbb{V}_\vartheta(S_n) \tag{11.4}$$

this fact is not surprising. This decomposition of the mean squared deviation of an estimator into the sum of the square of its bias and its variance reveals a phenomenon that is often referred to as the *bias-variance-tradeoff*. To keep the mean squared deviation small, it is thus sometimes not enough to search for an estimator with the smallest variance within the set of unbiased estimators (note that the latter set may even be empty). The

---

[2] This inequality was first published by M. Fréchet (1943) (see [FRE], p. 185); however, it is usually only referred to as *Cramér–Rao inequality*.

message is that the mean squared deviation depends on the sum of the two terms on the right-hand side of (11.4).

Suppose now that ϑ is a k-dimensional parameter, that is, $\Theta \subset \mathbb{R}^k$, and that T_n is an estimator of ϑ with a sample size n. Moreover, we assume throughout that $\mathbb{E}_\vartheta \| T_n \|^2 < \infty$ for each $\vartheta \in \Theta$. Then the concentration of the distribution of T_n is generally measured using the covariance matrix $\mathbb{C}\mathrm{ov}_\vartheta (T_n)$ of T_n. If S_n and T_n are two unbiased estimators of ϑ, that is, if $\mathbb{E}_\vartheta (S_n) = \mathbb{E}_\vartheta (T_n) = \vartheta$ for each $\vartheta \in \Theta$, one will prefer the estimator S_n over T_n if for each $\vartheta \in \Theta$ the inequality

$$\mathbb{C}\mathrm{ov}_\vartheta (T_n) \geq_L \mathbb{C}\mathrm{ov}_\vartheta (S_n) \tag{11.5}$$

holds (a motivation follows). Here, the notation $A \geq_L B$ for $k \times k$ symmetric matrices A and B generally means that

$$c^\top A c \geq c^\top B c \quad \text{for each } c \in \mathbb{R}^k, \tag{11.6}$$

that is, the matrix $A - B$ is positive semidefinite. The letter L as an index of the greater-than-or-equal sign stands for K. Löwner,[3] who introduced the partial order $\geq_L$ between symmetric matrices.

The inequality (11.5) is equivalent to $\mathbb{V}_\vartheta (c^\top T_n) \geq \mathbb{V}_\vartheta (c^\top S_n)$ for every $c \in \mathbb{R}^k$ (see Self-question 1). The estimator S_n is then better in the sense that the variance of any linear combination of the components of S_n is less than or equal to the variance of the corresponding components of T_n. For the word *better* to be justified, there should be strict inequality $\mathbb{V}_\vartheta (c^\top T_n) > \mathbb{V}_\vartheta (c^\top S_n)$ for at least one c and at least one $\vartheta \in \Theta$. In the case $k = 1$, (11.5) is equivalent to $\mathbb{V}_\vartheta (T_n) \geq \mathbb{V}_\vartheta (S_n)$.

Self-question 1: Why is (11.5) equivalent to $\mathbb{V}_\vartheta (c^\top T_n) \geq \mathbb{V}_\vartheta (c^\top S_n), c \in \mathbb{R}^k$?

We now want to generalize the information inequality (11.2) to the case that $\vartheta =: (\vartheta_1, \ldots, \vartheta_k)^\top$ is a k-dimensional parameter. To this end, let Θ be an open subset of $\mathbb{R}^k$, and suppose $T_n =: (T_{n,1}, \ldots, T_{n,k})^\top$ is an estimator of ϑ based on $X_1, \ldots, X_n$. We write

$$D_n(\vartheta) := \frac{\mathrm{d}}{\mathrm{d}\vartheta} \mathbb{E}_\vartheta (T_n^\top) = \left(\frac{\partial}{\partial \vartheta_i} \mathbb{E}_\vartheta (T_{n,j}) \right)_{1 \leq i,j \leq k}$$

---

[3] Karl Löwner (1893–1968), Czech-American mathematician, among others, professor at Charles University in Prague, emigrated to the U.S. in 1939, lastly professor at Stanford University. Main areas of work: Complex and real analysis.

for the Jacobian matrix of the mapping $\Theta \ni \vartheta \mapsto \mathbb{E}_\vartheta(T_n)$, which is assumed to exist. Note that $D_n(\vartheta) = I_k$ if T_n is unbiased for ϑ. We further assume that the estimator T_n is *regular* in the sense that for any choice of $i, j \in \{1, \ldots, k\}$

$$\int_{\mathcal{X}_0^n} \frac{\partial}{\partial \vartheta_i} f_n(x_1, \ldots, x_n, \vartheta) \, T_{n,j}(x_1, \ldots, x_n) \, \nu(\mathrm{d}x_1) \ldots \nu(\mathrm{d}x_n)$$

$$= \frac{\partial}{\partial \vartheta_i} \int_{\mathcal{X}_0^n} T_{n,j}(x_1, \ldots, x_n) \, f_n(x_1, \ldots, x_n, \vartheta) \, \nu(\mathrm{d}x_1) \ldots \nu(\mathrm{d}x_n).$$

Thus, the chain of equations ending in (11.1) shows that the score vector

$$U_n(\vartheta) = \frac{\mathrm{d}}{\mathrm{d}\vartheta} \log f_n(X_1, \ldots, X_n, \vartheta) =: \left(U_{n,1}(\vartheta), \ldots, U_{n,k}(\vartheta)\right)^\top$$

satisfies

$$\mathrm{Cov}_\vartheta\left(U_{n,i}(\vartheta), T_{n,j}\right) = \frac{\partial}{\partial \vartheta_i} \mathbb{E}_\vartheta(T_{n,j}), \quad i, j \in \{1, \ldots, k\}, \ \vartheta \in \Theta.$$

In matrix notation, we therefore have

$$D_n(\vartheta) = \mathbb{E}_\vartheta\left[U_n(\vartheta)\left(T_n - \mathbb{E}_\vartheta(T_n)\right)^\top\right], \quad \vartheta \in \Theta. \tag{11.7}$$

Self-question 2: Why can $\mathbb{E}_\vartheta(T_n)$ stand on the right-hand side?

With these preparations, we can state and prove a generalization of the inequality (11.2).

11.1 Theorem (Multivariate Information Inequality)
Under the standing assumptions, we have

$$\mathrm{Cov}_\vartheta(T_n) \geq_L \frac{1}{n} D_n^\top(\vartheta) I_1(\vartheta)^{-1} D_n(\vartheta) \quad \text{for each } \vartheta \in \Theta. \tag{11.8}$$

The equal sign holds if and only if

$$T_n = \mathbb{E}_\vartheta(T_n) + \frac{1}{n} D_n^\top(\vartheta) I_1(\vartheta)^{-1} U_n(\vartheta) \ \mathbb{P}_\vartheta\text{-almost surely.}$$

Proof For fixed $\vartheta \in \Theta$ we set $\widetilde{T}_n := T_n - \mathbb{E}_\vartheta(T_n)$, $D_n := D_n(\vartheta)$, $I_n := nI_1(\vartheta)$ and $U_n := U_n(\vartheta)$. Moreover, let $Y := \widetilde{T}_n - D_n^\top I_n^{-1} U_n$. If $0_{k \times k}$ denotes the $k \times k$ zero matrix

then, putting $A \leq_L B :\Longleftrightarrow B \geq_L A$ and using $\mathbb{E}(YY^\top) \geq_L 0_{k \times k}$ (cf. Problem 11.4 (a),
it follows that

$$
\begin{aligned}
0_{k \times k} \leq_L \mathbb{E}_\vartheta (YY^\top) \\
&= \mathbb{E}_\vartheta \left[\left(\tilde{T}_n - D_n^\top I_n^{-1} U_n \right) \left(\tilde{T}_n^\top - U_n^\top I_n^{-1} D_n \right) \right] \\
&= \mathbb{C}\mathrm{ov}_\vartheta (T_n) - D_n^\top I_n^{-1} \mathbb{E}_\vartheta \left[U_n \tilde{T}_n^\top \right] - \mathbb{E}_\vartheta \left[\tilde{T}_n U_n^\top \right] I_n^{-1} D_n \\
&\quad + D_n^\top I_n^{-1} \mathbb{E}_\vartheta \left[U_n U_n^\top \right] I_n^{-1} D_n .
\end{aligned}
$$

Since $\mathbb{E}_\vartheta \left[U_n U_n^\top \right] = I_n$ and $\mathbb{E}_\vartheta \left[U_n \tilde{T}_n^\top \right] = D_n$ (cf. (11.7)), the assertion follows. The
addition about the equals sign follows from

$$
\mathbb{E}(YY^\top) = 0_{k \times k} \Longleftrightarrow Y = 0_d \quad \mathbb{P}\text{-almost surely}
$$

(cf. Problem 11.4 (b)).   □

Under certain regularity assumptions regarding the density $f(\cdot, \vartheta)$ and the estimator
T_n, the covariance matrix of T_n thus has a lower bound in the sense of the Löwner partial
order $\geq_L$. If T_n is unbiased for ϑ, then, in generalization of (11.3), (11.8) takes the special
form

$$
\mathbb{C}\mathrm{ov}_\vartheta (T_n) \geq_L \frac{I_1(\vartheta)^{-1}}{n}, \quad \vartheta \in \Theta,
$$

and it follows that

$$
\mathbb{C}\mathrm{ov}_\vartheta \left(\sqrt{n}(T_n - \vartheta) \right) \geq_L I_1(\vartheta)^{-1}, \quad \vartheta \in \Theta. \tag{11.9}
$$

With regard to asymptotic aspects, we now assume that (T_n) is a sequence of estimators
of ϑ with the property

$$
\sqrt{n}(T_n - \vartheta) \xrightarrow{\mathcal{D}_\vartheta} N_k(0_k, \Sigma(\vartheta)), \quad \vartheta \in \Theta. \tag{11.10}
$$

Here, $\Sigma(\vartheta)$ is the *covariance matrix of the asymptotic normal distribution* of $\sqrt{n}(T_n - \vartheta)$
and thus a symmetric positive-semidefinite matrix.

In view of (11.9) and (11.10), the question arises as to whether the *asymptotic
information inequality*

$$
\Sigma(\vartheta) \geq_L I_1(\vartheta)^{-1}, \quad \vartheta \in \Theta, \tag{11.11}
$$

holds for asymptotically normally distributed estimators. Theorem 10.15 states that (11.11) is valid for every asymptotic MLE $(\widehat{\vartheta}_n)$, if the sequence $(\widehat{\vartheta}_n)$ is consistent, and that for $(\widehat{\vartheta}_n)$ the lower bound in (11.11) is attained for every $\vartheta \in \Theta$.

A sequence of estimators (T_n) is called *asymptotically efficient* for ϑ, if

$$\sqrt{n}(T_n - \vartheta) \xrightarrow{\mathcal{D}_\vartheta} N_k(0_k, I_1(\vartheta)^{-1}), \quad \vartheta \in \Theta. \tag{11.12}$$

In this case, the sequence (T_n) is also said to be a *best asymptotically normally distributed estimator* or shortly BAN-*estimator*. In the setting of Theorem 10.15, the sequence $(\widehat{\vartheta}_n)$ is therefore asymptotically efficient or a BAN estimator. However, Problem 11.5 shows that an asymptotically normally distributed estimator can fall below the asymptotic Cramér–Rao bound $I_1(\vartheta)^{-1}$ for individual values of ϑ, so that for such values there is so-called "superefficiency". The following fact, discovered by L. LeCam,[4] states that those values of ϑ for which superefficiency holds are a null set with respect to the Borel–Lebesgue measure λ^k. The original proof by LeCam was simplified by R.R. Bahadur[5] (see [BAH], or Section 5.16 of [PF2]).

11.2 Theorem (LeCam–Bahadur)
Under the regularity conditions 10.12, let (T_n) be a sequence of estimators of ϑ satisfying (11.10). Then there exists a set $N \in \mathcal{B}^k$ with $\lambda^k(N) = 0$ and

$$\Sigma(\vartheta) \geq I_1(\vartheta)^{-1} \text{ for every } \vartheta \in \Theta \cap N^c.$$

We now turn to a second important method of estimation of unknown parameters. To this end, let $X_1, X_2, \ldots$ be a sequence of i.i.d. real-valued random variables and assume $\mathbb{E}(X_1^{2k}) < \infty$. So, for an integer k, the $2k$-th moment of the underlying distribution exists. We do not assume, however, that X_1 has a density with respect to some σ-finite dominating measure on the σ-field $\mathcal{B}_0$. Since $\mathbb{E}(X_1^{2k}) < \infty$, all moments

$$m_\ell := \mathbb{E}(X_1^\ell), \quad \ell \in \{1, \ldots, 2k\},$$

exist.

---

[4] Lucien LeCam (1924–2000), French mathematician and statistician. After an industrial activity at the *Electricité de France*, LeCam was a professor at the University of California, Berkeley from 1960. He is considered the founder of modern asymptotic statistics.

[5] Raghu Raj Bahadur (1924–1997), Indian statistician, among others professor at the University of Chicago (1961–1991) and *Distinguished Visiting Professor* at the Indian Statistical Institute (1972–1997). Bahadur made fundamental contributions to mathematical statistics; the term *Bahadur efficiency* goes back to him.

Self-question 3: Why does $\mathbb{E}|X_1|^\ell < \infty$ $(\ell \in \{1, \ldots, 2k-1\})$ follow from $\mathbb{E}(X_1^{2k}) < \infty$?

Suppose we are interested in a k-dimensional parameter $\vartheta =: (\vartheta_1, \ldots, \vartheta_k)^\top \in \mathbb{R}^k$, which can be expressed as a function of $m_1, \ldots, m_k$. More precisely, there are subsets D and Θ of $\mathbb{R}^k$ and a bijective continuously differentiable mapping $g : D \to \Theta$ with the property

$$\vartheta = g(m_1, \ldots, m_k).$$

With the so-called *sample moments*

$$\widetilde{m}_{n,\ell} := \frac{1}{n} \sum_{j=1}^n X_j^\ell, \quad \ell \in \{1, \ldots, k\}, \tag{11.13}$$

it is natural to estimate ϑ by

$$\widetilde{\vartheta}_n := g\big(\widetilde{m}_{n,1}, \ldots, \widetilde{m}_{n,k}\big). \tag{11.14}$$

The estimator $\widetilde{\vartheta}_n = \widetilde{\vartheta}_n(X_1, \ldots, X_n)$ is called the *moment estimator* of ϑ (for the sample size n).

The sequence of estimators $(\widetilde{\vartheta}_n)$ is strongly consistent for ϑ, and we will soon see that $\sqrt{n}(\widetilde{\vartheta}_n - \vartheta)$ has a centered asymptotic normal distribution as $n \to \infty$.

Self-question 4: Why is the sequence of estimators $(\widetilde{\vartheta}_n)$ strongly consistent?

The asymptotics of the moment estimator can be derived fairly quickly: For this purpose, put

$$Y_j := \big(X_j, X_j^2, \ldots, X_j^k\big)^\top, \quad j \geq 1,$$

so $Y_1, Y_2, \ldots$ is a sequence of i.i.d. k-dimensional random vectors with expectation

$$a := \mathbb{E}(Y_1) = \big(m_1, m_2, \ldots, m_k\big)^\top \tag{11.15}$$

and covariance matrix

$$T := \mathbb{E}\big[(Y_1 - a)(Y_1 - a)^\top\big] = \Big(\mathbb{E}\big[(X_1^i - m_i)(X_1^j - m_j)\big]\Big)_{1 \le i,j \le k}$$

$$= \big(m_{i+j} - m_i m_j\big)_{1 \le i,j \le k}. \tag{11.16}$$

If we set $\overline{Y}_n := \frac{1}{n} \sum_{j=1}^n Y_j$, the multivariate central limit theorem 6.19 gives

$$\sqrt{n}(\overline{Y}_n - a) = \frac{1}{\sqrt{n}} \Big(\sum_{j=1}^n Y_j - na\Big) \xrightarrow{\mathcal{D}} N_k(0_k, T).$$

Since $\overline{Y}_n = \big(\widetilde{m}_{n,1}, \ldots, \widetilde{m}_{n,k}\big)^\top$, $g(\overline{Y}_n) = \widetilde{\vartheta}_n$, and $g(a) = \vartheta$, using the delta method (Theorem 6.25) we obtain the following result:

11.3 Theorem (Asymptotic Distribution of the Moment Estimator)
Let a be as in (11.15) and T as in (11.16). If $g'(a)$ denotes the Jacobian matrix of g at the point $a = g^{-1}(\vartheta)$, then, under the assumptions made for the moment estimator $\widetilde{\vartheta}_n$ defined in (11.14), we have

$$\sqrt{n}(\widetilde{\vartheta}_n - \vartheta) \xrightarrow{\mathcal{D}_\vartheta} N_k(0_k, \Sigma(\vartheta)), \tag{11.17}$$

where $\Sigma(\vartheta) = g'(a)\, T\, g'(a)^\top$.

An example of (11.17) is the convergence in distribution stated in (10.6), because $\widehat{\mu}_n$ and $\widehat{\sigma}_n^2$ given in (10.5) are the moment estimators of μ and σ^2, respectively. In this case, the moment estimator is a BAN estimator, but such a statement does not hold in general. We will see that using an estimator of ϑ, which satisfies certain conditions, but is not necessarily a BAN estimator, one can construct an asymptotically efficient estimator. Of course, one would like to use the MLE, but the likelihood equations (10.3), written in the form $U_n(\vartheta) = 0_k$, are often highly nonlinear and can therefore only be solved approximately by numerical methods (see for example Problem 10.3).

At this point, we fall back on the Newton[6] method, which is a powerful tool for approximating the zeros of a function. For this purpose, let $G \subset \mathbb{R}^k$ be an open set, $h : G \to \mathbb{R}^k$ a continuously differentiable function and $x^* \in G$ with $h(x^*) = 0_k$. Suppose the Jacobian matrix $h'(x)$ is invertible at the point x^*. By continuity, h' is then invertible

---

[6] Sir Isaac Newton (1643–1727), English scientist and civil servant. With his works on optics, celestial mechanics, mathematics, physics, and chemistry, Newton is one of the most significant scientists in human history. Very important for mathematics is his foundation of analysis, which has entered the history of science as fluxion calculus.

in some neighborhood of x^*. In this situation, starting from an initial value x_0 close to x^*, one can try to construct a so-called *Newton sequence* (x_n) that converges to x^*. Such a Newton sequence is defined via the recursion formula

$$x_{n+1} := x_n - h'(x_n)^{-1} h(x_n), \qquad n = 0, 1, 2, \ldots \tag{11.18}$$

(see, e.g., [ZEI], Section 5.1). The book [DEU] is entirely dedicated to Newton methods.

11.4 Construction of BAN Estimators

In what follows, we assume that the regularity conditions stated in 10.12 hold, and that the parameter space Θ is an open set. Suppose that we have any sequence $(\widetilde{\vartheta}_n)$ of estimators of ϑ that is strongly consistent and asymptotically normally distributed. Thus, for every $\vartheta \in \Theta$

$$\widetilde{\vartheta}_n \to \vartheta \;\; \mathbb{P}_\vartheta\text{-a.s.} \;\; \text{and} \;\; \sqrt{n}(\widetilde{\vartheta}_n - \vartheta) \xrightarrow{\mathcal{D}_\vartheta} N_k(0, \Sigma(\vartheta)).$$

Here, $\Sigma(\vartheta)$ is a symmetric positive-semidefinite matrix. As we have seen, these properties apply to the moment estimator under certain conditions. With the score vector $U_n(\vartheta) = \sum_{j=1}^n \frac{d}{d\vartheta} \log f(X_j, \vartheta)$ of $(X_1, \ldots, X_n)$ (cf. (10.23)) and the Jacobian matrix $H_n(\vartheta) = \frac{d}{d\vartheta^\top} U_n(\vartheta)$ of U_n at the point ϑ introduced in (10.26), we set

$$\widetilde{\vartheta}_n^{(1)} := \widetilde{\vartheta}_n - H_n(\widetilde{\vartheta}_n)^{-1} U_n(\widetilde{\vartheta}_n). \tag{11.19}$$

In view of (11.18), this definition amounts to a one-step Newton iteration.

Now, let $(\widehat{\vartheta}_n)$ be a strongly consistent sequence of MLE's of ϑ. We claim that

$$\sqrt{n}(\widetilde{\vartheta}_n^{(1)} - \widehat{\vartheta}_n) \xrightarrow{\mathbb{P}_\vartheta} 0 \quad \text{for each } \vartheta \in \Theta, \tag{11.20}$$

and we will provide a sketch of the proof in a moment (for a detailed proof see, e.g., [FER], p. 138). The striking consequence of (11.20) is that, with (11.20), the main theorem for MLE's (Theorem 10.15) and Slutsky's lemma we obtain

$$\sqrt{n}(\widetilde{\vartheta}_n^{(1)} - \vartheta) = \sqrt{n}(\widehat{\vartheta}_n - \vartheta) + \sqrt{n}(\widetilde{\vartheta}_n^{(1)} - \widehat{\vartheta}_n) \xrightarrow{\mathcal{D}_\vartheta} N_k(0_k, I_1(\vartheta)^{-1}).$$

Thus, the estimator $(\widetilde{\vartheta}_n^{(1)})$ is asymptotically efficient, that is, a BAN estimator.

A sketch of the proof of (11.20) starts with the Taylor expansion

$$U_n(\widetilde{\vartheta}_n) = U_n(\widehat{\vartheta}_n) + H_n(\widehat{\vartheta}_n)(\widetilde{\vartheta}_n - \widehat{\vartheta}_n) + R_n$$

of $U_n(\cdot)$ around $\widehat{\vartheta}_n$. Here, $R_n = o_{\mathbb{P}_\vartheta}(n^{1/2})$. In view of the likelihood equations, we have $U_n(\widehat{\vartheta}_n) = 0_k$, and (11.19) yields

$$\widetilde{\vartheta}_n^{(1)} - \widehat{\vartheta}_n = \widetilde{\vartheta}_n - \widehat{\vartheta}_n - H_n(\widetilde{\vartheta}_n)^{-1} U_n(\widetilde{\vartheta}_n)$$

$$= \left[I_k - H_n(\widetilde{\vartheta}_n)^{-1} H_n(\widehat{\vartheta}_n)\right](\widetilde{\vartheta}_n - \widehat{\vartheta}_n) - H_n(\widetilde{\vartheta}_n)^{-1} R_n$$

and thus

$$\sqrt{n}\big(\widetilde{\vartheta}_n^{(1)} - \widehat{\vartheta}_n\big) = \left[I_k - \left(\frac{H_n(\widetilde{\vartheta}_n)}{n}\right)^{-1}\frac{H_n(\widehat{\vartheta}_n)}{n}\right]\sqrt{n}\big(\widetilde{\vartheta}_n - \widehat{\vartheta}_n\big) - \left(\frac{H_n(\widetilde{\vartheta}_n)}{n}\right)^{-1}\frac{R_n}{\sqrt{n}}.$$

Due to (10.29) and the fact that both $\widetilde{\vartheta}_n$ and $\widehat{\vartheta}_n$ converge $\mathbb{P}_\vartheta$-almost surely to ϑ, a uniform strong law of large numbers (see, e.g., [FER], Chapter 16)) implies that $\frac{1}{n}H_n(\widetilde{\vartheta}_n) \to -I_1(\vartheta)$ $\mathbb{P}_\vartheta$-a.s. and $\frac{1}{n}H_n(\widehat{\vartheta}_n) \to -I_1(\vartheta)$ $\mathbb{P}_\vartheta$-a.s. Thus, the expression within the square bracket converges $\mathbb{P}_\vartheta$-a.s. to the $k \times k$ zero matrix. Taking differences, it follows from $\sqrt{n}(\widetilde{\vartheta}_n - \vartheta) = O_{\mathbb{P}_\vartheta}(1)$ and $\sqrt{n}(\widehat{\vartheta}_n - \vartheta) = O_{\mathbb{P}_\vartheta}(1)$ that $\sqrt{n}(\widetilde{\vartheta}_n - \widehat{\vartheta}_n) = O_{\mathbb{P}_\vartheta}(1)$. Since $R_n/\sqrt{n} = o_{\mathbb{P}_\vartheta}(1)$, (11.20) is proved.

If the Fisher information matrix $I_1(\vartheta)$ is available as a function of ϑ, the estimator

$$\widetilde{\vartheta}_n^{(2)} := \widetilde{\vartheta}_n - \frac{1}{n}I_1(\widetilde{\vartheta}_n)^{-1} U_n(\widetilde{\vartheta}_n)$$

can be used instead of (11.19). (11.20) applies also for this estimator, where the superscript (1) has to be replaced by (2).

We conclude this chapter with the notion of *Pitman[7]-efficiency*. The Pitman efficiency is used to compare the relative quality of two sequences of estimators of a real-valued aspect of an unknown distribution. Let $X_1, X_2, \ldots$ be a sequence of i.i.d. random variables or random vectors with distribution Q_ϑ, where $\vartheta \in \Theta$ with $\Theta \subset \mathbb{R}^k$ is an unknown parameter. Of interest is $\gamma(\vartheta)$, where $\gamma : \Theta \to \mathbb{R}$. Further, let (S_n) and (T_n) with $S_n = S_n(X_1, \ldots, X_n)$ and $T_n = T_n(X_1, \ldots, X_n)$ be two sequences of estimators of $\gamma(\vartheta)$ with

$$\sqrt{n}\big(S_n - \gamma(\vartheta)\big) \xrightarrow{\mathcal{D}_\vartheta} N\big(0, \sigma^2(\vartheta)\big), \tag{11.21}$$

$$\sqrt{n}\big(T_n - \gamma(\vartheta)\big) \xrightarrow{\mathcal{D}_\vartheta} N\big(0, \tau^2(\vartheta)\big) \tag{11.22}$$

as $n \to \infty$, where $0 < \sigma^2(\vartheta) < \infty, 0 < \tau^2(\vartheta) < \infty$ for each $\vartheta \in \Theta$.

---

[7] Edwin James George Pitman (1897–1993), Australian mathematician, from 1926 Professor of Mathematics at the University of Tasmania. Pitman made significant contributions to statistics (Pitman permutation test, Pitman Closeness Criterion, Pitman Efficiency).

11.5 Definition (Asymptotic Relative Efficiency, Pitman Efficiency)
In the above setting,

$$\mathrm{ARE}_\vartheta\big((T_n):(S_n)\big) := \frac{\sigma^2(\vartheta)}{\tau^2(\vartheta)}$$

is called the *asymptotic relative efficiency or Pitman efficiency of* (T_n) *with respect to* (S_n).

This notion is not limited to the setting of a parametric model. If the distribution function of X_1, denoted by F, belongs to a certain set $\mathcal{F}$ of distribution functions, and the functional $\gamma : \mathcal{F} \to \mathbb{R}$ captures an interesting real-valued aspect of F, which is to be estimated based on $X_1, \ldots, X_n$, then (S_n) and (T_n) are sequences of estimators of $\gamma(F)$. If then (11.21) and (11.22) apply, where ϑ is throughout replaced with F, we write

$$\mathrm{ARE}_F\big((T_n):(S_n)\big) = \frac{\sigma^2(F)}{\tau^2(F)}$$

for the asymptotic relative efficiency of (T_n) with respect to (S_n).

The Pitman efficiency can be interpreted as follows: Suppose $(m_n)_{n\geq 1}$ is a sequence of integers (depending on ϑ) with $m_n \to \infty$ and

$$\sqrt{n}\big(T_{m_n} - \gamma(\vartheta)\big) \xrightarrow{\ \mathcal{D}_\vartheta\ } \mathrm{N}\big(0, \sigma^2(\vartheta)\big) \quad \text{as } n \to \infty. \tag{11.23}$$

The comparison with (11.21) shows that the estimator T which employs a sample of size m_n has the same asymptotic variance as the estimator S, which uses the sample size n. If we rewrite (11.23) according to

$$\sqrt{\frac{n}{m_n}}\,\sqrt{m_n}\big(T_{m_n} - \gamma(\vartheta)\big) \xrightarrow{\ \mathcal{D}_\vartheta\ } \mathrm{N}\big(0, \sigma^2(\vartheta)\big) \tag{11.24}$$

and consider the convergence in distribution

$$\sqrt{m_n}\big(T_{m_n} - \gamma(\vartheta)\big) \xrightarrow{\ \mathcal{D}_\vartheta\ } \mathrm{N}\big(0, \tau^2(\vartheta)\big) \quad \text{as } n \to \infty,$$

which holds due to (11.22), then this can only occur together with (11.24) if

$$\lim_{n\to\infty}\sqrt{\frac{n}{m_n(\vartheta)}} = \frac{\sigma(\vartheta)}{\tau(\vartheta)}.$$

We therefore obtain

$$\mathrm{ARE}_\vartheta\big((T_n):(S_n)\big) = \lim_{n\to\infty}\frac{n}{m_n(\vartheta)}.$$

The $\text{ARE}_\vartheta\ ((T_n):(S_n))$ thus indicates the approximate proportion of samples—in relation to the sample size $m_n(\vartheta)$ used by T—that the estimator S requires in order to achieve the same asymptotic variance as T. For example, an ARE of 2 means that T is twice as efficient as S, because T requires about half the sample size compared to S to achieve the same accuracy measured by the variance.

11.6 Example (Estimation of the Center of a Symmetric Distribution)

As an example, we consider the frequently studied question of how to estimate the center of a symmetric distribution. For this purpose, let $X_1, X_2, \ldots$ be a sequence of i.i.d. random variables with an unknown distribution function F. We assume that the distribution of X_1 is symmetric around an unknown value $a = a(F)$; so $X_1 - a \overset{D}{=} -(X_1 - a) = a - X_1$. We further assume $\mathbb{E}_F\big(X_1^2\big) < \infty$, which entails that the variance of X_1, denoted by $\sigma^2(F) = \mathbb{V}_F(X_1)$, exists, and we suppose $\sigma^2(F) > 0$. For the expectation of X_1, we then have $\mathbb{E}_F(X_1) = a$.

Self-question 5: Why does $\mathbb{E}_F(X_1) = a$ hold?

A natural sequence of estimators (S_n) of a is given by $S_n := \overline{X}_n := \frac{1}{n}\sum_{j=1}^n X_j$, and by the Lindeberg–Lévy central limit theorem it follows that

$$\sqrt{n}(S_n - a) \overset{D_F}{\longrightarrow} \mathrm{N}\big(0, \sigma^2(F)\big) \quad \text{as } n \to \infty.$$

We now additionally assume that the distribution function F is differentiable at the point a and has a positive derivative denoted by $f(a)$. Note that under this condition a is the only value for which $F(a) = \frac{1}{2}$; that is, $a = F^{-1}\big(\frac{1}{2}\big)$ is the median of F or of X_1. To estimate this median, we use the *order statistics* of $X_1, \ldots, X_n$, denoted by $X_{n:1} \le X_{n:2} \le \ldots \le X_{n:n}$, and consider as a second estimator of a the so-called *sample median of $X_1, \ldots, X_n$*, defined by

$$T_n := \begin{cases} X_{n:\frac{n+1}{2}}, & \text{if } n \text{ is odd}, \\[2mm] \frac{1}{2}\left(X_{n:\frac{n}{2}} + X_{n:\frac{n}{2}+1}\right), & \text{if } n \text{ is even}. \end{cases}$$

The asymptotic behavior of T_n follows from

$$X_{n:r} \le t \iff \sum_{j=1}^n \mathbf{1}\{X_j \le t\} \ge r, \qquad (11.25)$$

which holds for each $r \in \{1, \ldots, n\}$, and from the fact that $\sum_{j=1}^n \mathbf{1}\{X_j \le t\}$ has the binomial distribution $\mathrm{Bin}(n, F(t))$.

Self-question 6: Why does the above equivalence hold?

According to Problem 11.9 we have

$$\sqrt{n}(T_n - a) \xrightarrow{D_r} N\left(0, \frac{1}{4f^2\left(F^{-1}(1/2)\right)}\right) \tag{11.26}$$

and therefore

$$\mathrm{ARE}_F\big((T_n) : (S_n)\big) = 4f^2\Big(F^{-1}\big(\tfrac{1}{2}\big)\Big)\,\sigma^2(F). \tag{11.27}$$

As a numerical example, we examine the case where X_1 has a Student's t-distribution with s degrees of freedom and thus the density

$$f_s(x) = \frac{\Gamma\left(\frac{s+1}{2}\right)}{\sqrt{\pi s}\,\Gamma\left(\frac{s}{2}\right)} \cdot \left(1 + \frac{x^2}{s}\right)^{-(s+1)/2}, \quad x \in \mathbb{R}$$

(see, e.g., [BD1], Section B.3.1.). The associated distribution function is denoted by F_s. In order for the variance of X_1 to exist, we assume $s \geq 3$. We have $\sigma^2(F_s) = \frac{s}{s-2}$ (see, e.g., [BD1], p. 530), and it follows that

$$\mathrm{are}_s := \mathrm{ARE}_{F_s}\big((T_n) : (S_n)\big) = \frac{4\Gamma^2\left(\frac{s+1}{2}\right)}{\pi s \Gamma^2\left(\frac{s}{2}\right)} \cdot \frac{s}{s-2}.$$

The following table shows these asymptotic relative efficiencies for various values of s. The entry $s = \infty$ corresponds to the case $X_1 \sim N(0, 1)$.

s	3	4	5	6	∞
are_s	1.621	1.125	0.961	0.879	0.637 $(= 2/\pi)$

For the small degrees of freedom 3 and 4, for which the t-distribution has heavy tails, the sample median is therefore a better estimator of the center of symmetry compared to the sample mean. Note that only the local behavior of the distribution function at the location of the median enters into the asymptotic variance of T_n.

If the distribution function F is known, one can check the conditions for the modification from Sect. 11.4 for each of the two sequences of estimators considered above and, if necessary, obtain an asymptotically efficient estimation sequence. Surprisingly, there is even a *universal* sequence that asymptotically efficiently estimates the center of a symmetric univariate distribution in the sense of (11.12) in the presence of any distribution function F. Such a sequence of estimators can be found in [STO]. Well readable classifications of this result into the more general so-called *semiparametric* or *adaptive estimation theory* are given by [PF2], Section 5.7, and [VW2].

Further information on the central limit theorem in (11.26) for sample medians and for more general order statistics in Problem 11.9 can be found in [REI].

Answers to the Self-Questions

Answer 1 The notation (11.5) is defined by (11.6) with $A := \mathbb{C}\mathrm{ov}(T_n)$ and $B := \mathbb{C}\mathrm{ov}(S_n)$, and it follows that $\mathbb{V}(c^\top T_n) = \mathbb{C}\mathrm{ov}(c^\top T_n, c^\top T_n) = c^\top \mathbb{C}\mathrm{ov}(T_n)c$ (cf. Remark 5.2 (b) with $A := c$ and $X := T_n$).

Answer 2 Because of $\mathbb{E}_\vartheta\left(U_n(\vartheta)\right) = 0_k$ and thus also $\mathbb{E}_\vartheta\left(U_n(\vartheta)\mathbb{E}_\vartheta(T_n)^\top\right) = 0_{k \times k}$.

Answer 3 Because of the inequality $|x|^\ell \leq 1 + |x|^{2k}$, which holds for every real number x and every $\ell \in \{1, \ldots, 2k - 1\}$.

Answer 4 By the strong law of large numbers, $\widetilde{m}_{n,j}$ converges almost surely to m_j for each $j \in \{1, \ldots, k\}$. Since the function g is continuous, $g\left(\widetilde{m}_{n,1}, \ldots, \widetilde{m}_{n,k}\right)$ converges almost surely to $g\left(m_1, \ldots, m_k\right) = \vartheta$.

Answer 5 Because of $X_1 - a \overset{\mathcal{D}}{=} a - X_1$, it follows that $\mathbb{E}_F(X_1) - a = a - \mathbb{E}_F(X_1)$ and thus $\mathbb{E}_F(X_1) = a$.

Answer 6 The r-smallest value of $X_1, \ldots, X_n$ is less than or equal to t if and only if at least r of the $X_1, \ldots, X_n$ are less than or equal to t.

Problems

11.1 Problem The random variables $X_1, \ldots, X_n$ are independent and have the same Poisson distribution $\mathrm{Po}(\vartheta)$, where $\vartheta \in \Theta := (0, \infty)$ is unknown. Show that, among all unbiased estimators of ϑ, the estimator $T_n := \frac{1}{n}\sum_{j=1}^n X_j$ has the uniformly smallest variance.

11.2 Problem The random variables $X_1, \ldots, X_n$ are i.i.d. with $X_1 \sim \mathrm{Bin}(1, \vartheta)$, where $\vartheta \in \Theta := [0, 1]$ is unknown. Show that, among all unbiased estimators of ϑ, the estimator $T_n := \frac{1}{n}\sum_{j=1}^n X_j$ has the uniformly smallest variance.

11.3 Problem Let $X_1, \ldots, X_n$ be i.i.d. random variables with the normal distribution $N(\mu, \sigma^2)$. Here, μ is *known* and $\vartheta := \sigma^2 \in \Theta := (0, \infty)$ is unknown. Prove the following claims:

(a) The Fisher information is given by $I_1(\vartheta) = \frac{1}{2\vartheta^2}$.
(b) The estimator defined by $T_n := \frac{1}{n} \sum_{j=1}^{n}(X_j - \mu)^2$ is unbiased for ϑ, and $\mathbb{V}_\vartheta(T_n) = \frac{2\vartheta^2}{n} = \frac{1}{nI_1(\vartheta)}$, $\vartheta \in \Theta$.
(c) Let $S_n := \frac{1}{n+2} \sum_{j=1}^{n}(X_j - \mu)^2$. Then

$$\mathbb{E}_\vartheta\big[(S_n - \vartheta)^2\big] = \frac{2\vartheta^2}{n+2} < \mathbb{E}_\vartheta\big[(T_n - \vartheta)^2\big], \quad \vartheta \in \Theta.$$

11.4 Problem Let Y be a k-dimensional random column vector on a probability space $(\Omega, \mathcal{A}, \mathbb{P})$ and $0_{k \times k}$ the $k \times k$ zero matrix. Show:

(a) $\mathbb{E}(YY^\top) \geq_L 0_{k \times k}$,
(b) $\mathbb{E}(YY^\top) = 0_{k \times k} \Longleftrightarrow Y = 0_d$ $\mathbb{P}$-almost surely.

11.5 Problem Suppose $X_1, X_2, \ldots$ are i.i.d. random variables with the normal distribution $N(\vartheta, 1)$, where $\vartheta \in \Theta := \mathbb{R}$.

(a) Show that the Fisher information is given by $I_1(\vartheta) = 1$.
(b) Let $\overline{X}_n = \frac{1}{n} \sum_{j=1}^{n} X_j$ be the MLE of ϑ. Since $\sqrt{n}(\overline{X}_n - \vartheta) \overset{\mathcal{D}_\vartheta}{\longrightarrow} N(0, 1)$, the sequence of estimators $(\overline{X}_n)$ is asymptotically efficient. Put

$$T_n := \overline{X}_n \mathbb{1}\big\{|\overline{X}_n| > n^{-1/4}\big\} + \frac{1}{2}\overline{X}_n \mathbb{1}\big\{|\overline{X}_n| \leq n^{-1/4}\big\}.$$

Show that $\sqrt{n}(T_n - \vartheta) \overset{\mathcal{D}_\vartheta}{\longrightarrow} N(0, \sigma^2(\vartheta))$, where $\sigma^2(\vartheta) = 1$ if $\vartheta \neq 0$ and $\sigma^2(\vartheta) = \frac{1}{4}$ if $\vartheta = 0$. Consequently, (T_n) is "super efficient" in the case $\vartheta = 0$.

11.6 Problem Let $X_1, X_2, \ldots$ be a sequence of i.i.d. random variables with $X_1 \sim Po(\vartheta)$, where $\vartheta \in (0, \infty)$ is unknown. Determine the MLE $\widehat{\vartheta}_n = \widehat{\vartheta}_n(X_1, \ldots, X_n)$ and directly prove consistency, asymptotic normality, and asymptotic efficiency.

11.7 Problem Show that the moment estimator satisfies a representation of the form (10.30) (with $\widetilde{\vartheta}_n$ instead of $\widehat{\vartheta}_n$). What does the influence function $\ell(\cdot, \vartheta)$ look like?

11.8 Problem Let $X_1, X_2, \ldots$ be a sequence of i.i.d. random variables with the Gamma distribution $\Gamma(\alpha, \lambda)$, that is, with the density

$$f(x, \alpha, \lambda) = \frac{\lambda^\alpha}{\Gamma(\alpha)} x^{\alpha-1} e^{-\lambda x}, \quad x > 0,$$

and $f(x, \alpha, \lambda) := 0$, otherwise. Here, $\alpha > 0$ and $\lambda > 0$ are unknown parameters.

(a) Show that $\mathbb{E}(X_1) = \frac{\alpha}{\lambda}$, $\mathbb{V}(X_1) = \frac{\alpha}{\lambda^2}$.
(b) With the notations in (11.13), show that the moment estimator $\tilde{\vartheta}_n := (\tilde{\alpha}_n, \tilde{\lambda}_n)^\top$ of $\vartheta := (\alpha, \lambda)^\top$ based on $X_1, \ldots, X_n$ is

$$\tilde{\alpha}_n = \frac{\tilde{m}_{n,1}^2}{\tilde{m}_{n,2} - \tilde{m}_{n,1}^2}, \quad \tilde{\lambda}_n = \frac{\tilde{m}_{n,1}}{\tilde{m}_{n,2} - \tilde{m}_{n,1}^2}.$$

(c) By Theorem 11.3, $\sqrt{n}(\tilde{\vartheta}_n - \vartheta)$ has a bivariate normal distribution with covariance matrix $g'(a)Tg'(a)^\top$ as $n \to \infty$. Can you specify $g'(a)$ and T?

11.9 Problem Let $X_1, X_2, \ldots$ be a sequence of i.i.d. random variables with distribution function F and $p \in (0, 1)$. Suppose F is differentiable at the point $F^{-1}(p)$, and we have $f(F^{-1}(p)) := F'(F^{-1}(p)) > 0$. Let r_n be a sequence of integers with $r_n \in \{1, \ldots, n\}$ for each n and

$$\frac{r_n}{n} = p + o\left(\frac{1}{\sqrt{n}}\right) \tag{11.28}$$

as $n \to \infty$. Furthermore, let $X_{n:1} \le X_{n:2} \le \cdots \le X_{n:n}$ be the order statistics of $X_1, \ldots, X_n$.

(a) Prove the following central limit theorem for order statistics:

$$\sqrt{n}\left(X_{n:r_n} - F^{-1}(p)\right) \xrightarrow{D} N\left(0, \frac{p(1-p)}{f^2(F^{-1}(p))}\right).$$

(b) Prove the statement (11.26).

Hint for (a) Use Problems 1.5 and 1.6 as well as (11.25).

Hint for (b) Use the inequalities $X_{n:\frac{n}{2}} \le \frac{1}{2}\left(X_{n:\frac{n}{2}} + X_{n:\frac{n}{2}+1}\right) \le X_{n:\frac{n}{2}+1}$.

11.10 Problem Let $X_1, X_2, \ldots$ be a sequence of i.i.d. random variables with the so-called *Laplace distribution*, also known as *double exponential distribution*, which has the density

$$f(x, a, \sigma) := \frac{1}{2\sigma} \exp\left(-\frac{|x-a|}{\sigma}\right), \quad x \in \mathbb{R},$$

where $a \in \mathbb{R}$ and $\sigma > 0$ are unknown parameters. Since the distribution is symmetric around a and the variance exists, both the sample mean and the sample median of $X_1, \ldots, X_n$ are suggested as estimators of a as in Example 11.6. Which of these estimators is better in terms of asymptotic relative efficiency?

$$\sigma(x, \mu) = \frac{1}{\mu} \left(\frac{1}{x} - \frac{1}{x} \right)$$

Likelihood Ratio Tests

12

This chapter deals with the testing of hypotheses in parametric models, specifically within the framework outlined in Chaps. 9 and 10. Let $X_1, X_2, \ldots$ be a sequence of i.i.d. random variables that take on values in $\mathcal{X}_0$, and which have an unknown distribution Q_ϑ, where $\vartheta \in \Theta$. The parameter space Θ is an open subset of $\mathbb{R}^k$, and there exists a σ-finite measure ν defined on the σ-field $\mathcal{B}_0$, which dominates each Q_ϑ with $\vartheta \in \Theta$. As in Chap. 10 we denote $f(\cdot, \vartheta)$ the density of Q_ϑ with respect to ν. We assume that this density satisfies the regularity conditions 10.12 in connection with the main theorem on MLE's. We also continue to use the canonical model stated in Chap. 9, that is, the set $\mathcal{X} := \{\mathbf{x} = (x_1, x_2, \ldots) : x_j \in \mathcal{X}_0 \text{ for each } j \geq 1\}$, where the random variable X_j assigns the value x_j to such a sequence $\mathbf{x}$.

In contrast to an estimation problem, in a *testing problem* the parameter space Θ is decomposed into two non-empty, disjoint subsets. Thus, we have

$$\Theta = \Theta_0 \uplus \Theta_1,$$

where $\Theta_0 \neq \emptyset$ and $\Theta_1 \neq \emptyset$. The

$$hypothesis\ H_0 : \vartheta \in \Theta_0$$

is to be tested versus the

$$alternative\ H_1 : \vartheta \in \Theta_1.$$

A *statistical test* of H_0 versus H_1 with sample size n is a measurable mapping $\varphi_n : \mathcal{X} \to [0, 1]$ with the requirement that $\varphi_n(\mathbf{x})$ depends on $\mathbf{x} = (x_1, x_2, \ldots)$ only through

$x_1, \ldots, x_n$. Here, $x_1, \ldots, x_n$ can be regarded as the data available for carrying out the test. If $\varphi_n(\mathbf{x}) = 1$ then the hypothesis H_0 is rejected, and in the case $\varphi_n(\mathbf{x}) = 0$, no objection is raised to H_0. If $0 < \varphi_n(\mathbf{x}) < 1$, one decides with probability $\varphi_n(\mathbf{x})$ against H_0, and with probability $1 - \varphi_n(\mathbf{x})$, no objection is raised to H_0. In the latter case, a pseudo-random number generator is needed to make a test decision. This *randomization*, which does not take into account the data $x_1, \ldots, x_n$, is necessary to obtain optimal tests in certain cases (see the explanations before Self-question 1). However, in practice, randomization is undesirable, because the decision of the test would then not depend solely on the data. Moreover, for asymptotic studies in connection with tests, randomization has no significance.

In what follows, we only consider so-called *non-randomized* tests. These are characterized by the fact that the case $0 < \varphi_n(\mathbf{x}) < 1$ is excluded. As a $\{0, 1\}$-valued mapping, φ_n is then an indicator function of a measurable subset $\mathcal{K}_n$ of $\mathcal{X}$, which is called the *critical region of the test*. The set $\mathcal{K}_n$ also only makes use of $x_1, \ldots, x_n$, so it is of the form $\mathcal{K}_n = \mathcal{K}_n^* \times (\mathcal{X}_0 \times \mathcal{X}_0 \times \cdots)$ with a subset $\mathcal{K}_n^*$ of $\mathcal{X}_0^n$. The test φ_n rejects the hypothesis H_0 if the data falls into the critical region, that is, if $(x_1, \ldots, x_n) \in \mathcal{K}_n^*$ or, equivalently, if $\mathbf{x} \in \mathcal{K}_n$. Otherwise, no objection is raised to H_0. The critical region $\mathcal{K}_n$ is often of the form

$$\mathcal{K}_n = \{T_n > c\} = \{\mathbf{x} \in \mathcal{X} : T_n(\mathbf{x}) > c\}.$$

Here, $T_n : \mathcal{X} \to \mathbb{R}$ is a measurable mapping, which, like $\mathcal{K}_n$, depends only on $x_1, \ldots, x_n$, and which is called the *test statistic*. The constant c is referred to as the *critical value*. The hypothesis H_0 is therefore rejected if T_n exceeds such a critical value. An example is provided by the test statistic T_n, given in (6.7), of the chi-square test. We will return to this test later.

In general, given $x_1, \ldots, x_n$, $f_n(x_1, \ldots, x_n, \vartheta) = \prod_{j=1}^n f(x_j, \vartheta)$ will be strictly positive for each ϑ in Θ. Therefore, it is inevitable that we may commit errors when performing a statistical test. A *type I error* is a mistaken rejection of H_0. In other words, this type of error, which sometimes is also called an *error of the first kind*, occurs if we reject H_0 and H_0 is true, that is, when $\vartheta \in \Theta_0$ and $(x_1, \ldots, x_n) \in \mathcal{K}_n^*$. A *type II error* is a mistaken acceptance of H_0, which means that $\vartheta \in \Theta_1$ and $(x_1, \ldots, x_n) \in \mathcal{X}_0^n \setminus \mathcal{K}_n^*$. This type of error is sometimes called an *error of the second kind.*. Regarded as a function of ϑ, the probability that the test φ_n rejects the hypothesis H_0 is called the *power function* of the test φ_n and is given by

$$\Theta \ni \vartheta \mapsto g_{\varphi_n}(\vartheta) := \mathbb{E}_\vartheta(\varphi_n) = \mathbb{P}_\vartheta((X_1, \ldots, X_n) \in \mathcal{K}_n^*).$$

For $\alpha \in (0, 1)$, a test φ_n is said to have *level (of significance)* α and we speak of a *level α test*, if $g_{\varphi_n}(\vartheta) \le \alpha$ for each $\vartheta \in \Theta_0$. With such a test, a type I error is thus committed with a probability of at most α. Since a test of level α is also of level α^* for each $\alpha^* \in (\alpha, 1)$, one usually gives a name to the smallest level of signifiance, which is then called the *size* of a test. It is common to limit oneself to level α tests with small α, such as $\alpha = 0.05$. This limitation *serves to protect the alternative* H_1. If such a test rejects H_0, one can be "practically sure that H_0 does not hold", since otherwise one would have arrived at this test result with a maximum (small) probability of at most α. If $\vartheta \in \Theta_1$, the test φ_n should reject H_0 if possible; the power function should therefore take as large values as possible on Θ_1. This vaguely formulated optimality criterion is obvious, and it can be specified, but for a fixed sample size n, optimal tests exist only in comparatively simple cases.

In connection with sequences of tests of H_0 versus H_1, the following concepts, based on the power function, are fundamental.

12.1 Definition (Asymptotic Level, Consistency)
Let $\alpha \in (0, 1)$ and (φ_n) be a sequence of tests $\varphi_n : \mathcal{X} \to [0, 1]$ of $H_0 : \vartheta \in \Theta_0$ versus $H_1 : \vartheta \in \Theta_1$. The sequence (φ_n)

(a) *is of (pointwise) asymptotic level* α, if

$$\limsup_{n \to \infty} g_{\varphi_n}(\vartheta) \le \alpha \quad \text{for every } \vartheta \in \Theta_0, \tag{12.1}$$

(b) *is consistent against any fixed alternative*, if

$$\lim_{n \to \infty} g_{\varphi_n}(\vartheta) = 1 \quad \text{for every } \vartheta \in \Theta_1.$$

Property (12.1) controls the probability of a type I error for large sample sizes. A stronger requirement than (12.1) is

$$\limsup_{n \to \infty} \sup_{\vartheta \in \Theta_0} g_{\varphi_n}(\vartheta) \le \alpha.$$

In this case, we say that the sequence (φ_n) of tests *has a uniform asymptotic level* α. It is often the case that

$$\lim_{n \to \infty} g_{\varphi_n}(\vartheta) = \alpha \quad \text{for every } \vartheta \in \Theta_0,$$

which is also stronger than (12.1).

The property of consistency is a weak requirement, because one should be able to detect any fixed alternative to H_0 with ever greater probability as the sample size grows beyond all limits.

12.2 Example (One-Sided Binomial Test)
Suppose $X_1, X_2, \ldots$ are i.i.d. $\{0, 1\}$-valued random variables with $\mathbb{P}_\vartheta(X_1 = 1) = \vartheta$, where $\vartheta \in \Theta := [0, 1]$. Suppose further that $\Theta_0 = [0, \vartheta_0]$ and $\Theta_1 = (\vartheta_0, 1]$, where

$0 < \vartheta_0 < 1$. A test of $H_0 : \vartheta \le \vartheta_0$ versus $H_1 : \vartheta > \vartheta_0$, that is, $H_0 : \vartheta \in \Theta_0$ versus $H_1 : \vartheta \in \Theta_1$, is called a *one-sided binomial test*. Note that $S_n := X_1 + \ldots + X_n$ has the binomial distribution $\mathrm{Bin}(n, \vartheta)$ under $\mathbb{P}_\vartheta$. To obtain a sequence (φ_n) of tests of asymptotic level α, we put

$$\varphi_n(\mathbf{x}) := \mathbf{1}\left\{ \sum_{j=1}^n x_j > n\vartheta_0 + \Phi^{-1}(1-\alpha)\sqrt{n\vartheta_0(1-\vartheta_0)} \right\}, \quad \mathbf{x} = (x_j)_{j\ge 1} \in \{0, 1\}^{\mathbb{N}},$$

where $\Phi^{-1}(1-\alpha)$ is the $(1-\alpha)$-quantile of the standard normal distribution. From the Lindeberg–Lévy central limit theorem, it follows that, as $n \to \infty$,

$$g_{\varphi_n}(\vartheta_0) = \mathbb{E}_{\vartheta_0}(\varphi_n) = \mathbb{P}_{\vartheta_0}\left(\frac{S_n - n\vartheta_0}{\sqrt{n\vartheta_0(1-\vartheta_0)}} > \Phi^{-1}(1-\alpha) \right)$$

$$\to 1 - \Phi(\Phi^{-1}(1-\alpha)) = \alpha.$$

Since for each $\ell \in \{0, 1, \ldots, n-1\}$

$$\mathbb{P}_\vartheta(S_n > \ell) = \sum_{j=\ell+1}^n \binom{n}{j} \vartheta^j (1-\vartheta)^{n-j} = \frac{1}{\ell!(n-\ell-1)!} \int_0^\vartheta t^\ell (1-t)^{n-\ell-1} \, dt, \quad (12.2)$$

the probability $\mathbb{P}_\vartheta(S_n > \ell)$ is a strictly increasing function of ϑ, and thus the sequence (φ_n) is of pointwise asymptotic level α and even of uniform asymptotic level α.

Now fix any $\vartheta_1 \in (\vartheta_0, 1)$. Putting $t_n = ((\sqrt{n}(\vartheta_0 - \vartheta_1) + \sqrt{\vartheta_0(1-\vartheta_0)}))/\sqrt{\vartheta_1(1-\vartheta_1)}$, we have

$$g_{\varphi_n}(\vartheta_1) = \mathbb{E}_{\vartheta_1}(\varphi_n) = \mathbb{P}_{\vartheta_1}\left(\frac{S_n - n\vartheta_1}{\sqrt{n\vartheta_1(1-\vartheta_1)}} > t_n \right).$$

Since $t_n \to -\infty$ as $n \to \infty$, again by the central limit theorem, it follows that $g_{\varphi_n}(\vartheta_1) \to 1$ as $n \to \infty$. Thus, the sequence (φ_n) is consistent against each alternative in Θ_1.

Self-question 1: Why does the second equals sign in (12.2) hold?

A hypothesis H_0 is called *simple*, if $|\Theta_0| = 1$, otherwise it is said to be *composite*. The same terminology is used analogously for the alternative H_1. To test a simple hypothesis $H_0 : \vartheta = \vartheta_0$ versus a simple alternative $H_1 : \vartheta = \vartheta_1$, that is, for the case $\Theta_0 = \{\vartheta_0\}$ and

$\Theta_1 = \{\vartheta_1\}$, there is an optimal test, dating back to J. Neyman[1] and E.S. Pearson.[2] This so-called *Neyman–Pearson test* uses as a test statistic the so-called *likelihood ratio*

$$L_n(x_1, \ldots, x_n) = \prod_{j=1}^{n} \frac{f(x_j, \vartheta_1)}{f(x_j, \vartheta_0)} \tag{12.3}$$

and rejects the hypothesis H_0 if $L_n(x_1, \ldots, x_n) > c$. Here, c is a critical value which, putting $L_n := L_n(X_1, \ldots, X_n)$, is obtained from the equation $\mathbb{P}_{\vartheta_0}(L_n > c) = \alpha$. If this equation does not hold for any c, then we set $c := \inf\{c^* > 0 : \mathbb{P}_{\vartheta_0}(L_n > c^*) \le \alpha\}$ and

$$\gamma := \frac{\alpha - \mathbb{P}_{\vartheta_0}(L_n > c)}{\mathbb{P}_{\vartheta_0}(L_n = c)}.$$

Moreover, we reject H_0 not only in the case $L_n > c$, but also with the probability γ if $L_n = c$. In the latter case, a pseudorandom number generator is needed. The idea of this so-called *randomization* is to fully exhaust the permitted maximum probability α of a type I error.

Self-question 2: Why does $0 < \gamma < 1$ hold under the given conditions?

The likelihood ratio in (12.3) serves as a guideline for a frequently used test principle in the case that H_0 and/or H_1 are composite. This strategic principle for the design of test statistics is based on the following notion.

12.3 Definition (Generalized Likelihood Ratio)
The quotient

$$\Lambda_n(x_1, \ldots, x_n) := \frac{\sup_{\vartheta \in \Theta_0} \prod_{j=1}^{n} f(x_j, \vartheta)}{\sup_{\vartheta \in \Theta} \prod_{j=1}^{n} f(x_j, \vartheta)} \tag{12.4}$$

is called *generalized likelihood ratio* (for short: GLR).

---

[1] Jerzy Neyman (1894–1981), from 1938 Professor at the University of California, Berkeley. Through Neyman, Berkeley became a world-famous center of statistics. Neyman supported the civil rights movement of *Martin Luther King* (1929–1968) and was part of the anti-Vietnam War movement. In 1968 he received the Medal of Science, the highest scientific award in the USA.

[2] Egon Sharpe Pearson (1895–1980), son of Karl Pearson, from 1933 head of the statistical department at Univ. College London. Main areas of work: Mathematical statistics, biometrics. Between 1928 and 1938 there were 10 joint publications with J. Neyman.

Note that the numerator figuring in (12.4) is $\prod_{j=1}^{n} f(x_j, \vartheta_0)$, if $\Theta_0 = \{\vartheta_0\}$, that is, if H_0 is a simple hypothesis. Otherwise, one has to find a MLE of ϑ *under the constraint* $\vartheta \in \Theta_0$. Regarding the denominator in (12.4), there is the familiar task of finding a MLE of ϑ without any constraint, which was a topic of Chap. 10. Even before Definition 10.14, it was emphasized that the supremum in the denominator of (12.4) does not necessarily have to be attained, and the same applies to the supremum in the numerator. Regarding $\Lambda_n(x_1, \ldots, x_n)$, we also suppose that asymptotic MLE's exist for the optimization problems in the numerator and denominator of (12.4) in the sense of Definition 10.14. So we assume both (10.24) and

$$f_n\big(\mathbf{x}, \widehat{\vartheta}_{n,0}(\mathbf{x})\big) = \sup_{t \in \Theta_0} f_n(\mathbf{x}, t) \quad \text{for each } \mathbf{x} \in M_n''.$$

Here, (M_n'') is a sequence of measurable subsets of $\mathcal{X}$ with $\mathbb{P}_\vartheta(M_n'') \to 1$ as $n \to \infty$. The sequence $(\widehat{\vartheta}_{n,0})$ is thus an asymptotic MLE in the sense of Definition 10.14 under the constraint $\vartheta \in \Theta_0$. On the set $M_n' \cap M_n''$, we thus have

$$\Lambda_n := \Lambda_n(X_1, \ldots, X_n) = \frac{\prod_{j=1}^{n} f(X_j, \widehat{\vartheta}_{n,0})}{\prod_{j=1}^{n} f(X_j, \widehat{\vartheta}_n)}. \qquad (12.5)$$

By definition, $\Lambda_n(x_1, \ldots, x_n) \le 1$. If H_0 does not hold, we may expect that the denominator in (12.4) is significantly larger than the numerator. One would therefore reject the hypothesis H_0 for "small" values of $\Lambda_n(x_1, \ldots, x_n)$, and such a test is subsequently called a *generalized likelihood ratio test* (for short: GLRT). Of course, it must be specified what the pleasantly vague wording *small values* means, and this will be done soon. We will see that, under the hypothesis H_0, the random variable M_n resulting from Λ_n by means of the transformation

$$M_n := -2 \log \Lambda_n \qquad (12.6)$$

can be approximated by a suitable quadratic form of an asymptotically normally distributed random vector and thus has an asymptotic chi-square distribution as $n \to \infty$. For simplicity, we first examine the case that H_0 is simple, that is, $\Theta_0 = \{\vartheta_0\}$ for some ϑ_0 in Θ. In this case, there is no need to find a supremum in the numerator of (12.4).

12.4 Theorem (GLRT, Simple Hypothesis)
Let $\Theta_0 = \{\vartheta_0\}$ and M_n as in (12.6). Under the standing assumptions,

$$M_n \xrightarrow{\mathcal{D}_{\vartheta_0}} \chi_k^2 \quad \text{as } n \to \infty.$$

Proof Since

$$\Lambda_n = \frac{\sup_{\vartheta \in \Theta_0} \prod_{j=1}^n f(X_j, \vartheta)}{\sup_{\vartheta \in \Theta} \prod_{j=1}^n f(X_j, \vartheta)} = \prod_{j=1}^n \frac{f(X_j, \vartheta_0)}{f(X_j, \widehat{\vartheta}_n)},$$

it follows that

$$M_n = -2 \log \Lambda_n = 2 \sum_{j=1}^n \left[\log f(X_j, \widehat{\vartheta}_n) - \log f(X_j, \vartheta_0) \right].$$

With $U_n(t)$ as in (10.25) and $H_n(t)$ as in (10.26), a Taylor expansion of $\log f(X_j, \vartheta)$ around $\vartheta = \vartheta_0$ yields

$$M_n = 2 \left(U_n^\top(\vartheta_0)(\widehat{\vartheta}_n - \vartheta_0) + \frac{1}{2}(\widehat{\vartheta}_n - \vartheta_0)^\top H_n(\vartheta_0)(\widehat{\vartheta}_n - \vartheta_0) + o_{\mathbb{P}_{\vartheta_0}}(1) \right).$$

Since $0_k = U_n(\widehat{\vartheta}_n) = U_n(\vartheta_0) + H_n(\vartheta_0)(\widehat{\vartheta}_n - \vartheta_0) + o_{\mathbb{P}_{\vartheta_0}}(\sqrt{n})$ and $\sqrt{n}(\widehat{\vartheta}_n - \vartheta_0) = O_{\mathbb{P}_{\vartheta_0}}(1)$, putting $Z_n := \sqrt{n}(\widehat{\vartheta}_n - \vartheta_0)$ gives

$$M_n = 2 \left(-\frac{1}{2}(\widehat{\vartheta}_n - \vartheta_0)^\top H_n(\vartheta_0)(\widehat{\vartheta}_n - \vartheta_0) + o_{\mathbb{P}_{\vartheta_0}}(1) \right)$$

$$= Z_n^\top \left(-\frac{1}{n} H_n(\vartheta_0) \right) Z_n + o_{\mathbb{P}_{\vartheta_0}}(1).$$

By the main Theorem 10.15 about MLE's, $Z_n \overset{\mathcal{D}_{\vartheta_0}}{\longrightarrow} Z$, where $Z \sim N_k(0_k, I_1(\vartheta_0)^{-1})$. Finally, because $-\frac{1}{n} H_n(\vartheta_0) \to I_1(\vartheta_0)$ $\mathbb{P}_{\vartheta_0}$-a.s., the mapping theorem and Theorem 5.16 imply

$$M_n \overset{\mathcal{D}_{\vartheta_0}}{\longrightarrow} Z^\top I_1(\vartheta_0) Z \sim \chi_k^2.$$

$\square$

An important message of Theorem 12.4 is that the limit distribution of M_n under H_0 does not depend on the specific value of ϑ_0. Since the mapping $\Lambda_n \mapsto M_n = -2 \log \Lambda_n$ is a continuous and strictly decreasing transformation, we thus obtain a GLRT of $H_0 : \vartheta = \vartheta_0$ versus $H_1 : \vartheta \neq \vartheta_0$ at asymptotic level α by rejecting H_0 if

$$M_n > \chi_{k;1-\alpha}^2.$$

Here, $\chi^2_{\ell;1-\alpha}$ is the general notation for the $(1-\alpha)$-quantile of the chi-square distribution with ℓ degrees of freedom.

Regarding the consistency of the sequence of tests (φ_n) defined by $\varphi_n := 1\{M_n > \chi^2_{k;1-\alpha}\}$, we have the following result:

12.5 Theorem (Consistency of the GLRT)

In the setting of Theorem 12.4, the sequence (φ_n) of GLRT's defined by $\varphi_n := 1\{M_n > \chi^2_{k;1-\alpha}\}$ is consistent against any fixed alternative, that is, we have

$$\lim_{n\to\infty} \mathbb{E}_\vartheta(\varphi_n) = 1 \quad \text{for every } \vartheta \in \Theta \text{ with } \vartheta \neq \vartheta_0.$$

Proof Fix any $\vartheta_1 \in \Theta$ with $\vartheta_1 \neq \vartheta_0$. Then

$$\Lambda_n = \prod_{j=1}^n \frac{f(X_j, \vartheta_0)}{f(X_j, \vartheta_1)} \cdot \prod_{j=1}^n \frac{f(X_j, \vartheta_1)}{f(X_j, \widehat{\vartheta}_n)}$$

and thus

$$M_n = -2\log \Lambda_n = Y_n + 2n V_n, \tag{12.7}$$

where

$$Y_n := 2 \sum_{j=1}^n \left\{ \log f(X_j, \widehat{\vartheta}_n) - \log f(X_j, \vartheta_1) \right\},$$

$$V_n := \frac{1}{n} \sum_{j=1}^n \log \frac{f(X_j, \vartheta_1)}{f(X_j, \vartheta_0)}.$$

By Theorem 12.4, $Y_n \xrightarrow{\mathcal{D}_{\vartheta_1}} \chi^2_k$ as $n \to \infty$, and the strong law of large numbers yields

$$V_n \to \mathbb{E}_{\vartheta_1}\left[\log \frac{f(X_1, \vartheta_1)}{f(X_1, \vartheta_0)} \right] \quad \mathbb{P}_{\vartheta_1}\text{-almost surely.}$$

According to (10.9), the above expectation is the Kullback–Leibler information $I_{KL}(\vartheta_1 : \vartheta_0)$. Since $I_{KL}(\vartheta_1 : \vartheta_0) > 0$, it follows that $\lim_{n\to\infty} \mathbb{P}_{\vartheta_1}(M_n > c) = 1$ for every $c > 0$. □

Self-question 3: Can you prove $\lim_{n\to\infty} \mathbb{P}_{\vartheta_1}(M_n > c) = 1$?

12.6 Example (Multinomial Distribution, Chi-Square Goodness-of-Fit Test)

As in the setting of Example 6.20, we consider repeated independent trials under the same condition. Each of these trials has s ($s \geq 2$) possible outcomes, which are denoted $1, 2, \ldots, s$. The outcome j, referred to as *hit of the j-th kind*, has the unknown probability $p_j, j \in \{1, \ldots, s\}$. Here, $p_1, \ldots, p_s$ are positive numbers with $p_1 + \ldots + p_s = 1$. Because of this equation, p_s is redundant, and so our parameter space is the open subset

$$\Theta := \left\{ \vartheta := (p_1, \ldots, p_{s-1}) : p_1 > 0, \ldots, p_{s-1} > 0, \ p_1 + \ldots + p_{s-1} < 1 \right\}$$

of $\mathbb{R}^k$, where $k := s - 1$.

This situation is modeled by a sequence $X_1, X_2, \ldots$ of i.i.d. s-dimensional random vectors, where $\mathbb{P}_\vartheta (X_1 = e_j) = p_j$, $j \in \{1, \ldots, s\}$. Here, e_j denotes the j-th canonical unit vector in $\mathbb{R}^s$. If ν is the counting measure on $\{e_1, \ldots, e_s\}$, then X_1 has the density

$$f(t, \vartheta) := \begin{cases} p_j, & \text{if } t = e_j, \quad j = 1, \ldots, s, \\ 0, & \text{otherwise,} \end{cases}$$

with respect to ν. Consequently, the random vector $(N_{n,1}, N_{n,2}, \ldots, N_{n,s}) := \sum_{j=1}^n X_j$ has the multinomial distribution $\mathrm{Mult}(n; p_1, \ldots, p_s)$. We thus have

$$\mathbb{P}_\vartheta (N_{n,1} = k_1, \ldots, N_{n,s} = k_s) = \frac{n!}{k_1! \cdot \ldots \cdot k_s!} \, p_1^{k_1} \cdot \ldots \cdot p_s^{k_s}$$

for any choice of $(k_1, \ldots, k_s) \in \mathbb{N}_0^s$ with $k_1 + \ldots + k_s = n$. Moreover, the joint density of $X_1, \ldots, X_n$ is given by

$$\prod_{j=1}^n f(X_j, \vartheta) = p_1^{N_{n,1}} p_2^{N_{n,2}} \cdot \ldots \cdot p_s^{N_{n,s}}. \tag{12.8}$$

To state the testing problem, let $\Theta_0 := \{\vartheta_0\} = \{(q_1, \ldots, q_{s-1})\}$, where $\vartheta_0 \in \Theta$, and we set $q_s := 1 - q_1 - \ldots - q_{s-1}$. The problem is to test the simple hypothesis $H_0 : \vartheta = \vartheta_0$— that is, $p_j = q_j$ for each $j \in \{1, \ldots, s\}$—against the general alternative $H_1 : \vartheta \notin \Theta_0$. The MLE of ϑ is given by the vector

$$\widehat{\vartheta}_n := \left(\frac{N_{n,1}}{n}, \ldots, \frac{N_{n,s-1}}{n} \right) \tag{12.9}$$

of individual relative hit frequencies (cf. Problem 12.1). We briefly set $\widehat{p}_{n,j} := \frac{1}{n} N_{n,j}$ for $j \in \{1, \ldots, s\}$, which in particular yields $\widehat{p}_{n,1} + \ldots + \widehat{p}_{n,s} = 1$. With these notations, the GLR occurring in (12.5) takes the form

$$\Lambda_n = \frac{q_1^{N_{n,1}} \cdot \ldots \cdot q_s^{N_{n,s}}}{\widehat{p}_{n,1}^{N_{n,1}} \cdot \ldots \cdot \widehat{p}_{n,s}^{N_{n,s}}} = \prod_{j=1}^{s} \left(\frac{q_j}{\widehat{p}_{n,j}} \right)^{N_{n,j}},$$

where $0^0 := 1$ as usual. Thus, the test statistic M_n of the GLRT becomes

$$M_n = 2 \sum_{j=1}^{s} N_{n,j} \log \left(\frac{N_{n,j}}{n q_j} \right). \tag{12.10}$$

By Theorem 12.4, $M_n \xrightarrow{D_{\vartheta_0}} \chi^2_{s-1}$. To obtain a sequence of tests having asymptotic level α, we thus reject H_0 if $M_n > \chi^2_{s-1;1-\alpha}$.

In the setting of this example, the more familiar test statistic is the so-called *Chi-square test statistic* T_n given in (6.7). Intuitively, T_n is more appealing than M_n because it directly compares the hit numbers $N_{n,j}$ with the numbers $n q_j$ expected under H_0. Using the multivariate central limit theorem and the mapping theorem, we proved in Example 6.20 that the limit distribution of T_n under H_0 is the χ^2_{s-1}-distribution and thus the same limit distribution as that of M_n. This is by no means accidental since $T_n - M_n \xrightarrow{P_{\vartheta_0}} 0$ (Problem 12.2). Thus, M_n and T_n are closely related.

The most important applications of the GLR (12.4) are cases in which the hypothesis H_0 is composite. We assume that the set Θ_0 is of the form $\Theta_0 = h(U)$ for an $\ell \in \{1, \ldots, k - 1\}$, where U is an open subset of $\mathbb{R}^\ell$ and $h : U \to \mathbb{R}^k$ is a twice continuously differentiable, injective function.

12.7 Example (Discrete Bivariate Random Vector and Independence)
Let (X, Y) be a discrete two-dimensional random vector, where

$$p_{i,j} := \mathbb{P}(X = x_i, Y = y_j) > 0, \quad i \in \{1, \ldots, r\}, \ j \in \{1, \ldots, s\}, \quad \sum_{i=1}^{r} \sum_{j=1}^{s} p_{i,j} = 1,$$

and $r, s \geq 2$. If we define Θ as the set of all vectors

$$\vartheta := (p_{1,1}, \ldots, p_{1,s}, p_{2,1}, \ldots, p_{2,s}, \ldots, p_{r,1}, \ldots p_{r,s-1})$$

with positive components, whose sum is less than 1, then Θ is an open subset of $\mathbb{R}^k$, where $k = rs - 1$. If ν denotes the counting measure on the set $\mathcal{X}_0 := \{x_1, \ldots, x_r\} \times \{y_1, \ldots, y_s\}$, then with $z := (x_i, y_j) \in \mathcal{X}_0$ and $f(z, \vartheta) := p_{i,j}$, the function $f(\cdot, \vartheta)$ is the density of (X, Y) with respect to ν under Q_ϑ, $\vartheta \in \Theta$.

The hypothesis H_0 of independence of X and Y is characterized by the equations

$$\mathbb{P}(X = x_i, Y = y_j) = \mathbb{P}(X = x_i)\,\mathbb{P}(Y = y_j), \quad \text{for all } i, j.$$

This is synonymous with the existence of positive numbers $p_1, \ldots, p_r$ with $p_1 + \ldots + p_r = 1$ and $q_1, \ldots, q_s$ with $q_1 + \ldots + q_s = 1$ such that $p_{i,j} = p_i q_j$ for each pair (i, j).

Self-question 4: Why does $p_{i,j} = p_i q_j \ \forall (i, j)$ provide independence of X and Y?

Because of $p_r = 1 - p_1 - \ldots - p_{r-1}$ and $q_s = 1 - q_1 - \ldots - q_{s-1}$, the hypothesis H_0 corresponds to the subset

$$\Theta_0 := \left\{ \vartheta \in \Theta : \exists p_1 > 0, \ldots, p_{r-1} > 0 \text{ with } \sum_{i=1}^{r-1} p_i < 1, \right.$$

$$\left. \exists q_1 > 0, \ldots, q_{s-1} > 0 \text{ with } \sum_{j=1}^{s-1} q_j < 1 \text{ and } p_{i,j} = p_i q_j \ \forall (i, j) \neq (r, s) \right\} \tag{12.11}$$

of Θ. We write U for the set of all vectors $u := (p_1, \ldots, p_{r-1}, q_1, \ldots, q_{s-1})$ with positive components and $\sum_{i=1}^{r-1} p_i < 1$ as well as $\sum_{j=1}^{s-1} q_j < 1$. Then $\Theta_0 = h(U)$, where

$$h := \left(h_{1,1}, \ldots, h_{1,s}, h_{2,1}, \ldots, h_{2,s}, \ldots, h_{r,1}, \ldots, h_{r,s-1} \right),$$

and $h_{i,j}(u) = p_i q_j$ for each $i \in \{1, \ldots, r\}$ and each $j \in \{1, \ldots, s\}$ with $(i, j) \neq (r, s)$. Here, we put $p_r := 1 - p_1 - \ldots - p_{r-1}$ and $q_s := 1 - q_1 - \ldots - q_{s-1}$. The set U is an open subset of $\mathbb{R}^\ell$, where $\ell = r - 1 + s - 1$.

In generalization of Theorem 12.4, we have the following result. Here, we recall the definition of the Fisher information matrix $I_1(\vartheta)$ in 10.12 (e).

12.8 Theorem (GLRT, Composite Hypothesis)
Let $X_1, X_2, \ldots$ be a sequence of i.i.d. random variables with ν-density $f(x, \vartheta)$, where $\vartheta \in \Theta \subset \mathbb{R}^k$. Suppose $f(x, \vartheta)$ satisfies the regularity conditions (a)–(e) stated in 10.12 in each point $\vartheta = h(u)$ with $u \in U$. The hypothesis to be tested is $H_0 : \vartheta \in \Theta_0 := \{h(u) : u \in U\}$. If $h'(u)$ denotes the $k \times \ell$ Jacobian matrix of h at u, then the matrix

$$\tilde{I}_1(u) := h'(u)^\top I_1(\vartheta) h'(u) \tag{12.12}$$

is invertible for each $u \in U$. Furthermore, let $(\widehat{\vartheta}_n)$ and $(\widehat{u}_n)$ be consistent sequences of asymptotic MLE's of ϑ and u, respectively. We then have

$$M_n = -2 \log \Lambda_n \xrightarrow{D_\vartheta} \chi^2_{k-\ell} \quad \text{for every } \vartheta \in \Theta_0.$$

Proof The sequence $(h(\widehat{u}_n))$ is a consistent MLE of ϑ *within the submodel given by* Θ_0, and it follows that

$$\Lambda_n = \prod_{j=1}^{n} \frac{f(X_j, h(\widehat{u}_n))}{f(X_j, \widehat{\vartheta}_n)} \tag{12.13}$$

and

$$M_n = 2 \sum_{j=1}^{n} \Big\{ \log f(X_j, \widehat{\vartheta}_n) - \log f(X_j, h(\widehat{u}_n)) \Big\}. \tag{12.14}$$

If you look at the expression within curly brackets, it is almost compelling to perform Taylor expansions, namely of $\log f(X_j, \widehat{\vartheta}_n)$ around ϑ and of $\log f(X_j, h(\widehat{u}_n))$ around u. If we set

$$Z_n := \sqrt{n}(\widehat{\vartheta}_n - \vartheta),$$

then—with $U_n(\vartheta)$ as in (10.25) and the convergence $n^{-1} H_n(\vartheta) \rightarrow -I_1(\vartheta)$ $\mathbb{P}_\vartheta$-a.s. (cf. (10.29)) for the Hessian matrix $H_n(\vartheta)$ defined in (10.26)—it follows that

$$\sum_{j=1}^{n} \log f(X_j, \widehat{\vartheta}_n) = \sum_{j=1}^{n} \log f(X_j, \vartheta) + \frac{1}{\sqrt{n}} U_n(\vartheta)^\top Z_n - \frac{1}{2} Z_n^\top I_1(\vartheta) Z_n + o_{\mathbb{P}_\vartheta}(1).$$

$$\tag{12.15}$$

The densities $\widetilde{f}(x, u) := f(x, h(u))$, $u \in U$, satisfy regularity conditions that correspond to those in 10.12 (a)–(e). By the chain rule, we have

$$\frac{\mathrm{d}}{\mathrm{d}u} \log \widetilde{f}(x, u) = h'(u)^\top \frac{\mathrm{d}}{\mathrm{d}\vartheta} \log f(x, \vartheta) \Big|_{\vartheta = h(u)}. \tag{12.16}$$

Putting

$$\widetilde{U}_n(u) := \sum_{j=1}^{n} \frac{\mathrm{d}}{\mathrm{d}u} \log \widetilde{f}(X_j, u),$$

$$\widetilde{Z}_n := \sqrt{n}(\widehat{u}_n - u),$$

then with $\vartheta = h(u)$ we obtain analogous to (12.15)

$$\sum_{j=1}^{n} \log f(X_j, h(\widehat{u}_n)) = \sum_{j=1}^{n} \log f(X_j, \vartheta) + \frac{1}{\sqrt{n}} \widetilde{U}_n(u)^\top \widetilde{Z}_n$$

$$- \frac{1}{2} \widetilde{Z}_n^\top h'(u)^\top I_1(\vartheta) h'(u) \widetilde{Z}_n + o_{\mathbb{P}_\vartheta}(1). \qquad (12.17)$$

From (12.16), it follows that

$$\widetilde{U}_n(u) = h'(u)^\top U_n(h(u)), \qquad (12.18)$$

and by analogy with (10.30), we get

$$\widetilde{Z}_n = \widetilde{I}_1(u)^{-1} \widetilde{U}_n(u) + o_{\mathbb{P}_\vartheta}(1). \qquad (12.19)$$

If we take into account $\vartheta = h(u)$ then (10.30), (12.18), and (12.19) imply

$$\widetilde{Z}_n = \widetilde{I}_1(u)^{-1} h'(u)^\top U_n(\vartheta)$$

$$= \widetilde{I}_1(u)^{-1} h'(u)^\top I_1(\vartheta) Z_n + o_{\mathbb{P}_\vartheta}(1). \qquad (12.20)$$

Putting

$$A := I_1(\vartheta) \left(I_k - h'(u) \widetilde{I}_1(u)^{-1} h'(u)^\top I_1(\vartheta) \right) \qquad (12.21)$$

and using (12.12), (12.15), (12.17), (12.18), and (12.20) as well as $n^{-1/2} U_n(\vartheta) = I_1(\vartheta) Z_n + o_{\mathbb{P}_\vartheta}(1)$, a direct calculation shows that M_n in (12.14) takes the form

$$M_n = Z_n^\top A Z_n + o_{\mathbb{P}_\vartheta}(1).$$

By the main Theorem 10.15 on MLE's, we have

$$Z_n \xrightarrow{\mathcal{D}_\vartheta} Z \sim N_k(0, \Sigma),$$

where $\Sigma := I_1(\vartheta)^{-1}$. The matrix A in (12.21) is symmetric, and the matrix $A\Sigma$ is idempotent, and it has rank $k - \ell$ (cf. Problem 12.4). By Problem 12.3, $Z^\top A Z \sim \chi^2_{k-\ell}$, so that the mapping theorem and Slutsky's lemma yield the assertion. $\qquad \square$

Since the limit distribution of M_n under H_0 does not depend on the specific value ϑ_0 in Θ_0, the GLRT for testing the composite null hypothesis H_0 is *asymptotically distribution-free*. As a corollary, it follows that the sequence of tests (φ_n) defined by

$$\varphi_n := \mathbf{1}\{ M_n > \chi^2_{k-\ell;1-\alpha} \}$$

has asymptotic level α, that is, we have

$$\lim_{n\to\infty} \mathbb{E}_\vartheta(\varphi_n) = \alpha \quad \text{for every } \vartheta \in \Theta_0.$$

In case of a composite hypothesis H_0 the consistency of the GLRT can often be shown directly (see, e.g., Problem 12.7 for the case of testing for independence in contingency tables considered in the next example). The following reasoning conveys the general idea why, under certain conditions, the GLRT is consistent against any fixed alternative. To this end, we assume the setting of Theorem 12.8, where $\Theta_0 = h(U)$, cf. the assumptions before Example 12.7. If we fix any $\vartheta_1 \in \Theta \setminus \Theta_0$, it follows with (12.13) that

$$\Lambda_n = \prod_{j=1}^{n} \frac{f(X_j, h(\widehat{u}_n))}{f(X_j, \vartheta_1)} \cdot \prod_{j=1}^{n} \frac{f(X_j, \vartheta_1)}{f(X_j, \widehat{\vartheta}_n)},$$

and thus

$$M_n = -2\log \Lambda_n = 2 \sum_{j=1}^{n} \log \frac{f(X_j, \widehat{\vartheta}_n)}{f(X_j, \vartheta_1)} + 2n \cdot \frac{1}{n} \sum_{j=1}^{n} \log \frac{f(X_j, \vartheta_1)}{f(X_j, h(\widehat{u}_n))}.$$

By Theorem 12.4, the first summand on the right-hand side converges in distribution under $\mathbb{P}_{\vartheta_1}$ to a χ_k^2-distribution. If $h(\widehat{u}_n)$ converges in probability under $\mathbb{P}_{\vartheta_1}$ to some $\vartheta^* \in \Theta_0$, it follows that

$$\frac{1}{n} \sum_{j=1}^{n} \log \frac{f(X_j, \vartheta_1)}{f(X_j, h(\widehat{u}_n))} \xrightarrow{\mathbb{P}_{\vartheta_1}} \mathbb{E}_{\vartheta_1}\left[\log \frac{f(X_1, \vartheta_1)}{f(X_1, \vartheta^*)} \right]$$

(see, e.g., [WIM], p. 235). In view of Lemma 10.10, the expectation is a Kullback–Leibler information and thus is (strictly) positive. It follows that $M_n \xrightarrow{\mathbb{P}_{\vartheta_1}} \infty$, which implies the consistency of the test.

12.9 Example (Test of Independence in Contingency Tables)
We assume the setting and the notations of Example 12.7 and assume $(X_j, Y_j)_{j\geq 1}$ is a sequence of i.i.d. random vectors with the same distribution as (X, Y). As we will soon see, the GLRT of the hypothesis H_0 of independence of X and Y, based on a sample of size n, employs the random variables

$$N_{ij} := \sum_{m=1}^{n} \mathbf{1}\{X_m = x_i, Y_m = y_j\}, \quad i \in \{1, \ldots, r\}, \ j \in \{1, \ldots, s\}.$$

Table 12.1 $r \times s$
Contingency Table

		1	$\cdots$	$\cdots$	$\cdots$	s	Σ
	1	N_{11}	$\cdots$	$\cdots$	$\cdots$	N_{1s}	N_{1+}
	2	N_{21}	$\cdots$	$\cdots$	$\cdots$	N_{2s}	N_{2+}
i	$\vdots$	$\vdots$	$\cdots$	$\cdots$	$\cdots$	$\vdots$	$\vdots$
	$\vdots$	$\vdots$	$\cdots$	$\cdots$	$\cdots$	$\vdots$	$\vdots$
	r	N_{r1}	$\cdots$	$\cdots$	$\cdots$	N_{rs}	N_{r+}
	Σ	N_{+1}	$\cdots$	$\cdots$	$\cdots$	N_{+s}	n

(column header: j)

To simplify the notation, we have suppressed the dependence of N_{ij} on the sample size n. Note that N_{ij} has the binomial distribution $\text{Bin}(n, p_{ij})$, and the rs-dimensional random vector $(N_{11}, \ldots, N_{rs})$ has the multinomial distribution $\text{Mult}(n; p_{11}, \ldots, p_{rs})$. It is common to arrange the N_{ij} as shown below in a so-called $r \times s$ *contingency table*. The row totals $N_{i+} := \sum_{j=1}^{s} N_{ij}$, $i \in \{1, \ldots, r\}$, and the column totals $N_{+j} := \sum_{i=1}^{r} N_{ij}$, $j \in \{1, \ldots, s\}$, are also listed in the margins (Table 12.1).

Self-question 5: What is the distribution of N_{i+}?

The density $f(\cdot, \cdot, \vartheta)$ of (X, Y) with respect to the counting measure ν on $\mathcal{X}_0 = \{x_1, \ldots, x_r\} \times \{y_1, \ldots, y_s\}$ is $f(x, y, \vartheta) = p_{ij}$, if $(x, y) = (x_i, y_j)$, and thus

$$\prod_{m=1}^{n} f(X_m, Y_m, \vartheta) = \prod_{i=1}^{r} \prod_{j=1}^{s} p_{ij}^{N_{ij}}.$$

With the definition of Θ_0 in (12.11) and Problem 12.1, the GLR in (12.4) takes the form

$$\Lambda_n = \frac{\sup_{p_i, q_j} \prod_i \prod_j (p_i q_j)^{N_{ij}}}{\prod_i \prod_j \left(\frac{N_{ij}}{n}\right)^{N_{ij}}}.$$

Here and in the following, i runs from 1 to r and j from 1 to s for all unspecified sums and products. Moreover, the supremum in the numerator extends over all positive $p_1, \ldots, p_r$ and $q_1, \ldots, q_s$ with $\sum_i p_i = 1$ and $\sum_j q_j = 1$.

Self-question 6: In which form does Problem 12.1 apply here?

Since

$$\sup_{p_i,q_j} \prod_i \prod_j (p_i q_j)^{N_{ij}} = \sup_{p_i} \prod_i p_i^{N_{i+}} \sup_{q_j} \prod_j q_j^{N_{+j}},$$

and since the random vectors $(N_{1+}, \ldots, N_{r+})$ and $(N_{+1}, \ldots, N_{+s})$ have the multinomial distributions $\mathrm{Mult}(n, p_1, \ldots, p_r)$ and $\mathrm{Mult}(n, q_1, \ldots, q_s)$, respectively, again with Problem 12.1, the GLR is

$$\Lambda_n = \frac{\prod_i \left(\frac{N_{i+}}{n}\right)^{N_{i+}} \prod_j \left(\frac{N_{+j}}{n}\right)^{N_{+j}}}{\prod_i \prod_j \left(\frac{N_{ij}}{n}\right)^{N_{ij}}}. \tag{12.22}$$

Using $n = \sum_j N_{+j}$ and $n = \sum_i N_{i+}$, $M_n = -2 \log \Lambda_n$ takes the form

$$M_n = 2n \sum_{i,j} \left\{ \frac{N_{ij}}{n} \log \frac{N_{ij}}{n} - \frac{N_{i+}}{n} \frac{N_{+j}}{n} \log \left(\frac{N_{i+}}{n} \frac{N_{+j}}{n} \right) \right\} \tag{12.23}$$

(Problem 12.5). By Theorem 12.8, M_n has a chi-square distribution with $(r-1)(s-1)$ degrees of freedom as $n \to \infty$ under H_0 (that is, for each $\vartheta \in \Theta_0$).

Self-question 7: Why is the number of degrees of freedom equal to $(r-1)(s-1)$?

The GLRT of independence in contingency tables (that is, for the independence of discrete random variables X and Y, each taking finitely many values) rejects the hypothesis H_0 if $M_n > \chi^2_{(r-1)(s-1);1-\alpha}$. By construction, the sequence of tests (φ_n) defined by

$$\varphi_n := \mathbf{1}\{M_n > \chi^2_{(r-1)(s-1);\, 1-\alpha}\}$$

has asymptotic level α.

A test statistic for testing H_0 that is more obvious compared to (12.23) is

$$T_n := \sum_{i=1}^r \sum_{j=1}^s \frac{\left(N_{ij} - n\frac{N_{i+}}{n}\frac{N_{+j}}{n}\right)^2}{n\frac{N_{i+}}{n}\frac{N_{+j}}{n}}, \tag{12.24}$$

because $\frac{N_{i+}}{n}$ is an estimator of $\mathbb{P}(X = x_i)$, and similarly, $\frac{N_{+j}}{n}$ is an estimator of $\mathbb{P}(Y = y_j)$. Since $\frac{N_{ij}}{n}$ estimates the probability $\mathbb{P}(X = x_i, Y = y_j)$, the numerator in (12.24) should be relatively small under H_0. Using the Taylor expansion

$$t \log t = g(t) = t - 1 + \frac{1}{2}(t-1)^2 - \frac{(t-1)^3}{6\rho^2}, \quad t > 0, \tag{12.25}$$

where $|\rho - 1| \leq |t - 1|$, it follows that $M_n - T_n = o_{\mathbb{P}_\vartheta}(1)$ for each $\vartheta \in \Theta_0$ (Problem 12.6), and thus also $T_n \overset{\mathcal{D}_\vartheta}{\longrightarrow} \chi^2_{(r-1)(s-1)}$ holds for each $\vartheta \in \Theta_0$. In the most important special case of a 2×2 contingency table, T_n in (12.24) takes the form

$$T_n = \frac{n\,(N_{11}N_{22} - N_{12}N_{21})^2}{N_{1+}N_{+1}N_{2+}N_{+2}}$$

(Problem 12.8).

Theorems 12.4 and 12.8 are *mathematical limit theorems*. With regard to applications, they make clear, among other things, how the critical values in connection with GLRT's should be chosen so that a given nominal level is maintained for large sample sizes. In the following, we show that there is an alternative to using critical values from limit distributions for test statistics beyond the class of GLRT's.

12.10 The Parametric Bootstrap

Let $X_1, X_2, \ldots$ be a sequence of i.i.d. random variables that take on values in $\mathcal{X}_0$, and which have an unknown distribution $\mathbb{P}^{X_1}$. The hypothesis to be tested is

$$H_0 : \mathbb{P}^{X_1} \in \{Q_\vartheta : \vartheta \in \Theta_0\},$$

where Θ_0 is a subset of $\mathbb{R}^k$. Let (T_n) with $T_n = T_n(X_1, \ldots, X_n)$ be *any* sequence of test statistics for H_0 with an *upper* rejection region, that is, rejection of H_0 if $T_n > c$. Here, c is a suitable critical value, which depends on the chosen level of significance $\alpha \in (0, 1)$ and on n, a fact that is emphasized by the notation $c = c(n, \alpha)$. However, the critical value generally also depends on the unknown distribution of T_n under H_0 if H_0 is composite, that is, if the set Θ_0 has at least two elements. The wish that the equation

$$\mathbb{P}_\vartheta (T_n > c) = \alpha$$

should apply to every ϑ in Θ_0 is therefore usually not feasible.

At this point, an idea attributed to B. Efron[3] enters the stage, which could only bear fruit with the advent of powerful computers. This idea is the so-called *bootstrap*, in this case more precisely the *parametric bootstrap*. The phrase *to pull oneself over a fence by one's bootstraps* means in a figurative sense that one frees oneself from an awkward situation using one's own bootstraps, much like Baron von Münchhausen allegedly pulled himself out of the swamp by his own topknot. The awkward situation in which one finds oneself with regard to finding a critical value when testing a hypothesis is described above, and the computer will prove to be the bootstrap.

The bootstrap requires some consistent sequence of estimators $(\widehat{\vartheta}_n)$ of ϑ, which, despite the chosen notation, does not necessarily have to consist of MLE's. The idea of the bootstrap is the following: If (a realization of) $\widehat{\vartheta}_n$ is "close to ϑ", then the probability measure $Q_{\widehat{\vartheta}_n}$ should "be close to Q_ϑ", which is vividly expressed by

$$\widehat{\vartheta}_n \approx \vartheta \implies Q_{\widehat{\vartheta}_n} \approx Q_\vartheta.$$

If $X_1^*, \ldots, X_n^*$ are independent random variables, each with the distribution $Q_{\widehat{\vartheta}_n}$, then— again expressed in a pleasantly vague way—we should have

$$\mathbb{P}_\vartheta(T_n(X_1, \ldots, X_n) > c) \approx \mathbb{P}_{\widehat{\vartheta}_n}\left(T_n(X_1^*, \ldots, X_n^*) > c\right) \tag{12.26}$$

(remember: $\mathbb{P}_\vartheta$ is the infinite product probability measure $Q_\vartheta \otimes Q_\vartheta \otimes \cdots$ on the product σ-field $\mathcal{B} = \mathcal{B}_o^{\mathbb{N}}$ over $\mathcal{X} = \mathcal{X}_0^{\mathbb{N}}$). The probability on the right-hand side of (12.26) can be estimated using a parametric *bootstrap algorithm* as follows:

Given $\widehat{\vartheta}_n = \widehat{\vartheta}_n(X_1, \ldots, X_n)$, that is, under the condition $X_1, \ldots, X_n$, using pseudo-random numbers, one generates realizations of random variables $X_1^*, \ldots, X_n^*$ which are independent and each have the distribution $Q_{\widehat{\vartheta}_n}$. This is used to calculate $T_n(X_1^*, \ldots, X_n^*)$. These two steps are carried out a total of b times. Here, b is the number of so-called *bootstrap replications*.

The computer thus generates a total of nb pseudorandom numbers which, conditionally on $X_1, \ldots, X_n$, are regarded as realizations of independent random variables $X_{i,j}^*$ ($i \in \{1, \ldots, n\}$, $j \in \{1, \ldots, b\}$) that are distributed according to $Q_{\widehat{\vartheta}_n}$. This then generates b

---

[3] Bradley Efron (*1938), American statistician, professor emeritus at Stanford University. Main areas of work: Bootstrap methods, empirical Bayes methods, multiple testing, variable selection, model choice. In 2007, he received the *National Medal of Science*, the highest scientific award in the USA.

realizations of the test statistic under $Q_{\widehat{\vartheta}_n}$ which, again conditionally on $X_1, \ldots, X_n$, are considered as realizations of the independent and identically distributed random variables

$$T_{n,1}^* := T_n(X_{1,1}^*, \ldots, X_{n,1}^*),$$

$$T_{n,2}^* := T_n(X_{1,2}^*, \ldots, X_{n,2}^*),$$

$$\vdots \quad \vdots \qquad \qquad \vdots$$

$$T_{n,b}^* := T_n(X_{1,b}^*, \ldots, X_{n,b}^*).$$

Let

$$\widehat{H}_{n,b}^*(t) := \frac{1}{b} \sum_{j=1}^{b} \mathbf{1}\{T_{n,j}^* \leq t\}, \quad t \in \mathbb{R},$$

be the empirical distribution function of $T_{n,1}^*, \ldots, T_{n,b}^*$, and denote $T_{(n:1)}^* \leq \cdots \leq T_{(n:b)}^*$ the order statistics of $T_{n,1}^*, \ldots, T_{n,b}^*$. The bootstrap test rejects H_0 at a desired level of significance α if $T_n > c_{n,b}^*(\alpha)$. Here,

$$c_{n,b}^*(\alpha) := \widehat{H}_{n,b}^{*-1}(1 - \alpha) = \begin{cases} T_{(n:b(1-\alpha))}^*, & \text{if } b(1 - \alpha) \text{ is an integer,} \\ T_{(n:\lfloor b(1-\alpha)+1 \rfloor)}^*, & \text{otherwise,} \end{cases}$$

is the $(1 - \alpha)$-quantile of the empirical distribution of $T_{n,1}^*, \ldots, T_{n,b}^*$.

In the specific case $b = 10.000$ and $\alpha = 0.05$, one therefore rejects H_0 if the observed value of the test statistic T_n is greater than $T_{(n:9.500)}^*$. The following result shows that, under certain assumptions, the above bootstrap procedure maintains a nominal level of significance asymptotically as $n \to \infty$ and $b \to \infty$. Here we generally use the notation $\|G\|_\infty := \sup_{x \in \mathbb{R}} |G(x)|$, where $G : \mathbb{R} \to \mathbb{R}$.

12.11 Theorem *Let $H_{n,\vartheta}(t) := \mathbb{P}_\vartheta(T_n \leq t)$, $t \in \mathbb{R}$, be the distribution function of T_n under $\mathbb{P}_\vartheta$, $\vartheta \in \Theta$. Suppose for each $\vartheta \in \Theta$, there exists a continuous distribution function H_ϑ, which strictly increases on the set $\{t \in \mathbb{R} : 0 < H_\vartheta(t) < 1\}$. Assume further that for every sequence (ϑ_n) in Θ with $\lim_{n \to \infty} \vartheta_n = \vartheta \in \Theta$, we have*

$$\lim_{n \to \infty} \|H_{n,\vartheta_n} - H_\vartheta\|_\infty = 0. \tag{12.27}$$

Finally, suppose that

$$\widehat{\vartheta}_n \xrightarrow{\mathbb{P}_\vartheta} \vartheta, \qquad \vartheta \in \Theta, \tag{12.28}$$

and that the mapping $\Omega \ni \omega \mapsto H_{n,\widehat{\vartheta}_n(\omega)}(t)$ is measurable for each $t \in \mathbb{R}$. For each $\vartheta \in \Theta$ and each $\alpha \in (0, 1)$, we then have the following:

(a) $\|\widehat{H}_{n,b}^* - H_\vartheta\|_\infty \xrightarrow{\mathbb{P}_\vartheta} 0$ as $n, b \to \infty$,

(b) $c_{n,b}^*(\alpha) \xrightarrow{\mathbb{P}_\vartheta} H_\vartheta^{-1}(1 - \alpha)$ as $n, b \to \infty$,

(c) $\lim_{n,b\to\infty} \mathbb{P}_\vartheta (T_n > c_{n,b}^*(\alpha)) = \alpha$.

Proof (a) The conditions (12.27) and (12.28) and the subsequence criterion 1.1 for convergence in probability yield

$$\|H_{n,\widehat{\vartheta}_n} - H_\vartheta\|_\infty \xrightarrow{\mathbb{P}_\vartheta} 0 \text{ as } n \to \infty. \tag{12.29}$$

W.l.o.g., let the underlying probability space be so rich that this space carries a sequence $(U_n)_{n\geq1}$ of i.i.d. random variables with a uniform distribution on the unit interval $(0, 1)$, and these random variables are independent of $(X_n)_{n\geq1}$.

Self-question 8: Why can we make this assumption without loss of generality?

The crucial idea is that, pointwise on this probability space, we can set

$$T_{n,j}^* := H_{n,\widehat{\vartheta}_n}^{-1}(U_j) := \inf \{t : H_{n,\widehat{\vartheta}_n}(t) \geq U_j\}, \quad j \geq 1.$$

As a quantile transformation of U_j, $T_{n,j}^*$ has the distribution function $H_{n,\widehat{\vartheta}_n}$, and *conditionally on* $\widehat{\vartheta}_n$, $T_{n,1}^*, T_{n,2}^*, \ldots$ are independent and identically distributed with distribution function $H_{n,\widehat{\vartheta}_n}$. Due to

$$T_{n,j}^* \leq t \iff U_j \leq H_{n,\widehat{\vartheta}_n}(t),$$

it follows that

$$\|\widehat{H}_{n,b}^* - H_{n,\widehat{\vartheta}_n}\|_\infty = \sup_{t\in\mathbb{R}} \left|\frac{1}{b} \sum_{j=1}^{b} \mathbf{1}\{U_j \leq H_{n,\widehat{\vartheta}_n}(t)\} - H_{n,\widehat{\vartheta}_n}(t)\right|$$

$$\leq \sup_{0\leq u\leq 1} \left|\frac{1}{b} \sum_{j=1}^{b} \mathbf{1}\{U_j \leq u\} - u\right|. \tag{12.30}$$

Self-question 9: Why does inequality (12.30) hold?

By the Glivenko-Cantelli theorem, the upper bound converges to zero almost surely as $b \to \infty$. The assertion now follows from the triangle inequality

$$\|\widehat{H}^*_{n,b} - H_\vartheta\|_\infty \leq \|H_{n,\widehat{\vartheta}_n} - H_\vartheta\|_\infty + \|\widehat{H}^*_{n,b} - H_{n,\widehat{\vartheta}_n}\|_\infty.$$

Assertion (b) follows from (a) and the continuity as well as the strict monotonicity of H_ϑ, and claim (c) is an immediate consequence of (b). $\square$

We will revisit the topic of the parametric bootstrap in Chap. 17 in connection with weighted L^2-statistics.

The tests presented in this chapter are based on an intuitively obvious principle, namely using the GLR for testing hypotheses within parametric models. We have limited ourselves to deriving the limit distribution of the test statistic defined in (12.6) if the null hypothesis is valid. In addition, we have examined questions of consistency of the GLRT and outlined connections to chi-square tests. For further reading on asymptotic tests in parametric models see, e.g., [SHA], Section 6.4, or [WIM], Section 6.2.

Answers to the Self-Questions

Answer 1 This equation follows from differentiation with respect to ϑ.

Answer 2 By definition of c, $\mathbb{P}_{\vartheta_0}(L_n > c) \leq \alpha$ and $\mathbb{P}_{\vartheta_0}(L_n \geq c) > \alpha$, since the probability measure $\mathbb{P}_{\vartheta_0}$ is continuous from above. The assertion follows from the definition of γ.

Answer 3 With (12.7) and $W_n := \frac{1}{n}(c - Y_n)$, we have $\mathbb{P}_{\vartheta_1}(M_n > c) = \mathbb{P}_{\vartheta_1}(V_n > W_n)$. Since $V_n \xrightarrow{\mathbb{P}_{\vartheta_1}} I_{KL}(\vartheta_1 : \vartheta_0) > 0$ and $W_n \xrightarrow{\mathbb{P}_{\vartheta_1}} 0$, the assertion follows.

Answer 4 Since $\sum_{j=1}^s p_{ij} = p_i$ and $\sum_{i=1}^r p_{ij} = q_j$, we have $\mathbb{P}(X = x_i) = p_i$ and $q_j = \mathbb{P}(Y = y_j)$.

Answer 5 Note that $N_{i+} = \sum_{m=1}^n \mathbf{1}\{X_m = x_i\}$. Since $\mathbb{P}(X = x_i) = p_{i1} + \ldots + p_{is} =: p_i$, the random variable N_{i+} has the binomial distribution $\mathrm{Bin}(n, p_i)$, since it is a sum of indicators of n independent events $\{X_1 = x_i\}, \ldots, \{X_n = x_i\}$ that have the same probability.

Answer 6 More generally, if $(M_1, \ldots, M_s)$ is a random vector with the multinomial distribution $\mathrm{Mult}(n, \rho_1, \ldots, \rho_s)$, then, according to Problem 12.1, the probability

$$\mathbb{P}(M_1 = k_1, \ldots, M_s = k_s) = \frac{n!}{k_1! \cdot \ldots \cdot k_s!} \rho_1^{k_1} \cdot \ldots \cdot \rho_s^{k_s} \qquad (k_j \in \mathbb{N}_0,\ k_1 + \ldots + k_s = n),$$

when regarded as a function of $(\rho_1, \ldots, \rho_s) \in [0, 1]^s$, attains its maximum under the constraint $\rho_1 + \ldots + \rho_s = 1$ for $\rho_j = \frac{1}{n} k_j$, $j = 1, \ldots, s$.

Answer 7 We have $k = rs - 1$ and $\ell = r - 1 + s - 1$, which entails $k - \ell = (r-1)(s-1)$.

Answer 8 If necessary, one forms the product of the underlying probability space with a probability space on which $U_1, U_2, \ldots$ are defined. A canonical choice for the latter space is the countably infinite product of the probability space $\big((0,1), \mathcal{B}^1_{|(0,1)}, \lambda^1_{|(0,1)}\big)$ with itself. Here, $\mathcal{B}^1_{|(0,1)}$ and $\lambda^1_{|(0,1)}$ denote the restrictions of the Borel σ-field and of the Borel–Lebesgue measure to $(0, 1)$, respectively.

Answer 9 Because $\{H_{n,\widehat{\vartheta}_n}(t) : t \in \mathbb{R}\} \subset [0, 1]$.

Problems

12.1 Problem Show that the likelihood function in (12.8), as a function of $\vartheta = (p_1, \ldots, p_{s-1})$ (for reminder: $p_s = 1 - p_1 - \ldots - p_{s-1}$), attains its maximum value for the vector $\widehat{\vartheta}_n$ specified in (12.9).

Hint After taking logarithms, only the summands with $N_{n,j} > 0$ need to be considered. Use the inequality $\log t \le t - 1$, $t > 0$.

12.2 Problem Let M_n and T_n be as in (12.10) or in (6.7). Show that the difference $T_n - M_n$ converges to zero in probability under H_0.

Hint Perform a Taylor expansion of the function $t \mapsto t \log t$ at the point $t = 1$.

12.3 Problem Let X be a k-dimensional random vector with $X \sim N_k(0_k, \Sigma)$, where the covariance matrix Σ is positive definite. Further, let A be a symmetric $k \times k$ matrix such that the matrix $A\Sigma$ is idempotent and has rank $r \ge 1$. Prove the following assertiona:

$$X^\top A X \sim \chi_r^2.$$

Hint First consider the case $\Sigma = I_k$.

12.4 Problem Let $A := I_1(\vartheta)\big(I_k - h'(u)\tilde{I}_1(u)^{-1}h'(u)^\top I_1(\vartheta)\big)$ be the matrix defined in (12.21) and $\Sigma := I_1(\vartheta)^{-1}$. Prove the following assertions:

(a) A is symmetric,
(b) $(A\Sigma)^2 = A\Sigma$,
(c) $A\Sigma$ has rank $k - \ell$.

12.5 Problem Prove the representation (12.23).

12.6 Problem Let M_n and T_n be the test statistics defined in (12.23) and (12.24), respectively. Show that $M_n - T_n = o_{\mathbb{P}_\vartheta}(1)$ for every $\vartheta \in \Theta_0$.

Hint Apply the equation

$$a \log a - b \log b = b \cdot \frac{a}{b} \log \frac{a}{b} + (a - b) \log b \qquad (a, b > 0)$$

to $a = \frac{1}{n}N_{i,j}$ and $b = n^{-2}N_{i+}N_{+j}$, and consider (12.25).

12.7 Problem

(a) Show that for the GLR test statistic M_n occurring in (12.23)

$$\frac{M_n}{n} \xrightarrow{\mathbb{P}} 2\sum_{i=1}^{r}\sum_{j=1}^{s}\mathbb{P}(X = x_i, Y = y_j) \log \frac{\mathbb{P}(X = x_i, Y = y_j)}{\mathbb{P}(X = x_i)\mathbb{P}(Y = y_j)}.$$

(b) What does the above stochastic limit have to do with Kullback–Leibler information?
(c) Show that the GLRT for independence in contingency tables is consistent against any bivariate distribution in which the components are not independent.

12.8 Problem Show that in the special case $r = s = 2$ T_n in (12.24) takes the form

$$T_n = \frac{n\,(N_{11}N_{22} - N_{12}N_{21})^2}{N_{1+}N_{+1}N_{2+}N_{+2}}.$$

12.9 Problem Let $\Theta := \{\vartheta := (\vartheta_1, \vartheta_2) \in \mathbb{R}^2 : \vartheta_1 \geq 0, \vartheta_2 \in \mathbb{R}\}$ and $(X_1, Y_1), (X_2, Y_2), \ldots$ be a sequence of i.i.d. bivariate random vectors with the bivariate normal distribution $N_2(\vartheta, I_2)$. The hypothesis to be tested is $H_0 : \vartheta_1 = \vartheta_2 = 0$, that is,

$\vartheta \in \Theta_0 := \{(0, 0)\}$, against the general alternative $H_1 : \vartheta \in \Theta \setminus \Theta_0$. Let $M_n = -2 \log \Lambda_n$ with Λ_n as in (12.5). Prove that

$$M_n \xrightarrow{\mathcal{D}} \frac{1}{2} Z_1 + \frac{1}{2} Z_2$$

under H_0. Here, Z_1 and Z_2 are independent random variables, where $Z_1 \sim \chi_1^2$ and $Z_2 \sim \chi_2^2$. The limit distribution is thus a mixture of two chi-square distributions.

12.10 Problem Let $X_1, \ldots, X_n$ be independent d-dimensional random vectors with the same normal distribution $N_d(\vartheta, \Sigma)$. Suppose the positive definite matrix Σ is known and $\vartheta \in \Theta := \mathbb{R}^d$ is unknown. The hypothesis to be tested is $H_0 : \vartheta = \vartheta_0$, against the alternative $H_1 : \vartheta \neq \vartheta_0$. Here, $\vartheta_0 \in \mathbb{R}^d$ is a given value. Show that in this case, the test statistic $M_n = -2 \log \Lambda_n$ resulting from the generalized likelihood ratio Λ_n takes the form

$$M_n = n (\overline{X}_n - \mu_0)^\top \Sigma^{-1} (\overline{X}_n - \mu_0), \quad \text{where} \quad \overline{X}_n := \frac{1}{n} \sum_{j=1}^n X_j,$$

and that M_n has a χ_d^2-distribution under H_0.

Probability Measures on Metric Spaces 13

So far, the focus has been on real-valued random variables or d-dimensional random vectors. In this chapter, we generalize this setting, which is characterized by a *finite dimension*. To motivate this departure from the finite-dimensional case, we consider a sequence $X_1, X_2, \ldots$ of independent random variables, each uniformly distributed on the interval $[0, 1]$. In Chap. 7, we introduced the empirical distribution function

$$\widehat{F}_n(t) = \frac{1}{n} \sum_{j=1}^{n} \mathbf{1}\{X_j \le t\}, \quad 0 \le t \le 1,$$

of $X_1, \ldots, X_n$. By the Glivenko–Cantelli theorem (Theorem 7.2),

$$\lim_{n \to \infty} \sup_{t \in [0,1]} \left| \widehat{F}_n(t) - t \right| = 0 \quad \mathbb{P}\text{-almost surely.}$$

If we equip each of the differences $\widehat{F}_n(t) - t$, $0 \le t \le 1$, with a "$\sqrt{n}$-magnifying glass" and put

$$B_n(t) := \sqrt{n}\big(\widehat{F}_n(t) - t\big), \quad 0 \le t \le 1, \tag{13.1}$$

we obtain a collection $(B_n(t))_{0 \le t \le 1}$ of random variables, which is referred to as the *uniform empirical process*. Figure 13.1 shows a realization of a uniform empirical process for the case $n = 25$.

The ticks on the t-axis (except for the one at $t = 1$) mark the realizations of $X_1, \ldots, X_n$. Because of the term $-t$ in (13.1), the values $B_n(t)$ decrease linearly between each two directly consecutive values of $X_1, \ldots, X_n$. At each of the points $t = X_j$, $B_n(t)$ jumps by

N. Henze, *Asymptotic Stochastics*, Mathematics Study Resources 10, https://doi.org/10.1007/978-3-662-68923-3_13

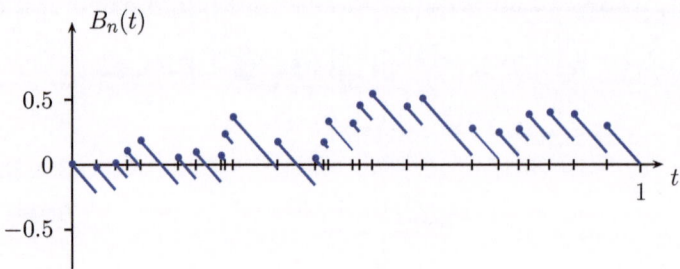

Fig. 13.1 Realization of a uniform empirical process ($n = 25$)

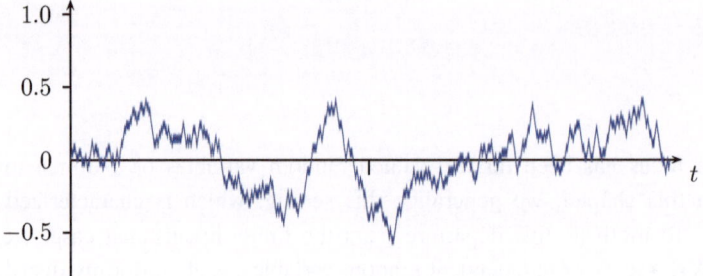

Fig. 13.2 Rough impression of a realization of the Brownian bridge

the value $1/\sqrt{n}$. In addition, B_n is also right-continuous with $\widehat{F}_n$, which is highlighted by the filled circles. Moreover, $\mathbb{P}(B_n(0) = 0) = 1 = \mathbb{P}(B_n(1) = 0)$.

Self-question 1: Why does $\mathbb{P}(B_n(0) = 0) = 1 = \mathbb{P}(B_n(1) = 0)$ hold?

For each $k \geq 1$ and for each choice of $t_1, \ldots, t_k$ with $0 \leq t_1 < t_2 < \ldots < t_k \leq 1$, the multivariate central limit theorem for i.i.d. sequences (Theorem 6.19) yields

$$\bigl(B_n(t_1), \ldots, B_n(t_k)\bigr) \xrightarrow{\;\mathcal{D}\;} N_k\bigl(0_k, \bigl(t_i \wedge t_j - t_i t_j\bigr)_{1 \leq i, j \leq k}\bigr) \quad \text{as } n \to \infty$$

(cf. Problem 7.4). This fact makes one curious whether perhaps even (B_n), as a *sequence of random functions* $B_n(\cdot)$ that take on values in a suitable function space, converges in distribution in a sense to be specified to a *random function*, denoted by B. As shown in Chap. 16, this is indeed the case, and the "limit object" B is called the *Brownian bridge*.

As we will see in Chap. 15, such a Brownian bridge also shows up as a limit object, as $n \to \infty$, if one starts with an i.i.d. sequence $(Z_n)_{n \geq 1}$ with $\mathbb{P}(Z_j = \pm 1) = \frac{1}{2}$ and considers the partial sums $S_{\lfloor nt \rfloor} := \sum_{j=1}^{\lfloor nt \rfloor} Z_j$, $0 \leq t \leq 1$. If we interpolate linearly between S_j and S_{j+1} for each $j \in \{0, \ldots, n-1\}$, scale with the factor $n^{-1/2}$ and subtract $S_n/\sqrt{n}$, we obtain a random continuous function on $[0, 1]$. Figure 13.2 shows a realization of such a random function for the case $n = 1000$. We will see that realizations of the Brownian bridge are almost surely continuous and nowhere differentiable functions. In this respect, the graph in Fig. 13.2 gives a rough impression of such a realization.

Now, suppose that, with a notion of convergence in distribution to be defined, and in a suitable function space D (say), we have $B_n \xrightarrow{\mathcal{D}} B$ as $n \to \infty$. Does it then follow that $T(B_n) \xrightarrow{\mathcal{D}} T(B)$, where $T : D \to \mathbb{R}$ is a suitable real-valued functional? The answer will be "yes", and it turns out that the function space D to be specified consists of bounded and measurable functions. Interesting functionals are then, among others, T_1 and T_2, where

$$T_1(x) := \|x\|_\infty := \sup_{0 \le t \le 1} |x(t)|, \qquad T_2(x) := \int_0^1 x^2(t)\, dt, \qquad x \in D.$$

Here,

$$T_1(B_n) = \|B_n\|_\infty = \sqrt{n} \sup_{0 \le t \le 1} \left| \widehat{F}_n(t) - t \right|$$

is the so-called *Kolmogorov–Smirnov statistic*. Furthermore,

$$T_2(B_n) := \int_0^1 B_n^2(t)\, dt = n \int_0^1 \left(\widehat{F}_n(t) - t \right)^2 dt$$

is the *Cramér–von Mises statistic* examined in Example 8.18 and denoted there by ω_n^2 (see (8.35)). Both $T_1(B_n)$ and $T_2(B_n)$ are celebrated statistics to test the hypothesis H_0 that the distribution underlying a sequence of i.i.d. random variables $X_1, X_2, \ldots$ is the uniform distribution on $[0, 1]$.

The above considerations show that it makes sense to study random variables that take on values in spaces that are much more general than $\mathbb{R}^d$. It has been found that metric spaces are suitable for this purpose. In this chapter, we first give a brief overview of metric spaces and related results that are important in the sequel. For background information on metric spaces, see, e.g., [HEI], Chapter 2, or [HAA], Chapter 3–4. A classic on the topic of probability measures on metric spaces is the book [PAR].

13.1 Definition (Metric Space)
Let S be a non-empty set. A *metric* on S is a function $\rho : S \times S \to \mathbb{R}$ with the following properties, valid for all $x, y, z \in S$:

(a) *Non-negativity*: $\rho(x, y) \ge 0$,
(b) *Uniqueness*: $\rho(x, y) = 0$ if and only if $x = y$,
(c) *Symmetry*: $\rho(x, y) = \rho(y, x)$,
(d) *Triangle Inequality*: $\rho(x, z) \le \rho(x, y) + \rho(y, z)$.

If these conditions hold then the pair (S, ρ) (or only S, if the metric is unambiguous) is called a *metric space*. The number $\rho(x, y)$ is called the *distance* between x and y.

13.2 Example

(a) If we set $\rho(x, y) := 1$ if $x \neq y$ and $\rho(x, y) := 0$ if $x = y$, then ρ is a metric on S. This metric is the so-called *discrete metric*, and the pair (S, ρ) is a *discrete metric space*.

(b) For $x = (x_1, \ldots, x_d) \in \mathbb{R}^d$ and $y = (y_1, \ldots, y_d) \in \mathbb{R}^d$, the *Euclidean metric* induced by the Euclidean norm $\| \cdot \|$ is defined by

$$\rho(x, y) := \|x - y\| = \left((x_1 - y_1)^2 + \ldots + (x_d - y_d)^2\right)^{1/2}.$$

The pair $(\mathbb{R}^d, \rho)$ is the *d-dimensional Euclidean space*.

(c) Let $S \subset \{0, 1\}^n$ be a set of binary codewords of length n. For $x = (x_1, \ldots, x_n)$ and $y = (y_1, \ldots, y_n)$ in S, let $\rho(x, y) := \mathbf{1}\{x_1 \neq y_1\} + \ldots + \mathbf{1}\{x_n \neq y_n\}$ be the number of positions in which the codewords x and y do not match. Then (S, ρ) is a metric space, and $\rho(x, y)$ is the *Hamming distance* between x and y.

13.3 Metric Spaces: Basic Concepts and Notations

Before we consider further examples of metric spaces, we introduce some general notations. If (S, ρ) is a metric space, then for $x \in S$ and $r > 0$

$$B(x, r) := B_\rho(x, r) := \{y \in S : \rho(x, y) < r\}$$

is the *open ball with radius r centered at x*. Strictly speaking, the naming *ball* only applies to the Euclidean metric in $\mathbb{R}^3$.

Self-question 2: Can you specify $B(x, r)$ in Example 13.2 (a) for $r \in \{1, 1.001\}$?

We say that a subset O of S is *open*, if for each $x \in O$ there is some $r > 0$ such that $B(x, r) \subset O$. This includes the case $O = \emptyset$. The collection of all open subsets of S is denoted by $\mathcal{O}$. A set $A \subset S$ is called *closed*, if its complement $A^c := S \setminus A$ is open. The notation $\mathfrak{A}$ stands for the collection of all closed sets. If M is a subset of S then

$$\overset{\circ}{M} := \bigcup \{O \in \mathcal{O} : O \subset M\}$$

denotes the *interior* of M, and

$$\overline{M} := \bigcap \{A \in \mathfrak{A} : A \supset M\}$$

is the *closure* of M. Thus, $\overset{\circ}{M}$ is the union of all open sets that are contained in M, and $\overline{M}$ is the intersection of all closed sets that contain M. The set-theoretic difference $\partial M :=$ $\overline{M} \setminus \overset{\circ}{M}$ is called the *boundary* of M, and each point of ∂M is said to be a *boundary point* of M. For each x in S and each non-empty subset M of S

$$\rho(x, M) := \inf\{\rho(x, y) : y \in M\}$$

denotes the *distance from x to M*. If $\varepsilon > 0$ and $M \subset S$, the notation

$$M^\varepsilon := \{x \in S : \rho(x, M) < \varepsilon\}$$

stands for the *ε-neighborhood* of M. Alternatively, M^ε is said to be the *parallel set* of M at a distance ε. In generalization to (6.3), the distance function $S \ni x \mapsto \rho(x, M)$ satisfies

$$|\rho(x, M) - \rho(z, M)| \leq \rho(x, z), \quad x, z \in S. \tag{13.2}$$

Self-question 3: Can you prove inequality (13.2)?

A sequence (x_n) in S *converges* to $x \in S$ if $\lim_{n \to \infty} \rho(x_n, x) = 0$. In this case, x is called the *limit* of (x_n). A sequence (x_n) in S is a *Cauchy sequence* if $\lim_{m,n \to \infty} \rho(x_n, x_m) = 0$. The metric space (S, ρ) is called *complete* if every Cauchy sequence in S has a limit in S. We say that a non-empty subset M of S is *complete*, if the metric space $(M, \rho_{|M})$ is complete. Here, $\rho_{|M}$ denotes the restriction of ρ to M.

A set $M \subset S$ is called *dense* (in S), if its closure equals S, that is, if $\overline{M} = S$. A metric space (S, ρ) is *separable*, if S contains a countable and dense subset.

A collection $\mathcal{O}_0 \subset \mathcal{O}$ of open sets is called a *basis* of $\mathcal{O}$, if every set O in $\mathcal{O}$ is a union of sets from $\mathcal{O}_0$. If $M \subset S$ and $\tilde{\mathcal{O}} \subset \mathcal{O}$ is a collection of open sets, then $\tilde{\mathcal{O}}$ is called an *open cover of M*, if $M \subset \bigcup\{O : O \in \tilde{\mathcal{O}}\}$. For separable metric spaces, we have the following result (see, e.g., [BI2], p. 237):

13.4 Theorem (Characterization of Separable Metric Spaces)
For a metric space (S, ρ), the following statements are equivalent:

(a) (S, ρ) *is separable.*
(b) (S, ρ) *has a countable basis.*
(c) *Every open cover of any subset of S has a countable subcover.*

A set $M \subset S$ is said to be *nowhere dense*, if its closure contains no interior points, that is, if $(\overline{M})^\circ = \emptyset$. For example, the integers as a subset of $\mathbb{R}$ are a nowhere dense set. The following result, named after R. Baire,[1] is one of the basic theorems of functional analysis (see, e.g., [HEI], Theorem 2.11.3, or [HAA], Theorem 15.1).

13.5 Theorem (Baire's Category Theorem)

Let (S, ρ) be a complete metric space. If $S = \bigcup_{n=1}^{\infty} A_n$ with closed sets $A_1, A_2, \ldots$, then at least one A_n contains an open ball.

A complete metric space can therefore *not* be written as a countable union of nowhere dense sets.

A special role will be assigned to compact subsets of a metric space. A set $M \subset S$ is called *compact* if every open cover of M has a *finite subcover*. Every compact set is closed and bounded in the sense that it is contained in a suitable ball $B(x, r)$ with $x \in S$ and $r > 0$ (Problem 13.2). By the Heine[2]–Borel theorem, the converse holds in the special case $S = \mathbb{R}^d$ with the Euclidean metric, that is, every bounded closed set is compact (see, e.g., [HEI], Theorem 2.8.4).

Self-question 4: Which sets are compact in a discrete metric space?

If M and N are subsets of S, then N is called an *ε-net for M*, if for every $x \in M$ there is some $y \in N$ with $\rho(x, y) < \varepsilon$. A set $M \subset S$ is said to be *totally bounded* if for every $\varepsilon > 0$ there is a *finite ε-net* $N \subset S$ for M. In this case, for every $\varepsilon > 0$ there exist finitely many points $x_1, \ldots, x_n$ in S such that every $x \in M$ has a distance from a suitable x_j that is less than ε.

A subset M of S is *relatively compact* if its closure $\overline{M}$ is compact. M is called *sequentially compact*, if every sequence in M has a convergent subsequence (whose limit does not necessarily have to belong to M).

By the Heine–Borel theorem, a set $M \subset \mathbb{R}^d$ is relatively compact if and only if it is bounded. The following result is a generalization of this fact to general metric spaces (see, e.g., [BI2], p. 239). The proof of part "(b) $\Longrightarrow$ (c)" is the subject of Problem 13.4.

13.6 Theorem (Characterization of Relative Compactness)

For a set $M \subset S$ the following statements are equivalent:

---

[1] René Louis Baire (1874–1932), French mathematician. Baire is considered the founder of the modern theory of real functions.

[2] Eduard Heine (1821–1881), Professor at the universities of Bonn (from 1848) and Halle (from 1856). Main areas of research: Real analysis, trigonometric series.

(a) M is relatively compact.
(b) M is sequentially compact.
(c) M is totally bounded, and $\overline{M}$ is complete.

If (S, ρ) is a metric space, then the *σ-field of Borel sets* on S, denoted by $\mathcal{B}(S)$, is the smallest σ-field on S that contains all open sets. In other words, we have $\mathcal{B}(S) = \sigma(\mathcal{O})$. This σ-field suggests itself as a natural domain for probability measures on metric spaces. If (S, ρ) is separable, the σ-field of Borel sets is already generated by the collection $\{B(x, \varepsilon) : x \in S, \ \varepsilon > 0\}$ of open balls (Problem 13.3), otherwise the latter σ-field is generally a proper subclass of all Borel sets (Problem 13.8).

If (S, ρ) is a metric spaces then

$$\mathcal{P} := \{P : \mathcal{B}(S) \to [0, 1] \,\big|\, P \text{ is a probability measure}\}$$

denotes the set of all probability measures on S (more precisely: on $\mathcal{B}(S)$).

In the special case $S = \mathbb{R}$, there is a large arsenal of probability measures, such as the normal distributions, the exponential distributions, or the Poisson distributions. In Chap. 5 we got to know the family of d-variate normal distributions. The naming *distribution* comes from the fact that, starting with a probability space $(\Omega, \mathcal{A}, \mathbb{P})$ and a random variable $X : \Omega \to \mathbb{R}$, the image (probability) measure of $\mathbb{P}$ under X, defined by $\mathbb{P}^X(B) := \mathbb{P}(X^{-1}(B))$, $B \in \mathcal{B}$, is commonly referred to as a *distribution*.

If (S, ρ) is a metric space, where S is an infinite set, it is an easy task to define a *discrete* probability measure on $\mathcal{B}(S)$. To this end, one chooses a countable subset $T =: \{x_1, x_2, \ldots\}$ of S and non-negative real numbers $p_1, p_2, \ldots$ with $\sum_{j \geq 1} p_j = 1$ and sets $P := \sum_{j \geq 1} p_j \delta_{x_j}$, where δ_x denotes the Dirac measure at the point x. Thus, in particular, $P(\{x_j\}) = p_j$, $j \geq 1$.

Self-question 5: Why are the singleton subsets of S Borel sets?

If S is an uncountable set, an interesting question is whether probability measures P can be defined on $\mathcal{B}(S)$ that are *nonatomic* in the sense that $P(\{x\}) = 0$ for every $x \in S$. The answer is "yes". Thus, in Chap. 15, we will get to know the *Wiener measure*, named after Norbert Wiener,[3] a nonatomic probability measure on the Borel sets of the space of continuous real-valued functions defined on the unit interval, equipped with the supremum

---

[3] Norbert Wiener (1894–1964), American mathematician and philosopher, PhD 1912 in mathematical logic. Wiener is known as the founder of cybernetics; he thus laid the foundations for control theory and control engineering. Named after him are, among others, the Wiener process, the Wiener filter, the Wiener–Khinchin theorem, and the Paley–Wiener–Zygmund theorem.

distance. This measure is accompanied by one of the most important stochastic processes, the so-called *Wiener process*, also known as *Brownian motion*, named after R. Brown.[4]

In order to identify probability measures, the following notion is useful.

13.7 Definition (Separating Class)

A collection $\mathcal{M} \subset \mathcal{B}(S)$ of Borel sets is called a *separating class* for $\mathcal{P}$, if any two probability measures in $\mathcal{P}$ that agree on $\mathcal{M}$ necessarily also agree on $\mathcal{B}(S)$, that is,

$$\text{if } P(A) = Q(A) \text{ for every } A \in \mathcal{M} \text{ then } P = Q.$$

Thus, if $\mathcal{M}$ is a separating class and P is any probability measure on $\mathcal{B}(S)$, then you know P if you know the values $P(A)$ with $A \in \mathcal{M}$. By the uniqueness theorem for measures (Theorem 1.28), if $\mathcal{M}$ is a π-system and $\sigma(\mathcal{M}) = \mathcal{B}(S)$, then $\mathcal{M}$ is a separating class for $\mathcal{P}$. In particular, the collection $\mathcal{O}$ of open sets is a separating class for $\mathcal{P}$. The same applies to the collection $\mathfrak{A}$ of closed sets.

A probability measure P on $\mathcal{B}(S)$ is also determined by integrals over certain classes of functions. For this purpose, let

$$\mathcal{C}_b(S) := \{f : S \to \mathbb{R} : f \text{ is bounded and continuous}\}$$

be the collection of bounded and continuous functions on S and

$$\mathcal{C}_b^0(S) := \{f \in \mathcal{C}_b(S) : f \text{ is uniformly continuous}\}$$

be the subcollection of functions that are even uniformly continuous. Here, a function $f : S \to \mathbb{R}$ is called *continuous at the point* $x \in S$, if for every $\varepsilon > 0$ there exists $\delta > 0$, such that, for each y in S, $\rho(x, y) < \delta$ implies $|f(x) - f(y)| < \varepsilon$. A function that is continuous at every point $x \in S$ is called *continuous*. A function $f : S \to \mathbb{R}$ is called *uniformly continuous*, if for every $\varepsilon > 0$ there is a $\delta > 0$ such that for all $x, y \in S$: If $\rho(x, y) < \delta$ then $|f(x) - f(y)| < \varepsilon$.

13.8 Theorem (The Integrals $\int f\,dP$ with $f \in \mathcal{C}_b^0(S)$ Determine P)

Let P and Q be arbitrary probability measures on $\mathcal{B}(S)$. We then have:

$$P = Q \iff \int f\,dP = \int f\,dQ \quad \text{for every } f \in \mathcal{C}_b^0(S).$$

---

[4] Robert Brown (1773–1858), Scottish physician and botanist, 1810 Fellow of the Royal Society and 1849–1853 President of the Linnean Society. In 1827, Brown observed that pollen grains in a liquid seemed to perform completely erratic, stochastic movements.

Proof Only the implication "⟸" needs to be shown, and it suffices to prove $P(A) = Q(A)$ for each closed set A. To this end, we approximate the indicator function $\mathbf{1}_A$ of A by uniformly continuous functions. So we fix any $A \in \mathfrak{A}$ and $\varepsilon > 0$. We define a function $f_\varepsilon : S \to \mathbb{R}$ by

$$f_\varepsilon(x) := \max\left(0, 1 - \frac{\rho(x, A)}{\varepsilon}\right), \quad x \in S. \tag{13.3}$$

Then $0 \le f_\varepsilon(x) \le 1$ for each $x \in S$, and using (13.2) we get

$$|f_\varepsilon(x) - f_\varepsilon(y)| \le \frac{\rho(x, y)}{\varepsilon}.$$

Hence f_ε is uniformly continuous. Furthermore, we have $\mathbf{1}_A \le f_\varepsilon \le \mathbf{1}_{A^\varepsilon}$.

Self-question 6: Why do these inequalities hold?

It follows that

$$P(A) = \int \mathbf{1}_A \, dP \le \int f_\varepsilon \, dP = \int f_\varepsilon \, dQ \le \int \mathbf{1}_{A^\varepsilon} \, dQ = Q(A^\varepsilon).$$

Since A is closed, each $x \in S$ belongs to A if and only if $\rho(x, A) = 0$. Thus, $A^\varepsilon \downarrow A$ as $\varepsilon \downarrow 0$. Since the probability measure Q is continuous from above, it follows that $P(A) \le Q(A)$. The reverse inequality $Q(A) \le P(A)$ holds for reasons of symmetry. $\quad\square$

13.9 The Space C[0, 1]

Let $C := C[0, 1] := \{x : [0, 1] \to \mathbb{R} \,|\, x \text{ continuous}\}$ be the collection of all continuous real-valued functions on $[0, 1]$. If we set

$$\|x\|_\infty := \sup_{0 \le t \le 1} |x(t)|, \quad x \in C,$$

and

$$\rho(x, y) := \|x - y\|_\infty = \max_{0 \le t \le 1} |x(t) - y(t)|,$$

then (C, ρ) becomes a metric space. Note that the convergence $\rho(x_n, x) \to 0$ of a sequence (x_n) to x means that the sequence of functions (x_n) converges uniformly to x.

By the Weierstrass[5] approximation theorem (see, e.g., [WEI], Theorem 4.6.2), or [HAA], Theorem 3.22), there is for every $x \in C$ and every $\varepsilon > 0$ a polynomial p such that $\rho(x, p) \leq \varepsilon$. Since p can be approximated with arbitrary precision by a polynomial of the same degree with *rational* coefficients, the countable set

$$\left\{ [0, 1] \ni t \mapsto \sum_{k=0}^{n} a_k t^k \,\middle|\, n \in \mathbb{N}_0, \, a_0, \ldots, a_n \in \mathbb{Q} \right\}$$

is dense in C, and thus the space (C, ρ) is separable.

Self-question 7: How do you approximate p by a "rational polynomial"?

The metric space (C, ρ) is also complete: If (x_n) is a Cauchy sequence in C, then $\varepsilon_n := \sup_{m \geq n} \|x_n - x_m\|_\infty \to 0$ as $n \to \infty$. Thus, for each fixed $t \in [0, 1]$, $(x_n(t))$ is a Cauchy sequence in $\mathbb{R}$. Since $\mathbb{R}$ is complete, the limit $x(t) := \lim_{n \to \infty} x_n(t)$ exists, and since $|x_n(t) - x_m(t)| \leq \varepsilon_n$ for every m with $m \geq n$, letting $m \to \infty$ gives $|x_n(t) - x(t)| \leq \varepsilon_n$, and thus

$$\lim_{n \to \infty} \|x_n - x\|_\infty = 0,$$

since t is arbitrary. Because x is a continuous function, the space (C, ρ) is complete.

Self-question 8: Why is x a continuous function?

We now show that, in contrast to $\mathbb{R}^d$, no closed ball in (C, ρ) is compact. To this end, we define for every integer $n \geq 2$ a function $z_n : [0, 1] \to [0, 1]$ by

$$z_n(t) := nt\mathbf{1}_{[0,1/n]}(t) + (2 - nt)\mathbf{1}_{(1/n,2/n]}(t), \qquad 0 \leq t \leq 1 \tag{13.4}$$

(see Fig. 13.3). For any fixed $\varepsilon > 0$, the sequence $(\varepsilon z_n)_{n \geq 2}$ has no convergent subsequence, since the assumption $\rho(\varepsilon z_{n_k}, z) \to 0$ for some $z \in C$ implies that z is the zero function.

---

[5] Karl Theodor Wilhelm Weierstrass (1815–1897), 1842–1855 activity as a teacher, 1856–1864 professor at the Gewerbeinstitut in Berlin, from 1864 professor at the University of Berlin. Weierstrass significantly influenced the development of mathematics ("Weierstrass school"). Main areas of research: complex analysis, algebra.

Fig. 13.3 Graph of the function z_n

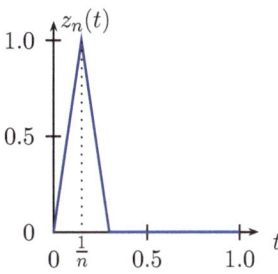

On the other hand, $\rho(\varepsilon z_{n_k}, 0) = \varepsilon$ for every k. By Theorem 13.6, the closed ball $\overline{B}(0, \varepsilon)$ is therefore not compact. If $x \in C$ is arbitrary, the transition from εz_n to $x + \varepsilon z_n$ shows that *no* closed ball $\overline{B}(x, \varepsilon)$ is compact. Thus, every compact set is nowhere dense, and since (C, ρ) is complete, it follows from Baire's category theorem (Theorem 13.5) that C, in contrast to $\mathbb{R}^d$, cannot be written as a countable union of compact sets and is thus not σ-*compact*.

These considerations show that compact subsets of C must be, loosely speaking, "rather thin". The following definition serves to characterize relatively compact subsets of C in a succinct way.

13.10 Definition (Modulus of Continuity)
Let $x \in C[0, 1]$. The function $w_x : (0, 1] \to \mathbb{R}$, defined by

$$w_x(\delta) := w(x, \delta) := \sup_{|s-t| \leq \delta} |x(s) - x(t)|, \quad 0 < \delta \leq 1, \tag{13.5}$$

is called the *modulus of continuity* of x.

Obviously, $\lim_{\delta \to 0} w_x(\delta) = 0$, because x is a continuous function on a compact interval and thus uniformly continuous. If $x, y \in C$ and $0 < \delta \leq 1$, then

$$|w_x(\delta) - w_y(\delta)| \leq 2\rho(x, y) \tag{13.6}$$

(Problem 13.9). Thus, the function $w(\cdot, \delta) : C \to \mathbb{R}$ is continuous.

Regarding the relative compactness in the space (C, ρ), the following theorem, named after C. Arzelà[6] and G. Ascoli,[7] applies (for a more general version see, e.g., [HEI], Theorem 4.9.3).

13.11 Theorem (Arzelà–Ascoli)
A subset M of $C[0, 1]$ is relatively compact if and only if

$$\sup_{x \in M} |x(0)| < \infty, \tag{13.7}$$

$$\lim_{\delta \to 0} \sup_{x \in M} w_x(\delta) = 0. \tag{13.8}$$

Property (13.7) states that the collection M of functions is *pointwise bounded* at the point $t = 0$, and (13.8) means that M is *equicontinuous*. Note that the set $\{z_n : n \geq 2\}$ of functions z_n defined in (13.4) violates the condition of equicontinuity and is therefore not relatively compact.

Proof of Theorem 13.11 It is easy to see that both (13.7) and (13.8) hold if $\overline{M}$ is compact: The function $\pi_0 : C \to \mathbb{R}$, defined by $\pi_0(x) := x(0)$, $x \in C$, is continuous, and since the image of a compact set under a continuous mapping is also compact and therefore bounded, (13.7) follows. To show (13.8), let $f_n(x) := w(x, 1/n)$, $x \in C$. Due to (13.6), f_n is continuous, and $f_n(x) \downarrow 0$ as $n \to \infty$. Fix any $\varepsilon > 0$, and put $O_n := \{x : f_n(x) < \varepsilon\}$, $n \geq 1$. Then $O_1 \subset O_2 \subset \ldots$ is an increasing sequence of open sets with $C = \cup_{n=1}^{\infty} O_n$. Since $\overline{M}$ is compact, $\overline{M} \subset O_n$ for some n, which yields (13.8). We now prove that (13.7) and (13.8) imply the relative compactness of M. By Theorem 13.6, we only need to show that M is totally bounded, that is, for every $\varepsilon > 0$ there is a finite ε-net for M, because $\overline{M}$ is also complete due to the completeness of C. According to (13.8), there is an integer k with $\sup_{x \in M} w_x(1/k) < \infty$. Since

$$|x(t)| \leq |x(0)| + \sum_{j=1}^{k} \left| x\left(\frac{jt}{k}\right) - x\left(\frac{(j-1)t}{k}\right) \right| \leq |x(0)| + k w_x\left(\frac{1}{k}\right),$$

(13.7) shows that $a < \infty$, where $a := \sup_{0 \leq t \leq 1} \sup_{x \in M} |x(t)|$. The graph $\{(t, x(t)) : 0 \leq t \leq 1\}$ of each function $x \in M$ is therefore a subset of $[0, 1] \times [-a, a]$. For a given $\varepsilon > 0$, let H be a finite ε-net for the interval $[-a, a]$. If we choose k using (13.8) so large that $w_x(1/k) < \varepsilon$ for each $x \in M$, and if we set $I_{kj} := [(j-1)/k, j/k]$, $j \in \{1, \ldots, k\}$, then the set B (say) of all continuous functions $y : [0, 1] \to \mathbb{R}$ with $y(\frac{j}{k}) \in H$ for

---

[6] Cesare Arzelà (1847–1912), Italian mathematician. Main field of research: Real functions.

[7] Guido Ascoli (1887–1957), Italian mathematician, after many years of work as a teacher, professor at various Italian universities. Main field of research: Partial differential equations.

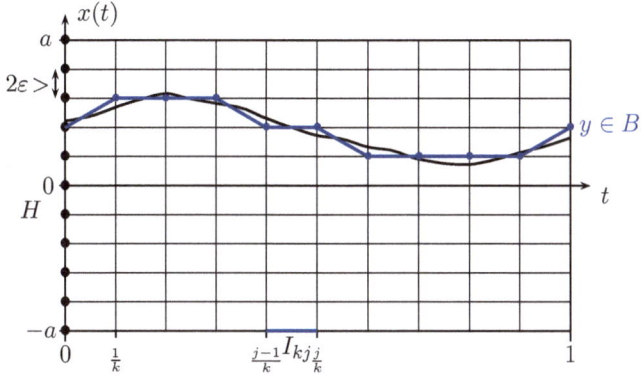

Fig. 13.4 Approximation of a function $x \in C$ by a function $y \in B$

each $j \in \{0, \ldots, k\}$, and which are linear on each of the intervals I_{kj}, is a 2ε-net for M. Figure 13.4 illustrates this fact. $\qquad\square$

We now get to know a separating class for probability measures on $\mathcal{B}(C)$.

13.12 Definition (Coordinate Projections, Finite-dimensional Sets)
For $k \in \mathbb{N}$ and $t_1, \ldots, t_k$ with $0 \le t_1 < \ldots < t_k \le 1$, let the mapping $\pi_{t_1, \ldots, t_k} : C \to \mathbb{R}^k$ be defined by

$$\pi_{t_1, \ldots, t_k}(x) := (x(t_1), \ldots, x(t_k)), \quad x \in C.$$

The mappings $\pi_{t_1, \ldots, t_k}$ $(k \in \mathbb{N}, \ 0 \le t_1 < \ldots < t_k \le 1)$ are called *coordinate projections*. Because

$$\max_{j=1,\ldots,k} \left| x(t_j) - y(t_j) \right| \le \|x - y\|_\infty \quad (x, y \in C),$$

$\pi_{t_1, \ldots, t_k} : C \to \mathbb{R}^k$ is a continuous mapping.

The collection

$$\mathcal{C}_f := \left\{ \pi^{-1}_{t_1, \ldots, t_k}(H) \,\middle|\, k \in \mathbb{N}, \ 0 \le t_1 < \ldots < t_k \le 1, \ H \in \mathcal{B}^k \right\}$$

is called the class of *finite-dimensional sets*.

Self-question 10: Why is every finite-dimensional set a Borel set in C?

We claim that the collection $\mathcal{C}_f$ is a *field*, that is, $C \in \mathcal{C}_f$, and $A^c \in \mathcal{C}_f$ whenever $A \in \mathcal{C}_f$. Moreover, $\mathcal{C}_f$ is a π-system. Because $C = \pi_1^{-1}(\mathbb{R})$, it follows that $C \in \mathcal{C}_f$, and $\mathcal{C}_f \ni A = \pi_{t_1,\ldots,t_k}^{-1}(H)$ with $H \in \mathcal{B}^k$ implies that $A^c = \pi_{t_1,\ldots,t_k}^{-1}(\mathbb{R}^k \setminus H) \in \mathcal{C}_f$. To show that $\mathcal{C}_f$ is a π-system, let $A, B \in \mathcal{C}_f$. This means there are integers k and ℓ as well as $t_1, \ldots, t_k, s_1, \ldots, s_\ell \in [0, 1]$ with $0 \le t_1 < \ldots < t_k \le 1$ and $0 \le s_1 < \ldots < s_\ell \le 1$, and Borel sets $H \in \mathcal{B}^k$ and $K \in \mathcal{B}^\ell$ with $A = \pi_{t_1,\ldots,t_k}^{-1}(H)$ and $B = \pi_{s_1,\ldots,s_\ell}^{-1}(K)$. We set $T := \{t_1, \ldots, t_k\} \cup \{s_1, \ldots, s_\ell\}$ and $m := |T|$, hence $\max(k, \ell) \le m \le k + \ell$. Suppose $T =: \{u_1, \ldots, u_m\}$, where $0 \le u_1 < \ldots < u_m \le 1$. Further, let $\alpha_1, \ldots, \alpha_k$ as well as $\beta_1, \ldots, \beta_\ell$ be integers with $1 \le \alpha_1 < \ldots < \alpha_k \le m$ and $1 \le \beta_1 < \ldots < \beta_\ell \le m$, such that $t_i = u_{\alpha_i}$ for $i \in \{1, \ldots, k\}$, and $s_j = u_{\beta_j}$ for $j \in \{1, \ldots, \ell\}$. With $L := \{(z_1, \ldots, z_m) \in \mathbb{R}^m : (z_{\alpha_1}, \ldots, z_{\alpha_k}) \in H, (z_{\beta_1}, \ldots, z_{\beta_\ell}) \in K\}$ it follows that

$$
\begin{aligned}
A \cap B &= \{x \in C : (x(u_{\alpha_1}), \ldots, x(u_{\alpha_k})) \in H, (x(u_{\beta_1}), \ldots, x(u_{\beta_\ell})) \in K\} \\
&= \{x \in C : (x(u_1), \ldots, x(u_m)) \in L\} \\
&= \pi_{u_1,\ldots,u_m}^{-1}(L).
\end{aligned}
$$

This shows that $A \cap B$ belongs to $\mathcal{C}_f$.

We will now see that the collection $\mathcal{C}_f$ of finite-dimensional sets is a separating class for $\mathcal{P}$ in the sense of Definition 13.7. For this purpose, let $x \in C$ and $\varepsilon > 0$ be arbitrary. Since x is a continuous function,

$$
\begin{aligned}
\overline{B(x, \varepsilon)} &= \bigcap_{r \in \mathbb{Q} \cap [0,1]} \{y \in C : |y(r) - x(r)| \le \varepsilon\} \\
&= \bigcap_{r \in \mathbb{Q} \cap [0,1]} \pi_r^{-1}([x(r) - \varepsilon, x(r) + \varepsilon]),
\end{aligned}
$$

and thus $\overline{B(x, \varepsilon)}$, being a countable intersection of sets in $\mathcal{C}_f$, belongs to the σ-field $\sigma(\mathcal{C}_f)$. Since $B(x, \varepsilon) = \bigcup_{n=1}^\infty \overline{B(x, \varepsilon - \frac{1}{n})}$, the collection $\mathcal{C}_f$ also contains all open balls. Because C is separable, the class of all open balls generates the σ-field $\mathcal{B}(C)$ (see Problem 13.3), Thus, $\sigma(\mathcal{C}_f) = \mathcal{B}(C)$, and since $\mathcal{C}_f$ is a π-system, it is a separating class for $\mathcal{P}$.

13.13 The Space $\mathbb{R}^\infty$

As a further example of a metric space, we consider the set

$$
S := \mathbb{R}^\infty := \{x = (x_k)_{k \ge 1} : x_k \in \mathbb{R} \text{ for every } k \ge 1\}
$$

of sequences of real numbers. With the metric $\rho : S \times S \to \mathbb{R}$ defined by

$$
\rho(x, y) := \sum_{k=1}^\infty \frac{\min(1, |x_k - y_k|)}{2^k}, \quad x = (x_k)_{k \ge 1}, \ y = (y_k)_{k \ge 1}, \tag{13.9}
$$

(S, ρ) becomes a complete separable metric space, and for $x^n = (x_k^n)_{k \geq 1}$ we have as $n \to \infty$:

$$\rho\left(x^n, x\right) \to 0 \Longleftrightarrow x_j^n \to x_j \quad \text{for every } j \geq 1 \tag{13.10}$$

(see Problems 13.5 and 13.6). Convergence in the space $(\mathbb{R}^\infty, \rho)$ is therefore colloquially "component-wise convergence". As a consequence, if we define for each integer k a mapping $\pi_k : S \to \mathbb{R}^k$ by $\pi_k(x) := (x_1, \dots, x_k)$, then π_k, which is the projection of $x = (x_j)_{j \geq 1}$ onto its first k components, is continuous and thus also $(\mathcal{B}(S), \mathcal{B}^k)$-measurable. Hence, the collection

$$\mathcal{R}_f^\infty := \left\{ \pi_k^{-1}(H) : k \in \mathbb{N}, \ H \in \mathcal{B}^k \right\} \tag{13.11}$$

of so-called *finite-dimensional sets* in $\mathbb{R}^\infty$ is a subcollection of $\mathcal{B}(\mathbb{R}^\infty)$. The class $\mathcal{R}_f^\infty$ of sets is a field (Problem 13.10). If we define for $x \in \mathbb{R}^\infty$, $\varepsilon > 0$ and each integer k a subset of $\mathbb{R}^\infty$ by

$$O_{k,\varepsilon}(x) := \pi_k^{-1}\left(\times_{j=1}^k (x_j - \varepsilon, x_j + \varepsilon) \right)$$
$$= \left\{ y = (y_j)_{j \geq 1} \in \mathbb{R}^\infty : |y_j - x_j| < \varepsilon \text{ for each } j = 1, \dots, k \right\},$$

then $O_{k,\varepsilon}(x)$ is an open set due to the continuity of π_k, and $y \in O_{k,\varepsilon}(x)$ implies $\rho(x, y) < \varepsilon + 1/2^k$.

Self-question 11: Why does the last implication hold?

For a given $r > 0$, one can choose $\varepsilon > 0$ and an integer k such that $\varepsilon + 2^{-k} < r$, which implies $O_{k,\varepsilon}(x) \subset B(x, r)$. Therefore, the subclass $\{ O_{k,\varepsilon}(x) : x \in \mathbb{R}^\infty, \ k \in \mathbb{N}, \ \varepsilon > 0 \}$ of $\mathcal{R}_f^\infty$ is a basis of the collection $\mathcal{O}$ of open sets in $\mathbb{R}^\infty$. Since $\mathbb{R}^\infty$ is separable, it follows that $\sigma(\mathcal{R}_f^\infty) = \mathcal{B}(\mathbb{R}^\infty)$. Therefore, the collection of finite-dimensional sets is a separating class for $\mathcal{P}$.

As in the space C[0, 1], no closed ball is compact in the space $\mathbb{R}^\infty$: If we define for $x = (x_j)_{j \geq 1} \in \mathbb{R}^\infty$, $k \in \mathbb{N}$, and $\varepsilon > 0$ a sequence $y^n = (y_j^n)_{j \geq 1}$ by $y_j^n := x_j$ if $j \leq k$, and $y_j^n := n$ if $j > k$, then $\{ y^n : n \geq 1 \} \subset O_{k,\varepsilon}(x) \subset B(x, \varepsilon + 2^{-k})$. Since the sequence $(y^n)_{n \geq 1}$ does not contain a convergent subsequence, no closed ball $\overline{B(x, r)}$ is compact. As in the space C[0, 1], every compact set is nowhere dense in $\mathbb{R}^\infty$, and by Baire's category theorem (Theorem 13.5), it follows that $\mathbb{R}^\infty$ cannot be written as a countable union of compact sets. If $A \subset \mathbb{R}^\infty$ is a closed set, then:

$$A \text{ compact} \Longleftrightarrow \forall k \in \mathbb{N} : \{ x_k : x = (x_j)_{j \geq 1} \in A \} \subset \mathbb{R} \text{ is bounded.} \tag{13.12}$$

(Problem 13.11).

We emphasize that, mutatis mutandis, the above considerations remain valid for the sequence space $\mathbb{R}^{\mathbb{N}_0} = \{x = (x_k)_{k\geq 0} : x_j \in \mathbb{R} \text{ for every } j \geq 0\}$ instead of $\mathbb{R}^\infty$.

Finally, we mention in particular [DUD] and [ALB] among the many recommendable further introductions to the field of topology of metric and more general spaces with a view towards stochastics.

Answers to the Self-Questions

Answer 1 We have $\widehat{F}_n(0) = 0$ if and only if $X_j > 0$ for each $j = 1, \ldots, n$. The probability of this event is 1, and thus $\mathbb{P}(B_n(0) = 0) = 1$. Likewise, $\widehat{F}_n(1) = 1$ if and only if $X_j \leq 1$ for each j, which also happens with probability 1. Consequently, $\mathbb{P}(B_n(1) = 0) = 1$.

Answer 2 Since $\rho(x, y) = 1$ if $x \neq y$, it follows that $B(x, 1) = \{x\}$ and $B(x, 1.001) = S$.

Answer 3 For $z \in M$ and $y \in S$, $\rho(x, M) \leq \rho(x, z) \leq \rho(x, y) + \rho(y, z)$. Hence $\rho(x, M) \leq \rho(x, y) + \rho(y, M)$ and thus $\rho(x, M) - \rho(y, M) \leq \rho(x, y)$. The assertion now follows for reasons of symmetry.

Answer 4 Exactly the finite subsets M of S are compact. For ε with $0 < \varepsilon \leq 1$, $B(x, \varepsilon) = \{x\}$ and thus $M = \cup_{x \in M}\{x\}$. For $\cup_{x \in M}\{x\}$ to have a finite subcover, M must be finite. Conversely, if $M =: \{x_1, \ldots, x_k\}$ is a finite set and $M \subset \cup_{i \in I} O_i$ with open sets O_i, then at most k of the O_i are sufficient to cover M.

Answer 5 For $x \in S$, $\{x\} = \cap_{k=1}^\infty B\left(x, \frac{1}{k}\right)$. As a countable intersection of open sets, $\{x\}$ is a Borel set. An alternative justification is that $\{x\}$, being the complement of the open (!) set $S \setminus \{x\}$, is closed.

Answer 6 If $x \in A$, then $\rho(x, A) = 0$ and thus $f_\varepsilon(x) = 1$. Since $f_\varepsilon(x) \geq 0$, this means $\mathbf{1}_A(x) \leq f_\varepsilon(x)$. To show $f_\varepsilon \leq \mathbf{1}_{A^\varepsilon}$ let w.l.o.g. $f_\varepsilon(x) > 0$. It follows that $\rho(x, A) < \varepsilon$ and thus $x \in A^\varepsilon$, so $f_\varepsilon(x) \leq 1 = \mathbf{1}_{A^\varepsilon}$.

Answer 7 Let p with $p(t) = \sum_{k=0}^n a_k t^t$ be a polynomial with real coefficients and $\varepsilon > 0$. For a_k there is a rational number q_k with $|a_k - q_k| < \varepsilon 2^{-(k+1)}$, $k \in \{0, \ldots, n\}$. For $q \in C$ with $q(t) = \sum_{k=0}^n q_k t^k$ it then follows that $\rho(p, q) \leq \sum_{k=0}^n |a_k - b_k| \leq \varepsilon$.

Answer 8 Because x is the limit of a *uniformly* convergent sequence of continuous functions on $[0, 1]$.

Answer 9 If $\rho(\varepsilon z_{n_k}, z) \to 0$ as $k \to \infty$, then $\varepsilon z_{n_k}(t) \to z(t)$, $0 \le t \le 1$. Since $z_{n_k}(0) = 0$ for each k and $z_{n_k}(t) = 0$ if $n_k \ge \frac{2}{t}$ for $0 < t \le 1$, it follows that $z \equiv 0$.

Answer 10 Let $\mathcal{O}^k$ and $\mathcal{B}^k$ be the class of open sets and Borel sets in $\mathbb{R}^k$, respectively. Further, let $\mathcal{G} := \{B \in \mathcal{B}^k : \pi_{t_1,...,t_k}^{-1}(B) \in \mathcal{B}(S)\}$. Since $\pi_{t_1,...,t_k}$ is continuous, $\pi_{t_1,...,t_k}^{-1}(\mathcal{O}^k) \subset \mathcal{O} \subset \mathcal{B}(S)$. Because $\mathcal{G}$ is a σ-field, it follows that $\mathcal{B}^k = \sigma(\mathcal{O}^k) \subset \mathcal{G}$.

Answer 11 According to the definition of $\rho(x, y)$, if $y \in O_{k,\varepsilon}(x)$ then

$$\rho(x, y) = \sum_{j=1}^{\infty} \frac{\min(1, |x_j - y_j|)}{2^j} \le \varepsilon \sum_{j=1}^{k} \frac{1}{2^j} + \sum_{j=k+1}^{\infty} \frac{1}{2^j} < \varepsilon + \frac{1}{2^k}.$$

Problems

13.1 Problem Let (S, ρ) be a metric space, $x \in S$, and $r > 0$. Show that

$$\partial B(x, r) \subset \{y \in S : \rho(x, y) = r\}.$$

Is strict inclusion possible?

13.2 Problem Let (S, ρ) be a metric space. Prove that every compact subset of S is bounded and closed.

13.3 Problem Suppose (S, ρ) is a separable metric space. Show that the Borel σ-field $\mathcal{B}(S)$ is generated by the system $\mathcal{M} := \{B(x, \varepsilon) : x \in S, \ \varepsilon > 0\}$ of open balls.

13.4 Problem Prove part "(b) $\Rightarrow$ (c)" of Theorem 13.6.

13.5 Problem Let (S, ρ) be a metric space and $\tilde{\rho}(x, y) := \min(1, \rho(x, y))$, $x, y \in S$. Prove the following claims:

(a) $\tilde{\rho}$ is a metric on S.
(b) ρ and $\tilde{\rho}$ generate the same class of open sets.
(c) With (S, ρ), $(S, \tilde{\rho})$ is also separable.
(d) With (S, ρ), $(S, \tilde{\rho})$ is also complete.

13.6 Problem Let (S_j, ρ_j), $j \in \mathbb{N}$, be metric spaces and $S := \times_{j=1}^{\infty} S_j$. For $x = (x_j)_{j \geq 1}$, $y = (y_j)_{j \geq 1} \in S$, let

$$\rho(x, y) := \sum_{j=1}^{\infty} \frac{\min(1, \rho_j(x_j, y_j))}{2^j}.$$

Prove the following statements:

(a) ρ is a metric on S, and for $x^n := \left(x_j^{(n)}\right)_{j \geq 1} \in S$, $n \geq 1$, and $x = (x_j)_{j \geq 1} \in S$, we have as $n \to \infty$:

$$\rho\left(x^{(n)}, x\right) \to 0 \iff \rho_j\left(x_j^{(n)}, x_j\right) \to 0 \text{ for every } j \geq 1.$$

(b) If S_j is separable for every j then S is separable.
(c) If S_j is complete for every j then S is complete.
(d) If $A_j \subset S_j$ is compact for every $j \geq 1$ then $A := \times_{j=1}^{\infty} A_j \subset S$ is compact.

Hint for (d) Use Theorem 13.6 and a suitable subsequence.

13.7 Problem Let (S', ρ') and (S'', ρ'') be metric spaces with Borel σ-fields $\mathcal{B}'$ and $\mathcal{B}''$, respectively. Further let $S := S' \times S''$ and

$$\rho((x', x''), (y', y'')) := \max\left(\rho'(x', y'), \rho''(x'', y'')\right), \quad (x', x''), (y', y'') \in S.$$

(a) Show that (S, ρ) is a metric space.
(b) Let $\mathcal{B}$ be the Borel σ-field of S. Prove the following claims:
  (b1) (S, ρ) is separable if and only if (S', ρ') and (S'', ρ'') are separable.
  (b2) If (S, ρ) is separable then $\mathcal{B} = \mathcal{B}' \otimes \mathcal{B}''$.

13.8 Problem Let (S, ρ) be a discrete metric space, where S is uncountable. What do the Borel σ-field and the σ-field generated by the open balls look like?

13.9 Problem Let $x, y \in C$ and $0 < \delta \leq 1$, and let $w_x(\delta)$ be the modulus of continuity defined in (13.5). Show that

$$|w_x(\delta) - w_y(\delta)| \leq 2\rho(x, y).$$

13.10 Problem Show that the system $\mathcal{R}_f^{\infty}$ defined in (13.11) is a field.

13.11 Problem Prove the equivalence (13.12).

Hint for "⟸" Cantor's diagonal argument.

13.12 Problem Let (S, ρ) be a metric space and P a probability measure on $\mathcal{B}(S)$. Prove the following *regularity property* of P: For every $B \in \mathcal{B}(S)$ and every $\varepsilon > 0$ there exist a closed set A and an open set O with $A \subset B \subset O$ and $P(O \setminus A) < \varepsilon$.

Hint Let $\mathcal{G}$ be the class of Borel sets B with the above properties. Show that $\mathcal{G}$ includes the closed sets, and that it is a σ-field over S.

Among other things, Chap. 6 dealt with the convergence in distribution $X_n \xrightarrow{\mathcal{D}} X$ of a sequence $X_1, X_2, \ldots$ of d-dimensional random vectors to a random vector X. This type of convergence was defined by

$$X_n \xrightarrow{\mathcal{D}} X :\Longleftrightarrow \lim_{n \to \infty} \mathbb{E}\big[f(X_n)\big] = \mathbb{E}\big[f(X)\big] \quad \text{for every } f \in \mathcal{C}_b(\mathbb{R}^d), \tag{14.1}$$

where $\mathcal{C}_b(\mathbb{R}^d)$ denotes the set of all continuous and bounded functions $f : \mathbb{R}^d \to \mathbb{R}$. Writing $Q_n := \mathbb{P}^{X_n}$ and $Q := \mathbb{P}^X$ for the distributions of X_n and X, respectively, (14.1) is the same as

$$\lim_{n \to \infty} \int_{\mathbb{R}^d} f \, dQ_n = \int_{\mathbb{R}^d} f \, dQ \quad \text{for every } f \in \mathcal{C}_b(\mathbb{R}^d). \tag{14.2}$$

Since both continuity and boundedness are defined also for real-valued functions whose domain is a metric space (S, ρ), (14.1) and (14.2) can be generalized in a straightforward manner, and we make these generalizations in this chapter. In the case of version (14.2), and completely detached from the idea that Q_n and Q are distributions, that is, image probability measures of $\mathbb{P}$ under X_n and under X, respectively, we speak of *weak convergence* of a sequence of probability measures Q_n to a probability measure Q.

In the following, let (S, ρ) be a metric space with Borel σ-field $\mathcal{B}(S)$, and let $\mathcal{P}$ denote the collection of all probability measures on $\mathcal{B}(S)$. The notations $\mathcal{C}_b(S)$ and $\mathcal{C}_b^0(S)$ stand for the sets of all continuous and bounded, and all uniformly continuous, bounded real-valued functions defined on S, respectively. Unless otherwise agreed, each unspecified integral is over S.

© The Author(s), under exclusive license to Springer-Verlag GmbH, DE, part of Springer Nature 2024
N. Henze, *Asymptotic Stochastics*, Mathematics Study Resources 10, https://doi.org/10.1007/978-3-662-68923-3_14

14.1 Definition (Weak Convergence)

If $P, P_1, P_2, \ldots$ are probability measures on $\mathcal{B}(S)$, we define

$$P_n \xrightarrow{\mathcal{D}} P \text{ as } n \to \infty :\Longleftrightarrow \lim_{n\to\infty} \int f \, dP_n = \int f \, dP \quad \text{for every } f \in C_b(S).$$

In this case, we say that the sequence (P_n) *converges weakly* to P.

Self-question 1: Why do $P_n \xrightarrow{\mathcal{D}} P$ and $P_n \xrightarrow{\mathcal{D}} Q$ imply $P = Q$?

14.2 Example (Dirac Measure)

Let $x_0, x_1, x_2, \ldots \in S$. If δ_x denotes the Dirac measure at $x \in S$, then:

$$\delta_{x_n} \xrightarrow{\mathcal{D}} \delta_{x_0} \Longleftrightarrow x_n \to x_0.$$

Proof "$\Longleftarrow$": Suppose $x_n \to x_0$. If $f \in C_b(S)$ then $\int f \, d\delta_{x_n} = f(x_n) \to f(x_0) = \int f \, d\delta_{x_0}$ and thus $\delta_{x_n} \xrightarrow{\mathcal{D}} \delta_{x_0}$.

We prove the implication "$\Longrightarrow$" by contraposition and assume that $x_n \not\to x_0$. Then there is an $\varepsilon > 0$ with $\rho(x_n, x_0) > \varepsilon$ for infinitely many n. We define

$$f_\varepsilon(x) := \max\left(0, 1 - \frac{\rho(x, x_0)}{\varepsilon}\right), \quad x \in S,$$

so $f_\varepsilon \in C_b(S)$ (cf. (13.3) with $A = \{x_0\}$) and $f_\varepsilon(x_0) = 1$. Since $\rho(x_n, x) > \varepsilon$ implies $f_\varepsilon(x_n) = 0$, it follows that $f_\varepsilon(x_n) = 0$ for infinitely many n, and hence we get $\int f_\varepsilon \, d\delta_{x_n} = f_\varepsilon(x_n) \not\to f_\varepsilon(x_0) = \int f_\varepsilon \, d\delta_{x_0}$. Thus, the sequence (δ_{x_n}) does not converge weakly to δ_{x_0}. $\square$

The Portmanteau Theorem (Theorem 6.3) also applies almost unchanged to the weak convergence of probability measures in general metric spaces. Part (e) of Theorem 6.3 is not applicable, since the notion of a distribution function presupposes an ordering structure.

14.3 Theorem (Portmanteau Theorem)

For probability measures $P, P_1, P_2, \ldots$ on $\mathcal{B}(S)$, the following statements are equivalent:

(a) $P_n \xrightarrow{\mathcal{D}} P$,

(b) $\int f \, dP_n \to \int f \, dP$ *for every* $f \in C_b^0(S)$,

(c) $\lim\sup_{n\to\infty} P_n(A) \leq P(A)$ *for every closed set A,*
(d) $\lim\inf_{n\to\infty} P_n(O) \geq P(O)$ *for every open set O,*
(e) $\lim_{n\to\infty} P_n(B) = P(B)$ *for every Borel set B with $P(\partial B) = 0$.*

A set $B \in \mathcal{B}(S)$ with the property $P(\partial B) = 0$ is called a *P-continuity set*. We denote the collection of all *P*-continuity sets by

$$\mathcal{C}(P) := \{B \in \mathcal{B}(S) : P(\partial B) = 0\}. \tag{14.3}$$

Proof of Theorem 14.3 The proof of the Portmanteau Theorem largely follows the proof of Theorem 6.3. The implication "(a) $\Longrightarrow$ (b)" holds because $\mathcal{C}_b^0(S) \subset \mathcal{C}_b(S)$.

"(b) $\Longrightarrow$ (c)" follows completely analogously to part "(a) $\Longrightarrow$ (b)" of the proof of Theorem 6.3, by replacing $\|x - A\|$ occurring in (6.4) with $\rho(x, A)$. The parts "(c) $\Longleftrightarrow$ (d)" and "(c)+ (d) $\Longrightarrow$ (e)" can be taken verbatim from the corresponding parts of the proof of Theorem 6.3.

"(e) $\Longrightarrow$ (a)": Fix any f in $\mathcal{C}_b(S)$. Since f is bounded, there is an $L > 0$ with $|f| < L$, so $|f(x)| < L$ for every $x \in S$. This implies $0 < (\frac{f}{L} + 1) \cdot \frac{1}{2} < 1$. Due to the linearity of the integral, we may assume w.l.o.g. that $0 < f < 1$. Let $P_n^f = P_n f^{-1}$ denote the image measure of P_n under f and λ^1 the Borel–Lebesgue measure in $\mathbb{R}^1$. By the change of variables formula (see, e.g., [DUR], p. 30) and Tonelli's theorem, we obtain

$$\int f \, dP_n = \int_0^1 t \, P_n^f(dt) = \int_0^1 \left(\int_0^t 1 \, \lambda^1(du)\right) P_n^f(dt) = \int_0^1 \left(\int_u^1 P_n^f(dt)\right) \lambda^1(du)$$

$$= \int_0^1 P_n(f > u) \, du.$$

Here, $P_n(f > u)$ stands for $P_n(\{x \in S : f(x) > u\})$. In the same way and with the same convention, $\int f \, dP = \int_0^1 P(f > u) \, du$. Since f is continuous, it follows that $\partial\{f > u\} \subset \{f = u\}$.

Self-question 2: Can you derive $\partial\{f > u\} \subset \{f = u\}$?

We thus obtain $P_n(\partial\{f > u\}) \leq P_n(f = u), n \geq 1$, and $P(\partial\{f > u\}) \leq P(f = u)$. Since $P(f = u) = 0$ for all but at most countably many values of u, we have $P_n(f > u) \to P(f > u)$ λ^1-almost everywhere. The claim $\int f \, dP_n \to \int f \, dP$ now follows from the dominated convergence theorem.

$\square$

Self-question 3: Why can $P(f = u) > 0$ hold only for at most countably many u?

As we will see, criterion (c) of the Portmanteau theorem is the most important for proving weak convergence. This also applies to the proof of the following result, which is a far-reaching generalization of Theorem 6.6.

14.4 Theorem (Mapping Theorem)
Let (S, ρ) and (S', ρ') be metric spaces and $h : S \to S'$ a $(\mathcal{B}(S), \mathcal{B}(S'))$-measurable mapping. The set of continuity points of h is denoted by C_h. If $P, P_1, P_2, \ldots$ are probability measures on $\mathcal{B}(S)$, then:

$$\text{If } P_n \xrightarrow{\mathcal{D}} P \text{ and } P(C_h) = 1 \text{ then } P_n^h \xrightarrow{\mathcal{D}} P^h.$$

Proof Let $A' \subset S'$ be a closed set. Since $\overline{h^{-1}(A')} \subset S \setminus C_h \cup h^{-1}(A')$ and $P_n^h = P_n h^{-1}$ as well as $P^h = P h^{-1}$, we obtain

$$\limsup_{n \to \infty} P_n\left(h^{-1}(A')\right) \leq \limsup_{n \to \infty} P_n\left(\overline{h^{-1}(A')}\right)$$

$$\leq P\left(\overline{h^{-1}(A')}\right)$$

$$\leq P(S \setminus C_h) + P\left(h^{-1}(A')\right)$$

$$= 0 + P\left(h^{-1}(A')\right).$$

The assertion now follows from Theorem 14.3 (c). □

Thus, under general conditions, weak convergence $P_n \xrightarrow{\mathcal{D}} P$ of probability measures is inherited under measurable mappings that are almost everwhere continuous with respect to P. The following subsequence criterion provides a necessary and sufficient condition for weak convergence.

14.5 Theorem (Subsequence Criterion for Weak Convergence)
Let $P, P_1, P_2, \ldots$ be probability measures on $\mathcal{B}(S)$. Then the following statements are equivalent:

(a) $P_n \xrightarrow{\mathcal{D}} P$.

(b) Every subsequence $(P_{n_k})_{k \geq 1}$ of (P_n) contains a sub-subsequence $(P_{n'_k})_{k \geq 1}$ with $P_{n'_k} \xrightarrow{\mathcal{D}} P$ as $k \to \infty$.

Proof Only the implication "(b) $\Longrightarrow$ (a)" needs to be shown, and the proof is by contraposition. To this end, suppose (P_n) does not converge weakly to P. Then there is some $f \in C_b(S)$ and some $\varepsilon > 0$ with $|\int f \, dP_{n_k} - \int f \, dP| > \varepsilon$ for a subsequence (P_{n_k}). Then, however, no subsequence of (P_{n_k}) can converge weakly to P.   □

We now rewrite Definition 14.1 and the results obtained so far for distributions of random variables with range S. If $(\Omega, \mathcal{A}, \mathbb{P})$ is a probability space, then any $(\mathcal{A}, \mathcal{B}(S))$-measurable mapping $X : \Omega \rightarrow S$ is called a *random element in S* or *S-valued random element*. In this respect, there are various specific namings, depending on S. For example, if $S = \mathbb{R}$ then one speaks of a *(real-valued) random variable*, and in the case $S = \mathbb{R}^d$ of a *d-dimensional random vector*. If S is the sequence space $\mathbb{R}^\infty$ introduced in Example 13.13, then X is is said to be a *random sequence*, and in the case $S = C[0, 1]$ (cf. Example 13.9), the term *random (continuous) function* is also common, see Sect. 14.9.

The *distribution of X* is the probability measure $\mathbb{P}^X = \mathbb{P}X^{-1}$ on $\mathcal{B}(S)$, that is, the image measure of $\mathbb{P}$ under the mapping X. If P is a probability measure on $\mathcal{B}(S)$, one can always specify a probability space $(\Omega, \mathcal{A}, \mathbb{P})$ and a random element $X : \Omega \rightarrow S$ such that X has the distribution P. One way to do this is the so-called *canonical construction* $\Omega := S$, $\mathcal{A} := \mathcal{B}(S)$ and $\mathbb{P} := P$, and one chooses X as the identity map on Ω. The message conveyed with the canonical construction is therefore that the underlying probability space is of little importance when one is interested in distributions.

The definition corresponding to Definition 14.1 for S-valued random elements that are all defined on the same probability space $(\Omega, \mathcal{A}, \mathbb{P})$ is as follows:

14.6 Definition (Convergence in Distribution)
Let $X, X_1, X_2, \ldots$ be random elements in S. Then we define

$$X_n \xrightarrow{\mathcal{D}} X :\Longleftrightarrow \mathbb{E}\big[f(X_n)\big] \rightarrow \mathbb{E}\big[f(X)\big] \text{ for every } f \in C_b(S)$$

and say that (X_n) *converges in distribution to X*. The distribution $\mathbb{P}^X$ of X is called *limit distribution* or *asymptotic distribution* of (X_n). Instead of $X_n \xrightarrow{\mathcal{D}} X$ there is also the hybrid notation $X_n \xrightarrow{\mathcal{D}} \mathbb{P}^X$.

A comparison with Definition 14.1 shows that, putting $P_n := \mathbb{P}^{X_n}$ and $P := \mathbb{P}^X$, the convergence in distribution $X_n \xrightarrow{\mathcal{D}} X$ is equivalent to weak convergence $P_n \xrightarrow{\mathcal{D}} P$. Stated in terms of random elements, the Portmanteau theorem (Theorem 14.3) reads as follows:

14.7 Theorem (Portmanteau Theorem)
For S-valued random elements $X, X_1, X_2, \ldots$ the following statements are equivalent:

(a) $X_n \xrightarrow{\mathcal{D}} X$,

(b) $\mathbb{E}[f(X_n)] \rightarrow \mathbb{E}[f(X)]$ *for every* $f \in C_b^0(S)$,
(c) $\limsup_{n\to\infty} \mathbb{P}(X_n \in A) \leq \mathbb{P}(X \in A)$ *for every closed set A,*
(d) $\liminf_{n\to\infty} \mathbb{P}(X_n \in O) \geq \mathbb{P}(X \in O)$ *for every open set O,*
(e) $\lim_{n\to\infty} \mathbb{P}(X_n \in B) = \mathbb{P}(X \in B)$ *for every set* $B \in \mathcal{B}(S)$ *with* $\mathbb{P}(X \in \partial B) = 0$.

In part (e), convergence $\mathbb{P}(X_n \in B) \rightarrow \mathbb{P}(X \in B)$ is for each Borel set B which is a $\mathbb{P}^X$-continuity set, that is, which belongs to the set $\mathcal{C}(\mathbb{P}^X)$ defined in (14.3). In this context, it is common to speak of an *X-continuity set*.

The analogue of the mapping theorem (Theorem 14.4) for random elements is as follows:

14.8 Theorem (Mapping Theorem)
Let (S, ρ) *and* (S', ρ') *be metric spaces and* $h : S \rightarrow S'$ *a* $(\mathcal{B}(S), \mathcal{B}(S'))$*-measurable mapping. The set of continuity points of h is denoted by* C_h. *Furthermore, let* $X, X_1, X_2, \ldots$ *be S-valued random elements defined on a common probability space. We then have:*

$$\text{If } X_n \xrightarrow{\mathcal{D}} X \text{ and } \mathbb{P}(X \in C_h) = 1 \text{ then } h(X_n) \xrightarrow{\mathcal{D}} h(X).$$

14.9 Random Functions
Let $(\Omega, \mathcal{A}, \mathbb{P})$ be a probability space and $X : \Omega \rightarrow C$ an $(\mathcal{A}, \mathcal{B}(C))$-measurable mapping, where for short $C := C[0, 1]$. The C-valued random element is called a *random function*. For a fixed $\omega \in \Omega$, $X(\omega)$ is a continuous function on $[0, 1]$, which is called the *path of X* (to ω). For the value of the function $X(\omega)$ at $t \in [0, 1]$ there are several notations, such as

$$X(\omega)(t) =: X_t(\omega) =: X(\omega, t), \qquad 0 \leq t \leq 1.$$

If we evaluate the random function X at the point $t \in [0, 1]$, that is, if we take the composition of X with the coordinate projection $\pi_t : C \rightarrow \mathbb{R}$ (cf. Definition 13.12), we obtain the random variable $X(t) := \pi_t \circ X$. We thus have

$$X(t) := \begin{cases} \Omega \rightarrow \mathbb{R}, \\ \omega \mapsto X(t)(\omega). \end{cases}$$

The mapping X is $(\mathcal{A}, \mathcal{B}(C))$-measurable if and only if the mapping $X(t)$ is $(\mathcal{A}, \mathcal{B}^1)$-measurable for each $t \in [0, 1]$ (Problem 14.4).

More generally, for every $k \geq 1$ and any choice of $t_1, \ldots, t_k \in [0, 1]$ with $0 \leq t_1 < \ldots < t_k \leq 1$, we obtain the k-dimensional random vector $\pi_{t_1,\ldots,t_k} \circ X$, which we denote by $(X(t_1), \ldots, X(t_k))$. The distributions of $(X(t_1), \ldots, X(t_k))$, where $k \geq 1$ and $0 \leq t_1 < \ldots < t_k \leq 1$, are called *finite-dimensional distributions* of X. We will use the common acronym *fidis* for these finite-dimensional distributions.

At the end of Sect. 13.12, we saw that the collection

$$\mathcal{C}_f = \left\{ \pi_{t_1,\dots,t_k}^{-1}(H) \,\middle|\, k \in \mathbb{N},\ 0 \le t_1 < \dots < t_k \le 1,\ H \in \mathcal{B}^k \right\}$$

of finite-dimensional sets in $\mathcal{B}(\mathrm{C})$ is a separating class for $\mathcal{P}$. This means that the distribution $\mathbb{P}^X$ of X, which is a probability measure on the σ-field of Borel sets of C, is uniquely determined by the values $\mathbb{P}^X(B)$ with $B \in \mathcal{C}_f$. By the definition of $\mathcal{C}_f$, any B in $\mathcal{C}_f$ has the representation $B = \pi_{t_1,\dots,t_k}^{-1}(H)$ with the meaning of $k, t_1, \dots, t_k$ and H as above. Since

$$\mathbb{P}^X(B) = \mathbb{P}^{\pi_{t_1,\dots,t_k} \circ X}(H) = \mathbb{P}\big((X(t_1), \dots, X(t_k)) \in H\big),$$

the distribution of X is thus uniquely determined by the fidis of X.

14.10 Partial Sum Processes

An important class of random functions are so-called *partial sum processes*. Here, the word *process* generally means a collection of random variables that are defined on a probability space. The starting point for partial sum processes is an i.i.d. sequence $Z_1, Z_2, \dots$ of real-valued random variables, where we assume $\mathbb{E}(Z_1) = 0$ and $\mathbb{E}(Z_1^2) = 1$.

We define $S_0 := 0$ and $S_k := Z_1 + \dots + Z_k$ for $k \ge 1$. For $t \in [0, 1]$, let

$$X_n(t) := \frac{S_{\lfloor nt \rfloor}}{\sqrt{n}} + (nt - \lfloor nt \rfloor) \cdot \frac{Z_{\lfloor nt \rfloor + 1}}{\sqrt{n}}. \tag{14.4}$$

The random function X_n is called the *n-th partial sum process* of the sequence $(Z_n)_{n \ge 1}$. If we interpret the variable t as *time*, the second summand in (14.4) causes a linear interpolation of the realizations of X_n between the "time points" $\frac{j}{n}$ and $\frac{j+1}{n}$, where $j \in \{0, 1, \dots, n-1\}$. As a result, the paths of X_n are continuous functions on $[0, 1]$.

Figure 14.1 shows a realization of such a partial sum process, with the distribution of Z_1 being the uniform distribution on the two values $+1$ and -1. This choice leads to the so-called *simple symmetric random walk* on the set $\mathbb{Z}$ of integers. After k steps, such a random walk is at the point S_k. In the next chapter, we will study the limit behavior of X_n as $n \to \infty$. Thereby, we get to know a far-reaching generalization of the multivariate central limit theorem.

Self-question 4: What is the limit distribution of $X_n(1)$ as $n \to \infty$?

Fig. 14.1 Realizations of
X_{100} (Here: $\mathbb{P}(Z_1 = 1) =$
$\mathbb{P}(Z_1 = -1) = 1/2$)

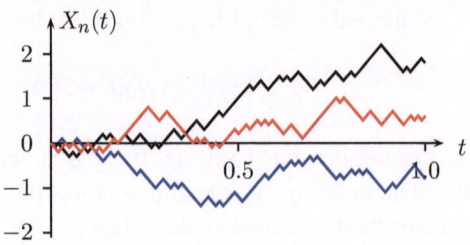

In the following, we will investigate how to prove weak convergence of probability measures or convergence in distribution of S-valued random elements.

14.11 Theorem (Criterion for Weak Convergence 1)

Let P be a probability measure on $\mathcal{B}(S)$ and $\mathcal{M}_P \subset \mathcal{B}(S)$ a π-system that possibly depends on P. Suppose that every open set is a countable union of sets in $\mathcal{M}_P$.

$$\text{If } P_n(A) \to P(A) \quad \text{for each } A \in \mathcal{M}_P \text{ then } P_n \xrightarrow{\mathcal{D}} P.$$

Proof Fix any open set $O \subset S$. We show that $P(O) \leq \liminf_{n\to\infty} P_n(O)$. The assertion would then follow from the Portmanteau theorem (Theorem 14.3). To this end, suppose $A_1, \ldots, A_k$ are sets in $\mathcal{M}_P$. Since $\mathcal{M}_P$ is a π-system, the inclusion-exclusion principle and the condition $P_n(A) \to P(A)$ for each $A \in \mathcal{M}_P$ yield

$$P_n\left(\bigcup_{j=1}^{k} A_j\right) = \sum_{j=1}^{k} P_n(A_j) - \sum_{1 \leq i < j \leq k} P_n(A_i \cap A_j) \pm \ldots + (-1)^{k-1} P_n(A_1 \cap \ldots \cap A_k)$$

$$\to P\left(\bigcup_{j=1}^{k} A_j\right) \quad \text{as } n \to \infty. \tag{14.5}$$

By assumption, there are sets $A_1, A_2, \ldots$ in $\mathcal{M}_P$ with $O = \bigcup_{j=1}^{\infty} A_j$. For any $\varepsilon > 0$, we choose an integer k with $P(O) - \varepsilon \leq P\left(\bigcup_{j=1}^{k} A_j\right)$.

Self-question 5: Why can we choose k like this?

From (14.5) it follows that

$$P(O) - \varepsilon \leq \lim_{n\to\infty} P_n\left(\bigcup_{j=1}^{k} A_j\right) \leq \liminf_{n\to\infty} P_n(O)$$

and thus the assertion, since ε was arbitrary. $\square$

The following criterion for weak convergence applies when the metric space (S, ρ) is separable.

14.12 Theorem (Criterion for Weak Convergence 2)

Let P and $\mathcal{M}_P$ be as in Theorem 14.11, and suppose (S, ρ) is separable. Furthermore, assume that the following condition is satisfied:

For every $x \in S$ and $\varepsilon > 0$ there is some $A \in \mathcal{M}_P$ with $x \in \overset{\circ}{A} \subset A \subset B(x, \varepsilon)$.   (14.6)

If $P_n(A) \to P(A)$ for each $A \in \mathcal{M}_P$ then $P_n \overset{\mathcal{D}}{\longrightarrow} P$.

Proof Let $O \subset S$ be any non-empty open set. In view of (14.6), for each $x \in O$ there is a set A_x in $\mathcal{M}_P$ with $x \in \overset{\circ}{A}_x \subset A_x \subset O$. Thus, $O = \bigcup_{x \in O} \overset{\circ}{A}_x$, that is, $\{\overset{\circ}{A}_x : x \in O\}$ is an open cover of O. Since (S, ρ) is separable, by Theorem 13.4 there is a countable subcover, that is, there exist $x_1, x_2, \ldots \in O$ with

$$O = \bigcup_{j=1}^{\infty} \overset{\circ}{A}_{x_j} = \bigcup_{j=1}^{\infty} A_{x_j}.$$

The assertion now follows from Theorem 14.11.   □

The criteria Theorems 14.11 and 14.12 are useful for proving weak convergence of a sequence (P_n) to *a specific* probability measure P. More flexibility, however, is gained if one can find a collection $\mathcal{M}$ of Borel sets such that *for every sequence (P_n) in $\mathcal{P}$ and every P in $\mathcal{P}$* the following implication holds:

If $\lim_{n \to \infty} P_n(B) = P(B)$ for every P-continuity set B in $\mathcal{M}$ then $P_n \overset{\mathcal{D}}{\longrightarrow} P$.

Such a collection $\mathcal{M}$ is subsequently called a *convergence-determining class* (for short: CDC). Such a CDC is always a separating class for $\mathcal{P}$. However, a separating class need not necessarily be a CDC, as we will soon see.

Self-question 6: Why is a CDC a separating class for $\mathcal{P}$?

To put it casually, a collection $\mathcal{M}$ must be "sufficiently rich" to be a CDC. To this end, let $\mathcal{M}$ be any non-empty collection of Borel sets. For any x in S and any $\varepsilon > 0$, we define

$$\mathcal{M}_{x,\varepsilon} := \{A \in \mathcal{M} : x \in \overset{\circ}{A} \subset A \subset B(x, \varepsilon)\},$$
$$\partial\mathcal{M}_{x,\varepsilon} := \{\partial A : A \in \mathcal{M}_{x,\varepsilon}\}.$$

If $\partial M_{x,\varepsilon}$ contains uncountably many pairwise disjoint sets then, regardless of $P \in \mathcal{P}$, at least one of these sets must have probability 0 under P (cf. Self-question 3).

14.13 Theorem (Sufficient Condition for a CDC)
Let (S, ρ) be separable and $\mathcal{M} \subset \mathcal{B}(S)$ be a π-system. If for each x in S and each $\varepsilon > 0$, the collection $\partial M_{x,\varepsilon}$ either contains $\emptyset$ or uncountably many pairwise disjoint sets, then $\mathcal{M}$ is a CDC.

Proof Fix any $P \in \mathcal{P}$, and let $\mathcal{M}_P := \mathcal{M} \cap \mathcal{C}(P)$ be the set of P-continuity sets in $\mathcal{M}$. Since

$$\partial(A \cap B) \subset \partial A \cup \partial B \tag{14.7}$$

$\mathcal{M}_P$ is a π-system.

Self-question 7: Why does (14.7) hold?

Suppose $P_n(A) \to P(A)$ for each $A \in \mathcal{M}_P$. Because of the condition imposed on $\partial M_{x,\varepsilon}$, the collection $\mathcal{M}_{x,\varepsilon}$ contains a set from $\mathcal{M}_P$ for each $x \in S$ and each $\varepsilon > 0$. Thus, $\mathcal{M}_P$ satisfies the condition of Theorem 14.12, and it follows that $P_n \xrightarrow{\mathcal{D}} P$. □

14.14 Example

(a) Let $\mathcal{M}$ be the collection of all finite intersections of open balls. Since $\partial B(x, r) \subset \{y \in S : \rho(x, y) = r\}$ (see Problem 13.1), $\mathcal{M}$ is a CDC.
(b) Let $S = \mathbb{R}^d$ and

$$\mathcal{M} := \left\{ \times_{j=1}^{d} (a_j, b_j] : a_1, \dots, a_d, b_1, \dots, b_d \in \mathbb{R}, \ a_j < b_j \text{ for } j \in \{1, \dots, d\} \right\} \cup \{\emptyset\}.$$

   The collection $\mathcal{M}$ satisfies the conditions of Theorem 14.13 and is therefore a convergence-determining class.
(c) Let $S = \mathbb{R}^d$ and $\mathcal{M} := \{(-\infty, x] : x \in \mathbb{R}^d\} \cup \{\emptyset\}$. Here, for $x = (x_1, \dots, x_d) \in \mathbb{R}^d$ $(-\infty, x] := \{y = (y_1, \dots, y_d) \in \mathbb{R}^d : y_j \leq x_j \text{ for } j = 1, \dots, d\}$. The collection $\mathcal{M}$ is a CDC (Problem 14.6).

14.15 Example (The Space $\mathbb{R}^\infty$, Continuation of Example 13.13)
In Example 13.13 we saw that the collection $\mathcal{R}_f^\infty = \{\pi_k^{-1}(H) : k \in \mathbb{N}, H \in \mathcal{B}^k\}$ of finite-dimensional sets is a separating class for $\mathcal{P}$. Here, $\pi_k : \mathbb{R}^\infty \to \mathbb{R}^k$, $\pi_k(x) := (x_1, \dots, x_k)$,

stands for the projection of a sequence $x = (x_j)_{j\geq 1}$ onto the vector of the first k sequence members.

The collection C_f^∞ is also a CDC: To prove this claim, we choose for each $x = (x_j)_{j\geq 1}$ in $\mathbb{R}^\infty$ and for each $\varepsilon > 0$ an integer k such that $2^{-k} < \frac{\varepsilon}{2}$. For each η with $0 < \eta < \frac{\varepsilon}{2}$ and each finite-dimensional set $A_\eta := \{y = (y_j) \in \mathbb{R}^\infty : |y_j - x_j| < \eta \text{ for each } j \leq k\}$, it follows that $x \in \overset{\circ}{A}_\eta = A_\eta \subset B(x, \varepsilon)$.

Self-question 8: Why does $A_\eta \subset B(x, \varepsilon)$ hold?

Since ∂A_η consists of all $y = (y_j)_{j\geq 1} \in \mathbb{R}^\infty$ with $|y_j - x_j| \leq \eta$ for each $j = 1, \ldots, k$, where the equal sign occurs for at least one j, the sets ∂A_η are pairwise disjoint for different values of η. Since the metric space $(\mathbb{R}^\infty, \rho)$ is separable, we can apply Theorem 14.13. Thus, $P_n \overset{\mathcal{D}}{\longrightarrow} P$ if and only if $P_n(B) \to P(B)$ for each finite-dimensional P-continuity set B or, equivalently,

$$P_n \overset{\mathcal{D}}{\longrightarrow} P \iff P_n\pi_k^{-1} \overset{\mathcal{D}}{\longrightarrow} P\pi_k^{-1} \text{ for each } k \geq 1.$$

Restated in terms of random elements $X = (X_j)_{j\geq 1}$ and $X^{(n)} = (X^{(n)})_{j\geq 1}, n \geq 1$, in $\mathbb{R}^\infty$, this result reads as follows:

$$X^{(n)} \overset{\mathcal{D}}{\longrightarrow} X \iff \left(X_1^{(n)}, \ldots, X_k^{(n)}\right) \overset{\mathcal{D}}{\longrightarrow} (X_1, \ldots, X_k) \text{ for each } k \geq 1. \qquad (14.8)$$

14.16 Example (The Space C[0, 1], Continuation of Sect. 13.9)

In the space $C = C[0, 1]$, the collection

$$C_f = \left\{\pi_{t_1,\ldots,t_k}^{-1}(H) : k \in \mathbb{N}, 0 \leq t_1 < \ldots < t_k \leq 1, H \in \mathcal{B}^k\right\}$$

of finite-dimensional sets is a separating class for $\mathcal{P}$ (see the end of 13.12). In the terminology of random functions (see Sect. 14.9), this result means that the distribution of a C-valued random element X is uniquely determined by the collection of its fidis, that is, by the set of all distributions of $(X(t_1), \ldots, X(t_k))$, where $k \geq 1$ and $0 \leq t_1 < \ldots < t_k \leq 1$. We ask ourselves: Is the collection of fidis of X also a CDC? That is, if $X, X_1, X_2, \ldots$ are C-valued random elements, do we have

$$X_n \overset{\mathcal{D}}{\longrightarrow} X \iff \left(X_n(t_1), \ldots, X_n(t_k)\right) \overset{\mathcal{D}}{\longrightarrow} \left(X(t_1), \ldots, X(t_k)\right) \qquad (14.9)$$

$$\text{for each } k \geq 1 \text{ and for each } 0 \leq t_1 < \ldots < t_k \leq 1$$

or, what is the same,

$$X_n \xrightarrow{\mathcal{D}} X \iff \pi_{t_1,\ldots,t_k} \circ X_n \xrightarrow{\mathcal{D}} \pi_{t_1,\ldots,t_k} \circ X \qquad (14.10)$$

$$\text{for each } k \geq 1 \text{ and for each } 0 \leq t_1 < \ldots < t_k \leq 1?$$

Since the projections $\pi_{t_1,\ldots,t_k} : C \to \mathbb{R}$ are continuous mappings, the mapping Theorem 14.8 yields "$\Longrightarrow$" both in (14.9) and (14.10). *We define* the collections of convergence in distributions on the right of "$\iff$" in (14.9) and (14.10) as *fidi convergence* of (X_n) to X. In the sequel, the notation for this will be

$$X_n \xrightarrow{\mathcal{D}_{\text{fidi}}} X :\iff \pi_{t_1,\ldots,t_k} \circ X_n \xrightarrow{\mathcal{D}} \pi_{t_1,\ldots,t_k} \circ X \qquad (14.11)$$

$$\text{for each } k \geq 1 \text{ and for each } 0 \leq t_1 < \ldots < t_k \leq 1.$$

With this new notation, we thus have $X_n \xrightarrow{\mathcal{D}} X \Longrightarrow X_n \xrightarrow{\mathcal{D}_{\text{fidi}}} X$. Fidi convergence is therefore a *necessary condition* for convergence in distribution in the space C.

The definition corresponding to (14.11) for probability measures $P, P_1, P_2, \ldots$ on $\mathcal{B}(C)$ is

$$P_n \xrightarrow{\mathcal{D}_{\text{fidi}}} P :\iff P_n \pi_{t_1,\ldots,t_k}^{-1} \xrightarrow{\mathcal{D}} P \pi_{t_1,\ldots,t_k}^{-1} \ \forall k \geq 1 \ \forall 0 \leq t_1 < \ldots < t_k \leq 1. \qquad (14.12)$$

In this case, too, we speak of *fidi convergence* of (P_n) to P.

In contrast to the space $\mathbb{R}^\infty$, in which the right-hand side of (14.8) is defined as fidi convergence of X_n to X, fidi convergence in C[0, 1] is not sufficient to obtain $X_n \xrightarrow{\mathcal{D}} X$. To illustrate this important message, we consider the sequence of functions (z_n) defined in (13.4), that is,

$$z_n(t) = nt\mathbf{1}_{[0,1/n]}(t) + (2 - nt)\mathbf{1}_{(1/n,2/n]}(t), \qquad 0 \leq t \leq 1$$

(see also Fig. 13.3). If we set $P_n := \delta_{z_n}$ and $P := \delta_0$, the sequence (P_n) does *not* converge weakly to P, since $z_n \nrightarrow 0$ (cf. Example 14.2). On the other hand, fix any integer k and real numbers $t_1, \ldots, t_k$ with $0 \leq t_1 < \ldots < t_k \leq 1$. Then, for sufficiently large n,

$$\pi_{t_1,\ldots,t_k}(z_n) = (0, 0, \ldots, 0) = \pi_{t_1,\ldots,t_k}(0).$$

Indeed, if $t_1 > 0$ then the above equality holds for each n with $\frac{2}{n} < t_1$, and in the remaining case $0 = t_1 < t_2$ it holds for each n with $\frac{2}{n} < t_2$. Thus, we have fidi convergence $P_n \xrightarrow{\mathcal{D}_{\text{fidi}}} P$ or, equivalently, $X_n \xrightarrow{\mathcal{D}_{\text{fidi}}} X$, if the random elements X_n and X take on the values (that is, functions) z_n and 0 with probability one, respectively.

The following general concept was already used in Chap. 6. Among other things, it allows to conclude $X_n \xrightarrow{\mathcal{D}} X$ and $P_n \xrightarrow{\mathcal{D}} P$ from $X_n \xrightarrow{\mathcal{D}_{\text{fidi}}} X$ and $P_n \xrightarrow{\mathcal{D}_{\text{fidi}}} P$, respectively.

14.17 Definition (Relative Compactness)

Let (S, ρ) be a metric space and $\mathcal{Q} \subset \mathcal{P}$ a non-empty set of probability measures on $\mathcal{B}(S)$. The set $\mathcal{Q}$ is called *relatively compact*, if for each sequence $(P_n)_{n \geq 1}$ in $\mathcal{Q}$ there is a subsequence $(P_{n_k})_{k \geq 1}$ and a probability measure Q on $\mathcal{B}(S)$ such that

$$P_{n_k} \xrightarrow{\mathcal{D}} Q \text{ as } k \to \infty. \tag{14.13}$$

If $X_1, X_2, \ldots$ are S-valued random elements on a probability space $(\Omega, \mathcal{A}, \mathbb{P})$ then, by definition, the sequence $(X_n)_{n \geq 1}$ is *relatively compact* if the set $\mathcal{Q} := \{\mathbb{P}^{X_n} : n \geq 1\}$ is relatively compact.

It should be emphasized that the probability measure Q occurring in (14.13) does not necessarily have to belong to $\mathcal{Q}$.

Together with Theorem 14.5, there is the following criterion for weak convergence:

14.18 Corollary

Let $P, P_1, P_2, \ldots$ be probability measures on $\mathcal{B}(S)$. If the set $\{P_n : n \geq 1\}$ is relatively compact, and if every weakly convergent subsequence of (P_n) converges to P, then $P_n \xrightarrow{\mathcal{D}} P$ as $n \to \infty$.

By the subsequence criterion 14.5, $P_n \xrightarrow{\mathcal{D}} P$ implies that $\{P_n : n \in \mathbb{N}\}$ is relatively compact. Relative compactness is therefore a necessary condition for weak convergence. The following result states that in the case of the metric space $C[0, 1]$, relative compactness and fidi convergence imply weak convergence.

14.19 Theorem (Fidi convergence and relative compactness imply $P_n \xrightarrow{\mathcal{D}} P$ in C)

Let $S = \mathrm{C} = C[0, 1]$ and $P, P_1, P_2, \ldots$ be probability measures on $\mathcal{B}(\mathrm{C})$. If $\{P_n : n \in \mathbb{N}\}$ is relatively compact and $P_n \xrightarrow{\mathcal{D}_{\text{fidi}}} P$, then $P_n \xrightarrow{\mathcal{D}} P$.

Proof Due to relative compactness, every subsequence (P_{n_i}) contains a sub-subsequence $(P_{n_i'})$ with $P_{n_i'} \xrightarrow{\mathcal{D}} Q$ for some $Q \in \mathcal{P}$. Fix any integer k and $t_1, \ldots, t_k$ with $0 \leq t_1 < \ldots < t_k \leq 1$. By the mapping theorem, $P_{n_i'} \pi_{t_1,\ldots,t_k}^{-1} \xrightarrow{\mathcal{D}} Q \pi_{t_1,\ldots,t_k}^{-1}$. Due to the fidi convergence $P_n \xrightarrow{\mathcal{D}_{\text{fidi}}} P$, it follows that $P_{n_i'} \pi_{t_1,\ldots,t_k}^{-1} \xrightarrow{\mathcal{D}} P \pi_{t_1,\ldots,t_k}^{-1}$. Thus, $Q \pi_{t_1,\ldots,t_k}^{-1} = P \pi_{t_1,\ldots,t_k}^{-1}$ and therefore $P(B) = Q(B)$ for each $B \in \mathcal{C}_f$. Since $\mathcal{C}_f$ is a separating class, we obtain $P = Q$ and thus the assertion. □

Self-question 9: Why does the assertion, that is, $P_n \xrightarrow{\mathcal{D}} P$, follow?

Stated in terms of C-valued random elements $X, X_1, X_2, \ldots$, Theorem 14.19 reads as follows: If $X_n \xrightarrow{\mathcal{D}_{\text{fidi}}} X$ and the sequence (X_n) is relatively compact, then $X_n \xrightarrow{\mathcal{D}} X$.

As the following theorem shows, relative compactness is also a powerful tool for proving the existence of probability measures on $\mathcal{B}(C)$.

14.20 Theorem (Existence of Probability Measures on C[0, 1])
Suppose $P_1, P_2, \ldots$ are probability measures on $\mathcal{B}(C)$. Assume that $\{P_n : n \geq 1\}$ is relatively compact, and that for every $k \geq 1$ and for any choice of $t_1, \ldots, t_k$ with $0 \leq t_1 < \ldots < t_k \leq 1$, there is a probability measure $Q_{t_1,\ldots,t_k}$ on $\mathcal{B}^k$ with

$$P_n \pi_{t_1,\ldots,t_k}^{-1} \xrightarrow{\mathcal{D}} Q_{t_1,\ldots,t_k}. \tag{14.14}$$

Then there exists a probability measure P on $\mathcal{B}(C)$ such that

$$P \pi_{t_1,\ldots,t_k}^{-1} = Q_{t_1,\ldots,t_k} \quad \text{for all } k \text{ and } t_1, \ldots, t_k \text{ with } 0 \leq t_1 < \ldots < t_k \leq 1.$$

Proof Due to the relative compactness, there are a subsequence (P_{n_j}) and a $P \in \mathcal{P}$ with $P_{n_j} \xrightarrow{\mathcal{D}} P$ as $j \to \infty$. Fix any $k \geq 1$ and $t_1, \ldots, t_k$ with $0 \leq t_1 < \ldots < t_k \leq 1$. By the mapping theorem, $P_{n_j} \pi_{t_1,\ldots,t_k}^{-1} \xrightarrow{\mathcal{D}} P \pi_{t_1,\ldots,t_k}^{-1}$. By (14.14), it follows that $P_{n_j} \pi_{t_1,\ldots,t_k}^{-1} \xrightarrow{\mathcal{D}} Q_{t_1,\ldots,t_k}$ and thus the assertion. $\qquad \square$

The last two results show how important the property of relative compactness is in terms of proving convergence in distribution, but how do we verify relative compactness? For collections of probability measures in $\mathbb{R}^d$, Prokhorov's theorem (Theorem 6.11) applies, according to which relative compactness is equivalent to tightness. Tightness can also be defined for probability measures on general metric spaces.

14.21 Definition (Tightness)
Let (S, ρ) be a metric space and $\mathcal{Q} \subset \mathcal{P}$ a non-empty set of probability measures on $\mathcal{B}(S)$. The set $\mathcal{Q}$ is called *tight*, if for every $\varepsilon > 0$ there is a compact subset K of S, such that

$$Q(K) \geq 1 - \varepsilon \quad \text{for every } Q \in \mathcal{Q}. \tag{14.15}$$

A sequence $(X_n)_{n \geq 1}$ of S-valued random elements on a probability space $(\Omega, \mathcal{A}, \mathbb{P})$ is *tight* if the set $\mathcal{Q} := \{\mathbb{P}^{X_n} : n \geq 1\}$ of the associated distributions is tight. In this case, it is common to say that the sequence (X_n) is *bounded in probability*, and, using the Landau notation introduced in Sect. 6.14, one writes $X_n = O_{\mathbb{P}}(1)$.

As J. Prokhorov was able to show, Theorem 6.11 holds under general conditions.

14.22 Theorem (Prokhorov)

Let (S, ρ) be a metric space and $\mathcal{Q} \subset \mathcal{P}$ a non-empty collection of probability measures on $\mathcal{B}(S)$. We then have:

(a) If $\mathcal{Q}$ is tight then $\mathcal{Q}$ is relatively compact.
(b) The converse in (a) holds if (S, ρ) is separable and complete.

Proof We first prove part (b) and then give a sketch of the proof of part (a). Let $\mathcal{Q} \subset \mathcal{P}$ be relatively compact, and fix any $\varepsilon > 0$. To find a compact subset K of S with (14.15), let (O_n) be any increasing sequence of open sets with $S = \cup_{n=1}^{\infty} O_n$. We first claim that there is an integer n with $P(O_n) > 1 - \varepsilon$ for every $P \in \mathcal{Q}$. Otherwise, for each n there would be some $P_n \in \mathcal{Q}$ with $P_n(O_n) \leq 1 - \varepsilon$. Since $\mathcal{Q}$ is relatively compact, there exist a subsequence (P_{n_k}) of (P_n) and some $Q \in \mathcal{P}$ with $P_{n_k} \xrightarrow{D} Q$ as $k \to \infty$. By the Portmanteau Theorem (Theorem 14.3), we then obtain for each fixed n:

$$Q(O_n) \leq \liminf_{k \to \infty} P_{n_k}(O_n) \leq \liminf_{k \to \infty} P_{n_k}(O_{n_k}) \leq 1 - \varepsilon. \tag{14.16}$$

However, $O_n \uparrow S$ implies $\lim_{n \to \infty} Q(O_n) = 1$, which contradicts the fact that (14.16) is satisfied for every n.

Since S is separable, for each $k \geq 1$ there are open balls $B_{k,j}$, $j \geq 1$, with radius $\frac{1}{k}$ and $S = \cup_{j=1}^{\infty} B_{k,j}$. After what has already been shown, there is an integer n_k with

$$P\left(\bigcup_{j=1}^{n_k} B_{k,j} \right) > 1 - \frac{\varepsilon}{2^k} \quad \text{for every } P \in \mathcal{Q}.$$

The set $M := \cap_{k \geq 1} (\cup_{j \leq n_k} B_{k,j})$ has a finite η-net for every $\eta > 0$ and is thus totally bounded. Since S is complete, the set K defined by $K := \overline{M}$ is also complete. By Theorem 13.6, K is compact, and (14.15) applies. Thus, $\mathcal{Q}$ is tight. □

Proof of (a) (sketch): Let (P_n) be a sequence in $\mathcal{Q}$. We need to show that there is a subsequence (P_{n_k}) of (P_n) and a $P \in \mathcal{P}$ with $P_{n_k} \xrightarrow{D} P$ as $k \to \infty$. Let $(K_n)_{n \geq 1}$ be an increasing sequence of compact subsets of S with

$$P_n(K_j) > 1 - \frac{1}{j} \quad \text{for each } j \geq 1 \text{ and each } n \geq 1.$$

For each $m \geq 1$, K_j has a finite $\frac{1}{m}$-net $N_{j,m}$. The set $N := \bigcup_{j=1}^{\infty} \bigcup_{m=1}^{\infty} N_{j,m}$ is countable, and $\bigcup_{j=1}^{\infty} K_j \subset \overline{N}$. Thus, the set $\bigcup_{j=1}^{\infty} K_j$ is separable by definition. Consequently, there is a countable collection $\widetilde{\mathcal{O}} \subset \mathcal{O}$ of open sets with the property that for each $x \in S$ and each $O \in \mathcal{O}$ the following holds: if $x \in \left(\bigcup_{j=1}^{\infty} K_j \right) \cap O$ then there is some $G \in \widetilde{\mathcal{O}}$ with $x \in G \subset \overline{G} \subset O$.

Let $\mathcal{H}$ be the countable collection of all finite unions of sets $\overline{G} \cap K_j$, where $G \in \widetilde{\mathcal{O}}$ and $j \geq 1$, extended by $\{\emptyset\}$. According to Cantor's diagonal argument, there is a subsequence (P_{n_k}) such that $\alpha(H) := \lim_{k \to \infty} P_{n_k}(H)$ exists for each $H \in \mathcal{H}$. The aim now is to construct a $P \in \mathcal{P}$ with

$$P(O) = \sup\{\alpha(H) : H \in \mathcal{H}, H \subset O\}, \quad O \in \mathcal{O}. \tag{14.17}$$

If such a P exists we would have

$$\alpha(H) = \lim_{k \to \infty} P_{n_k}(H) \leq \liminf_{k \to \infty} P_{n_k}(O)$$

for each $H \in \mathcal{H}$ with $H \subset O$ and thus $P(O) \leq \liminf_{k \to \infty} P_{n_k}(O)$ due to (14.17). The Portmanteau theorem then yields $P_{n_k} \xrightarrow{\ \mathcal{D}\ } P$. Using Carathéodory's extension theorem, the construction of P is done via a suitable outer measure (see, e.g., [BI2], pp. 61–63). □

14.23 Corollary *If (S, ρ) is complete and separable, then every finite set $\mathcal{Q} \subset \mathcal{P}$ is tight.*

Self-question 10: Why does the last conclusion hold?

What are the implications of Prokhorov's theorem for weak convergence and convergence in distribution in C[0, 1]? By the Arzelà–Ascoli theorem (Theorem 13.11), a set $M \subset$ C[0, 1] is relatively compact if and only if both $\sup_{x \in M} |x(0)| < \infty$ and $\lim_{\delta \to 0} \sup_{x \in M} w_x(\delta) = 0$. We thus obtain the following result:

14.24 Theorem (Characterization of Tightness in C[0, 1])
A sequence $(P_n)_{n \geq 1}$ of probability measures on $\mathcal{B}(C)$ is tight if and only if these conditions hold:

(a) *For each $\eta > 0$, there is an a such that $P_n(\{x \in C : |x(0)| \geq a\}) \leq \eta$ for each $n \geq 1$.*
(b) *For each $\varepsilon > 0$ and for each $\eta > 0$ there exists a $\delta \in (0, 1)$, such that*
$$P_n(\{x \in C : w_x(\delta) \geq \varepsilon\}) \leq \eta \text{ for each } n \geq 1.$$

Proof We first prove that tightness of (P_n) implies both (a) and (b). To this end, fix any $\eta > 0$. Due to tightness, there is a compact set $K \subset C$ with $P_n(K) > 1 - \eta$ for every $n \geq 1$. By Theorem 13.11, there exists an a with $K \subset \{x \in C : |x(0)| < a\}$. Thus, condition (a) holds. To prove (b), fix any $\varepsilon > 0$. With Theorem 13.11, there is a $\delta \in (0, 1)$ with $K \subset \{x : w_x(\delta) < \varepsilon\}$. This implies $P_n(\{x : w_x(\delta) < \varepsilon\}) > 1 - \eta$ for every $n \geq 1$, so condition (b) also holds. To show that (a) and (b) imply the tightness of $(P_n)_{n \geq 1}$, let $\varepsilon > 0$ be arbitrary. To find a compact set $K \subset C$ with $P_n(K) \geq 1 - \varepsilon$ for each $n \geq 1$, we first set $\eta := \varepsilon/2$ in (a). We then put $B := \{x : |x(0)| \leq a\}$, where $a > 0$ is chosen so that

$$P_n(B) \geq 1 - \tfrac{\varepsilon}{2}, \quad n \geq 1.$$

Now we use (b) and set $\eta_k := \varepsilon/2^{k+1}$ for each $k \geq 1$. According to (b), we can choose $\delta_k \in (0, 1)$ such that, setting $B_k := \{x \in C : w_x(\delta_k) < 1/k\}$, we have

$$P_n(B_k) \geq 1 - \tfrac{\varepsilon}{2^{k+1}}, \quad k \geq 1, \ n \geq 1.$$

If we define $K := \overline{B \cap \bigcap_{k=1}^{\infty} B_k}$, then $P_n(K) \geq 1 - \varepsilon$ for every $n \geq 1$. In view of Theorem 13.11, the set $A := B \cap \bigcap_{k=1}^{\infty} B_k$ is relatively compact. □

With the help of the above characterization, we obtain the following criterion for convergence in distribution $X_n \xrightarrow{\mathcal{D}} X$ of random elements in C[0, 1]. Here, we recall the fidi convergence $X_n \xrightarrow{\mathcal{D}_{\text{fidi}}} X$ (convergence of all finite-dimensional distributions) defined in (14.11).

14.25 Theorem (Criterion for Convergence in Distribution in C[0, 1])
Let $X, X_1, X_2, \ldots$ be C[0, 1]-valued random elements on a probability space $(\Omega, \mathcal{A}, \mathbb{P})$. If $X_n \xrightarrow{\mathcal{D}_{\text{fidi}}} X$ and

$$\lim_{\delta \to 0} \limsup_{n \to \infty} \mathbb{P}\big(w(X_n, \delta) \geq \varepsilon\big) = 0 \quad \text{for every } \varepsilon > 0, \tag{14.18}$$

then $X_n \xrightarrow{\mathcal{D}} X$.

Proof Let $P := \mathbb{P}^X$ and $P_n := \mathbb{P}^{X_n}$, $n \geq 1$. In view of Theorem 14.19 and Prokhorov's theorem, we only need to show tightness of $\{P_n : n \geq 1\}$. Since $X_n(0) \xrightarrow{\mathcal{D}} X(0)$, the set $\{P_n \circ \pi_0^{-1} : n \geq 1\}$ is tight, and thus condition (a) of Theorem 14.24 is satisfied. Because (14.18) is the same as

$$\lim_{\delta \to 0} \limsup_{n \to \infty} P_n(\{x : w(x, \delta) \geq \varepsilon\}) = 0 \quad \text{for every } \varepsilon > 0$$

also condition (b) of Theorem 14.24 holds. □

In order to prove $X_n \xrightarrow{\mathcal{D}} X$ in C[0, 1], it is thus necessary to show fidi convergence $X_n \xrightarrow{\mathcal{D}_{\text{fidi}}} X$. In addition, it must be shown that the probability of the event that the maximum modulus of the fluctuations of X_n over intervals of length δ is at least equal to an ε, however small, converges uniformly with respect to n to 0 as $\delta \downarrow 0$. In the next chapter, we will tailor this criterion to the situation of partial sum processes.

We conclude this chapter with a far-reaching generalization of Theorem 6.8 on convergence in distribution and independence as well as with a Slutsky's lemma for random elements in metric spaces.

14.26 Theorem (Convergence in Distribution and Independence)

Let S' and S'' be separable metric spaces, and put $S := S' \times S''$. Further, let $(\Omega, \mathcal{A}, \mathbb{P})$ be a probability space and $X, X_1, X_2, \ldots$ be S'-valued as well as $Y, Y_1, Y_2, \ldots$ be S''-valued random elements on Ω. Suppose X_n and Y_n are independent for each $n \geq 1$ and $X_n \xrightarrow{\mathcal{D}} X$ as well as $Y_n \xrightarrow{\mathcal{D}} Y$. Then

$$(X_n, Y_n) \xrightarrow{\mathcal{D}} (X, Y),$$

where X and Y are independent. This result applies analogously for more than two separable metric spaces and corresponding sequences of random elements.

Proof According to Problem 13.7, S is separable, and the Borel σ-field on S is the product of the Borel σ-fields on S' and S''. Thus, (X_n, Y_n), $n \geq 1$, are S-valued random elements. For each $n \geq 1$, let $P_n := \mathbb{P}^{(X_n, Y_n)}$, $P_n' := \mathbb{P}^{X_n}$ and $P_n'' := \mathbb{P}^{Y_n}$ be the distributions of (X_n, Y_n), X_n, and Y_n, respectively. Furthermore, let $P' := \mathbb{P}^X$ and $P'' := \mathbb{P}^Y$ be the distributions of X and Y, respectively. Since X_n and Y_n are independent, we have $P_n = P_n' \otimes P_n''$. In view of $X_n \xrightarrow{\mathcal{D}} X$ and $Y_n \xrightarrow{\mathcal{D}} Y$, it follows that $P_n'(B') \to P'(B')$ for each P'-continuity set B' and $P_n''(B'') \to P''(B'')$ for each P''-continuity set B''. For such sets B' and B'', we then get

$$P_n(B' \times B'') = P_n'(B')P_n''(B'') \to P'(B')P''(B'').$$

If we define $P := P' \otimes P''$, then $P_n \xrightarrow{\mathcal{D}} P$ by part (d) of Problem 14.11. This is the same as $(X_n, Y_n) \xrightarrow{\mathcal{D}} (X, Y)$, where X and Y are independent. An extension to more than two metric spaces and sequences of random elements is straightforward, as forming the product of measures is an associative operation.   □

Slutsky's lemma for d-dimensional random vectors $X, X_1, X_2, \ldots$ and $Y_1, Y_2, \ldots$ states that $X_n \xrightarrow{\mathcal{D}} X$ and $Y_n \xrightarrow{\mathbb{P}} 0_d$ imply $X_n + Y_n \xrightarrow{\mathcal{D}} X$ (Theorem 6.9). However, care must be taken when defining convergence in probability $X_n \xrightarrow{\mathbb{P}} X$ for S-valued random elements $X, X_1, X_2, \ldots$. Since only $\rho(X_n, X)$ can be used to measure the "distance" between X_n and X, the mapping defined on Ω via $\omega \mapsto \rho(X_n(\omega), X(\omega))$ must be $(\mathcal{A}, \mathcal{B}^1)$-measurable, that is, a random variable on Ω. This is the case when the metric space (S, ρ) is separable. Then, by Problem 13.7, the pair (X_n, X) is a random element in $S \times S$, so due to $\mathcal{B}(S \times S) = \mathcal{B}(S) \otimes \mathcal{B}(S)$ a $(\mathcal{A}, \mathcal{B}(S \times S))$-measurable mapping on Ω with values in $S \times S$. Consequently, $\rho(X_n, X)$ is a real-valued random variable on Ω.

Self-question 11: Why does the last conclusion hold?

14.27 Definition (Convergence in Probability)
Let (S, ρ) be a separable metric space and $X, X_1, X_2, \ldots$ random elements in S. If

$$\lim_{n \to \infty} \mathbb{P}\big(\rho(X_n, X) \geq \varepsilon\big) = 0 \quad \text{for every } \varepsilon > 0,$$

then we say the sequence (X_n) *converges in probability to* X and write briefly $X_n \xrightarrow{\mathbb{P}} X$. If specifically $\mathbb{P}(X = a) = 1$ for some $a \in S$, then we also write $X_n \xrightarrow{\mathbb{P}} a$.

If $X_n \xrightarrow{\mathbb{P}} X$, the limit random element X is unique with probability 1 (Problem 14.5).

14.28 Theorem (Slutzky's Lemma)
Let (X_n, Y_n), $n \geq 1$, be random elements in $S \times S$ and X a random element in S.

$$\text{If } X_n \xrightarrow{\mathcal{D}} X \text{ and } \rho(X_n, Y_n) \xrightarrow{\mathbb{P}} 0, \text{ then } Y_n \xrightarrow{\mathcal{D}} X.$$

Proof Let A be a closed subset of S, and fix any $\varepsilon > 0$. With $A_\varepsilon := \{x \in S : \rho(x, A) \leq \varepsilon\}$ we then have

$$\mathbb{P}(Y_n \in A) = \mathbb{P}(Y_n \in A, \ \rho(X_n, Y_n) \geq \varepsilon) + \mathbb{P}(Y_n \in A, \ \rho(X_n, Y_n) < \varepsilon)$$

$$\leq \mathbb{P}(\rho(X_n, Y_n) \geq \varepsilon) + \mathbb{P}(X_n \in A_\varepsilon).$$

The set A_ε is closed. Since $X_n \xrightarrow{\mathcal{D}} X$, the Portmanteau theorem (Theorem 14.7) implies $\limsup_{n \to \infty} \mathbb{P}(Y_n \in A) \leq \mathbb{P}(X \in A_\varepsilon)$. As $\varepsilon \downarrow 0$, the corresponding sets A_ε are a descending sequence that converges to A, as A is closed. Since the probability measure $\mathbb{P}^X$ is continuous from above, we obtain

$$\limsup_{n \to \infty} \mathbb{P}(Y_n \in A) \leq \mathbb{P}(X \in A).$$

The assertion now follows from the Portmanteau theorem. $\square$

14.29 Corollary (Convergence in Probability Implies Convergence in Distribution)
If $X_n \xrightarrow{\mathbb{P}} X$ then $X_n \xrightarrow{\mathcal{D}} X$.

Proof The assertion follows if we set $X_n := X$ and $Y_n := X_n$, $n \geq 1$, in Slutsky's lemma.
$\square$

If, after studying this chapter, you would like to learn about further results related to convergence in distribution, valuable sources are [BI2] or Chapter 11 in [DUD].

Answers to the Self-Questions

Answer 1 If $P_n \xrightarrow{\mathcal{D}} P$ and $P_n \xrightarrow{\mathcal{D}} Q$ then $\int f \, dP = \int f \, dQ$ for every $f \in C_b(S)$ and thus $P = Q$ by Theorem 13.8.

Answer 2 If $x \in \partial\{f > u\}$, then there are sequences (x_n) in $\{f > u\}$ and (y_n) in $\{f \leq u\}$ with $x_n \to x$ and $y_n \to x$. Since f is continuous, we have $f(x_n) \to f(x)$ and $f(y_n) \to f(x)$. Because $f(x_n) > u$ and $f(y_n) \leq u$ for each n, it follows that $f(x) = u$.

Answer 3 Note that $\{u \in (0, 1) : P(f = u) > 0\} \subset \bigcup_{n=1}^{\infty} A_n$, where $A_n := \{u \in (0, 1) : P(f = u) \geq \frac{1}{n}\}$. Since A_n has at most n elements, the set $\{u \in (0, 1) : P(f = u) > 0\}$ is countable.

Answer 4 Since $X_n(1) = S_n/\sqrt{n}$, the Lindeberg–Lévy central limit theorem yields $X_n(1) \xrightarrow{\mathcal{D}} N(0, 1)$.

Answer 5 Because the probability measure P is continuous from below.

Answer 6 Let $P, Q \in \mathcal{P}$ with $P(B) = Q(B)$ for every $B \in \mathcal{M}$. The sequence (P_n) with $P_n := P$, $n \geq 1$, trivially satisfies $P_n \xrightarrow{\mathcal{D}} P$. Since $\mathcal{M}$ is a CDC, we also have $P_n \xrightarrow{\mathcal{D}} Q$. It follows that $P = Q$ (see Self-question 1).

Answer 7 If $x \in \partial(A \cap B)$, then there are sequences (x_n) in $A \cap B$ and (y_n) in $(A \cap B)^c = A^c \cup B^c$ with $x_n \to x$ and $y_n \to x$. Hence there is either a subsequence of (y_n) in A^c or in B^c that converges to x, and it follows that $x \in \partial A \cup \partial B$.

Answer 8 Fix any $y \in A_\eta$. Since $0 < \eta < \frac{\varepsilon}{2}$, the definition of the metric ρ in $\mathbb{R}^\infty$ (cf. (13.9)) implies

$$\rho(x, y) \leq \sum_{j=1}^{k} \frac{\min(1, |x_j - y_j|)}{2^j} + \sum_{j=k+1}^{\infty} \frac{1}{2^j} < \eta \sum_{j=1}^{k} \frac{1}{2^k} + \frac{\varepsilon}{2} < \varepsilon.$$

Answer 9 Because every subsequence of (P_n) contains a sub-subsequence that converges weakly to P, the assertion follows from Corollary 14.18.

Answer 10 Every finite collection Q is relatively compact, since there are only finitely many elements of Q for the members of a sequence (P_n) in Q. By Prokhorov's theorem, Q is tight.

Answer 11 The mapping defined on $S \times S$ that assigns $\rho(x, y)$ to (x, y) is continuous and thus $(\mathcal{B}(S \times S), \mathcal{B}^1)$-measurable. Note that $\rho(X_n, X)$ is the composition $\rho \circ (X_n, X)$ and therefore $(\mathcal{A}, \mathcal{B}^1)$-measurable.

Problems

14.1 Problem Show that the collection $\mathcal{C}(P) := \{ B \in \mathcal{B}(S) : P(\partial B) = 0 \}$ of P-continuity sets defined in (14.3) is a field.

14.2 Problem Let (S, ρ) be a metric space and $x_1, x_2, \dots$ a sequence in S. Show: If $\delta_{x_n} \xrightarrow{D} P$ for a $P \in \mathcal{P}$, then there is an $x \in S$ with $P = \delta_x$.

Hint First assume that the sequence (x_n) has a convergent subsequence. Then show that this assumption is necessary for the validity of the premise "$\delta_{x_n} \xrightarrow{D} P$ for a $P \in \mathcal{P}$".

14.3 Problem Let $\mathcal{P}$ be the set of probability measures on the Borel σ-field $\mathcal{B}(S)$ of a metric space (S, ρ). For P and Q in $\mathcal{P}$ we define

$$\varrho(P, Q) := \inf \{ \varepsilon > 0 : P(B) \le Q(B^\varepsilon) + \varepsilon \text{ and } Q(B) \le P(B^\varepsilon) + \varepsilon \text{ for every } B \in \mathcal{B}(S) \}.$$

Prove the following statements:

(a) ϱ defines a metric on $\mathcal{P}$ (so-called *Prokhorov metric*).
(b) If $P(B) \le Q(B^\varepsilon) + \varepsilon$ for every $B \in \mathcal{B}(S)$, then $Q(B) \le P(B^\varepsilon) + \varepsilon$ for every $B \in \mathcal{B}(S)$.
(c) If $\varrho(P_n, P) \to 0$ then $P_n \xrightarrow{D} P$.

Note: If S is separable, then the converse also holds in (c) (see, e.g., [BI2], p. 72).

14.4 Problem Assume the setting of Sect. 14.9. Prove that the mapping $X : \Omega \to C$ is $(\mathcal{A}, \mathcal{B}(C))$-measurable if and only if the mapping $X(t) : \Omega \to \mathbb{R}$ $(\mathcal{A}, \mathcal{B}^1)$-measurable for each $t \in [0, 1]$.

14.5 Problem Assume the setting of Definition 14.27. Show that $X_n \xrightarrow{\text{P}} X$ and $X_n \xrightarrow{\text{P}} Y$ imply $\mathbb{P}(X = Y) = 1$.

14.6 Problem Show that the collection $\mathcal{M} := \{(-\infty, x] : x \in \mathbb{R}^d\} \cup \{\emptyset\}$ (see Example 14.14 (c)) is a CDC for convergence in distribution in $\mathbb{R}^d$.

14.7 Problem Let (S, ρ) be a separable metric space, $(\Omega, \mathcal{A}, \mathbb{P})$ a probability space, and $\mathcal{L}^0 := \{X : \Omega \to S | X \ (\mathcal{A}, \mathcal{B}(S))\text{-measurable}\}$ the collection of all S-valued random elements defined on Ω. For $X, Y \in \mathcal{L}^0$, let

$$d_K(X, Y) := \inf \{\varepsilon \geq 0 : \mathbb{P}(\rho(X, Y) > \varepsilon) \leq \varepsilon\}.$$

Prove the following claims:

(a) The infimum is attained.
(b) $d_K(X, Y) = 0 \iff \mathbb{P}(X = Y) = 1$.
(c) $d_K(X, Z) \leq d_K(X, Y) + d_K(Y, Z)$ $(X, Y, Z \in \mathcal{L}^0)$.
(d) $d_K(X_n, X) \to 0 \iff X_n \xrightarrow{\text{P}} X$.

Note If we identify random elements X and Y with $X = Y$ $\mathbb{P}$-almost surely, then d_K is a metric, which is named after Ky Fan.[1]

14.8 Problem Let $Z, Z_1, Z_2, \ldots$ be i.i.d. $\mathbb{N}_0$-valued random variables on a probability space $(\Omega, \mathcal{A}, \mathbb{P})$. Further let $F(k) := \mathbb{P}(Z \leq k), k \in \mathbb{N}_0$, and

$$X_k^{(n)} := \sqrt{n}\left(\sum_{j=1}^{n} \mathbf{1}\{Z_j \leq k\} - F(k)\right), \quad k \in \mathbb{N}_0.$$

Show: The sequence $(X^{(n)})$ defined by $X^{(n)} := (X_k^{(n)})_{k \geq 0}$ converges in distribution in the sequence space $S = \mathbb{R}^{\mathbb{N}_0}$ (see the end of Sect. 13.13) to a random element $X = (X_k)_{k \geq 0}$ of $\mathbb{R}^{\mathbb{N}_0}$. What do the finite-dimensional distributions of X look like?

14.9 Problem Prove that a uniformly integrable sequence of random variables is tight.

14.10 Problem Let (S, ρ) be a metric space, $A \subset S$ and $\mathcal{Q} := \{\delta_x : x \in A\}$. Show that $\mathcal{Q}$ is relatively compact if and only if A is relatively compact.

---

[1] Ky Fan (1914–2010), Chinese-American mathematician. Ky Fan held various professorships, from 1965 at the University of California, Santa Barbara. Main areas of research: Nonlinear and convex analysis, functional analysis, optimization, topology.

14.11 Problem Suppose (S', ρ') and (S'', ρ'') are separable metric spaces with Borel σ-fields $\mathcal{B}'$ and $\mathcal{B}''$, respectively. Let $S := S' \times S''$ and

$$\rho\big((x', x''), (y', y'')\big) := \max\big(\rho'(x', y'), \rho''(x'', y'')\big), \quad (x', x''), (y', y'') \in S$$

as in Problem 13.7. Moreover, let P be a probability measure on $\mathcal{B}(S)$, and write P' and P'' the corresponding marginal distributions, that is, $P'(B') := P(B' \times S'')$ for each $B' \in \mathcal{B}'$ and $P''(B'') := P(S' \times B'')$ for each $B'' \in \mathcal{B}''$. Finally, let $\mathcal{M} := \{B' \times B'' : B' \in \mathcal{B}', B'' \in \mathcal{B}''\}$ be the collection of *measurable rectangles*. Prove the following statements:

(a) If $P_n \xrightarrow{\mathcal{D}} P$ then $P'_n \xrightarrow{\mathcal{D}} P'$ and $P''_n \xrightarrow{\mathcal{D}} P''$.
(b) The converse in part (a) does not generally hold.
(c) $\mathcal{M}$ is a convergence-determining class (CDC).
(d) $P_n \xrightarrow{\mathcal{D}} P \iff P_n(B' \times B'') \to P(B' \times B'')$ for all $B' \in \mathcal{C}(P')$ and all $B'' \in \mathcal{C}(P'')$.

14.12 Problem Let (S, ρ), (S', ρ') be metric spaces and $h : S \to S'$ an *arbitrary* mapping. Show:

(a) The set $A_{\varepsilon,\delta} := \{x \in S : \exists y, z \in S \text{ with } \rho(x, y) < \delta, \rho(x, z) < \delta \text{ and } \rho'(h(y), h(z)) \geq \varepsilon\}$ is open for every $\varepsilon > 0$ and every $\delta > 0$.
(b) The set C_h of continuity points of h is a Borel set.

Hint for (b) Write the set of discontinuity points in terms of countable unions of countable intersections of the sets $A_{\varepsilon,\delta}$.

Wiener Process, Donsker's Theorem, and Brownian Bridge

15

In this chapter, we get to know the Wiener process, which is the starting point for many other stochastic processes. This process is accompanied by the Wiener measure on the σ-field of Borel sets on the function space $C := C[0, 1]$. According to the title of this book, a limit theorem must not be missing, and that is Donsker's theorem, which represents a far-reaching generalization of the of Lindeberg–Lévy central limit theorem. With the help of the Wiener process, the Brownian bridge emerges, which plays an important role in nonparametric statistics.

The year was 1872, when Karl Weierstrass introduced a function that was continuous but not differentiable at any point. In a time when analysis was still strongly influenced by intuition, the appearance of such a pathological function seemed simply strange, and so it is no wonder that it was given the name *Weierstrass monster* (see, e.g., [VOL], Section 5). In this chapter, we will become acquainted with a probability measure on the Borel σ-field $\mathcal{B}(C)$ that puts all its probability mass on the continuous but nowhere differentiable functions in the unit interval. So, by chance, such "monster functions" are generated with probability one!

The ominous probability measure, which pays so much attention to continuous and at the same time nowhere differentiable functions, is named the *Wiener measure* in honor of Norbert Wiener. We will first define this probability measure, denoted by W, on $\mathcal{B}(C)$ and only later worry about its existence. If the Wiener measure W exists, then by virtue of the canonical construction $(\Omega, \mathcal{A}, \mathbb{P}) := (C, \mathcal{B}(C), W)$ and $W : C \to C$ with $W(x) := x$ for each $x \in C$, there is a probability space and a C-valued random element W, such that W has the distribution W (note the small notational difference between W and W). If we set $W(t) := \pi_t \circ W$, $0 \le t \le 1$, as in Sect. 14.9, we obtain a collection $(W(t))_{0 \le t \le 1}$ of random variables, that is, a stochastic process, and this process is called *Wiener process* or *Brownian motion*. Here, the name Brown stands for Robert Brown, who in 1827 noted that

© The Author(s), under exclusive license to Springer-Verlag GmbH, DE, part of Springer Nature 2024
N. Henze, *Asymptotic Stochastics*, Mathematics Study Resources 10,
https://doi.org/10.1007/978-3-662-68923-3_15

pollen grains in a liquid seemed to perform erratic, stochastic movements. The existence of the Wiener measure and the Wiener process W are thus two sides of the same coin.

From Sect. 14.9 it is known that the fidis of W, that is, the collection of distributions of $(W(t_1), \ldots, W(t_k))$, where $k \in \mathbb{N}$ and $t_1, \ldots, t_k \in [0, 1]$ with $0 \leq t_1 < \ldots < t_k \leq 1$, determine the distribution of W and thus the Wiener measure W. We can therefore try to define W by making demands on these fidis.

15.1 Definition (Wiener Measure, Wiener Process)

A probability measure W on $\mathcal{B}(C)$ is called *Wiener measure*, if the following holds:

(a) $\mathrm{W}\big(W(0) = 0\big) = 1$.
(b) For each t with $0 < t \leq 1$, $W(t)$ has the normal distribution $N(0, t)$ under W.
(c) For every $k \geq 2$ and any choice of $t_0, \ldots, t_k \in [0, 1]$ with $0 \leq t_0 \leq t_1 \leq \ldots \leq t_k \leq 1$, the random variables $W(t_1) - W(t_0)$, $W(t_2) - W(t_1), \ldots, W(t_k) - W(t_{k-1})$ are independent under W.

A stochastic process $W : \Omega \to C$ with the above properties is called a *Wiener process* or *Brownian motion*.

Self-question 1: What does the above independence imply in the case $t_1 = t_2$?

Requirement (a) means that the Wiener process starts at 0 with probability one. Condition (b) seems harmless, but the third requirement is tricky. It states that the *increments* $W(t) - W(s)$ of this process are independent over even the smallest disjoint time intervals. This requirement is very strong, and it is initially unclear whether it can be fulfilled at all. Why? If you want to make predictions about the realizations of $W(t) - W(s)$ for $t > s$ at a given $s \in (0, 1)$, knowing the realizations of $W(u)$ for $0 \leq u \leq s$, then property (c) states that the latter knowledge is completely irrelevant. The paths $W(\omega)$, $\omega \in \Omega$, must therefore be quite "pathological". They will indeed turn out to be "Weierstrass monsters".

We will now draw some conclusions from the definition, which in particular provide the finite-dimensional distributions necessary to determine the distribution W of W.

15.2 Corollary (Properties of the Wiener Measure or the Wiener Process)

Under the Wiener measure W, *we have:*

(a) $W(t) - W(s) \sim W(t - s) \sim N(0, t - s)$, $0 \leq s \leq t \leq 1$.
(b) $Cov\big(W(s), W(t)\big) = \min(s, t)$, $0 \leq s, t \leq 1$.

(c) For every $k \geq 1$ and $t_1, \ldots, t_k$ with $0 \leq t_1 \leq t_2 \leq \ldots \leq t_k \leq 1$,

$$\left(W(t_1), W(t_2), \ldots, W(t_k)\right)^{\top} \sim N_k(0_k, \Sigma),$$

where

$$\Sigma = \left(\min(t_i, t_j)\right)_{1 \leq i, j \leq k}. \tag{15.1}$$

Property (a) states that the Wiener process has *stationary increments*: If $[s_1, t_1]$ and $[s_2, t_2]$ are intervals of the same length $\ell := t_1 - s_1 = t_2 - s_2$, then

$$W(t_2) - W(s_2) \sim W(t_1) - W(s_1) \sim W(\ell) - W(0) = W(\ell).$$

Here, the equal sign applies with probability one. Note that property (c) determines all finite-dimensional distributions of the Wiener process and thus of the Wiener measure, provided the latter exists at all. Since these fidis are all normal distributions, the Wiener process is by definition a *Gaussian process*. In the interpretation of W as a C-valued random element, the term *Gaussian random element* is also common.

Proof of Corollary 15.2 (a) Fix any $s, t \in [0, 1]$ with $0 \leq s \leq t \leq 1$. Note that $W(t) = W(s) + \left(W(t) - W(s)\right)$, where $W(s) \left(= W(s) - W(0)\right)$ and $W(t) - W(s)$ are independent by Definition 15.1 (c). This implies

$$\mathbb{E}\left(e^{iu\,W(t)}\right) = \mathbb{E}\left(e^{iu\,W(s)}\right) \cdot \mathbb{E}\left(e^{iu(W(t)-W(s))}\right), \quad u \in \mathbb{R}.$$

With (1.5) and Definition 15.1 (b), we get

$$\mathbb{E}\left(e^{iu\,W(t)}\right) = \exp\left(-\tfrac{tu^2}{2}\right), \quad \mathbb{E}\left(e^{iu\,W(s)}\right) = \exp\left(-\tfrac{su^2}{2}\right), \quad u \in \mathbb{R},$$

and thus

$$\mathbb{E}\left(e^{iu(W(t)-W(s))}\right) = \exp\left(-\tfrac{(t-s)u^2}{2}\right), \quad u \in \mathbb{R}.$$

Since the characteristic function of a random variable uniquely determines its distribution, we obtain $W(t) - W(s) \sim N(0, t - s)$. Because, according to Definition 15.1 (b), $W(t - s) \sim N(0, t - s)$, the assertion follows.

To prove (b), let s, $t \in [0, 1]$ with $0 \le s \le t \le 1$. We have $W(s)W(t) = W(s)^2 + W(s)(W(t) - W(s))$, and thus

$$\mathrm{Cov}(W(s), W(t)) = \mathbb{E}[W(s)W(t)] = \mathbb{E}[W(s)^2] + \mathbb{E}[W(s)(W(t) - W(s))]$$
$$= \mathbb{E}[W(s)^2] + 0 = s = \min(s, t),$$

since $\mathbb{E}[W(s)] = \mathbb{E}[W(t)] = 0$.

The proof of (c) follows from the fact that the random vector $(W(t_1), \ldots, W(t_k))^\top$, written as a column vector according to

$$\begin{pmatrix} W(t_1) \\ W(t_2) \\ W(t_3) \\ \vdots \\ \vdots \\ W(t_k) \end{pmatrix} = \begin{pmatrix} 1\,0\,0\,0\cdots 0 \\ 1\,1\,0\,0\cdots 0 \\ 1\,1\,1\,0\cdots 0 \\ \vdots\ \vdots\ \vdots\ 1\cdots 0 \\ \vdots\ \vdots\ \vdots\ \vdots\ \ddots\ 0 \\ 1\,1\,1\,1\cdots 1 \end{pmatrix} \cdot \begin{pmatrix} W(t_1) \\ W(t_2) - W(t_1) \\ W(t_3) - W(t_2) \\ \vdots \\ \vdots \\ W(t_k) - W(t_{k-1}) \end{pmatrix}, \qquad (15.2)$$

arises from an affine transformation of a k-dimensional random vector, whose components are increments of the Wiener process and thus independent according to Definition 15.1 (c). The random vector on the right-hand side in (15.2) has thus a k-variate normal distribution with expectation 0_k and, due to the independence of its components, covariance matrix $D := \mathrm{diag}(t_1, t_2 - t_1, t_3 - t_2, \ldots, t_k - t_{k-1})$. If we briefly write A for the matrix of zeros and ones in (15.2), then $ADA^\top = \Sigma = (\min(t_i, t_j))_{1 \le i, j \le k}$ (Problem 15.1). $\qquad \square$

Self-question 2: Why has the vector on the right in (15.2) a normal distribution?

To prove the existence of the Wiener measure and thus also that of the Wiener process, we use the partial sum process introduced in Sect. 14.10. The following considerations also serve to prepare the fundamental theorem of Donsker.

Let $Z_1, Z_2, \ldots$ be a sequence of i.i.d. random variables that are defined on a probability space $(\Omega, \mathcal{A}, \mathbb{P})$. Regarding the distribution of Z_1, we only assume $\mathbb{E}(Z_1^2) < \infty$ as well as $\mathbb{E}(Z_1) = 0$ and $0 < \sigma^2 := \mathbb{V}(Z_1)$. As in Sect. 14.10 we define $S_0 := 0$, $S_n := Z_1 + \ldots + Z_n$ for each $n \ge 1$, and we set

$$X_n(t) := \frac{1}{\sigma \sqrt{n}} S_{\lfloor nt \rfloor} + (nt - \lfloor nt \rfloor) \frac{1}{\sigma \sqrt{n}} Z_{\lfloor nt \rfloor + 1}, \quad n \ge 1,\ 0 \le t \le 1. \qquad (15.3)$$

Fig. 15.1 The stochastic
behavior of X_n is determined
by $X_n\left(\frac{j}{n}\right)$, $j \in \{0, \ldots, n\}$

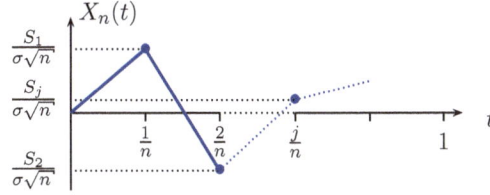

This n-th partial sum process $X_n = (X_n(t))_{0 \le t \le 1}$ of the sequence $(Z_n)_{n \ge 1}$ has the property that its stochastic behavior solely rests on the random variables

$$X_n\left(\frac{j}{n}\right) = \frac{S_j}{\sigma\sqrt{n}}, \quad j \in \{0, 1, \ldots, n\},$$

because only a linear interpolation takes place between the "points in time" $t = \frac{j-1}{n}$ and $t = \frac{j}{n}$, $j \in \{1, \ldots, n\}$ (see Fig. 15.1).

15.3 Theorem (Fidi Convergence of the Partial Sum Process)
Let $X_n = (X_n(t))_{0 \le t \le 1}$ be the partial sum process defined in (15.3). Furthermore, let k be any integer and $t_1, \ldots, t_k$ be any numbers with $0 \le t_1 < \ldots < t_k \le 1$. Then

$$\left(X_n(t_1), \ldots, X_n(t_k)\right) \xrightarrow{\mathcal{D}} N_k(0_k, \Sigma) \quad \text{as } n \to \infty,$$

where $\Sigma = \left(\min(t_i, t_j)\right)_{1 \le i, j \le k}$.

Proof Setting

$$R_n(t) := \frac{nt - \lfloor nt \rfloor}{\sigma\sqrt{n}} \cdot Z_{\lfloor nt \rfloor + 1}$$

we have

$$X_n(t) = \frac{1}{\sigma\sqrt{n}} S_{\lfloor nt \rfloor} + R_n(t).$$

Since $\mathbb{E}(R_n(t)) = 0$ and $\mathbb{V}(R_n(t)) \le \frac{1}{n}$, we get $R_n(t) \xrightarrow{\mathbb{P}} 0$ and thus $\left(R_n(t_1), \ldots, R_n(t_k)\right) \xrightarrow{\mathbb{P}} 0_k$. Due to Slutsky's lemma (Theorem 6.9), it is only necessary to show

$$\frac{1}{\sigma\sqrt{n}}\left(S_{\lfloor nt_1 \rfloor}, \ldots, S_{\lfloor nt_k \rfloor}\right) \xrightarrow{\mathcal{D}} N_k(0_k, \Sigma). \tag{15.4}$$

In doing so, we can assume w.l.o.g. that $t_1 > 0$.

Self-question 3: Why can we assume w.l.o.g. that $t_1 > 0$?

Analogous to (15.2), we have

$$
\frac{1}{\sigma\sqrt{n}}
\begin{pmatrix}
S_{\lfloor nt_1\rfloor}\\
S_{\lfloor nt_2\rfloor}\\
S_{\lfloor nt_3\rfloor}\\
\vdots\\
\vdots\\
S_{\lfloor nt_k\rfloor}
\end{pmatrix}
=
\begin{pmatrix}
1\,0\,0\,0\cdots0\\
1\,1\,0\,0\cdots0\\
1\,1\,1\,0\cdots0\\
\vdots\,\vdots\,\vdots\,1\cdots0\\
\vdots\,\vdots\,\vdots\,\vdots\,\ddots\,0\\
1\,1\,1\,1\cdots1
\end{pmatrix}
\cdot
\frac{1}{\sigma\sqrt{n}}
\begin{pmatrix}
S_{\lfloor nt_1\rfloor}\\
S_{\lfloor nt_2\rfloor}-S_{\lfloor nt_1\rfloor}\\
S_{\lfloor nt_3\rfloor}-S_{\lfloor nt_2\rfloor}\\
\vdots\\
\vdots\\
S_{\lfloor nt_k\rfloor}-S_{\lfloor nt_{k-1}\rfloor}
\end{pmatrix}.
$$

The components of the random vector on the right-hand side are independent, as they are formed by pairwise disjoint blocks of the independent random variables $Z_1, \ldots, Z_n$. The sequence of random vectors to the right of the matrix denoted by A in (15.2), which consists of ones and zeros, converges as $n \to \infty$ to the normal distribution $N_k(0_k, D)$, where $D = \mathrm{diag}(t_1, t_2 - t_1, \ldots, t_k - t_{k-1})$ (Problem 15.2). In view of $ADA^{\top} = \Sigma = (\min(t_i, t_j))_{1\leq i,j\leq k}$ (see Problem 15.1), and using the mapping Theorem 6.6 as well as the reproduction Theorem 5.7 for the normal distribution, the assertion follows.   □

The result of Theorem 15.3 provides a promising hint on how we can prove the existence of the Wiener measure W. If we translate statement (c) of Corollary 15.2 into the language of measure theory, W—provided this probability measure exists—has the property

$$
W \circ \pi_{t_1,\ldots,t_k}^{-1} = N_k(0_k, \Sigma)
$$

for each $k \geq 1$ and $t_1, \ldots, t_k$ with $0 \leq t_1 < \ldots < t_k \leq 1$, where Σ is given in (15.1). If we write $P_n := \mathbb{P}^{X_n}$ for the distribution of the partial sum process X_n in (15.3), then Theorem 15.3 states that

$$
P_n \circ \pi_{t_1,\ldots,t_k}^{-1} \xrightarrow{\mathcal{D}} W \circ \pi_{t_1,\ldots,t_k}^{-1} \tag{15.5}
$$

for every $k \geq 1$ and $t_1, \ldots, t_k$ with $0 \leq t_1 < \ldots, < t_k \leq 1$, because $P_n \circ \pi_{t_1,\ldots,t_k}^{-1}$ is the distribution of $(X_n(t_1), \ldots, X_n(t_k))$. We are now assuming purely hypothetically that the set $\{P_n : n \geq 1\}$ is relatively compact. Then there is a subsequence $(P_{n_j})_{j\geq1}$ of the sequence $(P_n)_{n\geq1}$ and a probability measure Q on $\mathcal{B}(C)$ with $P_{n_j} \xrightarrow{\mathcal{D}} Q$ as $j \to \infty$. By the mapping Theorem 14.8, this implies the fidi-convergence $P_{n_j} \xrightarrow{\mathcal{D}_{\text{fidi}}} Q$. Since the fidi-convergence (15.5) is inherited by the subsequence (P_{n_j}), Q and W have the same finite-dimensional distributions. Because these fidis uniquely determine a probability measure on $\mathcal{B}(C)$, we have found a probability measure Q that we can immediately rename to W. Thus, Wiener measure exists if the partial sum process is relatively compact.

Self-question 4: Does $P_n \xrightarrow{\mathcal{D}} W$ hold if the partial sum process is relatively compact?

Since, by Prokhorov's theorem, tightness and relative compactness are equivalent terms in the space C[0, 1], we can alternatively show the tightness of the partial sum process (more precisely: of the sequence $(X_n)_{n\geq 1}$). With $P_n = \mathbb{P}^{X_n}$, condition (a) of Theorem 14.24 is satisfied, since $X_n(0) = 0$ for each $n \geq 1$. Condition (b) of this theorem reads

$$\lim_{\delta \to 0} \limsup_{n \to \infty} \mathbb{P}\big(w(X_n, \delta) \geq \varepsilon\big) = 0 \quad \text{for every } \varepsilon > 0, \tag{15.6}$$

and we will now prove this condition, which has already shown up earlier. To this end, we start with a lemma regarding the modulus of continuity of an arbitrary C-valued random element X defined on a probability space $(\Omega, \mathcal{A}, \mathbb{P})$.

15.4 Lemma *Let $k \geq 3$ and $t_0, \ldots, t_k \in [0, 1]$ with $0 = t_0 < t_1 < \ldots < t_k = 1$. Furthermore, let $\varepsilon > 0$ and $0 < \delta < 1$. If*

$$\min_{2 \leq j \leq k-1} (t_j - t_{j-1}) \geq \delta, \tag{15.7}$$

then

$$\mathbb{P}\big(w(X, \delta) \geq 3\varepsilon\big) \leq \sum_{j=1}^{k} \mathbb{P}\left(\sup_{t_{j-1} \leq s \leq t_j} |X(s) - X(t_{j-1})| \geq \varepsilon\right).$$

The proof of this lemma is the subject of Problem 15.3. Note that the inequalities $t_1 \geq \delta$ and $1 - t_{k-1} \geq \delta$ are *not* required in (15.7).

Since the stochastic behavior of the partial sum process X_n solely depends on $X_n(\frac{j}{n})$, $j \in \{0, \ldots, n\}$, it is natural to choose the values $t_0, \ldots, t_k$ occurring in Lemma 15.4 as multiples of $\frac{1}{n}$. For this reason, we set

$$t_j := \frac{m_j}{n}, \quad j = 0, 1, \ldots, k,$$

where $m_0, \ldots, m_k$ are integers with $0 = m_0 < m_1 < \ldots < m_k = n$. If we require additionally

$$\frac{m_j}{n} - \frac{m_{j-1}}{n} \geq \delta \quad \text{for each } j \text{ with } 2 \leq j \leq k-1,$$

then condition (15.7) is satisfied. Due to the definition and the polygonal character of the partial sum process, Lemma 15.4 implies

$$\mathbb{P}\big(w(X_n, \delta) \geq 3\varepsilon\big) \leq \sum_{j=1}^{k} \mathbb{P}\left(\max_{m_{j-1} \leq \ell \leq m_j} \left| \frac{S_\ell - S_{m_{j-1}}}{\sigma\sqrt{n}} \right| \geq \varepsilon \right). \tag{15.8}$$

We now make a further specialization and choose the values $m_0, m_1, \ldots, m_k$ as equidistant as possible. For this purpose, given any $\delta \in (0, 1)$, we set $m := \lceil n\delta \rceil := \min\{r \in \mathbb{N} : r \geq n\delta\}$ and $m_j := jm$, if $0 \leq j < k$, as well as $m_k := n$. With this choice of $t_j = m_j/n$, the inequalities in (15.7) hold.

Self-question 5: Why do the inequalities in (15.7) hold with this choice?

Since $m_{k-1} = (k-1)m < m_k = n \leq km$ should also apply, we set $k := \lceil n/m \rceil$. Then $m_k - m_{k-1} \leq m$. Since the random variables $Z_1, \ldots, Z_n$ that define the partial sums occurring on the right-hand side of (15.8) are independent and identically distributed, it follows that, with the choice $m_j = jm$, $j \in \{1, \ldots, k-1\}$, the first $k-1$ terms in (15.8) are each equal to

$$\mathbb{P}\left(\max_{\ell \leq m} \left| \frac{S_\ell}{\sigma\sqrt{n}} \right| \geq \varepsilon \right).$$

Since $m_k - m_{k-1} \leq m$, the summand for $j = k$ is at most equal to the above probability, and thus (15.8) becomes

$$\mathbb{P}\big(w(X_n, \delta) \geq 3\varepsilon\big) \leq k\,\mathbb{P}\left(\max_{\ell \leq m} |S_\ell| \geq \varepsilon\sigma\sqrt{n} \right). \tag{15.9}$$

With these considerations, the following result is obtained quite quickly.

15.5 Lemma (Tightness of the Partial Sum Process)

The sequence $(X_n)_{n \geq 1}$ of partial sum processes defined in (15.3) is tight if

$$\lim_{\lambda \to \infty} \limsup_{n \to \infty} \lambda^2 \mathbb{P}\left(\max_{k \leq n} |S_k| \geq \lambda\sigma\sqrt{n} \right) = 0.$$

Proof Fix any $\varepsilon > 0$. We have to prove (15.6), where we can replace ε with 3ε. The starting point is inequality (15.9). With the above setting $k := \lceil n/m \rceil$, we write (15.9) in the form

$$\mathbb{P}\big(w(X_n, \delta) \geq 3\varepsilon\big) \leq \left\lceil \frac{n}{m} \right\rceil \mathbb{P}\left(\max_{\ell \leq m}|S_\ell| \geq \frac{\varepsilon}{\sqrt{2\delta}} \cdot \sigma\sqrt{m}\sqrt{2\delta}\sqrt{\frac{n}{m}}\right). \tag{15.10}$$

With the choice $m = m(n, \delta) := \lceil n\delta \rceil$ as above, $k = k(n, \delta) = \lceil n/m \rceil$ is also a function of n and δ. As $n \to \infty$, we obtain

$$\lim_{n\to\infty} k(n, \delta) = \frac{1}{\delta}, \quad \lim_{n\to\infty} \frac{n}{m(n, \delta)} = \frac{1}{\delta}.$$

Thus, there is a $n_0 = n_0(\delta)$ such that $k(n, \delta) < \frac{2}{\delta}$ and $\frac{n}{m(n,\delta)} > \frac{1}{2\delta}$ for each $n \geq n_0$. From (15.10) it then follows that

$$\mathbb{P}\big(w(X_n, \delta) \geq 3\varepsilon\big) \leq \frac{2}{\delta} \cdot \mathbb{P}\left(\max_{\ell \leq m}|S_\ell| \geq \frac{\varepsilon}{\sqrt{2\delta}} \cdot \sigma\sqrt{m}\right)$$

for each $n \geq n_0$, and we get

$$\limsup_{n\to\infty} \mathbb{P}\big(w(X_n, \delta) \geq 3\varepsilon\big) \leq \frac{2}{\delta} \limsup_{n\to\infty} \mathbb{P}\left(\max_{\ell \leq n}|S_\ell| \geq \frac{\varepsilon}{\sqrt{2\delta}} \cdot \sigma\sqrt{n}\right).$$

Now, setting $\lambda := \varepsilon/\sqrt{2\delta}$ yields

$$\limsup_{n\to\infty} \mathbb{P}\big(w(X_n, \delta) \geq 3\varepsilon\big) \leq \frac{4}{\varepsilon^2}\lambda^2\limsup_{n\to\infty} \mathbb{P}\left(\max_{\ell \leq n}|S_\ell| \geq \lambda \cdot \sigma\sqrt{n}\right)$$

and thus the assertion. $\square$

In order to apply this lemma, we need an upper bound for the probability

$$\mathbb{P}\left(\max_{\ell \leq n}|S_\ell| \geq \lambda\sigma\sqrt{n}\right).$$

For later purposes, it will be important to get rid of the maximum over ℓ *within* the bracket of the above expression. The following result serves this purpose.

15.6 Lemma (Etemadi's[1] Inequality)

Let $Z_1, \ldots, Z_n$ be independent random variables on a probability space $(\Omega, \mathcal{A}, \mathbb{P})$. If we set $S_0 := 0$ and $S_k := \sum_{j=1}^{k} Z_j$ for $k \in \{1, \ldots, n\}$, then for every $\alpha > 0$:

$$\mathbb{P}\left(\max_{k \leq n} |S_k| \geq 3\alpha\right) \leq 3 \max_{k \leq n} \mathbb{P}(|S_k| \geq \alpha).$$

Proof We briefly write

$$A := \left\{ \max_{k \leq n} |S_k| \geq 3\alpha \right\}$$

for the event whose probability is to be bounded from above and set

$$B_k := \{|S_k| \geq 3\alpha, \ |S_j| < 3\alpha \text{ for } j = 0, \ldots, k-1\}, \quad k \in \{1, \ldots, n\}.$$

If we regard k as a point in time then B_k is the event that the inequality $|S_k| \geq 3\alpha$ holds *for the first time* at time k. By definition, the events $B_1, \ldots, B_k$ are pairwise disjoint, and we have $A = B_1 \uplus \ldots \uplus B_k$. Furthermore,

$$A = A \cap \{|S_n| \geq \alpha\} \uplus A \cap \{|S_n| < \alpha\},$$

so that we get the first upper bound

$$\mathbb{P}(A) \leq \mathbb{P}(|S_n| \geq \alpha) + \sum_{k=1}^{n} \mathbb{P}(B_k \cap \{|S_n| < \alpha\}). \tag{15.11}$$

We now use the inequality

$$\mathbb{P}(B_k \cap \{|S_n| < \alpha\}) \leq \mathbb{P}(B_k \cap \{|S_n - S_k| > 2\alpha\}). \tag{15.12}$$

Self-question 6: Why does this inequality hold?

Since $Z_1, \ldots, Z_n$ are independent, also the events B_k and $\{|S_n - S_k| > 2\alpha\}$ are independent, since they depend only on $Z_1, \ldots, Z_k$ and $Z_{k+1} \ldots, Z_n$, respectively. As a

---

[1] Nasrollah Etemadi (*1945), Professor emeritus at the University of Illinois, Chicago. His name is also associated with a simple proof of the strong law of large numbers. Main field of research: probability theory.

consequence, $\mathbb{P}\big(B_k \cap \{|S_n - S_k| > 2\alpha\}\big) = \mathbb{P}(B_k)\mathbb{P}(|S_n - S_k| > 2\alpha)$. Since $\sum_{k=1}^{n} \mathbb{P}(B_k) \leq 1$, we use (15.11) and (15.12) to obtain

$$\mathbb{P}(A) \leq \mathbb{P}\big(|S_n| \geq \alpha\big) + \max_{k \leq n} \mathbb{P}\big(|S_n - S_k| > 2\alpha\big)$$

$$\leq \mathbb{P}\big(|S_n| \geq \alpha\big) + \max_{k \leq n} \big(\mathbb{P}\big(|S_n| \geq \alpha\big) + \mathbb{P}\big(|S_k| \geq \alpha\big)\big)$$

$$\leq 3 \max_{k \leq n} \mathbb{P}\big(|S_k| \geq \alpha\big).$$

□

Self-question 7: Why does the second inequality hold?

We are now able to prove the existence of the Wiener measure and thus also of the Wiener process. It should be emphasized that there are various other methods of proof of this result (see, e.g., [HID], [LEG], [MOP], or [SCH]).

15.7 Theorem *The Wiener measure* W *exists.*

Proof In (15.3), we choose a special partial sum process, in which Z_1 has the normal distribution $N(0, \sigma^2)$. Due to the addition theorem for the normal distribution, the distribution of $S_k/(\sigma\sqrt{k})$ is then the standard normal distribution $N(0, 1)$ for each $k \in \{1, \ldots, n\}$. If N is a random variable with this normal distribution, then for each $k \in \{1, \ldots, n\}$

$$\mathbb{P}\big(|S_k| \geq \lambda\sigma\sqrt{n}\big) = \mathbb{P}\left(|N| \geq \lambda\sqrt{\frac{n}{k}}\right) \leq \mathbb{P}(|N| \geq \lambda) \leq \frac{\mathbb{E}\left(N^4\right)}{\lambda^4} = \frac{3}{\lambda^4}.$$

Here, the last estimate used Markov's inequality (Theorem 1.5). We thus get

$$\lim_{\lambda \to \infty} \limsup_{n \to \infty} \lambda^2 \max_{1 \leq k \leq n} \mathbb{P}(|S_k| \geq \lambda\sigma\sqrt{n}) = 0,$$

and the assertion follows from Etemadi's inequality and Lemma 15.5. □

15.8 The Wiener Process on [0, 1]
As explained at the beginning of this chapter, the existence of the Wiener measure W on the σ-field of Borel sets of C also provides a probability space $(\Omega, \mathcal{A}, \mathbb{P})$ and an $(\mathcal{A}, \mathcal{B}(C))$-measurable mapping $W : \Omega \to C$ with $\mathbb{P}^W = W$. In the terminology of Sect. 14.9, W is a random function, called the *Wiener process* or *Brownian motion*. The realizations $W(\omega)$, $\omega \in \Omega$, of W are called *paths of* W.

With $W(t) := \pi_t \circ W$, $0 \le t \le 1$, one obtains a collection $(W(t))_{0 \le t \le 1}$ of random variables on Ω, which is also referred to as the Wiener process or Brownian motion. We recall once again the decisive properties of the Wiener process (cf. Definition 15.1): First, $\mathbb{P}(W(0) = 0) = 1$; the Wiener process thus starts with probability one at 0. Since generalizations can also be allowed here, this case is often called the *standard Wiener process*. Furthermore, $W(t)$ has the normal distribution $N(0, t)$ for each t with $0 < t \le 1$, and the increments $W(t) - W(s)$ over finitely many pairwise disjoint intervals (s, t) are independent random variables.

From this we concluded that W has stationary increments (cf. Corollary 15.2), which means that the distribution of the increment $W(t) - W(s)$ depends only on the length $\ell = t - s$ of the interval $[s, t]$ and thus has the same distribution as $W(\ell)$. Last but not least, we found that the Wiener process is a Gaussian process, because all finite-dimensional distributions are normal distributions. These normal distributions are centered, and the covariance between $W(s)$ and $W(t)$ is given by $\min(s, t)$.

The Wiener process is a fundamental stochastic process. It forms the starting point for many other processes, and its study fills entire books (see, e.g., [HID], [LEG], [MOP], or [SCH]). It has continuous paths, as W is a $C[0, 1]$-valued random element. However, almost all of these paths are "Weierstrass monsters", because one can prove: With probability one, the paths of W are

- nowhere differentiable,
- on no interval $[a, b]$ with $a < b$ increasing or decreasing (Problem 15.5),
- of unbounded variation on every interval $[a, b]$ with $a < b$.

Figure 15.2 shows three realizations of the partial sum process X_n with $n = 1000$, obtained from simulations. Each of these plots is based on normally distributed pseudorandom numbers for $Z_1, \ldots, Z_{1000}$. These plots can be considered as approximate paths of a Wiener process, and they convey at least a vague idea of the erratic behavior of these paths.

We now arrive at a basic result of this chapter.

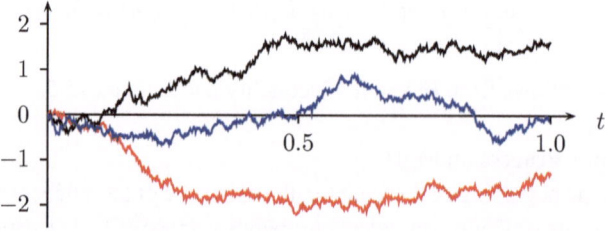

Fig. 15.2 Three (approximate) paths of a Wiener process

15.9 Theorem (Donsker,[2] 1952)

Suppose $Z_1, Z_2, \ldots$ are i.i.d. random variables with $\mathbb{E}(Z_1^2) < \infty$, $\mathbb{E}(Z_1) = 0$, and $0 < \sigma^2 := \mathbb{V}(Z_1) < \infty$. Let $S_0 := 0$ and $S_n := \sum_{j=1}^{n} Z_j$ for each $n \geq 1$. Then, for the sequence $(X_n)_{n \geq 1}$ of partial sum processes defined by

$$X_n(t) := \frac{1}{\sigma \sqrt{n}} S_{\lfloor nt \rfloor} + (nt - \lfloor nt \rfloor) \cdot \frac{Z_{\lfloor nt \rfloor + 1}}{\sigma \sqrt{n}}, \quad 0 \leq t \leq 1,$$

we have $X_n \xrightarrow{\mathcal{D}} W$ in C as $n \to \infty$.

Proof By Theorem 15.3, we have already shown the fidi convergence of X_n to W. In view of Etemadi's inequality and Lemma 15.5, it thus only remains to prove

$$\lim_{\lambda \to \infty} \limsup_{n \to \infty} \lambda^2 \max_{1 \leq k \leq n} \mathbb{P}\left(|S_k| > \lambda \sigma \sqrt{n}\right) = 0$$

(cf. the proof of Theorem 15.7). An application of Chebyshev's inequality yields

$$\mathbb{P}\left(|S_k| > \lambda \sigma \sqrt{n}\right) \leq \frac{k\sigma^2}{\lambda^2 \sigma^2 n} = \frac{k}{\lambda^2 n}. \tag{15.13}$$

This upper bound, however, only provides the factor λ^2 in the denominator, which is insufficient in view of what we need to show. We can, however, combine this upper bound with another one that follows from the Lindeberg–Lévy central limit theorem. If we set

$$Y_k := \frac{S_k}{\sigma \sqrt{k}}, \quad k \geq 1,$$

this theorem yields $Y_k \xrightarrow{\mathcal{D}} N$ as $k \to \infty$, where N is a standard normally distributed random variable. Together with Markov's inequality (Theorem 1.5), we therefore obtain

$$\lim_{k \to \infty} \mathbb{P}\left(|Y_k| > \lambda\right) = \mathbb{P}(|N| > \lambda) \leq \frac{\mathbb{E}(N^4)}{\lambda^4} = \frac{3}{\lambda^4}.$$

Consequently, there is an integer $k(\lambda)$ depending on λ with

$$\mathbb{P}\left(|Y_k| > \lambda\right) \leq \frac{4}{\lambda^4} \quad \text{for every } k > k(\lambda). \tag{15.14}$$

---

[2] Monroe David Donsker (1924–1991), American mathematician, from 1962 Professor at the Courant Institute of Mathematical Sciences of the New York University. Main field of research: Probability theory.

For each such k and each n with $k \leq n$, we thus get

$$\mathbb{P}(|S_k| > \lambda \sigma \sqrt{n}) \leq \mathbb{P}(|S_k| > \lambda \sigma \sqrt{k}) = \mathbb{P}(|Y_k| > \lambda) \leq \frac{4}{\lambda^4}.$$

A combination with (15.13) yields

$$\max_{1 \leq k \leq n} \mathbb{P}(|S_k| > \lambda \sigma \sqrt{n}) \leq \max\left(\frac{k(\lambda)}{\lambda^2 n}, \frac{4}{\lambda^4}\right),$$

and it follows that

$$\limsup_{n \to \infty} \lambda^2 \max_{1 \leq k \leq n} \mathbb{P}(|S_k| > \lambda \sigma \sqrt{n}) \leq \frac{4}{\lambda^2}$$

and thus the assertion.   □

Self-question 8: Why is the number 4 in the numerator in (15.14)?

Donsker's theorem contains the Lindeberg–Lévy central limit theorem as a special case, because

$$\frac{S_n}{\sigma \sqrt{n}} = X_n(1) \xrightarrow{\mathcal{D}} W(1) \sim N(0, 1),$$

according to the mapping theorem. Like the Lindeberg–Lévy theorem, also Donsker's theorem conveys the important message that, as $n \to \infty$, the specific distribution of Z_1 does not matter. In fact, we only require the existence of the second moment, that is, $\mathbb{E}(Z_1^2) < \infty$, and we assume that $\mathbb{V}(Z_1) > 0$ and $\mathbb{E}(Z_1) = 0$. The fact that the limit distribution (that is, the Wiener measure W) does not depend on the specific distribution of Z_1 is often referred to as the *invariance principle*.

Figure 15.3 illustrates this fact. The blue plot shows a realization of a partial sum process X_n with $n = 1000$, obtained from simulations, where the distribution of Z_1 is a uniform distribution on the values $+1$ and -1. Regarding the red plot, the distribution of Z_1 is an exponential distribution Exp(1) shifted by -1. Despite the fact that the plots were generated using very different initial distributions for Z_1, it is hard to discern which one corresponds to which distribution.

15.10 Functional Central Limit Theorem
Donsker's theorem proves particularly powerful when coupled with the mapping theorem. If $h : C \to \mathbb{R}^k$ is a $(\mathcal{B}(C), \mathcal{B}^k)$-measurable mapping with $W(\mathcal{C}(h)) = 1$, that is, if h is

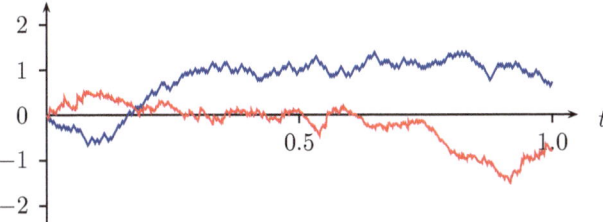

Fig. 15.3 Realizations of partial sum processes X_{1000} with $\mathbb{P}(Z_1 = \pm 1) = 1/2$ (blue) and $Z_1 \sim$ Exp$(1) - 1$ (red)

continuous almost everywhere with respect to W, then Donsker's theorem and the mapping theorem imply

$$h(X_n) \xrightarrow{\mathcal{D}} h(W).$$

Because this convergence links the (central limit) theorem $X_n \xrightarrow{\mathcal{D}} W$ of Donsker with a function, this fact is referred to as the *functional central limit theorem*.

The functional central limit theorem has, among other things, the following important consequence: If one can obtain the limit distribution of $h(X_n)$ for a *specific* partial sum process, that is, for a *specific* distribution of Z_1, then one knows the distribution of $h(W)$. Such an approach is often successful with a very simple partial sum process. This characteristic applies to the so-called *simple symmetric random walk*. In this case, the distribution of Z_1 is given by $\mathbb{P}(Z_1 = 1) = \mathbb{P}(Z_1 = -1) = \frac{1}{2}$. The following result is based on this idea of proof.

15.11 Theorem (Distribution of $\max_{0 \leq t \leq 1} W(t)$)
If $(W(t))_{0 \leq t \leq 1}$ *is the Wiener process, then*

$$\max_{0 \leq t \leq 1} W(t) \sim |N|,$$

where $N \sim N(0, 1)$. *Thus, for every* $u \geq 0$:

$$\mathbb{P}\left(\max_{0 \leq t \leq 1} W(t) \leq u\right) = 2\Phi(u) - 1, \quad u \geq 0.$$

Here, Φ *denotes the distribution function of the standard normal distribution.*

Proof Let X_n be the partial sum process corresponding to an i.i.d. sequence $Z_1, Z_2, \ldots$ with $\mathbb{P}(Z_1 = \pm 1) = \frac{1}{2}$. Putting $S_0 := 0$ and $S_n := \sum_{j=1}^{n} Z_j$ for $n \geq 1$, we have

$$\max_{0 \leq t \leq 1} X_n(t) = \frac{1}{\sqrt{n}} \max_{k=0,\ldots,n} S_k.$$

By Problem 15.4, the random variables on the right-hand side converge in distribution to $|N|$. Since the functional $h : C \to \mathbb{R}$ defined by $h(x) := \max_{0 \le t \le 1} x(t)$ is continuous, the mapping theorem implies

$$\max_{0 \le t \le 1} X_n(t) = h(X_n) \xrightarrow{\mathcal{D}} h(W) = \max_{0 \le t \le 1} W(t).$$

$\square$

Self-question 9: Why is the functional $h : C \to \mathbb{R}$, $x \mapsto \max_{0 \le t \le 1} x(t)$, continuous?

15.12 Corollary *Let $Z_1, Z_2, \ldots$ be i.i.d. random variables with $\mathbb{E}(Z_1^2) < \infty$, $\mathbb{E}(Z_1) = 0$, and $0 < \sigma^2 := \mathbb{V}(Z_1)$. Then*

$$\frac{1}{\sigma \sqrt{n}} \max_{k=0,\ldots,n} S_k \xrightarrow{\mathcal{D}} |N|,$$

where $N \sim N(0, 1)$.

We provide two more results regarding distributions of functionals of the Wiener process. These functionals are defined by

$$h_+(x) := \lambda^1 \left(\{ t \in [0, 1] : x(t) > 0 \} \right), \quad x \in C,$$
$$h_0(x) := \sup \{ t \in [0, 1] : x(t) = 0 \}, \quad x \in C.$$

In descriptive terms, $h_+(x)$ is the time span that the function x spends above the x-axis, and $h_0(x)$ stands for the time of the last zero of x (see Fig. 15.4).

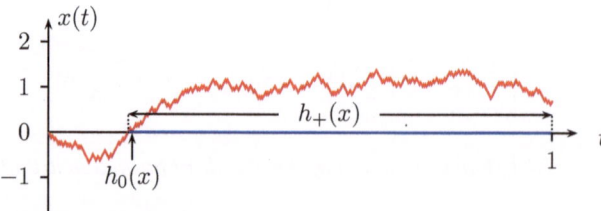

Fig. 15.4 The functionals h_+ and h_0

Both functionals meet the requirements of the mapping theorem (see, e.g., [BI2], pp. 246–248). For the partial sum process X_n based on the simple symmetric random walk (that is, $\mathbb{P}(Z_1 = \pm 1) = \frac{1}{2}$), we have (see, e.g., [HE0], pp. 23, 48):

$$\lim_{n \to \infty} \mathbb{P}\big(h_0(X_n) \le u\big) = \lim_{n \to \infty} \mathbb{P}\big(h_+(X_n) \le u\big) = \frac{2}{\pi} \arcsin \sqrt{u}, \quad 0 \le u \le 1. \quad (15.15)$$

With the functional central limit theorem, this leads to the famous *Arcsine law*:

15.13 Theorem (Arcsine Law for the Wiener Process)
If $(W(t))_{0 \le t \le 1}$ is the Wiener process, then

$$\mathbb{P}\big(h_0(W) \le u\big) = \mathbb{P}\big(h_+(W) \le u\big) = \frac{2}{\pi} \arcsin \sqrt{u}, \quad 0 \le u \le 1.$$

The distribution of $h_0(W)$ is therefore called the *Arcsine distribution*. The same applies to $h_+(W)$. Figure 15.5 shows the density and the distribution function of the Arcsine distribution. The density is U-shaped, and thus the probability mass is strongly concentrated near 0 and 1. With high probability, the last zero of the Wiener process W occurs either "quite early or quite late" in the course of time. A similar effect applies to the time span that W spends above the x-axis. With high probability, this is "either quite large or quite small". The noteworthy message of the Arcsine law is that the limit statements in (15.15) applies to *every* partial sum process X_n that meets the requirements of Donsker's theorem.

The next result is based on an orthogonal decomposition of the covariance function $K(s, t) = \min(s, t)$ of the Wiener process. This is accompanied by a general approach

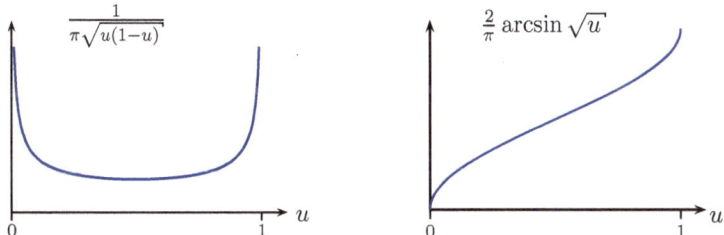

Fig. 15.5 Density (left) and distribution function (right) of the Arcsine distribution

to orthogonal expansions of stochastic processes, which is known under the keyword *Karhunen[3]–Loève[4] expansion.*

15.14 Theorem (Karhunen–Loève Expansion of the Wiener Process)
Let $W = (W(t))_{0 \le t \le 1}$ be a Wiener process and $N_1, N_2, \ldots$ be a sequence of i.i.d. standard normally distributed random variables on a probability space $(\Omega, \mathcal{A}, \mathbb{P})$. Then the series

$$\widetilde{W}(t) := \sum_{j=1}^{\infty} \frac{\sqrt{2} \sin\left(\left(j - \frac{1}{2}\right) \pi t\right)}{\left(j - \frac{1}{2}\right) \pi} N_j, \quad 0 \le t \le 1, \tag{15.16}$$

converges uniformly in t in $L^2(\Omega, \mathcal{A}, \mathbb{P})$, and we have $W \stackrel{\mathcal{D}_{fidi}}{=} \widetilde{W}$, that is, W and $\widetilde{W}$ have the same finite-dimensional distributions.

The proof uses a theorem by J. Mercer[5], see, e.g., [KOE], p. 145.

15.15 Theorem (Mercer, 1909)
Let $K : [0, 1]^2 \to \mathbb{R}$ be a non-vanishing, continuous, symmetric and positive semi-definite function, that is, for every square integrable function g on $[0, 1]$ we have

$$\int_0^1 \int_0^1 g(s)K(s, t)g(t) \, ds \, dt \ge 0.$$

Then

$$K(s, t) = \sum_{j=1}^{\infty} \lambda_j \varphi_j(s) \varphi_j(t), \quad 0 \le s, t \le 1. \tag{15.17}$$

Here, $\lambda_1, \lambda_2, \ldots$ are the positive eigenvalues and $\varphi_1, \varphi_2, \ldots$ the associated normalized eigenfunctions of the integral operator $g \mapsto Ag$, $(Ag)(s) = \int_0^1 K(s, t)g(t)dt$ associated with the kernel K. The series in (15.17) converges uniformly and absolutely.

---

[3] Kari Karhunen (1915–1992), Finnish statistician, graduated in 1947 with the thesis written in German *Über lineare Methoden in der Wahrscheinlichkeitsrechnung.* In 1963 he took over the management of the insurance company SUOMI.

[4] Michel Loève (1907–1979), PhD 1941 at the École Polytechnique, after being arrested by the German occupiers he worked at the Institut Henri Poincaré in Paris and 1946–1948 at the University of London, from 1948 Professor of Mathematics at the University of California, Berkeley. Main field of research: Probability theory.

[5] James Mercer (1883–1932), British mathematician. Main field of research: Analysis.

For the special case $K(s,t) = \min(s,t)$, that is, the covariance kernel of the Wiener process, Mercer's theorem takes the following explicit form. The proof is the subject of Problem 15.10.

15.16 Theorem (Mercer's theorem for $K(s,t) = \min(s,t)$)

$$We\ have\ \min(s,t) = \sum_{j=1}^{\infty} \lambda_j \, \varphi_j(s) \, \varphi_j(t), \quad 0 \le s, t \le 1,$$

$$where\ \lambda_j = \frac{1}{\pi^2 \left(j - \frac{1}{2}\right)^2}, \quad \varphi_j(t) = \sqrt{2} \sin\left(\left(j - \tfrac{1}{2}\right)\pi t\right), \quad j \ge 1. \tag{15.18}$$

Proof of Theorem 15.14 Let

$$\widetilde{W}_n(t) := \sum_{j=1}^{n} \sqrt{\lambda_j}\, \varphi_j(t)\, N_j, \quad n \ge 1. \tag{15.19}$$

Due to the independence of $N_1, N_2, \ldots$, we have for m, n with $1 \le m < n$

$$\mathbb{E}\left(\widetilde{W}_n(t) - \widetilde{W}_m(t)\right)^2 = \sum_{j=m+1}^{n} \lambda_j \varphi_j^2(t) \le \frac{2}{\pi^2} \sum_{j=m+1}^{\infty} \frac{1}{\left(j - \frac{1}{2}\right)^2} \to 0 \text{ as } m \to \infty.$$

Thus, $(W_n(t))_{n\ge 1}$ is a Cauchy sequence in $L^2(\Omega, \mathcal{A}, \mathbb{P})$. Since this space is complete, the limit of $(W_n(t))_{n\ge 1}$, which is denoted $\widetilde{W}(t)$ in (15.16), exists, and it follows that

$$\lim_{n\to\infty} \sup_{0\le t\le 1} \mathbb{E}\left[\left(\widetilde{W}(t) - \widetilde{W}_n(t)\right)^2\right] = 0. \tag{15.20}$$

To prove $W \stackrel{\mathcal{D}_{\mathrm{fidi}}}{=} \widetilde{W}$, fix any $k \ge 1$ and $t_1, \ldots, t_k$ with $0 \le t_1 < \ldots < t_k \le 1$. We have to show that $\left(\widetilde{W}(t_1), \ldots, \widetilde{W}(t_k)\right)$ has the normal distribution $N_k\left(0_k, \left(t_i \wedge t_j\right)_{1\le i, j\le k}\right)$. By Theorem 1.17, this is the same as

$$\sum_{\ell=1}^{k} c_\ell \widetilde{W}(t_\ell) \sim N\left(0, \sum_{\ell,m=1}^{k} c_\ell c_m \left(t_\ell \wedge t_m\right)\right) \tag{15.21}$$

for each choice of real numbers $c_1, \ldots, c_k$. Using the addition theorem for the normal distribution, we obtain

$$\sum_{\ell=1}^{k} c_\ell \widetilde{W}_n(t_\ell) = \sum_{\ell=1}^{k} c_\ell \left(\sum_{j=1}^{n} \sqrt{\lambda_j} \varphi_j(t_\ell) N_j \right) = \sum_{j=1}^{n} \sqrt{\lambda_j} \left(\sum_{\ell=1}^{k} c_\ell \varphi_j(t_\ell) \right) N_j$$

$$\sim N\left(0, \sum_{j=1}^{n} \lambda_j \sum_{\ell,m=1}^{k} c_\ell c_m \, \varphi_j(t_\ell) \varphi_j(t_m) \right)$$

$$= N\left(0, \sum_{\ell,m=1}^{k} c_\ell c_m \sum_{j=1}^{n} \lambda_j \varphi_j(t_\ell) \varphi_j(t_m) \right).$$

By Mercer's theorem, the variance of this normal distribution converges to the variance of the normal distribution in (15.21) as $n \to \infty$. Since $\sum_{\ell=1}^{k} c_\ell \widetilde{W}_n(t_\ell) \xrightarrow{L^2} \sum_{\ell=1}^{k} c_\ell \widetilde{W}(t_\ell)$, the claim follows.   □

Self-question 10: Why does this L^2-convergence imply the claim?

The Karhunen–Loève expansion from Theorem 15.14 yields the following result regarding the distribution of the integral of the squared Wiener process. Since the paths of the Wiener process are continuous with probability 1, this integral is to be understood as a pathwise Riemann integral (see also Problem 15.16).

15.17 Corollary (The Distribution of $\int_0^1 W^2(t)\,dt$)
For the Wiener process $W = (W(t))_{0 \le t \le 1}$, we have:

$$\int_0^1 W^2(t)\,dt \sim \sum_{j=1}^{\infty} \frac{N_j^2}{\left(j - \frac{1}{2} \right)^2}. \tag{15.22}$$

Here, $N_1, N_2, \ldots$ is a sequence of i.i.d. standard normal random variables.

Proof The integral in (15.22) only uses a sequence of finite-dimensional distributions of W, and by Theorem 15.14 the latter coincide with the corresponding fidis of $(\widetilde{W}(t))_{0 \le t \le 1}$ with $\widetilde{W}(t)$ as in (15.16). We therefore can prove (15.22) with $\widetilde{W}$ instead of W. From the definition of $\widetilde{W}_n(t)$ in (15.19), and due to the orthogonality and normalization of the functions φ_j, we obtain

$$\int_0^1 \widetilde{W}_n^2(t)\,dt = \sum_{j=1}^{n} \lambda_j N_j^2 \tag{15.23}$$

with λ_j as in (15.18) and the meaning of $N_1, \ldots, N_n$ as in the corollary. We prove

$$\lim_{n \to \infty} \mathbb{E}\left| \int_0^1 \left(\tilde{W}^2(t) - \tilde{W}_n^2(t) \right) dt \right| = 0. \tag{15.24}$$

By Markov's inequality, it herewith follows that, as $n \to \infty$, the sum in (15.23) converges in probability and thus also in distribution to the right-hand side of (15.22), which was to be shown. Using the triangle inequality for integrals, Fubini's theorem, the formula $a^2 - b^2 = (a - b)(a + b)$ and Chebyshev's inequality, the expectation occurring in (15.24) can be bounded from above by

$$\int_0^1 \sqrt{\mathbb{E}\left[\left(\tilde{W}(t) - \tilde{W}_n(t)\right)^2\right]} \sqrt{\mathbb{E}\left[\left(\tilde{W}(t) + \tilde{W}_n(t)\right)^2\right]} \, dt$$

(Problem 15.9). After taking the supremum over t, the first factor of the integrand can be put in front of the integral. Using (15.20), the claim follows if one takes into account the inequalities $(a + b)^2 \le 2a^2 + 2b^2$ and $\sqrt{u + v} \le \sqrt{u} + \sqrt{v}$ ($u, v \ge 0$), and notes that $\mathbb{E}[\tilde{W}^2(t)] = t$ and $\mathbb{E}[\tilde{W}_n^2(t)] \le 2 \sum_{j=1}^n \lambda_j$.   □

Self-question 11: Why does the last inequality hold?

There are many other processes that can be constructed using the Wiener process. A special role for statistics is played by the so-called *Brownian bridge*.

15.18 Definition (Brownian Bridge)
A C[0, 1]-valued random element $B = (B(t))_{0 \le t \le 1}$ is called a *Brownian bridge*, if the following conditions hold:

(a) $\mathbb{P}(B(0) = 0) = 1 = \mathbb{P}(B(1) = 0)$.
(b) For each $k \ge 1$ and $t_1, \ldots, t_k$ with $0 \le t_1 < \ldots < t_k \le 1$, we have

$$\left(B(t_1), \ldots, B(t_k)\right) \sim N_k\left(0_k, \left(\min(t_i, t_j) - t_i t_j\right)_{1 \le i, j \le k}\right).$$

Like the Wiener process, thus also the Brownian bridge is a Gaussian process. Property (a) states that a Brownian bridge not only starts at 0 (like a Wiener process), but also ends at 0 at time $t = 1$, whence the naming *bridge*. The covariances $\mathrm{Cov}(B(s), B(t)) = \min(s, t) - st$ differ from the corresponding covariances of the Wiener process by the minus term. In the special case of the uniform distribution on the unit interval, these covariances already showed up in Problem 7.4 in connection with the empirical distribution function.

The existence of the Brownian bridge is easily shown:

15.19 Theorem *The Brownian bridge exists.*

Proof For each $x \in C$, let $h(x)(t) := x(t) - t\,x(1)$, $0 \leq t \leq 1$. The mapping $h : C \to C$ is continuous, and we have $h(x)(1) = 0$.

Self-question 12: Why is the mapping h continuous?

Furthermore, if $x(0) = 0$ then $h(x)(0) = 0$. Now suppose $(W(t))_{0 \leq t \leq 1}$ is a Wiener process. We set

$$B(t) := W(t) - t W(1) = (h \circ W)(t), \quad 0 \leq t \leq 1,$$

so $B := h \circ W$. Then $\mathbb{P}(B(0) = 0) = 1 = \mathbb{P}(B(1) = 0)$, and thus condition 15.18 (a) holds. By definition, for each $k \geq 1$ and $t_1, \ldots, t_k$ with $0 \leq t_1 < \ldots < t_k \leq 1$, we have

$$
\begin{pmatrix} B(t_1) \\ B(t_2) \\ \vdots \\ B(t_k) \end{pmatrix}
=
\begin{pmatrix}
1 & 0 & \cdots & 0 & -t_1 \\
0 & 1 & 0 & 0 & -t_2 \\
\vdots & 0 & \ddots & 0 & \vdots \\
0 & 0 & \cdots & 1 & -t_k
\end{pmatrix}
\begin{pmatrix} W(t_1) \\ \vdots \\ W(t_k) \\ W(1) \end{pmatrix}.
$$

As an affine transformation of a $(k + 1)$-variate normally distributed random vector, the random vector $(B(t_1), \ldots, B(t_k))$ has a k-variate normal distribution with expectation 0_k. Furthermore, if $0 \leq s, t \leq 1$ then

$$
\begin{aligned}
\mathrm{Cov}(B(s), B(t)) &= \mathbb{E}\left[(W(s) - s W(1))(W(t) - t W(1))\right] \\
&= \mathbb{E}\left[W(s)W(t)\right] - s\,\mathbb{E}\left[W(1)W(t)\right] - t\,\mathbb{E}\left[W(s)W(1)\right] \\
&\quad + st\,\mathbb{E}\left[W(1)^2\right] \\
&= s \wedge t - st - ts + st \\
&= s \wedge t - st.
\end{aligned}
$$

Thus, condition 15.18 (b) also holds.   □

Figure 15.6 shows the transition from a function x to $h(x)$. Since the value $x(1)$ is visually "tied down to the x-axis" by subtracting $1 \cdot x(1)$, the Brownian bridge is often referred to as *tied down Brownian motion*. Figure 15.7 shows realizations of an

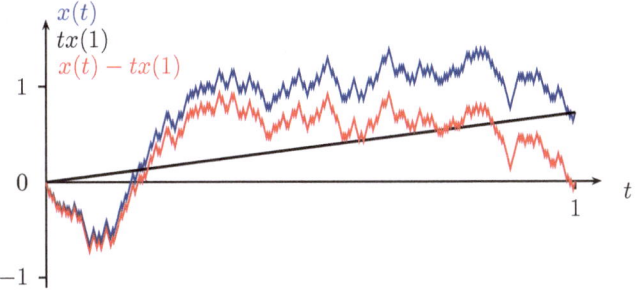

Fig. 15.6 Transition from a function $x(t)$ to $x(t) - tx(1)$

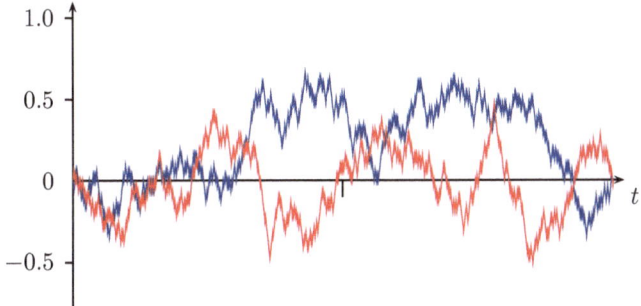

Fig. 15.7 Two realizations of an (approximate) Brownian bridge

approximate Brownian bridge, which have been obtained by tying down partial sum processes to the x-axis. Like the paths of the Wiener process, also the paths of the Brownian bridge are nowhere differentiable with probability 1.

The following result states, in colloquial terms, that the distribution of the Brownian bridge is equal to the conditional distribution of the Wiener process under the condition $W(1) = 0$. The adjective "colloquial" refers to the fact that the event $\{W(1) = 0\}$ has probability 0. We therefore consider a conditional distribution of W, in which we condition on the event $\{0 \leq W(1) \leq \varepsilon\}$ with $\varepsilon > 0$ and then let ε converge to zero.

15.20 Theorem *Let* $W = (W(t))_{0 \leq t \leq 1}$ *be a Wiener process and* $\varepsilon > 0$. *If we set*

$$P_\varepsilon(A) := \mathbb{P}(W \in A \mid 0 \leq W(1) \leq \varepsilon), \quad A \in \mathcal{B}(C),$$

then $P_\varepsilon \xrightarrow{\mathcal{D}} B$ *as* $\varepsilon \downarrow 0$. *Here,* B *is a Brownian bridge.*

Proof We assume that W is defined on a probability space $(\Omega, \mathcal{A}, \mathbb{P})$ and set $B := (B(t))_{0 \leq t \leq 1}$, where

$$B(t) := W(t) - tW(1), \quad 0 \leq t \leq 1. \tag{15.25}$$

By the Portmanteau theorem (Theorem 14.3), we need to prove

$$\limsup_{\varepsilon \downarrow 0} \mathbb{P}(W \in A | 0 \leq W(1) \leq \varepsilon) \leq \mathbb{P}(B \in A)$$

for every non-empty closed Borel set A of C. To this end, we use the fact that for any choice of $k \geq 1$ and $t_1, \ldots, t_k$ with $0 \leq t_1 < \ldots < t_k \leq 1$, the random vector $(W(1), B(t_1), \ldots, B(t_k))$ has a $(k+1)$-variate normal distribution, where, due to $\mathrm{Cov}(W(s), W(t)) = \min(s, t)$,

$$\mathbb{E}[W(1)B(t_j)] = \mathbb{E}\big[W(1)(W(t_j) - t_j W(1))\big] = t_j - t_j = 0, \quad j \in \{1, \ldots, k\}.$$

By Theorem 5.11, $W(1)$ and $\big(B(t_1), \ldots, B(t_k)\big)$ are independent for each choice of k and $t_1, \ldots, t_k$. Thus,

$$\mathbb{P}(W(1) \in C, B \in M) = \mathbb{P}(W(1) \in C) \cdot \mathbb{P}(B \in M) \quad \text{for every } C \in \mathcal{B}^1 \tag{15.26}$$

and every M in $\mathcal{C}_f = \big\{\pi_{t_1,\ldots,t_k}^{-1}(H) \big| k \in \mathbb{N}, 0 \leq t_1 < \ldots < t_k \leq 1, H \in \mathcal{B}^k\big\}$. For any fixed $C \in \mathcal{B}^1$, we set

$$\mathcal{D}_C := \big\{M \in \mathcal{B}(\mathrm{C}) : \mathbb{P}(W(1) \in C, B \in M) = \mathbb{P}(W(1) \in C)\mathbb{P}(B \in M)\big\}.$$

Due to (15.26), we have $\mathcal{C}_f \subset \mathcal{D}_C$, and as can be directly verified, $\mathcal{D}_C$ is a *Dynkin system*, that is, the following properties hold:

- $\mathrm{C}[0, 1] \in \mathcal{D}_C$.
- If $D, E \in \mathcal{D}_C$ with $D \subset E$ then $E \setminus D \in \mathcal{D}_C$.
- If $E_1, E_2, \ldots$ are pairwise disjoint sets in $\mathcal{D}_C$, then $\uplus_{j=1}^{\infty} E_j \in \mathcal{D}_C$.

Since $\mathcal{D}_C$ is a Dynkin system that contains $\mathcal{C}_f$, it also contains the smallest Dynkin system $\delta(\mathcal{C}_f)$ (say) on Ω that contains $\mathcal{C}_f$. Since $\mathcal{C}_f$ is a π-system, we have $\mathcal{B}(\mathrm{C}) = \sigma(\mathcal{C}_f) = \delta(\mathcal{C}_f) \subset \mathcal{D}_C$ and hence

$$\mathbb{P}(B \in M | 0 \leq W(1) \leq \varepsilon) = \mathbb{P}(B \in M) \quad \text{for every } M \in \mathcal{B}(\mathrm{C}).$$

Due to (15.25),

$$\|W - B\|_\infty = \sup_{0 \leq t \leq 1} |W(t) - (W(t) - tW(1))| = |W(1)|.$$

So, if $A \subset C$ is a closed set and $\delta > 0$, we can conclude from $|W(1)| \leq \delta$ and $W \in A$ that $B \in A_\delta := \{x \in C : \|x - A\|_\infty \leq \delta\}$. If $0 < \varepsilon < \delta$, then

$$\mathbb{P}(W \in A | 0 \leq W(1) \leq \varepsilon) \leq \mathbb{P}(B \in A_\delta | 0 \leq W(1) \leq \varepsilon) = \mathbb{P}(B \in A_\delta).$$

Since A is closed, the assertion now follows by taking the limit $\delta \to 0$.   □

The following result will prove important in the next chapter in connection with the nonparametric two-sample problem.

15.21 Theorem (Reproduction Theorem for the Brownian Bridge)
Let B_1, B_2 be independent Brownian bridges and $a_1, a_2 \in \mathbb{R}$ with $a_1^2 + a_2^2 = 1$. Then also

$$B := a_1 B_1 + a_2 B_2$$

is a Brownian bridge.

Proof First, $\mathbb{P}\big(B(0) = 0\big) = 1 = \mathbb{P}\big(B(1) = 0\big)$. For each $k \geq 1$ and $t_1, \ldots, t_k$ with $0 \leq t_1 < \ldots < t_k \leq 1$, the random vector $(B(t_1), \ldots, B(t_k))$ has a k-variate normal distribution.

Self-question 13: Why does the last statement hold?

We have $\mathbb{E}(B(t)) = 0, 0 \leq t \leq 1$. If $K(s, t) := \min(s, t) - st, 0 \leq s, t \leq 1$, denotes the covariance function of a Brownian bridge, the independence of B_1 and B_2 implies

$$\begin{aligned}
\mathbb{E}\big[B(s)B(t)\big] &= \mathbb{E}\big[(a_1 B_1(s) + a_2 B_2(s))(a_1 B_1(t) + a_2 B_2(t))\big] \\
&= a_1^2 K(s, t) + a_1 a_2 \mathbb{E}\big[B_1(s)B_2(t)\big] + a_2 a_1 \mathbb{E}\big[B_2(s)B_1(t)\big] + a_2^2 K(s, t) \\
&= (a_1^2 + a_2^2) K(s, t) \\
&= K(s, t).
\end{aligned}$$

Thus, B fulfills conditions (a) and (b) of Definition 15.18. Since B_1 and B_2 are defined on a probability space $(\Omega, \mathcal{A}, \mathbb{P})$ as C-valued random elements, $(B_1, B_2) : \Omega \to C \times C$ is a $C \times C$-valued random element due to the separability of C. The mapping $h : C \times C \to C$ defined by $h(x, y) := a_1 x + a_2 y$ is continuous, and thus $B = a_1 B_1 + a_2 B_2 = h(B_1, B_2)$ is a C-valued random element defined on Ω.   □

This result can be generalized to more than two independent Brownian bridges (Problem 15.8).

15.22 The Wiener Process on $[0, \infty)$

We conclude this chapter by extending the Wiener process to the half-axis $[0, \infty)$. For this purpose, let $C[0, \infty) := \{x : [0, \infty) \to \mathbb{R}, \ x \text{ continuous}\}$ denote the collection of all continuous real-valued functions on $[0, \infty)$. We set

$$\rho(x, y) := \sum_{j=1}^{\infty} \frac{1}{2^j} \cdot \frac{\max_{0 \le t \le j} |x(t) - y(t)|}{1 + \max_{0 \le t \le j} |x(t) - y(t)|}, \quad x, y \in C[0, \infty).$$

Thereby, $(C[0, \infty), \rho)$ becomes a complete separable metric space, and for every sequence (x_n) in $C[0, \infty)$ and every $x \in C[0, \infty)$ we have

$$\rho(x_n, x) \to 0 \iff \max_{t \in K} |x_n(t) - x(t)| \to 0 \text{ for every compact set } K \subset [0, \infty).$$

Convergence in the space $(C[0, \infty), \rho)$ is therefore uniform convergence on each compact subset of $[0, \infty)$.

A *Wiener process* $W = (W(t))_{t \ge 0}$ on $[0, \infty)$ is defined as a $C[0, \infty)$-valued random element on a probability space $(\Omega, \mathcal{A}, \mathbb{P})$ with properties (a), (b) and (c) of Definition 15.1, where (b) applies for every $t > 0$ and (c) without the restriction $t_k \le 1$. From this, similar to Definition 15.1, each of the properties (a)–(c) of Corollary 15.2 can be concluded, without the upper limit 1 for s, t in (a)–(b) and for t_k in (c). Therefore, a Wiener process $(W(t))_{t \ge 0}$ on $[0, \infty)$ is a centered Gaussian process with continuous paths, which starts with probability one in 0 and has independent and stationary increments. Moreover, $W(t) \sim N(0, t)$ for every $t > 0$.

We construct such a process using a Brownian bridge. The starting point is the function h defined by

$$h : \begin{cases} C[0, 1] \to C[0, \infty), \\ x \mapsto h(x), \quad h(x)(t) := (1 + t) \cdot x\left(\frac{t}{1+t}\right), \quad 0 \le t < \infty. \end{cases}$$

This function is continuous, and thus with every $C[0, 1]$-valued random element $X : \Omega \to C[0, 1]$ the mapping $h \circ X : \Omega \to C[0, \infty)$ is a $C[0, \infty)$-valued random element.

Self-question 14: Why is the above mapping $h : C[0, 1] \to C[0, \infty)$ continuous?

If $B := (B(t))_{0 \le t \le 1}$ is a Brownian bridge, we set $W := h \circ B$, so

$$W(t) := h(B)(t)$$

$$= (1+t)B\left(\frac{t}{1+t}\right), \quad t \ge 0. \tag{15.27}$$

Then W is a random element in $C[0, \infty)$, and we have $\mathbb{E}(W(t)) = 0$ for each $t \ge 0$. For arbitrary real numbers s and t with $0 \le s \le t$, it follows that

$$\mathrm{Cov}(W(s), W(t)) = (1+s)(1+t)\mathrm{Cov}\left(B\left(\frac{s}{1+s}\right), B\left(\frac{t}{1+t}\right)\right)$$

$$= (1+s)(1+t)\left(\min\left(\frac{s}{1+s}, \frac{t}{1+t}\right) - \frac{s}{1+s}\frac{t}{1+t}\right)$$

$$= s(1+t) - st = s = \min(s, t).$$

Furthermore, $\mathbb{P}(W(0) = 0) = 1$. Due to (15.27), the random vector $(W(t_1), \ldots, W(t_k))$ has the normal distribution $N_k\big(0_k, (\min(t_i, t_j))_{1 \le i, j \le k}\big)$ for each $k \ge 1$ and $t_1, \ldots, t_k$ with $0 \le t_1 < \ldots < t_k$. Since this collection ot finite-dimensional distributions determines the distribution of W, all other properties required for a standard Wiener process on $[0, \infty)$ also apply.

As already emphasized, the standard Wiener process $(W(t))_{t \ge 0}$ is the starting point for many other interesting stochastic processes. For $\mu \in \mathbb{R}$ and $\sigma > 0$, the process $X(t) := \mu t + \sigma W(t)$, $t \ge 0$, defines a *Wiener process with drift μ and volatility σ*. In financial mathematics, the so-called *geometric Brownian motion with drift μ and volatility σ*, defined by

$$S(t) := S(0)\exp\left(\left(\mu - \frac{\sigma^2}{2}\right)t + \sigma W(t)\right), \quad t \ge 0,$$

plays a major role. Here, for example, $S(t)$ stands for the price of a stock at time t (see, e.g., [BIK])). If $W_1, \ldots, W_n$ are independent standard Wiener processes, one can define an *n-dimensional standard Wiener process* via $\mathbf{W}(t) := \big(W_1(t), \ldots, W_n(t)\big)$, $t \ge 0$.

The following result allows, among other things, to prove the invariance in distribution of the Wiener process with respect to "projective reflections at $t = \infty$" (see property 15.24 (c)).

15.23 Theorem (Strong Law of Large Numbers for the Wiener Process)

Let $W = (W(t))_{t \ge 0}$ be a Wiener process on a probability space $(\Omega, \mathcal{A}, \mathbb{P})$. Then

$$\lim_{t \to \infty} \frac{W(t)}{t} = 0 \quad \mathbb{P}\text{-almost surely.}$$

Proof Since the increments $W(k) - W(k-1)$, $k = 1, 2, \ldots$, are independent and each standard normally distributed random variables, the strong law of large numbers (Theorem 1.2) yields

$$\lim_{n \to \infty} \frac{W(n)}{n} = 0 \quad \mathbb{P}\text{-almost surely.}$$

Thus, there exists a set $\Omega_1 \in \mathcal{A}$ with $\mathbb{P}(\Omega_1) = 1$ and the property that for every $\varepsilon > 0$ and every $\omega \in \Omega_1$ there is a n_1 depending on ω and ε, such that $|W(n, \omega)/n| \leq \varepsilon$ for every $n \geq n_1$. We therefore only need to tackle the maximum absolute fluctuation of W between each two integers. Using Kolmogorov's inequality (Theorem 1.4) we get

$$\mathbb{P}\left(\max_{0 \leq k \leq 2^m} \left| W\left(n + \frac{k}{2^m}\right) - W(n) \right| > n^{2/3} \right) \leq \frac{\mathbb{V}\big(W(n+1) - W(n)\big)}{n^{4/3}} = \frac{1}{n^{4/3}}$$

for each $m, n \geq 1$.

Self-question 15: To which random variables X_j is Theorem 1.4 applied here?

Putting $A_n := \big\{ \sup_{n \leq u \leq n+1} |W(u) - W(n)| > n^{2/3} \big\}$, and taking the limit $m \to \infty$ in the above inequality, it follows that $\mathbb{P}(A_n) \leq n^{-4/3}$. Since $\sum_{n=1}^{\infty} \mathbb{P}(A_n) < \infty$, the Borel–Cantelli lemma (Theorem 1.3) implies that with probability one, only finitely many of the events $A_1, A_2, \ldots$ occur. Thus, there exists a set $\Omega_2 \in \mathcal{A}$ with $\mathbb{P}(\Omega_2) = 1$ and the property that for each $\omega \in \Omega_2$ there is a n_2 (that depends on ω), such that

$$\frac{1}{n} \sup_{n \leq u \leq n+1} |W(\omega, u) - W(\omega, n)| \leq \frac{1}{n^{1/3}}$$

for each $n \geq n_2$. For each ω in $\Omega_1 \cap \Omega_2$ and each t with $t \geq \max(n_1, n_2)$ and $s := \lfloor t \rfloor$, we then have

$$\left| \frac{W(\omega, t)}{t} \right| \leq \left| \frac{W(\omega, s)}{s} \right| \frac{s}{t} + \frac{1}{s} \sup_{s \leq u \leq s+1} |W(\omega, u) - W(\omega, s)| \frac{s}{t} \leq \varepsilon + \frac{1}{s^{1/3}}.$$

Since $\mathbb{P}(\Omega_1 \cap \Omega_2) = 1$, the assertion follows. $\square$

Finally, we compile some properties of the Wiener process and note connections between Brownian motion and the Wiener process (see also Problems 15.11–15.15).

15.24 Properties of the Wiener Process, Connection to the Brownian Bridge

(a) Let $((W(t))_{t\geq 0}$ be a Wiener process and $a > 0$. Then the process W^* defined by

$$W^*(t) := \frac{1}{\sqrt{a}}\, W\,(at)\,, \quad t \geq 0,$$

is also a Wiener process. The Wiener process is therefore *self-similar under stretching of the time axis*.

(b) Let $((W(t))_{t\geq 0}$ be a Wiener process and $r > 0$. Then the process $\widetilde{W}$ defined by

$$\widetilde{W}(t) := W(t + r) - W(r), \quad t \geq 0,$$

is also a Wiener process. The Wiener process is therefore *invariant in distribution with respect to shifts of the time axis*.

(c) Suppose $((W(t))_{t\geq 0}$ is a Wiener process. Setting

$$\widehat{W}(t) := t\, W\left(\frac{1}{t}\right), \quad t > 0,$$

and $\widehat{W}(0) := 0$, then $\widehat{W} = (\widehat{W})_{t\geq 0}$ is also a Wiener process. The transition from W to $\widehat{W}$ is called the *projective reflection of W at $t = \infty$*.

(d) Let $((W(t))_{t\geq 0}$ be a Wiener process, and put

$$B(t) := (1 - t)W\left(\frac{t}{1-t}\right), \quad 0 \leq t < 1,$$

as well as $B(1) := 0$. Then the process $B = (B(t))_{0\leq t\leq 1}$ is a Brownian bridge.

(e) Suppose $B = (B(t))_{0\leq t\leq 1}$ is a Brownian bridge, and Z is a standard normally distributed random variable that is independent of B. Then

$$W(t) := B(t) + tZ, \quad 0 \leq t \leq 1,$$

is a Wiener process on $[0, 1]$.

Answers to the Self-Questions

Answer 1 If $t_1 = t_2$ then $W(t_2) - W(t_1) = 0$ is a constant random variable, whose generated σ-field is equal to $\{\emptyset, \Omega\}$. This σ-field is independent of all sub-σ-fields of $\mathcal{A}$.

Answer 2 Because the components are normally distributed according to Corollary 15.2 (a) and independent in view of Definition 15.1 (c).

Answer 3 If $t_1 = 0$, then (15.4) follows in the case $k = 1$, because there is the one-point distribution δ_0 on both sides of the convergence arrow. If $k \geq 2$, (15.4) follows using the Cramér–Wold device (Theorem 6.18), because the first component of the vector on the left-hand side of (15.4) is equal to 0, and the covariance matrix Σ has the property $\min(t_1, t_j) = \min(0, t_j) = 0$ for each $j \in \{1, \ldots, k\}$.

Answer 4 Yes, due to the subsequence criterion (Theorem 14.5).

Answer 5 Fix any $j \in \{2, \ldots, k-1\}$. Since $m \geq n\delta$, it follows that

$$t_j - t_{j-1} = \frac{jm - (j-1)m}{n} = \frac{m}{n} \geq \delta.$$

Answer 6 If each of the events B_k and $\{|S_n| < \alpha\}$ occurs, then $|S_n - S_k| > 2\alpha$. Otherwise, the triangle inequality would yield $|S_k| = |S_k - S_n + S_n| \leq |S_n - S_k| + |S_n| < 3\alpha$.

Answer 7 If $|S_n - S_k| > 2\alpha$, then at least one of the events $\{|S_n| \geq \alpha\}$ and $\{|S_k| \geq \alpha\}$ must occur because of the triangle inequality.

Answer 8 You may choose any number a greater than 3. It is only important that all but finitely many terms of the sequence $(\mathbb{P}(|Y_k| > \lambda))$ are less than or equal to a/λ^4.

Answer 9 If $x, y \in C$ with $\|x - y\| \leq \varepsilon$, then

$$h(x) = \max_{0 \leq t \leq 1} x(t) \leq \max_{0 \leq t \leq 1} \left(y(t) + \varepsilon \right) \leq h(y) + \varepsilon.$$

From reasons of symmetry, it follows that $|h(x) - h(y)| \leq \varepsilon$.

Answer 10 Because L^2-convergence implies convergence in distribution, and the limit distribution of that type of convergence is uniquely determined.

Answer 11 Using $\mathbb{E}(N_i N_j) = \delta_{i,j}$ we obtain

$$\mathbb{E}\left(\tilde{W}_n^2(t)\right) = \sum_{i=1}^{n} \sum_{j=1}^{n} \sqrt{\lambda_i \lambda_j} \varphi_i(t) \varphi_j(t) \delta_{i,j} = \sum_{i=1}^{n} \lambda_i \varphi_i^2(t).$$

Since $\varphi_i^2(t) \leq 2$, the assertion follows.

Answer 12 For $x, y \in C$ we have $|h(x)(t) - h(y)(t)| = |x(t) - y(t) - t(x(1) - y(1))| \leq 2\|x - y\|_\infty$.

Answer 13 The statement follows from the addition Theorem 5.13 for the multivariate normal distribution.

Answer 14 For $x, y \in C[0, 1]$ and $t \geq 0$ we have $|h(x)(t) - h(y)(t)| \leq (1+t)\left|x\left(\frac{t}{1+t}\right) - y\left(\frac{t}{1+t}\right)\right|$ and thus $\max_{0 \leq t \leq j} |h(x)(t) - h(y)(t)| \leq (1+j)\|x - y\|_\infty$. It follows that

$$\rho\big(h(x), h(y)\big) \leq \|x - y\|_\infty \sum_{j=1}^{\infty} \frac{j+1}{2^j} = 3\|x - y\|_\infty.$$

Answer 15 Note that $X_j = W\big(n + j2^{-m}\big) - W\big(n + (j-1)2^{-m}\big)$, $j \in \{1, \dots, 2^m\}$.

Problems

15.1 Problem Let A be the matrix of ones and zeros in (15.2) and D the diagonal matrix $\mathrm{diag}(t_1, t_2 - t_1, t_3 - t_2, \dots, t_k - t_{k-1})$. Show that

$$ADA^\top = (\min(t_i, t_j))_{1 \leq i, j \leq k}.$$

15.2 Problem Let X_n be the partial sum process defined in (15.3). Moreover, let $k \in \mathbb{N}$ and $t_1, \dots, t_k$ with $0 \leq t_1 < \dots < t_k \leq 1$. Prove:

$$\frac{1}{\sigma\sqrt{n}}\big(S_{\lfloor nt_1 \rfloor}, S_{\lfloor nt_2 \rfloor} - S_{\lfloor nt_1 \rfloor}, \dots, S_{\lfloor nt_k \rfloor} - S_{\lfloor nt_{k-1} \rfloor}\big) \xrightarrow{\mathcal{D}} N_k(0_k, D), \tag{15.28}$$

where $D = \mathrm{diag}(t_1, t_2 - t_1, t_3 - t_2, \dots, t_k - t_{k-1})$.

Hint Use the Lindeberg–Lévy central limit theorem, Slutsky's lemma, and Theorem 14.26.

15.3 Problem Let $k \geq 3$ and $t_0, \dots, t_k$ with $0 = t_0 < t_1 < \dots < t_k = 1$. Further, let $\varepsilon > 0$ and $0 < \delta < 1$. Show: If

$$\min_{2 \leq j \leq k-1} (t_j - t_{j-1}) \geq \delta, \tag{15.29}$$

then for every $x \in C[0, 1]$:

$$w_x(\delta) \leq 3 \max_{1 \leq j \leq k} \sup_{t_{j-1} \leq s \leq t_j} |x(s) - x(t_{j-1})|. \tag{15.30}$$

From this, conclude the statement of Lemma 15.4.

Hint Let $I_j := [t_{j-1}, t_j]$, $j \in \{1, \dots, k\}$. In which of the intervals $I_1, \dots, I_k$ can s and t lie, so that the inequality $|s - t| \leq \delta$ is fulfilled?

15.4 Problem Let $Z_1, Z_2, \dots$ be i.i.d. random variables with $\mathbb{P}(Z_1 = \pm 1) = \frac{1}{2}$. Further, let $S_0 := 0$, $S_n := \sum_{j=1}^{n} Z_j$ for $n \geq 1$ and $M_n := \max_{j=0,\dots,n} S_j$. Prove the following statements:

(a) $\mathbb{P}(S_n > k) + \mathbb{P}(S_n = k)$ for each $k \in \{0, \dots, n\}$.

> **Hint** Decompose the event $\{M_n \geq k\}$ according to whether, in addition, $S_n = k$, $S_n < k$, or $S_n > k$ holds. Consider $\{(j, S_j) : 0 \leq j \leq n\}$ as a random walk and use a symmetry argument.

(b) With a standard normally distributed random variable N, we have

$$\frac{M_n}{\sqrt{n}} \xrightarrow{\mathcal{D}} |N| \quad \text{as } n \to \infty.$$

15.5 Problem Let $a, b \in \mathbb{R}$ with $a < b$ and $(W(t))_{0 \leq t \leq 1}$ be a Wiener process on a probability space $(\Omega, \mathcal{A}, \mathbb{P})$. Show:

$$\mathbb{P}\left(\bigcap_{\{(s,t):a \leq s < t \leq b\}} \{W(s) \leq W(t)\} \right) = 0.$$

Why does this countable intersection belong to the σ-field $\mathcal{A}$?

Hint For u, v with $u < v$, we have $\mathbb{P}(W(u) \leq W(v)) = \frac{1}{2}$.

15.6 Problem On a common probability space $(\Omega, \mathcal{A}, \mathbb{P})$, let $W = (W(t))_{0 \leq t \leq 1}$ be a Wiener process and U be a random variable that is independent of W and uniformly distributed on $[0, 1]$. Further, let $Y(t) := W(t)$ if $U \neq t$, and $Y(t) := 0$ if $U = t$. Prove the following claims:

(a) The processes $Y := (Y(t))_{0 \leq t \leq 1}$ and W have the same finite-dimensional distributions.

(b) The paths $Y(\omega)$, $\omega \in \Omega$, of Y are discontinuous with probability one.

Conclusion: The fidis are not sufficient to assess the path behavior.

15.7 Problem Let $(W(t))_{0 \leq t \leq 1}$ be a Wiener process. Prove that

$$\min_{0 \leq t \leq 1} W(t) \sim -|N|,$$

where $N \sim N(0, 1)$.

15.8 Problem State and prove a more general reproduction theorem for Brownian bridges compared to Theorem 15.21.

15.9 Problem Show the inequality

$$\mathbb{E}\left| \int_0^1 \left(\widetilde{W}^2(t) - \widetilde{W}_n^2(t) \right) dt \right| \leq \int_0^1 \sqrt{\mathbb{E}\left[(\widetilde{W}(t) - \widetilde{W}_n(t))^2 \right]} \sqrt{\mathbb{E}\left[(\widetilde{W}(t) + \widetilde{W}_n(t))^2 \right]} \, dt$$

used in the proof of Corollary 15.17.

15.10 Problem Consider the integral operator A associated with the kernel $K(s, t) := \min(s, t)$, that is,

$$Af(s) := \int_0^1 K(s, t) f(t) \, dt, \quad 0 \leq s \leq 1,$$

on the space L^2 of square integrable functions on $[0, 1]$. Prove the following statements:

(a) For each $\ell \geq 1$, $\lambda_\ell := \left(\left(\ell - \frac{1}{2} \right) \pi \right)^{-2}$ is an eigenvalue of A with associated normalized eigenfunction

$$\varphi_\ell(s) := \sqrt{2} \sin \left(\left(\ell - \frac{1}{2} \right) \pi s \right), \quad 0 \leq s \leq 1.$$

Hint Differentiate both sides of the equation $\lambda f(s) = \int_0^1 \min(s, t) f(t) \, dt$ twice.

(b) All non-zero eigenvalues of A are listed in (a).

Hint Proceed analogously as after (8.42).

15.11 Problem Let $((W(t))_{t \geq 0}$ be a Wiener process and $a > 0$. Prove: The process defined by

$$W^*(t) := \frac{1}{\sqrt{a}} W(at), \quad t \geq 0,$$

W^* is also a Wiener process.

15.12 Problem Let $(W(t))_{t \geq 0}$ be a Wiener process and $r > 0$. Show that the process $\widetilde{W}$ defined by

$$\widetilde{W}(t) := W(t + r) - W(r), \quad t \geq 0,$$

is also a Wiener process.

15.13 Problem Let $(W(t))_{t \geq 0}$ be a Wiener process and

$$\widehat{W}(t) := t W\left(\frac{1}{t}\right), \quad t > 0,$$

as well as $\widehat{W}(0) := 0$. Prove that $\widehat{W} = (\widehat{W}(t))_{t \geq 0}$ is also a Wiener process.

Hint Use Theorem 15.23.

15.14 Problem Let $(W(t))_{t \geq 0}$ be a Wiener process and

$$B(t) := (1 - t) W\left(\frac{t}{1 - t}\right), \quad 0 \leq t < 1,$$

as well as $B(1) := 0$. Show that the process $B = (B(t))_{0 \leq t \leq 1}$ is a Brownian bridge.

Hint Use Theorem 15.23.

15.15 Problem Let $B = (B(t))_{0 \leq t \leq 1}$ be a Brownian bridge and Z a standard normally distributed random variable that is independent of B. Prove that

$$W(t) := B(t) + t Z, \quad 0 \leq t \leq 1,$$

defines a Wiener process on $[0, 1]$.

15.16 Problem Let $W = (W(t))_{t \geq 0}$ be a Wiener process on a probability space $(\Omega, \mathcal{A}, \mathbb{P})$. For $a > 0$ and $\omega \in \Omega$, let

$$\left(\int_0^a W(t)\, dt\right)(\omega) := \int_0^a W(t, \omega)\, dt$$

be the integral of W from 0 to a, defined as a pathwise Riemann integral. Show that

$$\int_0^a W(t)\,dt \sim \mathrm{N}\!\left(0, \frac{a^3}{3}\right).$$

The stochastic process $[0, \infty) \ni a \mapsto \int_0^a W(t)\,dt$ is called the *integrated Wiener process*.

Hint Consider Riemann sums $\sum_{j=1}^{k_n} W(t_{n,j})(t_{n,j+1} - t_{n,j})$.

The Space D[0, 1], Empirical Processes

16

As we have already seen at the beginning of Chap. 13, the space $C = C[0, 1]$ is unsuitable as a range for random functions whose realizations exhibit discontinuities. An example is the uniform empirical process

$$B_n(t) = \sqrt{n}\big(\widehat{F}_n(t) - t\big), \quad 0 \le t \le 1, \tag{16.1}$$

where $\widehat{F}_n(t) = \frac{1}{n} \sum_{j=1}^{n} \mathbf{1}\{X_j \le t\}, 0 \le t \le 1$, is the empirical distribution function of independent random variables $X_1, \ldots, X_n$, each having a uniform distribution on the unit interval $[0, 1]$, cf. Chap. 7. Consider these random variables as the initial segment of length n of a sequence $(X_j)_{j \ge 1}$ defined on a common probability space $(\Omega, \mathcal{A}, \mathbb{P})$. According to Problem 7.4, for each $k \ge 1$ and any choice of $t_1, \ldots, t_k$ with $0 \le t_1 < \ldots < t_k \le 1$, we have $\big(B_n(t_1), \ldots, B_n(t_k)\big) \xrightarrow{\mathcal{D}} N_k\big(0_k, (t_i \wedge t_j - t_i t_j)_{1 \le i, j \le k}\big)$. The variances and covariances of the asymptotic normal distribution are thus those of a Brownian bridge.

If we add the suppressed argument $\omega \in \Omega$ in (16.1), that is, if we write

$$B_n(\omega, t) := \sqrt{n}\big(\widehat{F}_n(\omega, t) - t\big), \quad 0 \le t \le 1, \ \omega \in \Omega,$$

where $\widehat{F}_n(\omega, t) = \frac{1}{n} \sum_{j=1}^{n} \mathbf{1}\{X_j(\omega) \le t\}$, then $B_n(\omega, \cdot)$ is a right-continuous function defined on $[0, 1]$, whose left-hand limits exist at every point $t \in (0, 1]$ (see Fig. 13.1). We will find that this collection of functions, denoted D[0, 1] in the following, can be equipped with a suitable metric that renders D[0, 1] a complete separable metric space. Moreover, $\Omega \ni \omega \mapsto B_n(\omega, \cdot)$ is a $(\mathcal{A}, \mathcal{B}(D[0, 1]))$-measurable mapping. Here, $\mathcal{B}(D[0, 1])$ denotes the σ-field of Borel sets of this metric space. We will further see that in the space D[0, 1] the sequence (B_n) of uniform empirical processes converges in distribution to a suitably defined Brownian bridge B on D[0, 1], and we will get to know

N. Henze, *Asymptotic Stochastics*, Mathematics Study Resources 10,
https://doi.org/10.1007/978-3-662-68923-3_16

various statistical applications of this fact to nonparametric one- and two-sample problems. Another important result of this chapter will be Donsker's theorem in D[0, 1].

The convergence in distribution in the space D[0, 1] involves considerable technical effort. We will state various auxiliary results without proof. Details can be found, e.g., in [BI2], pp. 121–146, or in [PAR], pp. 231–254.

For a function $x : [0, 1] \to \mathbb{R}$ and any t with $0 \le t < 1$, let

$$x(t+) := \lim_{s \downarrow t} x(s)$$

be the *right-hand limit* of x at the point t. Similarly, for any t with $0 < t \le 1$,

$$x(t-) := \lim_{s \uparrow t} x(s)$$

is the *left-hand limit* of x at t. As the examples $x(t) := \sin\left(\frac{1}{t}\right)$ for $0 < t \le 1$ and $x(0) := 0$ as well as $x(t) := \sin\left(\frac{1}{1-t}\right)$ for $0 \le t < 1$ and $x(1) := 0$ show, such limits do not necessarily have to exist. If we demand the existence of all these limits as well as the right-continuity of x at every point $t \in [0, 1)$, we arrive at the so-called *Càdlàg space* D[0, 1]. The acronym *Càdlàg* is derived from the French: continue **à d**roite, limites **à g**auche.

16.1 Definition (Càdlàg Space D[0, 1])
Let

$$D[0, 1] := \left\{ x : [0, 1] \to \mathbb{R} \,\middle|\, x(t+) = x(t) \;\forall\, t \in [0, 1),\; x(t-) \text{ exists } \forall\, t \in (0, 1] \right\}$$

be the collection of all right-continuous functions with existing left-hand limits. In the following, we mostly write $D := D[0, 1]$.

Obviously, C is a subset of D. We will soon see that every function in D is Borel measurable and bounded. To quantify the fluctuations of the functions in D over subsets of [0, 1], we set for $x \in D$ and a non-empty subset T of [0, 1]

$$w_x(T) := w(x, T) := \sup_{s,t \in T} |x(s) - x(t)|.$$

The connection with the modulus of continuity of a function $x \in C[0, 1]$ as defined in (13.5) is given by

$$w_x(\delta) = \sup_{|u-v| \le \delta} |x(u) - x(v)| = \sup_{0 \le t \le 1-\delta} w_x([t, t + \delta]).$$

Note the small notational difference between w and w, because both functions have different arguments.

Fig. 16.1 In each of the
intervals (t_j, t_{j+1}),
$j \in \{0, \ldots, k-1\}$, the
function values $x(t)$ are within
the respective gray strip

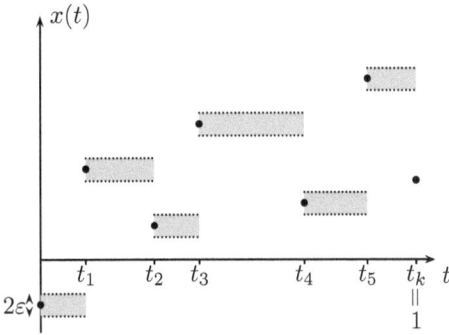

The following result is important to understand properties of Càdlàg functions.

16.2 Lemma *For each $x \in D$ and each $\varepsilon > 0$ there is a decomposition of $[0, 1]$ of the form $0 = t_0 < t_1 < \ldots < t_k = 1$ such that*

$$\mathrm{w}_x\big([t_{j-1}, t_j)\big) < \varepsilon, \quad j \in \{1, 2, \ldots, k\}. \tag{16.2}$$

Proof Let s be the supremum of those $t \in [0, 1]$ for which the half-open interval $[0, t)$ can be decomposed into finitely many intervals of the form $[t_{j-1}, t_j)$ satisfying (16.2). Because $x(0) = x(0+)$, it follows that $s > 0$. Since the left-hand limit $x(s-)$ exists, the interval $[0, s)$ itself can be decomposed in that way. The case $s < 1$ is not possible, because then we would have $x(s) = x(s+)$. □

Figure 16.1 illustrates the statement of Lemma 16.2. In each of the open intervals (t_j, t_{j+1}) with $j \in \{0, \ldots, k-1\}$, the function values $x(t)$ are within the respective gray strip. The values of x at the points $t_0, \ldots, t_k$ are emphasized by means of small filled circles. From Lemma 16.2 it follows that a function x in D can have at most countably many jumps (Problem 16.1). Furthermore, x is the uniform limit of a sequence of functions of the form

$$z(t) := \sum_{j=1}^{k-1} x(t_j)\mathbf{1}_{[t_j, t_{j+1})}(t) + x(1)\mathbf{1}_{\{t_k\}}(t), \quad 0 \le t \le 1$$

$(0 = t_0 < \ldots < t_k = 1)$ and thus Borel measurable. Last but not least, it turns out that every function in D is bounded, that is, we have

$$\|x\| := \sup_{0 \le t \le 1} |x(t)| < \infty.$$

Here, $\|x\|$ has the meaning just assigned for the entire chapter.

Self-question 1: Why is every function in D bounded?

A function $x : [0, 1] \to \mathbb{R}$ is continuous and thus belongs to C[0, 1] if and only if $\lim_{\delta \to 0} w_x(\delta) = 0$. A corresponding characterization of the functions in D[0, 1] is achieved by a modification of the modulus of continuity w_x. For this purpose, we set for each δ with $0 < \delta < 1$

$$w'_x(\delta) := \inf\left\{ \max_{1 \leq i \leq k} w_x([t_{i-1}, t_i)) \,\Big|\, k \in \mathbb{N},\ 0 = t_0 < \ldots < t_k = 1,\ \min_{1 \leq i \leq k} (t_i - t_{i-1}) > \delta \right\}.$$

The function $(0, 1) \ni \delta \mapsto w'_x(\delta)$ is called the *Càdlàg modulus* (of x). Note that the infimum runs over all "δ-sparse decompositions" (that is, decompositions with $t_i - t_{i-1} > \delta$ for each i) of [0, 1], and that $w'_x(\delta)$ does not depend on the value $x(1)$.

According to Lemma 16.2, we have $\lim_{\delta \to 0} w'_x(\delta) = 0$ for each $x \in D$. Conversely, if $x : [0, 1] \to \mathbb{R}$ is any function, then $\lim_{\delta \to 0} w'_x(\delta) = 0$ implies $x \in D$ (Problem 16.2).

There are some relations between the Càdlàg modulus w'_x and the modulus of continuity w_x (Problem 16.3): First, we have

$$w'_x(\delta) \leq w_x(2\delta), \quad \text{if } \delta < \frac{1}{2}.$$

If

$$H(x) := \sup_{0 < t \leq 1} |x(t) - x(t-)| \tag{16.3}$$

denotes the maximum (absolute) jump in x, then

$$w_x(\delta) \leq 2w'_x(\delta) + H(x), \quad 0 < \delta < 1.$$

Self-question 2: Why is the supremum in (16.3) attained?

Since each of the functions x in D is bounded, we could equip also the set D with the supremum metric

$$\rho(x, y) := \|x - y\| := \sup_{0 \leq s, t \leq 1} |x(s) - x(t)|, \quad x, y \in D,$$

just as was done in Sect. 13.9 for C[0, 1]. With regard to convergence in distribution, the metric space (D, ρ) is extensively studied in Chapter V of [POL]. It is not separable, because $\rho(x_u, x_v) = 1$ if $u \neq v$ for the functions x_u defined by $x_u := \mathbf{1}_{[u,1]}, 0 \leq u \leq 1$. According to an idea of the Ukrainian mathematician A.V. Skorokhod, the functions x_u and x_v should have a small distance in the case $u \approx v$, We should therefore allow uniformly small deformations of the time scale. To this end, let

$$\mathcal{G} := \{g : [0, 1] \rightarrow [0, 1] : g \text{ continuous, increasing, bijective}\}$$

be the set of all continuous, increasing and bijective mappings of the unit interval onto itself. The set $\mathcal{G}$ is a group with respect to forming compositions (denoted by $\circ$), and we have $g(0) = 0$ and $g(1) = 1$. We write I for the identity map on [0, 1], that is, $\mathrm{I}(t) := t$, $0 \leq t \leq 1$, and (as a reminder) $\|x\| = \sup_{0 \leq t \leq 1} |x(t)|$ for each bounded function $x :$ $[0, 1] \rightarrow \mathbb{R}$. With these preliminaries, we set

$$d_S(x, y) := \inf_{g \in \mathcal{G}} \max \left(\|x \circ g - y\|, \|g - \mathrm{I}\| \right), \quad x, y \in \mathrm{D}. \tag{16.4}$$

For each $\varepsilon > 0$, we thus have: If $d_S(x, y) < \varepsilon$, then there is a $g \in \mathcal{G}$ such that:

$$\sup_{0 \leq t \leq 1} |x(g(t)) - y(t)| < \varepsilon, \quad \sup_{0 \leq t \leq 1} |g(t) - t| < \varepsilon.$$

16.3 Definition and Theorem (Skorokhod Metric)

The function $d_S : \mathrm{D} \times \mathrm{D} \rightarrow \mathbb{R}$ is a metric on D (so-called *Skorokhod metric*). The metric space (D, d_S) is separable, but not complete.

Proof That d_S is indeed a metric on D is the subject of Problem 16.4. A countable subset of D that is dense with respect to d_S is given by all functions of the type

$$\sum_{j=1}^{k} q_j \mathbf{1}_{[t_j, t_{j+1})}(t) + q_0 \mathbf{1}_{\{t_k\}}(t), \quad 0 \leq t \leq 1.$$

Here, k is an integer, and $0 = t_0 < \ldots < t_k = 1$ is a decomposition of [0, 1] with rational numbers $t_1, \ldots, t_{k-1}$. Furthermore, $q_0, \ldots, q_k$ are rational numbers (see, e.g., [BI2], pp. 127–128). The following example shows that the metric space (D, d_S) is not complete. To this end, we define a sequence (x_n) of functions in D by $x_n := \mathbf{1}_{[0,a_n)}$ and $a_n := 1/2^n, n \geq 1$. If we choose a function g_n from $\mathcal{G}$ so that g_n is linear on each of the intervals $[0, a_n]$ and $[a_n, 1]$ and $g(a_n) := a_{n+1}$ (see Fig. 16.2), then $\|g_n - \mathrm{I}\| = a_{n+1}$ and $x_{n+1} = \mathbf{1}_{[0,a_{n+1})}$ as well as

$$x_{n+1} \circ g_n = \mathbf{1}_{[0,a_{n+1})} \circ g_n = \mathbf{1}_{[0,a_n)} = x_n.$$

Fig. 16.2 (x_n) is a Cauchy sequence that does not have a limit in D

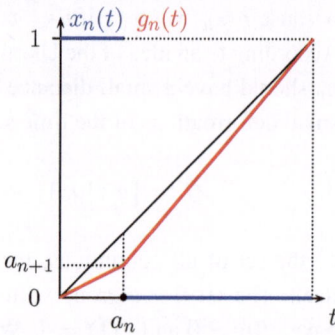

Therefore, $\|x_{n+1} \circ g_n - x_n\| = 0$ and hence $d_S(x_n, x_{n+1}) \le a_{n+1} = \frac{1}{2^{n+1}}$. The sequence (x_n) is thus a Cauchy sequence in (D, d_S).

Self-question 3: Why does $d_S(x_n, x_m) \to 0$ hold as $m, n \to \infty$?

Obviously, $\lim_{n \to \infty} x_n(t) = 0$ for each $t > 0$. If $x \in D$ denotes the zero function, that is, $x(t) := 0, 0 \le t \le 1$, we obtain $d_S(x_n, x) = 1$ for every $n \ge 1$, because $\|x_n \circ g - x\| = 1$ for each $g \in \mathcal{G}$ due to $g(0) = 0$. Thus, (x_n) does not have a limit in D.   □

If we choose $g = \mathrm{I}$ in (16.4), it follows that

$$d_S(x, y) \le \rho(x, y) = \|x - y\|. \tag{16.5}$$

Convergence in the Skorokhod metric is therefore weaker than uniform convergence. Because

$$\sup_{0 \le t \le 1} |x(g(t)) - y(t)| = \sup_{0 \le t \le 1} |x(t) - y(g^{-1}(t))|, \quad x, y \in D, \ g \in \mathcal{G},$$

and since the mapping $\mathcal{G} \ni g \mapsto g^{-1}$ is a bijection onto $\mathcal{G}$, we obtain

$$d_S(x, y) = \inf_{g \in \mathcal{G}} \max\left(\|x - y \circ g\|, \|g - \mathrm{I}\|\right), \quad x, y \in D.$$

If $\lim_{n \to \infty} d_S(x_n, x) = 0$ there is thus a sequence (g_n) in $\mathcal{G}$ with $\|x_n - x \circ g_n\| \to 0$ and $\|g_n - \mathrm{I}\| \to 0$. From the inequalities

$$|x_n(t) - x(t)| \le |x_n(t) - x(g_n(t))| + |x(g_n(t)) - x(t)| \tag{16.6}$$

$$\le \|x_n - x \circ g_n\| + w_x(\|g_n - \mathrm{I}\|),$$

which are valid for each $t \in [0, 1]$, we can thus draw the following conclusions:

16.4 Corollary *Let (x_n) be a sequence in D and $x \in$ D. If $d_S(x_n, x) \to 0$, then the following hold:*

(a) *$x_n(t) \to x(t)$ for every continuity point t of x,*
(b) *$x_n(t) \to x(t)$ for every t except for at most countably many exceptions,*
(c) *$\|x_n - x\| \to 0$, if x is continuous.*

Self-question 4: Why does Conclusion 16.4 (a) apply?

Conclusion 16.4 (c) shows that the Skorokhod topology, when restricted to the subset C, coincides with the topology of uniform convergence on C. Thus, the σ-field $\mathcal{B}(C)$ of Borel sets in C is equal to the trace σ-field $\mathcal{B}(D) \cap C$.

By the following modification d_S° of the Skorokhod metric d_S, D becomes a separable and at the same time *complete* metric space The metrics d_S and d_S° are topologically equivalent, that is, they generate the same collections of open sets. To define the metric d_S°, we first set for each g in $\mathcal{G}$

$$\|g\|^\circ := \sup_{s<t} \left| \log \frac{g(t) - g(s)}{t - s} \right|. \tag{16.7}$$

Whereas in the Definition (16.4) of d_S, the closeness of a time-deformation g in $\mathcal{G}$ to the identity I is measured by the supremum distance $\|g - I\|$, now the slopes $(g(t) - g(s))/(t - s)$ of chords are considered. If $g = I$ then all these slopes are equal to one and thus their logarithm is equal to zero, so that $\|I\|^\circ = 0$. It should be emphasized that the supremum in (16.7) can be infinite. However, such time-deformations do not enter into the following definition, because we set

$$d_S^\circ(x, y) := \inf_{g\in\mathcal{G}} \max \left(\|g\|^\circ, \|x \circ g - y\| \right).$$

Regarding $d_S(x, y)$ and $d_S^\circ(x, y)$, we have

$$d_S(x, y) \le e^{d_S^\circ(x,y)} - 1$$

(Problem 16.5). Therefore, $d_S^\circ(x_n, x) \to 0$ implies $d_S(x_n, x) \to 0$.

16.5 Theorem (The Space (D, d_S°)**)**

(a) d_S° *is a metric on* D *equivalent to* d_S.
(b) *The metric space* (D, d_S°) *is separable and complete.*

Proof That d_S° is a metric on D follows from Problem 16.6 by direct calculation. The proof of the equivalence of d_S and d_S° as well as the properties noted in (b) can be found, e.g., in [BI2], pp. 126–129.   □

The following two results characterize the relatively compact sets in the space D by means of the Càdlàg modulus. For a proof, see, e.g., [BI2], pp. 130–133.

16.6 Theorem ("Arzelà–Ascoli in D[0, 1]**")**
A set $A \subset D[0, 1]$ *is relatively compact with respect to the Skorokhod topology if and only if these two conditions hold:*

$$\sup_{x \in A} \|x\| < \infty, \tag{16.8}$$

$$\lim_{\delta \to 0} \sup_{x \in A} w_x'(\delta) = 0. \tag{16.9}$$

Note that, in contrast to condition (13.7) of the Arzelà–Ascoli theorem in the space C[0, 1] (Theorem 13.11), it is required that all graphs $\{(t, x(t)) : 0 \le t \le 1\}$ with $x \in A$ are subsets of the compact rectangle $[0, 1] \times [-M, M]$ for some $M \in (0, \infty)$. The example of the sequence (x_n) of functions with $x_n := n\mathbf{1}_{[0.5,1)}, n \ge 1$, shows that, instead of (16.8), the weaker condition $\sup_{x \in A} |x(t_0)| < \infty$ for only one $t_0 \in [0, 1]$ is not sufficient. With $A := \{x_n : n \ge 1\}$, both (16.9) and $\sup_{n \ge 1} |x_n(0.25)| < \infty$ hold, but the set A is not relatively compact.

Self-question 5: Why does condition (16.9) hold for the set A?

The second characterization of relatively compact sets of functions in D uses a modification of the Càdlàg modulus w_x'. For each $\delta \in (0, 1)$ and $x \in D$, let

$$w_x''(\delta) := \sup_{t_1 \le t \le t_2, t_2 - t_1 \le \delta} \left\{ |x(t) - x(t_1)| \wedge |x(t_2) - x(t)| \right\}. \tag{16.10}$$

Here, the supremum is over all triples $(t_1, t, t_2) \in [0, 1]^3$ with the stated restrictions.

16.7 Theorem (Characterization of Relatively Compact Sets of Functions in D**)**
A set $A \subset D[0, 1]$ *is relatively compact with respect to the Skorokhod topology if and only*

if (16.8) *holds and the following conditions are satisfied:*

$$\lim_{\delta \to 0} \sup_{x \in A} w_x''(\delta) = 0, \tag{16.11}$$

$$\lim_{\delta \to 0} \sup_{x \in A} |x(\delta) - x(0)| = 0, \tag{16.12}$$

$$\lim_{\delta \to 0} \sup_{x \in A} |x(1-) - x(1 - \delta)| = 0. \tag{16.13}$$

We have $w_x''(\delta) \le w_x'(\delta)$ (Problem 16.7), but an inequality of the type $w_x'(\delta) \le a w_x''(\delta)$ for some $a \in (0, \infty)$ (not depending on x and δ) cannot hold, because then (16.8) and (16.11) would already imply the relative compactness of A, and the conditions (16.12) and (16.13) would be unnecessary. For the sequence (x_n) defined by $x_n := 1_{[0,1/n)}, n \ge 1$, we have $w_{x_n}''(\delta) = 0$ for every $\delta \in (0, 1)$ and $w_{x_n}'(\delta) = 1$ if $n \ge 1/\delta$.

Self-question 6: Why does $w_{x_n}''(\delta) = 0$ hold for every $\delta \in (0, 1)$?

By Theorem 14.19, one can prove weak convergence $P_n \xrightarrow{\mathcal{D}} P$ of probability measures on the σ-field $\mathcal{B}(C)$ of Borel sets in the space $C[0, 1]$ by showing the relative compactness of the set $\{P_n : n \ge 1\}$ and the fidi convergence $P_n \xrightarrow{\mathcal{D}_{\text{fidi}}} P$, that is,

$$P_n \pi_{t_1,\dots,t_k}^{-1} \xrightarrow{\mathcal{D}} P \pi_{t_1,\dots,t_k}^{-1}$$

for each $k \ge 1$ and $t_1, \dots, t_k$ with $0 \le t_1 < t_2 \dots < t_k = 1$, cf. (14.12).

In the space $D[0, 1]$, the situation becomes more complicated because the projections

$$\pi_{t_1,\dots,t_k}(x) := \big(x(t_1), \dots, x(t_k)\big), \quad x \in D,$$

are generally no longer continuous mappings. Here, the word *continuous* always refers to the Skorokhod topology for the rest of this chapter.

16.8 Theorem (Properties of Projections in the Space D)

(a) *The projections π_0 and π_1 are continuous.*

(b) *If $0 < t < 1$ then π_t is continuous at $x \in D$ if and only if x is continuous at t.*

(c) *Every projection $\pi_{t_1,\dots,t_k} : D \to \mathbb{R}^k$ is $(\mathcal{B}(D), \mathcal{B}^k)$-measurable.*

Proof

(a) follows from the fact that every time-deformation $g \in \mathcal{G}$ satisfies $g(0) = 0$ and $g(1) = 1$.

(b) Let $0 < t < 1$. If t is a point of continuity of x, then $d_S(x_n, x) \to 0$ implies $\pi_t(x_n) = x_n(t) \to x(t) = \pi_t(x)$ according to Corollary 16.4 (a) . We prove the converse by contraposition and assume that x is not continuous at the point t. For each n with $t - \frac{1}{n} > 0$, let $g_n \in \mathcal{G}$ be defined by $g_n(t) := t - \frac{1}{n}$, and let g_n be linear on $[0, t]$ and on $[t, 1]$ (cf. Fig. 16.2 with the modification that a_n is replaced by t and a_{n+1} by $t - \frac{1}{n}$). If we define $x_n(s) := x(g_n(s))$, $0 \le s \le 1$, then $d_S(x_n, x) \to 0$, but $x_n(t)$ does not converge to $x(t)$.

(c) Since the Borel σ-field in $\mathbb{R}^k$ is generated by the collections of d-dimensional rectangles $[a_1, b_1] \times \cdots \times [a_k, b_k]$ with $a_j, b_j \in \mathbb{R}$ and $a_j < b_j$ for each $j \in \{1, \ldots, k\}$, it suffices to consider the case $k = 1$. Furthermore, we can assume $t < 1$ because π_1 is continuous and thus measurable. For each $\varepsilon > 0$ with $t + \varepsilon \le 1$, let

$$h_\varepsilon(x) := \int_t^{t+\varepsilon} x(s)\,ds, \qquad x \in D.$$

If $d_S(x_n, x) \to 0$ then $\lim_{n \to \infty} h_\varepsilon(x_n) = h_\varepsilon(x)$ (cf. Problem 16.8). The mapping $D \ni x \mapsto \int_t^{t+\varepsilon} x(s)\,ds$ is therefore continuous and thus $(\mathcal{B}(D), \mathcal{B}^k)$-measurable. Due to the right-continuity of x, we have $h_{k^{-1}}(x) \to \pi_t(x)$ as $k \to \infty$. Being the limit of $(\mathcal{B}(D), \mathcal{B}^k)$-measurable mappings, π_t is also $(\mathcal{B}(D), \mathcal{B}^k)$-measurable.

$\square$

Since all projections are measurable mappings, we can consider finite-dimensional sets, that is, subsets of D of the form $\pi_{t_1,\ldots,t_k}^{-1}(H)$, where $k \ge 1$, $0 \le t_1 < \ldots < t_k \le 1$, and $H \in \mathcal{B}^k$, as we already did in the space C. If T is any non-empty subset of $[0, 1]$, we set

$$\mathcal{D}_T := \left\{ \pi_{t_1,\ldots,t_k}^{-1}(H) : k \ge 1, \ 0 \le t_1 < \ldots < t_k \le 1, \ t_1, \ldots, t_k \in T, \ H \in \mathcal{B}^k \right\}$$

in what follows.

16.9 Theorem (Generating Systems of $\mathcal{B}(D)$)
Let $T \subset [0, 1]$. If $1 \in T$ and T is dense in $[0, 1]$, then:

(a) $\mathcal{D}_T$ is a separating class for $\mathcal{P}$ in the sense of Definition 13.7.
(b) $\mathcal{B}(D) = \sigma(\mathcal{D}_T)$.

Proof See, e.g., [BI2], pp. 134–135.

In particular, the σ-field $\mathcal{B}(D)$ of Borel sets in D is generated by the collection $\mathcal{D}_{[0,1]}$ of all finite-dimensional sets. Note that due to the right-continuity of the functions in D and the condition $\overline{T} = [0, 1]$, the projection π_0 is measurable with respect to the σ-field $\sigma(\mathcal{D}_T)$. Thus, we may assume w.l.o.g. that $0 \in T$.

Suppose $P, P_1, P_2, \ldots$ are probability measures on $\mathcal{B}(D)$. To state a criterion for weak convergence $P_n \xrightarrow{D} P$ in the space D, set

$$T_P := \{t \in [0, 1] : \pi_t \text{ is continuous on the complement of a } P\text{-null set}\}.$$

If $t \in T_P$, then there is a set $N \in \mathcal{B}(D)$ with $P(N) = 0$, and $\pi_t : D \to \mathbb{R}$ is continuous at every point $x \in D \setminus N$. Since π_0 and π_1 are continuous, we have $\{0, 1\} \subset T_P$. Let for $t \in (0, 1)$

$$J_t := \{x \in D : x(t) \neq x(t-)\}$$

denote the set of those $x \in D$ that are discontinuous at the point t. Then, by Theorem 16.8 (b), the equivalence

$$t \in T_P \iff P(J_t) = 0 \tag{16.14}$$

holds for each t with $0 < t < 1$.   □

16.10 Lemma *The set $[0, 1] \setminus T_P$ is countable.*

Proof According to (16.14), we have to show that $P(J_t) > 0$ can hold only for at most countably many t. If we define for $\varepsilon > 0$ and $t \in (0, 1)$ the set $J_t(\varepsilon) := \{x \in D : |x(t) - x(t-)| > \varepsilon\}$, then for each $\delta > 0$ there can be at most finitely many t with $P(J_t(\varepsilon)) \geq \delta$. Namely, suppose this inequality holds for infinitely many $t_1, t_2, \ldots$. But then $P(\limsup_{n \to \infty} J_{t_n}(\varepsilon)) \geq \delta$, which, however, contradicts the fact that for each $x \in D$ there are only finitely many t with $|x(t) - x(t-)| > \varepsilon$. Since the probability measure P is continuous from below we have $P(J_t) = \lim_{k \to \infty} P(J_t(1/k))$ and thus

$$\left\{ t \in (0, 1) : P(J_t) > 0 \right\} \subset \bigcup_{k=1}^{\infty} \left\{ t \in (0, 1) : P\left(J_t\left(\frac{1}{k}\right)\right) \geq \frac{1}{2}P(J_t) \right\}.$$

Since a countable union of finite sets is countable, the assertion follows.   □

Suppose $P, P_1, P_2, \ldots$ are probability measures on $\mathcal{B}(D)$. If k is any integer, and if $t_1, \ldots, t_k$ belong to T_P, then the projection $\pi_{t_1,\ldots,t_k}$ is almost everywhere continuous with respect to P, and from the mapping Theorem 14.4 it follows that

$$P_n \xrightarrow{D} P \implies P_n \pi_{t_1,\ldots,t_k}^{-1} \xrightarrow{D} P \pi_{t_1,\ldots,t_k}^{-1}. \tag{16.15}$$

With these preparations, we are now able to state an analogue to Theorem 14.19.

16.11 Theorem (First Criterion for Weak Convergence in D**)**
Let $P, P_1, P_2, \ldots$ *be probability measures on* $\mathcal{B}(\mathrm{D})$. *If the set* $\{P_n : n \geq 1\}$ *is relatively compact, and if*

$$P_n \pi_{t_1, \ldots, t_k}^{-1} \xrightarrow{\mathcal{D}} P \pi_{t_1, \ldots, t_k}^{-1}$$

for each $k \geq 1$ *and each choice of* $t_1, \ldots, t_k \in T_P$, *then* $P_n \xrightarrow{\mathcal{D}} P$.

Proof Let $(P_{n_j})_{j \geq 1}$ be any subsequence of (P_n) with $P_{n_j} \xrightarrow{\mathcal{D}} Q$ as $j \to \infty$ for a probability measure Q on $\mathcal{B}(\mathrm{D})$. By Corollary 14.18, we only need to show $P = Q$. Fix any integer k and $t_1, \ldots, t_k$ in T_P. By assumption, $P_{n_j} \pi_{t_1, \ldots, t_k}^{-1} \xrightarrow{\mathcal{D}} P \pi_{t_1, \ldots, t_k}^{-1}$. If $t_1, \ldots, t_k$ also belong to T_Q, (16.15) yields $P_{n_j} \pi_{t_1, \ldots, t_k}^{-1} \xrightarrow{\mathcal{D}} Q \pi_{t_1, \ldots, t_k}^{-1}$. Therefore, if $t_1, \ldots, t_k \in T_P \cap T_Q$ then $P \pi_{t_1, \ldots, t_k}^{-1} = Q \pi_{t_1, \ldots, t_k}^{-1}$. We have $1 \in T_P \cap T_Q$, and by Lemma 16.10, the set $T_P \cap T_Q$ is dense in $[0, 1]$. Using Theorem 16.9 (a) , it follows that $P = Q$. $\square$

By Prokhorov's theorem (Theorem 14.22), we can replace *relatively compact* with *tight* in the above theorem. The concept of tightness of a set of probability measures leads us to relatively compact subsets of D, and Theorem 16.7 contains a characterization of such subsets. With the help of this characterization, one can now state conditions analogous to Theorem 14.24 for tightness and also for convergence in distribution. We state a related result without proof (see, e.g., [BI2], p. 141).

16.12 Theorem (Second Criterion for Weak Convergence in D**)**
Let $P, P_1, P_2, \ldots$ *be probability measures on* $\mathcal{B}(\mathrm{D})$ *with the following properties:*

$$P_n \pi_{t_1, \ldots, t_k}^{-1} \xrightarrow{\mathcal{D}} P \pi_{t_1, \ldots, t_k}^{-1} \quad \textit{for all } k \geq 1, \textit{ for all } t_1, \ldots, t_k \in T_P, \tag{16.16}$$

$$\lim_{\delta \downarrow 0} P\big(\{x \in \mathrm{D} : |x(1) - x(1 - \delta)| \geq \varepsilon\}\big) = 0 \quad \textit{for every } \varepsilon > 0, \tag{16.17}$$

$$\lim_{\delta \downarrow 0} \limsup_{n \to \infty} P_n\big(\{x \in \mathrm{D} : w_x''(\delta) \geq \varepsilon\}\big) = 0 \quad \textit{for every } \varepsilon > 0.$$

Then $P_n \xrightarrow{\mathcal{D}} P$.

As was done in Chap. 14, both the results obtained so far and further results may be restated in terms of D-valued random elements. If $(\Omega, \mathcal{A}, \mathbb{P})$ is a probability space, then any $(\mathcal{A}, \mathcal{B}(\mathrm{D}))$-measurable mapping $X : \Omega \to \mathrm{D}$ is a (D-valued) *random element* or a (D-valued) *random function*. As in Sect. 14.9, for a fixed $\omega \in \Omega$ the function $X(\omega)$ from D is called a *path of* X (to ω). We also adopt the notations introduced there and write $X(\omega)(t) =: X_t(\omega) =: X(\omega, t)$ and $X(t) := \pi_t \circ X$ for the real-valued random variable, which assigns to each ω in Ω the value $X(t)(\omega)$. As in Sect. 14.9, X is $(\mathcal{A}, \mathcal{B}(D))$-measurable if and only if for each $t \in [0, 1]$ the mapping $X(t)$ is $(\mathcal{A}, \mathcal{B}^1)$-measurable and thus a real-valued random variable (see also Problem 14.4).

In the following, let $X, X_1, X_2, \ldots$ be D-valued random elements that are defined on a probability space $(\Omega, \mathcal{A}, \mathbb{P})$. With the abbreviation $T_X := T_{\mathbb{P}X}$, we have the following criterion for convergence in distribution in the space D[0, 1] (see [BI2], p. 142).

16.13 Theorem (Criterion for Convergence in Distribution in D[0, 1]**)**
Suppose that the following conditions hold:

(a) $(X_n(t_1), \ldots, X_n(t_k)) \xrightarrow{\mathcal{D}} (X(t_1), \ldots, X(t_k))$ *for each* $k \geq 1$ *and any choice of* $t_1, \ldots, t_k \in T_X$.

(b) $X(1) - X(1 - \delta) \xrightarrow{\mathbb{P}} 0$ *as* $\delta \downarrow 0$.

(c) *There exist a continuous, increasing function* $L : [0, 1] \to \mathbb{R}$ *and constants* $\alpha > 0$, $\beta \geq 0$, *such that for every* $\gamma > 0$ *and every* $n \geq 1$ *as well as for all* r, s, t *with* $0 \leq r \leq s \leq t \leq 1$ *we have:*

$$\mathbb{P}\big(|X_n(s) - X_n(r)| \wedge |X_n(t) - X_n(s)| \geq \gamma\big) \leq \frac{1}{\gamma^{4\beta}} (L(t) - L(r))^{2\alpha} . \qquad (16.18)$$

Then $X_n \xrightarrow{\mathcal{D}} X$.

Note that conditions (a) and (b) correspond to (16.16) and (16.17), respectively. Condition (c) is based on a technically complex maximal inequality (see [BI2], pp. 108–112). A usually easy to verify sufficient condition for (16.18) is

$$\mathbb{E}\left(|X_n(s) - X_n(r)|^{2\beta} \cdot |X_n(t) - X_n(s)|^{2\beta}\right) \leq (L(t) - L(r))^{2\alpha} . \qquad (16.19)$$

Self-question 7: Why does (16.18) follow from (16.19)?

In Chap. 15, we proved the existence of the Wiener measure D as a probability measure on the Borel σ-field $\mathcal{B}(C)$ in the space C[0, 1]. This was accompanied by the Wiener process $(W(t))_{0 \leq t \leq 1}$. Using the canonical embedding

$$\iota : \begin{cases} C \to D, \\ x \mapsto \iota(x) := x \end{cases}$$

of C into the space D, we define the *Wiener measure* on the σ-field $\mathcal{B}(D)$ by the image measure of D under the continuous and thus $(\mathcal{B}(C), \mathcal{B}(D))$-measurable mapping ι, that is, by the probability measure $D \circ \iota^{-1}$. By construction, $D \circ \iota^{-1}(C) = 1$, and D is the restriction of $D \circ \iota^{-1}$ to the trace σ-field $\mathcal{B}(C) = \mathcal{B}(D) \cap C$. In this context, it should be noted that C as a subset of D is closed and thus a Borel subset of D (see Problem 16.10).

In the sequel, we will briefly set $D := D \circ \iota^{-1}$ and speak of the *Wiener measure on* D. Using the canonical construction $(\Omega, \mathcal{A}, \mathbb{P}) := (D, \mathcal{B}(D), D)$ and $W : D \to D$ with $W(x) := x$, $x \in D$, there is then a probability space and a D-valued random element with $\mathbb{P}^W = D$. Putting $W(t) := \pi_t \circ W$, $0 \le t \le 1$, we also obtain a stochastic process $(W(t))_{0 \le t \le 1}$, which, completely unchanged to the case C, is referred to as the *Wiener process*. The only difference to Chap. 15 is that the paths of W can now formally also belong to $D \backslash C$. However, since $\mathbb{P}(W \in C) = 1$, these paths are continuous with probability one.

The partial sum process X_n introduced in (15.3) includes a linear interpolation that renders the paths of X_n continuous functions. If this interpolation is omitted, that is, if we define

$$X_n(t) := \frac{1}{\sigma\sqrt{n}} S_{\lfloor nt \rfloor}, \quad 0 \le t \le 1,$$

then X_n is a D-valued random element, whose realizations are constant on the intervals $\left[\frac{j-1}{n}, \frac{j}{n} \right)$, $j \in \{1, \dots, n\}$ (Fig. 16.3).

The following result is the version of Donsker's theorem (Theorem 15.9) in the space D[0, 1].

16.14 Theorem (Donsker)

Let $Z_1, Z_2, \dots$ *be i.i.d. random variables with* $\mathbb{E}(Z_1^2) < \infty$, $\mathbb{E}(Z_1) = 0$, *and* $0 < \sigma^2 := \mathbb{V}(Z_1)$. *Putting* $S_0 := 0$, $S_n := \sum_{j=1}^{n} Z_j$, $n \ge 1$, *and*

$$X_n(t) := \frac{1}{\sigma\sqrt{n}} S_{\lfloor nt \rfloor}, \quad 0 \le t \le 1, \tag{16.20}$$

we have $X_n \xrightarrow{\mathcal{D}} W$ *in* D[0, 1] *as* $n \to \infty$.

Fig. 16.3 Realization of a partial sum process in D[0, 1]

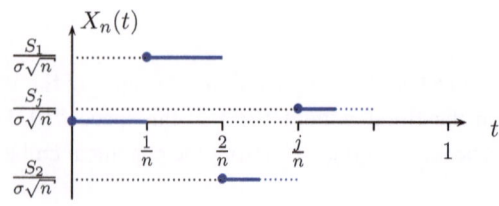

Proof We use Theorem 16.13. Since $\mathbb{P}(W \in C) = 1$, we have $T_W = [0, 1]$, and condition (a) of Theorem 16.13 follows from the multivariate central limit theorem (see also the proof of Theorem 15.3). Condition (b) is also fulfilled, as $W(1) - W(1 - \delta) \sim W(\delta) \sim N(0, \delta)$. We will now show that (16.19) holds with $L(t) := 2t$ and $\alpha = \beta = 1$, which completes the proof. The increments $X_n(r) - X_n(s)$ and $X_n(t) - X_n(s)$ over intervals $[r, s]$ and $[s, t]$ with $r \leq s \leq t$ are independent random variables since they are formed from disjoint blocks of the random variables $Z_1, \ldots, Z_n$. We thus obtain

$$\mathbb{E}\left[|X_n(s) - X_n(r)|^2 \cdot |X_n(t) - X_n(s)|^2\right] = \frac{\lfloor ns \rfloor - \lfloor nr \rfloor}{n} \cdot \frac{\lfloor nt \rfloor - \lfloor ns \rfloor}{n}.$$

Self-question 8: Is there a further argument that justifies this equation?

Since $r \leq s \leq t$ it follows that

$$\mathbb{E}\left[|X_n(s) - X_n(r)|^2 |X_n(t) - X_n(s)|^2\right] \leq \left(\frac{\lfloor nt \rfloor - \lfloor nr \rfloor}{n}\right)^2.$$

If $t - r < \frac{1}{n}$ then the left-hand side of this inequality is equal to zero, and if $t - r \geq \frac{1}{n}$ then the right-hand side is at most equal to $4(t - r)^2$.

Self-question 9: Why is the right side at most equal to $4(t - r)^2$ in the second case?

Thus, (16.19) holds with $\alpha = \beta = 1$ and $L(t) = 2t$.   □

Together with the mapping Theorem 14.4, Theorem 16.14 provides a functional central limit theorem on D[0, 1]: If $h : D \to \mathbb{R}$ is a $(\mathcal{B}(D), \mathcal{B}^1)$-measurable function that is almost everywhere continuous with respect to D, then $h(X_n) \xrightarrow{\mathcal{D}} h(W)$. For some functions like, for example, the one defined by

$$h_+(x) := \lambda^1(\{t \in [0, 1] : x(t) > 0\}), \quad x \in D,$$

(whose $(\mathcal{B}(D), \mathcal{B}^1)$-measurability is shown in [BI2], p. 247) takes a simple form for the partial sum process X_n defined in (16.20) (compared to the more complicated "interpolated process" from (15.3)), because $h_+(X_n)$ is the number of positive partial sums among $S_1, \ldots, S_{n-1}$, divided by n.

We conclude this chapter with statistical applications of Theorem 16.13. For this purpose, let $X_1, X_2, \ldots$ be a sequence of i.i.d. random variables on a common probability

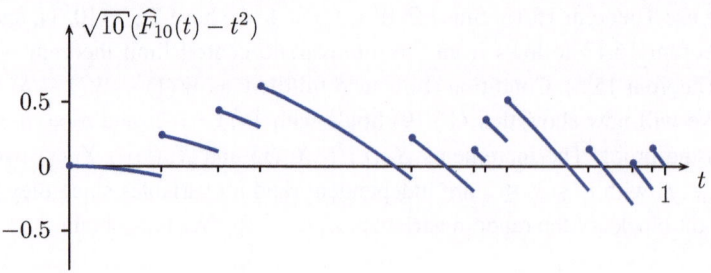

Fig. 16.4 Realization of an empirical process ($n = 10$, $F(t) = t^2$)

space $(\Omega, \mathcal{A}, \mathbb{P})$. Here we make the restrictive assumption[1] $\mathbb{P}(0 \leq X_1 \leq 1) = 1$. Let $F(t) := \mathbb{P}(X_1 \leq t)$, $0 \leq t \leq 1$, be the distribution function of X_1 and

$$\widehat{F}_n(t) := \frac{1}{n} \sum_{j=1}^{n} \mathbf{1}\{X_j \leq t\}, \quad 0 \leq t \leq 1,$$

be the empirical distribution function of $X_1, \ldots, X_n$. For each $t \in [0, 1]$ we then define a random variable $Y_n(t)$ on Ω by

$$Y_n(t) := \sqrt{n}\big(\widehat{F}_n(t) - F(t)\big), \quad 0 \leq t \leq 1. \tag{16.21}$$

The collection $(Y_n(t))_{0 \leq t \leq 1}$ of random variables is called the *empirical process* (associated with $X_1, \ldots, X_n$). An important special case is the *uniform empirical process*, which arises in the case $F(t) = t$, that is, $X_1 \sim U(0, 1)$ (see (13.1)). If we make the suppressed argument $\omega \in \Omega$ in (16.21) explicit and write

$$Y_n(\omega, t) := \sqrt{n}\left(\frac{1}{n} \sum_{j=1}^{n} \mathbf{1}\{X_j(\omega) \leq t\} - F(t)\right), \quad \omega \in \Omega, 0 \leq t \leq 1,$$

then, for fixed $\omega \in \Omega$, the function $Y_n(\omega, \cdot) : [0, 1] \to \mathbb{R}$ belongs to D[0, 1]. Moreover, since $Y_n(t)$ is a $(\mathcal{A}, \mathcal{B}^1)$-measurable mapping for each t, the mapping $Y_n : \Omega \to D$ that assigns to each $\omega \in \Omega$ the function $Y_n(\omega, \cdot) : [0, 1] \to \mathbb{R}$ is a D-valued random element on Ω.

Figure 16.4 shows the realization of an empirical process for the case $n = 10$ and $F(t) = t^2$, $0 \leq t \leq 1$. Thus, each of the random variables X_j has the same distribution

---

[1] Regarding the Càdlàg spaces D[0, ∞) and D(−∞, ∞) of right-continuous functions with existing left-hand limits defined on [0, ∞) and (−∞, ∞), respectively see, e.g., [POL], Chapters V and VI, or [BI2], Section 16.

as the maximum of two independent random variables with a uniform distribution on the interval [0, 1].

As in the space C[0, 1], also a D[0, 1]-valued random element Y on a probability space $(\Omega, \mathcal{A}, \mathbb{P})$ is called a *Gaussian random element*, if all fidis of Y are normally distributed, that is, if for each $k \geq 1$ and each choice of $t_1, \ldots, t_k$ with $0 \leq t_1 < \ldots < t_k \leq 1$, the random vector $(Y(t_1), \ldots, Y(t_k))$ has a k-variate normal distribution. As before, we set $Y(t) := \pi_t \circ Y, 0 \leq t \leq 1$. This manner of speaking is equivalent to the term *Gaussian process*, when referring to the stochastic process $(Y(t))_{0 \leq t \leq 1}$. A Gaussian random element or a Gaussian process is called *centered*, if the above normal distributions have the expectation 0_k for each k.

In connection with the limit distribution of the empirical process, the Brownian bridge on D[0, 1] will play a crucial role. This Brownian bridge is defined using the Wiener process W on D just as in the space C, namely according to

$$B(t) := W(t) - tW(1), \quad 0 \leq t \leq 1,$$

that is, by setting $B := h \circ W$ with $h : D \to D$, where $h(x)(t) := x(t) - tx(1), 0 \leq t \leq 1$. Like the Wiener process on D, the Brownian bridge B on D is also entirely concentrated on the subset C of D, that is, we have $\mathbb{P}(B \in C) = 1$.

16.15 Theorem (Convergence in Distribution of the Empirical Process)
Let $X_1, X_2, \ldots$ be a sequence of i.i.d. random variables with $\mathbb{P}(0 \leq X_1 \leq 1) = 1$ and distribution function F. For the sequence (Y_n) of empirical processes defined by (16.21), we have

$$Y_n \xrightarrow{\ \mathcal{D}\ } Y \text{ in } D[0, 1].$$

Here, Y is a D-valued centered Gaussian process with covariance function

$$Cov(Y(s), Y(t)) = F(s) \wedge F(t) - F(s)F(t), \quad 0 \leq s, t \leq 1.$$

Proof We apply Theorem 16.13 and first consider the special case $F(t) = t$, that is, we assume that $X_1 \sim U(0, 1)$. In this situation, Y is the Brownian bridge B. According to Problem 7.4, all finite-dimensional distributions of Y_n converge to the corresponding finite-dimensional distributions of B, so condition (a) of Theorem 16.13 is satisfied. Furthermore, we have

$$B(1) - B(1 - \delta) \sim N(0, \delta(1 - \delta))$$

and thus $B(1) - B(1 - \delta) \overset{\mathbb{P}}{\longrightarrow} 0$ as $\delta \downarrow 0$. Therefore, condition (b) of Theorem 16.13 also holds. We now prove

$$\mathbb{E}\left[\left(Y_n(s) - Y_n(r)\right)^2 \cdot \left(Y_n(t) - Y_n(s)\right)^2\right] \leq 6(t - r)^2, \quad 0 \leq r \leq s \leq t \leq 1. \quad (16.22)$$

If we can show (16.22), the assertion holds for the case $X_1 \sim U(0, 1)$, because then, putting $\alpha = \beta = 1$ and $L(t) = \sqrt{6}t$, (16.19) and thus condition (c) of Theorem 16.13 would also apply. To prove (16.22), we assume w.l.o.g. $0 \leq r < s < t \leq 1$. Since $X_1 \sim U(0, 1)$, it follows that

$$Y_n(u) = \sqrt{n}\left(\frac{1}{n}\sum_{i=1}^{n} \mathbf{1}\{X_i \leq u\} - u\right) = \frac{1}{\sqrt{n}}\sum_{i=1}^{n}\left(\mathbf{1}\{X_i \leq u\} - u\right), \quad 0 \leq u \leq 1.$$

Setting

$$\alpha_i := \mathbf{1}\{r < X_i \leq s\} - (s - r), \quad \beta_i := \mathbf{1}\{s < X_i \leq t\} - (t - s), \quad i \in \{1, \dots, n\},$$

we get

$$Y_n(s) - Y_n(r) = \frac{1}{\sqrt{n}}\sum_{i=1}^{n}\alpha_i, \quad Y_n(t) - Y_n(s) = \frac{1}{\sqrt{n}}\sum_{k=1}^{n}\beta_k. \quad (16.23)$$

If we abbreviate the expected value on the left-hand side of (16.22) with $\Delta_n(r, s, t)$ and write the squares of the differences appearing in (16.23) as double sums, the linearity of taking expectations yields

$$\Delta_n(r, s, t) = \frac{1}{n^2}\sum_{i,j=1}^{n}\sum_{k,\ell=1}^{n}\mathbb{E}[\alpha_i\alpha_j\beta_k\beta_\ell].$$

Since $\mathbb{E}(\alpha_i) = \mathbb{E}(\beta_i) = 0$ for each $i \in \{1, \dots, n\}$, symmetry arguments and the multiplication rule for expected values imply

$$\Delta_n(r, s, t) = \frac{1}{n^2}\left(n\mathbb{E}[\alpha_1^2\beta_1^2] + n(n-1)\mathbb{E}[\alpha_1^2]\mathbb{E}[\beta_2^2] + 2n(n-1)\mathbb{E}[\alpha_1\beta_1]\mathbb{E}[\alpha_2\beta_2]\right)$$

and thus

$$\Delta_n(r, s, t) \leq \mathbb{E}[\alpha_1^2\beta_1^2] + \mathbb{E}[\alpha_1^2]\mathbb{E}[\beta_2^2] + 2\mathbb{E}[\alpha_1\beta_1]\mathbb{E}[\alpha_2\beta_2].$$

Because each of the random variables α_1^2 and α_2^2 can only take on two values, and since $1\{r < X_1 \le s\}1\{s < X_1 \le t\} = 0$, we obtain

$$\mathbb{E}[\alpha_1^2\beta_1^2] = \mathbb{E}\left[\left(1\{r < X_1 \le s\} - (s-r)\right)^2\left(1\{s < X_1 \le t\} - (t-s)\right)^2\right]$$

$$= (s-r)\left(1 - (s-r)\right)^2(t-s)^2 + (t-s)(s-r)^2\left(1 - (t-s)\right)^2$$

$$+ \left(1 - (t-r)\right)(s-r)^2(t-s)^2$$

$$\le 3(s-r)(t-s)$$

$$\le 3(t-r)^2.$$

Here, the second equal sign follows by distinguishing the cases $r < X_1 \le s, s < X_1 \le t$, and $X_1 \in [0, 1] \setminus (r, t]$. In the same way, it follows that

$$\mathbb{E}[\alpha_1^2]\mathbb{E}[\beta_2^2] \le (t-r)^2,$$

$$\mathbb{E}[\alpha_1\beta_1]\mathbb{E}[\alpha_2\beta_2] \le (t-r)^2$$

and thus $\Delta_n(r, s, t) \le 6(t-r)^2$ (Problem 16.11). The assertion of the theorem is therefore valid if X_1 is uniformly distributed on the interval $[0, 1]$.

Building on what has already been shown, we now prove Theorem 16.15 for the general case that X_1 has the distribution function F. The crucial point is that we can generate a random variable X with the distribution function F via the quantile transformation $X := F^{-1}(U)$ from a random variable U with a uniform distribution on the interval $[0, 1]$. Since all is about distributions, we therefore assume w.l.o.g. that $X_j := F^{-1}(U_j), j \ge 1$, where $U_1, U_2, \ldots$ are i.i.d. random variables with a uniform distribution on $[0, 1]$. In the following, let

$$\widehat{G}_n(t) := \frac{1}{n}\sum_{j=1}^{n} 1\{U_j \le t\}, \quad 0 \le t \le 1,$$

be the empirical distribution function of $U_1, \ldots, U_n$, and write

$$Z_n(t) := \sqrt{n}\left(\widehat{G}_n(t) - t\right), \quad 0 \le t \le 1,$$

for the uniform empirical process based on $U_1, \ldots, U_n$. According to what has already been shown, $Z_n \xrightarrow{\mathcal{D}} B$ in D, where B is a Brownian bridge. Since $X_j \le t$ if and only if $U_j \le F(t)$, it follows that $\widehat{F}_n(t) = \widehat{G}_n(F(t))$ and thus $Y_n(t) = Z_n(F(t))$. We now define a mapping $\psi : \mathrm{D} \to \mathrm{D}$ by putting $(\psi(x))(t) := x(F(t))$, where $x \in \mathrm{D}$ and $0 \le t \le 1$.

> **Self-question 10**: Why does $\psi(x) = x \circ F$ belong to D?

By Corollary 16.4 (c), we know that in D $d_S(x_n, x) \to 0$ and $x \in C$ together give the convergence $\|x_n - x\| \to 0$. By the definition of ψ we have $\|\psi(x_n) - \psi(x)\| \le \|x_n - x\|$, and thus $\|\psi(x_n) - \psi(x)\| \to 0$. According to (16.5), the latter convergence in turn implies $d_S(\psi(x_n), \psi(x)) \to 0$. Since $\mathbb{P}(B \in C) = 1$, the already shown convergence $Z_n \xrightarrow{\mathcal{D}} B$ and the mapping Theorem 14.4 lead to $Y_n = \psi(Z_n) \xrightarrow{\mathcal{D}} \psi(B) =: Y$. The process Y has the properties claimed in the theorem (Problem 16.12).   □

We conclude this chapter with two statistical applications of the results obtained, namely with goodness-of-fit tests and with a nonparametric two-sample test.

16.16 The Kolmogorov–Smirnov[2] Goodness-of-Fit Test

Suppose that $X_1, X_2, \ldots$ is a sequence of i.i.d. random variables with an unknown, *continuous* distribution function F. If F_0 is a given *continuous* distribution function, then a nonparametric goodness-of-fit test is a statistical test of the hypothesis

$$H_0 : \ F = F_0$$

against the alternative $H_1 : F \ne F_0$. Here, the most important special case is $F(t) = t$, $0 \le t \le 1$, that is, a test of the hypothesis that the underlying distribution is uniform on $[0, 1]$. To test H_0 based on (realizations of) $X_1, \ldots, X_n$, it is almost obvious to estimate the unknown F by the empirical distribution function

$$\widehat{F}_n(x) := \frac{1}{n} \sum_{j=1}^{n} \mathbf{1}\{X_j \le x\}, \quad x \in \mathbb{R},$$

of $X_1, \ldots, X_n$. If the hypothesis $H_0 : F = F_0$ holds, then the Glivenko–Cantelli theorem (Theorem 7.2) states that the sequence (K_n) of random variables defined by

$$K_n := \sup_{x \in \mathbb{R}} \left| \widehat{F}_n(x) - F_0(x) \right| \tag{16.24}$$

converges to 0 almost surely as $n \to \infty$. K_n is called the *Kolmogorov statistic* for testing H_0, and a test of H_0 based on K_n rejects H_0 for "too large" values of K_n. What exactly

---

[2] Nikolai Vasilyevich Smirnov (1900–1966), was a leading Russian mathematical statistician. From 1938 he worked at the Steklov Institute, where in his last year of life he succeeded A.N. Kolmogorov as head of the Department of Mathematical Statistics.

"too large" means depends on the permitted probability of a type I error, which occurs when H_0 is falsely rejected.

We will now see that the distribution of K_n under H_0 (for short: null distribution of K_n) does not depend on the unknown distribution function as long as the latter is *continuous*. Why? Since F is continuous, $X_1, \ldots, X_n$ are pairwise distinct with probability one.

Self-question 11: Why does the last claim hold?

If $X_{(1)} < \ldots < X_{(n)}$ denote the order statistics of $X_1, \ldots, X_n$, we have $\widehat{F}_n(X_{(j)}) = \frac{j}{n}$ for each $j \in \{1, \ldots, n\}$. As a consequence, it follows that

$$K_n = \max_{j=1,\ldots,n} \left(\max \left(\left| F_0\left(X_{(j)}\right) - \frac{j}{n} \right|, \left| F_0\left(X_{(j)}\right) - \frac{j-1}{n} \right| \right) \right).$$

The Kolmogorov statistic thus depends on $X_1, \ldots, X_n$ only via $F_0(X_{(1)}), \ldots, F_0(X_{(n)})$. Now, suppose H_0 holds. If we set $U_j := F_0(X_j)$, $j \geq 1$, then $U_1, U_2, \ldots$ is a sequence of i.i.d. random variables, each having the uniform distribution $U(0, 1)$. Writing $U_{(1)} < \ldots < U_{(n)}$ for the order statistics of $U_1, \ldots, U_n$, we then obtain

$$\left(F_0(X_{(1)}), \ldots, F_0(X_{(n)})\right) \overset{\mathcal{D}}{=} \left(U_{(1)}, \ldots, U_{(n)}\right).$$

This is the very reason why the null distribution of K_n does not depend on F_0. When studying this distribution one can therefore assume w.l.o.g. that $X_1 \sim U(0, 1)$.

The following result shows that the limit distribution of $\sqrt{n} K_n$ as $n \to \infty$ under H_0 is linked with the Brownian bridge.

16.17 Theorem (Limit Distribution of the Kolmogorov Statistic under H_0)
Under the hypothesis $H_0 : F = F_0$, we have

$$\sqrt{n} K_n \overset{\mathcal{D}}{\longrightarrow} \|B\| = \max_{0 \leq t \leq 1} |B(t)| \quad as \ n \to \infty.$$

Here, B is a Brownian bridge.

Proof Let B_n be the uniform empirical process defined in (16.1). According to Problem 16.13 (a) , the mapping $h : D \to \mathbb{R}$ defined by $h(x) := \|x\|$, $x \in D$, is $(\mathcal{B}(D), \mathcal{B}^1)$-measurable. Therefore, $h(B_n) = \|B_n\|$ is a random variable, which has the same distribution as $\sqrt{n} K_n$. Because $\mathbb{P}(B \in C) = 1$, the mapping Theorem 14.4 is applicable according to Problem 16.13 (b). From $B_n \overset{\mathcal{D}}{\longrightarrow} B$ it follows that $h(B_n) \overset{\mathcal{D}}{\longrightarrow} h(B)$ and thus the assertion.   □

The distribution of the maximum absolute value $\|B\|$ of the Brownian bridge is called *Kolmogorov distribution*. The distribution function of $\|B\|$ is given by

$$\mathbb{P}\big(\|B\| \le \xi\big) = 1 - 2\sum_{j=1}^{\infty}(-1)^{j-1}e^{-2j^2\xi^2}, \quad 0 < \xi < \infty.$$

We will indicate at the end of this chapter how a proof of this result can be obtained with relatively elementary means (see Remark 16.23).

The Kolmogorov goodness-of-fit test is consistent against the general alternative H_1 : $F \ne F_0$, since $\sqrt{n}K_n \to \infty$ $\mathbb{P}$-almost surely if X_1 has the distribution function F (Problem 16.16).

16.18 The Cramér–von Mises Goodness-of-Fit Test

The test statistic K_n defined in (16.24) uses the maximum absolute difference between the empirical distribution function $\widehat{F}_n$ and the hypothetical distribution function F_0 as a measure of distance between $\widehat{F}_n$ and F_0. Alternatively, H. Cramér and R. von Mises suggested to employ the test statistic defined by

$$\omega_n^2 := n\int_{-\infty}^{\infty} \big(\widehat{F}_n(x) - F_0(x)\big)^2 \, \mathrm{d}F_0(x). \tag{16.25}$$

We encountered ω_n^2, referred to as the *Cramér–von Mises statistic*, already in Sect. 8.18 in connection with a test of $H_0 : F = F_0$ in the special case $F_0(t) = t, 0 \le t \le 1$.

The integration in (16.25) is done with respect to the probability measure μ_0 on $\mathcal{B}^1$ defined by F_0. If one transforms μ_0 using the probability integral transform $x \mapsto t := F_0(x)$ then, due to the continuity of F_0, the resulting image measure is the uniform distribution on $(0, 1)$, and the change of variables formula (see, e.g., [DUR], p. 30) yields

$$\omega_n^2 = n\int_0^1 \big(\widehat{F}_n(F_0^{-1}(t)) - t\big)^2 \, \mathrm{d}t. \tag{16.26}$$

Because $X_j \le F_0^{-1}(t)$ is equivalent to $F_0(X_j) \le t$, $\widehat{F}_n \circ F_0^{-1}$ is the empirical distribution function of $F_0(X_1), \ldots, F_0(X_n)$, and since these random variables have the uniform distribution $U(0, 1)$, the null distribution of ω_n^2 (like that of K_n) does not depend on the specific continuous distribution function F_0. This means that we can set $F_0^{-1}(t) = t$ in (16.26) if there is interest in the null distribution of ω_n^2. Thus, we have

$$\omega_n^2 \overset{\mathcal{D}}{=} \int_0^1 B_n^2(t) \, \mathrm{d}t$$

under H_0, where B_n is the uniform empirical process defined in (16.1). According to Problem 16.14, the mapping $h : \mathrm{D} \to \mathbb{R}$ defined by $h(x) := \int_0^1 x^2(t) \, \mathrm{d}t$ satisfies the prerequisites of the mapping Theorem 14.4. With Theorem 16.15 we obtain $h(B_n) \xrightarrow{\mathcal{D}} h(B)$, that is,

$$\omega_n^2 \xrightarrow{\mathcal{D}} \int_0^1 B^2(t)\, dt.$$

In Sect. 8.18 we derived the limit distribution of ω_n^2 in the case $X_1 \sim U(0, 1)$ using the theory of simply-degenerate U-statistics. From (8.43) we have

$$\int_0^1 B^2(t)\, dt \stackrel{\mathcal{D}}{=} \sum_{j=1}^{\infty} \frac{1}{\pi^2 j^2} \left(N_j^2 - 1 \right) + \frac{1}{6}.$$

Here, $N_1, N_2, \ldots$ is a sequence of i.i.d. standard normal random variables. Such a sequence also appears in connection with the distribution of $\int_0^1 W^2(t)\, dt$, where W is a Wiener process, see (15.22). The distribution of $\int_0^1 B^2(t)\, dt$ is called *Cramér–von Mises distribution*.

To be able to carry out the Cramér–von Mises goodness-of-fit test, representation (16.25) is not very helpful at first glance. Using the order statistics $X_{(1)} < \ldots < X_{(n)}$ of $X_1, \ldots, X_n$ and the fact that $\widehat{F}_n \circ F_0^{-1}$ is constant between the points $F_0(X_{(1)}), \ldots F_0(X_{(1)})$ of the interval $[0, 1]$, it can be shown by direct calculation that ω_n^2 can be written in the form

$$\omega_n^2 = \frac{1}{12n} + \sum_{j=1}^{n} \left(F_0(X_{(j)}) - \frac{2j-1}{2n} \right)^2$$

(Problem 16.14). Like the Kolmogorov test, also the Cramér–von Mises goodness-of-fit test is consistent against each alternative $H_1 : F \neq F_0$ (Problem 16.16).

16.19 The Nonparametric Two-Sample Problem

A further statistical application that we already have been acquainted with in Chap. 8 (see Example 8.25) is the classical *nonparametric two-sample problem*. The starting point of this problem are two sequences $(X_j)_{j\geq 1}$ and $(Y_j)_{j\geq 1}$ of real-valued random variables, which are all defined on the same probability space $(\Omega, \mathcal{A}, \mathbb{P})$. We assume that all random variables are independent. Furthermore, we suppose that $X_1, X_2, \ldots$ have the same distribution function F, and that $Y_1, Y_2, \ldots$ have the same distribution function G. So we have $F(t) = \mathbb{P}(X_j \leq t)$ and $G(t) = \mathbb{P}(Y_j \leq t), t \in \mathbb{R}$, for each $j \geq 1$. Furthermore, we assume that F and G are *continuous*, but otherwise unknown. The hypothesis to be tested is

$$H_0 : F = G,$$

against the general alternative $H_1 : F \neq G$. To carry out the test, we dispose of realizations of the random variables $X_1, \ldots, X_m$ and $Y_1, \ldots, Y_n$.

To test H_0, it suggests itself to use the empirical distribution functions

$$\widehat{F}_m(t) := \frac{1}{m} \sum_{j=1}^{m} \mathbf{1}\{X_j \leq t\}, \qquad \widehat{G}_n(t) := \frac{1}{n} \sum_{j=1}^{n} \mathbf{1}\{Y_j \leq t\}$$

of the two samples $X_1, \ldots, X_m$ and $Y_1, \ldots, Y_n$ as estimators of F and G, respectively, and to reject H_0 if a suitably defined distance between $\widehat{F}_m$ and $\widehat{G}_n$ is "too large". Of course, it must again be specified what exactly "too large" means.

Following a suggestion by A.N. Kolmogorov and N.W. Smirnov, we consider the supremum distance

$$K_{m,n} := \left\| \widehat{F}_m - \widehat{G}_n \right\|_{\infty} := \sup_{x \in \mathbb{R}} \left| \widehat{F}_m(x) - \widehat{G}_n(x) \right| \tag{16.27}$$

between $\widehat{F}_m$ and $\widehat{G}_n$, which is commonly referred to as the *Kolmogorov–Smirnov statistic*. The *Kolmogorov–Smirnov test* rejects H_0 for large values of $K_{m,n}$.

Since $\widehat{F}_m$ and $\widehat{G}_n$ are step functions, the supremum in (16.27) is attained at one of the points $X_1, \ldots, X_m, Y_1, \ldots, Y_n$. Because F and G are continuous, $X_1, \ldots, X_m, Y_1, \ldots, Y_n$ take on pairwise distinct values with probability one.

In the sequel, we will investigate the distribution of $K_{m,n}$ for the case that the hypothesis H_0 holds, that is, all random variables have the same unknown distribution function F. Since $K_{m,n}$ does not change when we change $X_1, \ldots, X_m, Y_1, \ldots, Y_n$ to $F(X_1), \ldots, F(X_m), F(Y_1), \ldots, F(Y_n)$, and since the latter random variables have a uniform distribution on $(0, 1)$, the null distribution of $K_{m,n}$ does not depend on the unknown continuous distribution function F. So from now on we assume w.l.o.g. that $X_1, \ldots, X_m, Y_1, \ldots, Y_n$ are uniformly distributed on the unit interval. When passing through the unit interval from left to right, the difference $\widehat{F}_m(t) - \widehat{G}_n(t)$ of the empirical distribution functions jumps upwards by the value $\frac{1}{m}$ and downwards by the value $\frac{1}{n}$, depending on whether an X_i or a Y_j occurs in the ordered sample of all $X_1, \ldots, X_m, Y_1, \ldots, Y_n$. Since thereby the indices i and j do not matter, in order to determine $K_{m,n}$ we only need a sequence of length $m + n$ of symbols x and y, where x and y appear m and n times, respectively.

In the case $m = 3$, $n = 2$ and the sequence $x\, y\, x\, x\, y$, the difference of the empirical distribution functions first jumps from zero to the value $\frac{1}{3}$, then down by the value $\frac{1}{2}$, then up twice by the value $\frac{1}{3}$, and finally down by the value $\frac{1}{2}$, so that the initial difference zero is reached again. In this case, the largest distance between $\widehat{F}_m$ and $\widehat{G}_n$ is equal to $\frac{1}{2}$. Since under H_0 for reasons of symmetry all permutations of the involved random variables are equally likely, it becomes clear that finding the distribution of $K_{m,n}$ under H_0 leads to a purely combinatorial problem: one occupies m of a total of $m + n$ places arranged in a row with an x and the remaining ones with a y. All $\binom{m+n}{n}$ such distinguishable orderings are equally likely, and for each such ordering one can determine the largest absolute value of the difference of the empirical distribution functions.

We now turn to the limit distribution of $K_{m,n}$ as m, $n \to \infty$. In contrast to a central limit theorem for two-sample U-statistics (Theorem 8.24), where we required that the quotient $\frac{m}{m+n}$ converges to a $\tau \in (0, 1)$ (cf. (8.45)), m and n may tend to infinity independently of each other in what follows.

16.20 Theorem (Limit Null Distribution of the Kolmogorov–Smirnov Statistic)
Under H_0, we have

$$\sqrt{\frac{mn}{m+n}} K_{m,n} \xrightarrow{\mathcal{D}} \|B\| \quad as \ m, \ n \to \infty,$$

where B denotes a Brownian bridge.

Proof We recall the assumption that all random variables are uniformly distributed on $[0, 1]$ and set

$$A_m(t) := \sqrt{m}(\widehat{F}_m(t) - t), \qquad C_n(t) := \sqrt{n}(\widehat{G}_n(t) - t), \quad 0 \le t \le 1.$$

Furthermore, we put

$$a_{m,n} := \sqrt{\frac{n}{m+n}}, \qquad c_{m,n} := -\sqrt{\frac{m}{m+n}}. \tag{16.28}$$

Under H_0, we have

$$\sqrt{\frac{mn}{m+n}} K_{m,n} \overset{\mathcal{D}}{=} \max_{0 \le t \le 1} \left| a_{m,n}\sqrt{m}(\widehat{F}_m(t) - t) + c_{m,n}\sqrt{n}(\widehat{G}_n(t) - t) \right|$$

$$= \|a_{m,n}A_m + c_{m,n}C_n\|.$$

By Theorem 16.15, the uniform empirical processes A_m and C_n satisfiy $A_m \xrightarrow{\mathcal{D}} A$ as $m \to \infty$ and $C_n \xrightarrow{\mathcal{D}} C$ as $n \to \infty$, where A and C are Brownian bridges. Since A_m and C_n are independent, Theorem 14.26 implies that, as $m, n \to \infty$, (A_m, C_n) converges in distribution to (A, C) in the space D $\times$ D, where A and C are independent.

Self-question 12: Why is Theorem 14.26 applicable (there, we have $m = n$)?

If a and c are non-zero real numbers, then the mapping Theorem 14.4, applied to the function $h : D \times D \to D$ defined by $h(x, y) := ax + c$, shows that $a A_m + c C_n$ converges in distribution to $a A + c C$ as m, $n \to \infty$. If specifically $a^2 + c^2 = 1$ applies, it follows from the reproduction theorem (Theorem 15.21) that $B := a A + c C$ is a Brownian bridge.

The sequences $(a_{m,n})$ and $(c_{m,n})$ defined in (16.28) are bounded, and we have $a_{m,n}^2 + c_{m,n}^2 = 1$. If $a_{m,n}$ converges to a value a and $c_{m,n}$ converges to a value c, then

$$\| a_{m,n} A_m + c_{m,n} C_n - (a A_m + c C_n) \| = \| (a_{m,n} - a) A_m + (c_{m,n} - c) C_n \|$$

$$\leq | a_{m,n} - a | \cdot \| A_m \| + | c_{m,n} - c | \cdot \| C_n \|.$$

Since $\| A_m \| \xrightarrow{\mathcal{D}} \| A \|$ and $\| C_n \| \xrightarrow{\mathcal{D}} \| C \|$, $(\| A_m \|)_{m \geq 1}$ and $(\| C_n \|)_{n \geq 1}$ are tight sequences, and it thus follows that

$$\| a_{m,n} A_m + c_{m,n} C_n - (a A_m + c C_n) \| = o_{\mathbb{P}}(1) \tag{16.29}$$

as m, $n \to \infty$. Since $a^2 + c^2 = 1$, the above reasoning yields $a A_m + c C_n \xrightarrow{\mathcal{D}} B$, where B is a Brownian bridge. The inequality $d_S(x, y) \leq \| x - y \|$, together with (16.29) and Slutsky's lemma (Theorem 14.28) implies

$$a_{m,n} A_m + c_{m,n} C_n \xrightarrow{\mathcal{D}} B \quad \text{as } m, n \to \infty. \tag{16.30}$$

We would have obtained the same convergence in distribution if, instead of $a_{m,n}$ and $b_{m,n}$, we had started with any subsequences of these sequences. The only decisive factor is that sum of the squares of the limits of these subsequences is equal to one. The assertion now follows from the subsequence criterion (Theorem 14.5).   □

In the nonparametric two-sample problem, one sometimes does not want to test the hypothesis $H_0 : F = G$ against the general alternative $H_1 : F \neq G$, but has more specific alternatives in mind. One such alternative is that Y_1 is *stochastically larger* than X_1. This concept of a *stochastic order* is defined by the condition $\mathbb{P}(Y_1 > t) \geq \mathbb{P}(X_1 > t)$ for each $t \in \mathbb{R}$, where the greater-than sign should stand for at least one t. From this perspective, the hypothesis $H_0 : F = G$ of equality of both distributions is tested against the *one-sided alternative*

$$H_1^+ : F \geq G \text{ and } F \neq G$$

that X_1 is stochastically larger than Y_1.

In this situation, it is fairly obvious to use the test statistic

$$K_{m,n}^+ := \sup_{x \in \mathbb{R}} \left(\widehat{F}_m(x) - \widehat{G}_n(x) \right)$$

instead of $K_{m,n}$ defined in (16.27), and to reject H_0 in favor of H_1^+ if $K_{m,n}^+$ exceeds a suitable critical value. A close examination ot the proof of Theorem 16.20 shows that the next result follows from the convergence in distribution stated in (16.30).

16.21 Theorem (Limit Distribution of $K_{m,n}^+$ under H_0)
As m, n → ∞, we have

$$\sqrt{\frac{mn}{m+n}}K_{m,n}^+ \xrightarrow{\mathcal{D}} \max_{0\le t\le 1} B(t),$$

where B is a Brownian bridge.

Proof The mapping $h : D \to \mathbb{R}$ defined by $h(x) := \sup_{0\le t\le 1} x(t)$ is $(\mathcal{B}(D), \mathcal{B}^1)$-measurable, since the supremum can be taken over all rational numbers in the interval [0, 1] due to the right-continuity, and for each fixed t the mapping $D \ni x \mapsto x(t) = \pi_t \circ x$ is measurable. Since h is continuous on the subset C of D, the assertion follows from (16.30) and the mapping theorem.   □

In contrast to the Kolmogorov distribution, that is, the distribution of $\|B\|$, the distribution of the maximum of the Brownian bridge has a simple form. As the following result shows, this maximum has a Weibull distribution.

16.22 Theorem (Distribution of $\max_{0\le t\le 1} B(t)$)
We have

$$\mathbb{P}\left(\max_{0\le t\le 1} B(t) \le \xi\right) = 1 - \exp\left(-2\xi^2\right), \quad \xi \ge 0.$$

Proof The proof uses Theorem 16.21 and combinatorial arguments. For the special case $m = n$ that theorem reads

$$\sqrt{\frac{n}{2}} \max_{0\le t\le 1} \left(\widehat{F}_n(t) - \widehat{G}_n(t)\right) \xrightarrow{\mathcal{D}} \max_{0\le t\le 1} B(t) \quad \text{as } n \to \infty,$$

where $\widehat{F}_n$ and $\widehat{G}_n$ are the empirical distribution functions of independent random variables $X_1, \ldots, X_n$ and $Y_1, \ldots, Y_n$, respectively, each of which has a uniform distribution on [0, 1]. An equivalent formulation is

$$\sqrt{\frac{1}{2n}} \max_{0\le t\le 1} \left(\sum_{j=1}^n \mathbf{1}\{X_j \le t\} - \sum_{j=1}^n \mathbf{1}\{Y_j \le t\}\right) \xrightarrow{\mathcal{D}} \max_{0\le t\le 1} B(t). \tag{16.31}$$

According to the considerations made before Theorem 16.20, the above maximum depends only on the order in which the X_i and Y_j occur when passing through the unit interval from left to right. The difference in the sums of indicators occurring in (16.31) jumps up or down by one in each case if an X_i or a Y_j is noted in the ordered sample of all $2n$ random variables according to ascending values. By reasons of symmetry, we can thus imagine ordered arrangements of $2n$ symbols, of which n are equal to x and n are equal to y. All $\binom{2n}{n}$ such ordered arrangements are equally likely. If we replace each x by 1 and each y by -1, we arrive at an equivalent model in which n of the $2n$ values $0, 1, \ldots, 2n$, regarded as "times" and plotted on a horizontal axis, are randomly chosen as times for upward steps and the remaining ones for downward steps. To this end, we set

$$W_{2n}^\circ := \left\{ (a_1, \ldots, a_{2n}) \in \{-1, 1\}^{2n} : a_1 + \ldots + a_{2n} = 0 \right\}.$$

Here, $a_j = 1$ or $a_j = -1$ indicates an upward or downward step at time $j - 1$, where $j \in \{1, \ldots, 2n\}$. Each $2n$-tuple $(a_1, \ldots, a_{2n})$ in W_{2n}° defines a polygonal line that starts in $(0, 0)$ in a Cartesian coordinate system and connects the points $(0, 0)$, $(1, a_1)$, $(2, a_1 + a_2)$, $\ldots$, $(2n - 1, a_1 + \ldots + a_{2n-1})$ and $(2n, 0)$. Conversely, each such polygonal line, also called a *path*, corresponds exactly to a sequence of $2n$ symbols, of which n are equal to x and n are equal to y. The information that the j-th symbol is an x or a y is interpreted as an upward or a downward step at time $j - 1$, respectively, where $j \in \{1, \ldots, 2n\}$, see Fig. 16.5 for an illustration in the case $n = 4$.

In the following, let $\mathbb{P}_n$ be the uniform distribution on W_{2n}°. We write

$$M_{2n}((a_1, \ldots, a_{2n})) := \max \left\{ \sum_{j=1}^{k} a_j : k \in \{1, \ldots, 2n\} \right\}$$

for the maximum level that a path given by $(a_1, \ldots, a_{2n}) \in W_{2n}^\circ$ attains. Then M_{2n} is a random variable defined on W_{2n}°, and we have the following equality in distribution:

$$\sqrt{\frac{1}{2n}} \max_{0 \le t \le 1} \left(\sum_{j=1}^{n} \mathbf{1}\{X_j \le t\} - \sum_{j=1}^{n} \mathbf{1}\{Y_j \le t\} \right) \overset{\mathcal{D}}{=} \frac{M_{2n}}{\sqrt{2n}}.$$

We now show

$$\lim_{n \to \infty} \mathbb{P}_n \left(\frac{M_{2n}}{\sqrt{2n}} \le \xi \right) = 1 - \exp\left(-2\xi^2\right), \quad \xi > 0, \tag{16.32}$$

Fig. 16.5 Illustrating a path corresponding to the arrangement $x\, x\, y\, x\, x\, y\, y\, y$

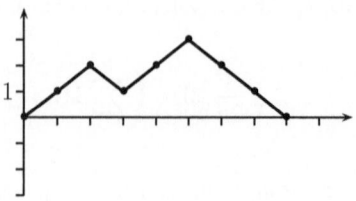

Fig. 16.6 Illustrating the reflection principle

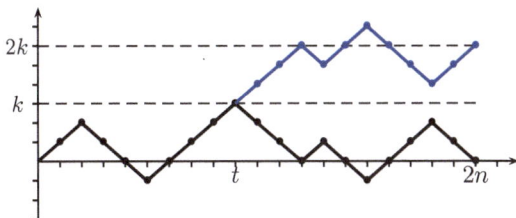

which proves Theorem 16.22. For this purpose, we first derive the equations

$$\mathbb{P}_n \left(M_{2n} \geq k \right) = \frac{\binom{2n}{n+k}}{\binom{2n}{n}}, \quad k \in \{0, 1, \ldots, n\}. \tag{16.33}$$

Since (16.33) is obviously valid for $k = 0$, we have to determine the number of all paths whose maximum level is at least equal to k, where $k \geq 1$. Each such path attains the level k at some abscissa t (say) for the *first time*. We now reflect the part of the path that starts at (t, k) across the line $y = k$. As a result, we obtain a path that ends at the point $(2n, 2k)$ and is marked blue in Fig. 16.6.

Conversely, given any path from $(0, 0)$ to $(2n, 2k)$, this path must attain the level k for the first time at some abscissa t. If we reflect the part of the path starting at (t, k) across the line $y = k$, we get a path starting at (t, k) and ending at $(2n, 0)$. It is obvious that this *reflection principle* defines a bijective mapping between all paths from $(0, 0)$ to $(2n, 0)$ whose maximum level is $\geq k$, and all paths from $(0, 0)$ to $(2n, 2k)$.

Since a path from $(0, 0)$ to $(2n, 2k)$ is determined by the choice of $n + k$ from the $2n$ points in time $0, 1, \ldots, 2n - 1$ at which the upward steps occur, the number of favorable paths is given by $\binom{2n}{n+k}$, which proves (16.33). For any $\xi > 0$ we now set $k_n := \lceil \xi \sqrt{2n} \rceil$. Since M_{2n} is integer-valued, (16.33) yields

$$\mathbb{P}_n \left(\frac{M_{2n}}{\sqrt{2n}} \geq x \right) = \mathbb{P}_n \left(M_{2n} \geq k_n \right) = \frac{\binom{2n}{n+k_n}}{\binom{2n}{n}} = \prod_{j=0}^{k_n-1} \left(1 - \frac{k_n}{n - j + k_n} \right).$$

Using the inequalities $1 - \frac{1}{t} \leq \log t \leq t - 1$, which hold for any $t > 0$, we obtain

$$\log \mathbb{P}_n \left(\frac{M_{2n}}{\sqrt{2n}} \geq x \right) \leq -k_n \sum_{j=0}^{k_n-1} \frac{1}{n - j + k_n} \leq -\frac{k_n^2}{n + k_n},$$

$$\log \mathbb{P}_n \left(\frac{M_{2n}}{\sqrt{2n}} \geq x \right) \geq -k_n \sum_{j=0}^{k_n-1} \frac{1}{n - j} \geq -\frac{k_n^2}{n - k_n + 1}.$$

Since

$$\lim_{n\to\infty} \frac{k_n^2}{n + k_n} = 2\xi^2 = \lim_{n\to\infty} \frac{k_n^2}{n - k_n + 1},$$

(16.32) follows.   □

16.23 Remark Using repeated reflections and the principle of inclusion and exclusion, one can also derive the Kolmogorov distribution (see, e.g., [HE0], Section 3.6).

Perhaps the concepts and results presented in this chapter have piqued your curiosity for more. On the one hand, a goodness-of-fit test that only tests a simple hypothesis of the type $H_0 : F = F_0$ is comparatively uninteresting. Of greater practical, but also theoretical interest are tests of a hypothesis of the type $H_0 : F \in \mathcal{F}_\Theta := \{F(\cdot, \vartheta) : \vartheta \in \Theta\}$. Here, $\mathcal{F}_\Theta$ is a class of distribution functions characterized by a finite-dimensional parameter. An important example in this respect is testing for normality. If the null hypothesis is $F \in \mathcal{F}_\Theta$, it makes sense to compare the empirical distribution function $\widehat{F}_n$ of $X_1, \ldots, X_n$ with the distribution function $F(\cdot, \widehat{\vartheta}_n)$. Here, $\widehat{\vartheta}_n = \vartheta_n(X_1, \ldots, X_n)$ is an estimator of the parameter $\vartheta \in \Theta$. If the estimation error $\widehat{\vartheta}_n - \vartheta$ allows of a representation of the form (10.30), and if the function that maps $\vartheta \in \Theta$ into $F(x, \vartheta)$ is "smooth" for each x, one can hope to make the approximation

$$\widehat{F}_n(t) - F(t, \widehat{\vartheta}_n) \approx \widehat{F}_n(t) - F(t, \vartheta) + \frac{\partial}{\partial \vartheta} F(t, \vartheta)\big(\widehat{\vartheta}_n - \vartheta\big)$$

watertight and to get a handle on the asymptotic behavior of test statistics based on functionals of the difference $\widehat{F}_n(\cdot) - F(\cdot, \widehat{\vartheta}_n)$. A treasure trove for related results is [SHW]. This book also contains a brief introduction to the general theory of empirical processes. The starting point of such a theory is a sequence of random elements on a probability space $(\Omega, \mathcal{A}, \mathbb{P})$ with values in a general set $\mathcal{X}$, which is equipped with a σ-field $\mathcal{C}$. The so-called *empirical measure* associated with $X_1, \ldots, X_n$ is $P_n := \frac{1}{n} \sum_{j=1}^n \delta_{X_j}$. Here, δ_x stands for the Dirac measure at the point $x \in \mathcal{X}$. If $P := \mathbb{P} \circ X_1^{-1}$ denotes the distribution of X_1, then $Z_n := \sqrt{n}(P_n - P)$ defines the *empirical process* of $X_1, \ldots, X_n$. If $\mathcal{F}$ is a collection of measurable real-valued functions on $\mathcal{X}$ with $\int_{\mathcal{X}} |f| \, dP < \infty$, and if we set $Q(f) := \int_{\mathcal{X}} f \, dQ$ for a general probability measure Q on $\mathcal{C}$, then

$$Z_n(f) := \sqrt{n}\big(P_n(f) - P(f)\big) = \sqrt{n}\Big(\frac{1}{n} \sum_{j=1}^n \mathbb{E}[f(X_j)] - \mathbb{E}[f(X_1)]\Big), \quad f \in \mathcal{F},$$

is called the *empirical process indexed by* $\mathcal{F}$. In the special case $\mathcal{X} = [0, 1], \mathcal{C} = \mathcal{B}^1 \cap [0, 1]$ and $\mathcal{F} := \{\mathbf{1}_{(-\infty, t]} : t \in [0, 1]\}$, we have

$$Z_n(t) := Z_n(\mathbf{1}_{(-\infty, t]}) = \sqrt{n}\left(\frac{1}{n}\sum_{j=1}^{n} \mathbf{1}\{X_j \leq t\} - F(t)\right), \quad 0 \leq t \leq 1,$$

where $F(t) := \mathbb{P}(X_1 \leq t)$. Thus $Z_n(\cdot)$ is the empirical process defined in (16.21).

A newer and comprehensive presentation of the general theory of empirical processes is provided by [VW1].

Answers to the Self-Questions

Answer 1 With the decomposition given in Lemma 16.2, $|x(t)| \leq \max_{0 \leq j \leq k} |x(t_j)| + \varepsilon$.

Answer 2 Because for each $\varepsilon > 0$ there are only finitely many $t \in (0, 1]$ with $|x(t) - x(t-)| > \varepsilon$. In this respect, the supremum is a maximum.

Answer 3 If $m > n$ the triangle inequality implies

$$d_S(x_n, x_m) \leq \sum_{j=0}^{m-n+1} d_S(x_{n+j}, x_{n+j-1}) \leq \sum_{j=0}^{m-n+1} \frac{1}{2^{n+j-1}} \leq \frac{1}{2^n}.$$

Answer 4 If t is a continuity point of x, then the second summand on the right-hand side of (16.6) converges to 0 because $g_n(t) \to t$. The same applies to the first summand, as $\|x_n - x \circ g_n\|$ converges to zero.

Answer 5 For the decomposition $t_0 := 0$, $t_1 := 0.5$, and $t_2 := 1$ of $[0, 1]$ we have

$$\sup_{s, t \in [t_0, t_1)} |x_n(s) - x_n(t)| = 0, \qquad \sup_{s, t \in [t_1, t_2)} |x_n(s) - x_n(t)| = 0, \quad n \geq 1.$$

Thus $\sup_{x \in A} w'_x(\delta) = 0$, if $\delta < 0.5$.

Answer 6 The crucial point is that there is a minimum on the right-hand side of (16.10). By definition of x_n this minimum is equal to zero, if $t_2 < \frac{1}{n}$ or $t_1 \geq \frac{1}{n}$, but also in the remaining case $t_1 < \frac{1}{n}, t_2 \geq \frac{1}{n}$.

Answer 7 If U and V are any random variables that are defined on the same probability space, then (pointwise on that space)

$$1\{|U| \wedge |V| \geq \gamma\} \leq \frac{|U|^{2\beta}|V|^{2\beta}}{\gamma^{4\beta}}.$$

If we take expectations, the assertion follows.

Answer 8 Because $X_n(t) = (\sigma\sqrt{n})^{-1} \sum_{j=1}^{\lfloor nt \rfloor} Z_j$ and $\mathbb{E}(Z_j) = 0$ as well as $\mathbb{V}(Z_j) = \sigma^2$ we have

$$\mathbb{E}\big(|X_n(s) - X_n(r)|^2\big) = \mathbb{V}\bigg(\frac{1}{\sigma\sqrt{n}} \sum_{j=\lfloor nr \rfloor+1}^{\lfloor ns \rfloor} Z_j\bigg) = \frac{\sigma^2(\lfloor ns \rfloor - \lfloor nr \rfloor)}{\sigma^2 n}.$$

Answer 9 If $t - r \geq \frac{1}{n}$, then $1 \leq nt - nr$ and thus $\lfloor nt \rfloor - \lfloor nr \rfloor \leq nt - (nr - 1) \leq n(t - r) + 1 \leq 2n(t - r)$.

Answer 10 Because both F and x are right-continuous functions with existing left-hand limits.

Answer 11 Let P_j be the distribution of X_j ($j = 1, 2$). Due to independence, $P_1 \otimes P_2$ is the joint distribution of X_1 and X_2. The continuity of F implies $P_j(\{x\}) = 0$ for each $x \in \mathbb{R}$ ($j = 1, 2$). Putting $\Delta := \{(x, x) : x \in \mathbb{R}\}$, Tonelli's theorem yields

$$\mathbb{P}(X_1 = X_2) = \int_\Delta P_1 \otimes P_2(\mathrm{d}x, \mathrm{d}y) = \int_\mathbb{R} P_2(\{x\})P_1(\mathrm{d}x) = 0.$$

Answer 12 Taking the limit $m, n \to \infty$ means that m and n are members of two increasing subsequences $(m_s)_{s\geq1}$ and $(n_s)_{s\geq1}$ and s tends to infinity.

Problems

16.1 Problem Show that every function in D has at most countably many discontinuity points.

16.2 Problem Let $x : [0, 1] \to \mathbb{R}$ be any function. Show: If $\lim_{\delta\to0} w'_x(\delta) = 0$, then $x \in D[0, 1]$.

16.3 Problem Let $x \in D$. Prove the following inequalities between the càdlàg-modulus w'_x and the modulus of continuity w_x:

(a) $w'_x(\delta) \le w_x(2\delta)$, if $\delta < \frac{1}{2}$,

(b) $w_x(\delta) \le 2w'_x(\delta) + H(x)$, if $0 < \delta < 1$. Here, $H(x) = \sup_{0 < t \le t} |x(t) - x(t-)|$.

16.4 Problem Let $\mathcal{G}$ be the set of all continuous, strictly increasing and bijective mappings $g : [0, 1] \to [0, 1]$. For $x, y \in D[0, 1]$ let

$$d_S(x, y) := \inf_{g \in \mathcal{G}} \max \left(\sup_{t \in [0,1]} |g(t) - t|, \; \sup_{t \in [0,1]} |x(t) - y(g(t))| \right).$$

Show that d_S is a metric on $D[0, 1]$.

16.5 Problem Prove the following claims:

(a) If $g \in \mathcal{G}$ then $\|g - \mathrm{I}\| \le e^{\|g\|^\circ} - 1$.

(b) $d_S(x, y) \le e^{d_S^\circ(x,y)} - 1$, $x, y \in D$.

Hint for (a) $|u - 1| \le e^{|\log u|} - 1$ if $u > 0$ and

$$|g(t) - t| = t \left| \frac{g(t) - g(0)}{t - 0} - 1 \right|, \quad t > 0.$$

Hint for (b) $v \le e^v - 1$, $v \in \mathbb{R}$.

16.6 Problem For $g \in \mathcal{G}$, let $\|g\|^\circ$ be defined as in (16.7). Prove the following statements:

(a) $\|g^{-1}\|^\circ = \|g\|^\circ$, $g \in \mathcal{G}$.

(b) $\|g_1 \circ g_2\|^\circ \le \|g_1\|^\circ + \|g_2\|^\circ$, $g_1, g_2 \in \mathcal{G}$.

16.7 Problem Show the inequality

$$w''_x(\delta) \le w'_x(\delta) \qquad x \in D, \; 0 < \delta < 1.$$

Hint Assume $w'_x(\delta) < \eta$ for some η and consider a δ-thin decomposition $0 = s_0 < s_1 < \dots < s_k = 1$ of $[0, 1]$ with $w_x([s_{i-1}, s_i)) < \eta$ for each $i \in \{1, \dots, k\}$. What do the conditions $t_2 - t_1 \le \delta$ and $t_1 \le t \le t_2$ occurring in (16.10) mean for this decomposition?

16.8 Problem Prove the following claim: If $x, x_1, x_2, \dots$ are functions in D with $d_S(x_n, x) \to 0$ then

$$\lim_{n \to \infty} \int_0^1 |x_n(t) - x(t)| \, dt = 0.$$

In particular, the functional $D \ni x \mapsto \int_0^1 x(t)\, dt$ defined on D is continuous with respect to the Skorokhod topology.

16.9 Problem For $x \in D$, let $H(x) := \sup_{0 < t \leq 1} |x(t) - x(t-)|$ be the greatest jump of x. Show that for $x, y \in D$ and $\varepsilon > 0$:

(a) If $\|x - y\| < \varepsilon$ then $|H(x) - H(y)| < 2\varepsilon$.
(b) If $d_S(x, y) < \varepsilon$ then $|H(x) - H(y)| < 2\varepsilon$.

The function $H : D \to \mathbb{R}$ is therefore continuous with respect to the Skorokhod topology.

16.10 Problem Prove that C[0, 1] is a closed and nowhere dense subset of D[0, 1].

Hint Consider Problem 16.9.

16.11 Problem Show that with the notations of the proof of Theorem 16.15:

(a) $\mathbb{E}[\alpha_1^2]\mathbb{E}[\beta_2^2] = (s - r)(1 - (s - r))(t - s)(1 - (t - s)) \leq (t - r)^2,$
(b) $\mathbb{E}[\alpha_1\beta_1]\mathbb{E}[\alpha_2\beta_2] = \{-(s - r)(t - s)\}^2 \leq (t - r)^2.$

16.12 Problem Prove that the process $Y = \psi(B)$ obtained in the proof of Theorem 16.15 is a centered Gaussian process with covariance function $\mathrm{Cov}(Y(s), Y(t)) = F(s) \wedge F(t) - F(s)F(t)$, where $0 \leq s, t \leq 1$.

16.13 Problem Show:

(a) The mapping $h : D \to \mathbb{R}$ defined by $h(x) := \|x\|$ is $(\mathcal{B}(D), \mathcal{B}^1)$-measurable.
(b) Let $x, x_1, x_2, \ldots \in D$ with $d_S(x_n, x) \to 0$ and $x \in C$. Then $\|x_n\| \to \|x\|$.

Hint for (a) It is necessary to show that $\{x \in D : \|x\| < a\} \in \mathcal{B}(D)$ for every $a > 0$. All projections $\pi_t \circ x, 0 \leq t \leq 1$, are measurable.

16.14 Problem Let $h : D \to \mathbb{R}$ be defined by $h(x) := \int_0^1 x^2(t)\, dt$. Show:

(a) h is $(\mathcal{B}(D), \mathcal{B}^1)$-measurable.
(b) h is continuous on C with respect to the Skorokhod topology.

16.15 Problem Show that the Cramér–von Mises statistic defined in (16.25) takes the form

$$n \int_0^1 \left(\widehat{F}_n\left(F_0^{-1}(t)\right) - t \right)^2 \mathrm{d}t = \frac{1}{12n} + \sum_{j=1}^n \left(F_0\left(X_{(j)}\right) - \frac{2j-1}{2n} \right)^2. \qquad (16.34)$$

16.16 Problem Consider the testing problem $H_0 : F = F_0$ against $H_1 : F \neq F_0$ from Sects. 16.16 and 16.18, and suppose that $F \neq F_0$. Prove that $\sqrt{n}K_n \to \infty$ $\mathbb{P}$-a.s. and $\omega_n^2 \to \infty$ $\mathbb{P}$-a.s. as $n \to \infty$.

16.15 Problem

$$\ldots$$

16.16 Problem

This chapter deals with random elements that take on values in a separable infinite-dimensional Hilbert space $\mathbb{H}$. Such $\mathbb{H}$-valued random elements play an important role in so-called *functional data analysis* (see, e.g., [HOK]). The crucial difference to the finite-dimensional case discussed in Chap. 6 is that balls with respect to the norm in $\mathbb{H}$ are not relatively compact. In generalization of the notions of an expectation and a covariance matrix of a d-dimensional random vector, we get to know how one defines an expectation and a covariance operator of an $\mathbb{H}$-valued random element, and what conditions must apply for them to exist. Topics of this chapter also include the characteristic functional, the normal distribution in a separable Hilbert space, and a central limit theorem for triangular arrays of row-wise independent $\mathbb{H}$-valued random elements. The chapter concludes with an application of the theory to a statistical problem. Background knowledge on Hilbert spaces and the required operator theory can be found, e.g., in [HAA], and [WEI]. A compendium of normal distributions in Hilbert spaces is [KUK].

17.1 Hilbert Spaces: Basics

In the following, let $\mathbb{H}$ denote a separable Hilbert space over $\mathbb{R}$ with inner product $\langle x, y \rangle$, $x, y \in \mathbb{H}$, and norm $\|x\| := \sqrt{\langle x, x \rangle}$. Thus, $\| \cdot \|$ stands for the norm on $\mathbb{H}$ (which renders $\mathbb{H}$ a Banach space) and not for the Euclidean norm in $\mathbb{R}^d$ in this chapter. If $\{e_1, e_2, \ldots\}$ is an orthonormal basis (in short: ONB) of $\mathbb{H}$, then for $x, y \in \mathbb{H}$:

$$x = \sum_{k=1}^{\infty} \langle x, e_k \rangle e_k \quad \left(:\Longleftrightarrow \lim_{n \to \infty} \left\| x - \sum_{k=1}^{n} \langle x, e_k \rangle e_k \right\| = 0 \right),$$

$$\|x\|^2 = \sum_{k=1}^{\infty} \langle x, e_k \rangle^2, \tag{17.1}$$

N. Henze, *Asymptotic Stochastics*, Mathematics Study Resources 10, https://doi.org/10.1007/978-3-662-68923-3_17

$$\langle x, y \rangle^2 \leq \|x\|^2 \|y\|^2, \tag{17.2}$$

$$\langle x, y \rangle = \sum_{k=1}^{\infty} \langle x, e_k \rangle \langle e_k, y \rangle. \tag{17.3}$$

Equation (17.1) is called *Parseval's*[1] *equation*, and (17.2) is the *Cauchy–Schwarz inequality*. Equation (17.3) is called the *generalized Parseval's equation* (see Problem 17.1). From (17.1) we obtain *Bessel's*[2] *inequality*

$$\sum_{k=1}^{\ell} \langle x, e_k \rangle^2 \leq \|x\|^2, \quad \ell \geq 1. \tag{17.4}$$

The metric $\rho(x, y) = \|x - y\|$ renders $(\mathbb{H}, \rho)$ a complete separable metric space. It should be emphasized that we occasionally also use a different notation for the elements of $\mathbb{H}$, namely f, g and h.

17.2 Example

(a) The d-dimensional Euclidean space $\mathbb{R}^d$ becomes a Hilbert space when equipped with the inner product $\langle x, y \rangle = x_1 y_1 + \ldots + x_d y_d$, where $x = (x_1, \ldots, x_d)$ and $y = (y_1, \ldots, y_d)$ are in $\mathbb{R}^d$.

(b) The set $\mathbb{H} := \ell^2 := \{x = (x_k)_{k \geq 1} \in \mathbb{R}^\infty : \sum_{k=1}^{\infty} x_k^2 < \infty\}$ of all square summable sequences of real numbers is a separable Hilbert space, if we define an inner product via $\langle x, y \rangle := \sum_{k=1}^{\infty} x_k y_k$, where $x = (x_k)_{k \geq 1}$ and $y = (y_k)_{k \geq 1}$ are in ℓ^2.

(c) In generalization of (b), let $(\Omega, \mathcal{A}, \mu)$ be a measure space, where $\mathcal{A} = \sigma(\mathcal{M})$ for a *countable* collection $\mathcal{M}$ of subsets of Ω. Then the set $\mathbb{H} := L^2(\Omega, \mathcal{A}, \mu)$ of all equivalence classes of measurable functions $f : \Omega \to \mathbb{R}$ with $\int_\Omega f^2 \, d\mu < \infty$ is a separable Hilbert space with the inner product

$$\langle f, g \rangle := \int_\Omega f g \, d\mu.$$

Here, we also follow the custom of noting the representative f of this equivalence class instead of the equivalence class $[f]$ of all functions that are equal to f almost everywhere with respect to μ.

---

[1] Marc-Antoine Parseval (1755–1836), French mathematician, known for the Parseval's equation named after him.

[2] Friedrich Wilhelm Bessel (1784–1846), astronomer, mathematician, geodesist and physicist, from 1810 professor of astronomy at the University of Königsberg. The *Bessel functions* named after him enable a mathematical description of many physical phenomena. He proved inequality (17.4) in 1828 in the special case of Fourier series.

Self-question 1: Why is (b) a special case of (c)?

(d) An important special case of (c) results from the choice $\Omega := [0, 1]$, $\mathcal{A} = \mathcal{B}^1 \cap [0, 1]$, and $\mu = \lambda^1_{|[0,1]}$. This Hilbert space is briefly denoted by $L^2[0, 1]$. Among other things, it has the following orthonormal bases (see, e.g., [HSE], Theorem 2.4.18):
   (i) $B_1 := \{e_0(t) = 1, \ e_n(t) = \sqrt{2}\cos(n\pi t), \ n \geq 1\}$,
   (ii) $B_2 := \{e_n(t) = \sqrt{2}\sin(n\pi t), \ n \geq 1\}$,
   (iii) $B_3 := \{e_0(t) := 1, \ e_{2n-1}(t) = \sqrt{2}\sin(2n\pi t), \ e_{2n}(t) = \sqrt{2}\cos(2n\pi t), \ n \geq 1\}$.

Every infinite-dimensional separable Hilbert space $\mathbb{H}$ is isomorphic to the Hilbert space ℓ^2 listed in Example 17.2 (b): If $\{e_1, e_2, \ldots\}$ is any ONB of $\mathbb{H}$, and if one defines a mapping $T : \mathbb{H} \to \ell^2$ by $T(x) := (\langle x, e_k\rangle)_{k\geq 1}$, then T indeed has the range ℓ^2 according to (17.2) and is bijective. If $\langle \cdot, \cdot\rangle_2$ denotes the inner product in ℓ^2, then due to (17.3) we also have

$$\langle T(x), T(y)\rangle_2 = \sum_{k=1}^{\infty}\langle x, e_k\rangle\langle e_k, y\rangle = \langle x, y\rangle.$$

Self-question 2: Why is T injective?

We have already encountered integral operators on the Hilbert space $L^2(\mathbb{R}^d, \mathcal{B}^d, \mathrm{d}F)$ in Chap. 8 (see the considerations after (8.25)). In the following, we compile some important properties of operators on Hilbert spaces.

17.3 Properties of Operators
A mapping $T : \mathbb{H} \to \mathbb{H}$ is called an *operator* (on $\mathbb{H}$). Usually one writes $Tx := T(x)$, $x \in \mathbb{H}$, for short. The operator T is called

- *linear*, if $T(ax + by) = a\,Tx + b\,Ty$, $x, y \in \mathbb{H}$, $a, b \in \mathbb{R}$,
- *bounded*, if there is a $K \in [0, \infty)$ with $\|Tx\| \leq K\,\|x\|$ for every $x \in \mathbb{H}$,
- *compact*, if $T(B)$ is relatively compact whenever B is a bounded subset of $\mathbb{H}$,
- *self-adjoint*, if $\langle Tx, y\rangle = \langle x, Ty\rangle$, $x, y \in \mathbb{H}$,
- *positive*, if $\langle Tx, x\rangle \geq 0$ for every $x \in \mathbb{H}$.

A compact linear operator is called a *trace-class operator*, if

$$\sum_{k=1}^{\infty}|\langle e_k, Te_k\rangle| < \infty. \tag{17.5}$$

Here, $\{e_1, e_2, \ldots\}$ is any ONB of $\mathbb{H}$. In this case, the number

$$\mathrm{tr}(T) := \sum_{k=1}^{\infty} \langle e_k, T e_k \rangle$$

is called the *trace of T*. Both condition (17.5) and the definition of the trace do not depend on the specific choice of an ONB of $\mathbb{H}$ (see, e.g., [WEI], p. 176).

In the case $\mathbb{H} = \mathbb{R}^d$, every $d \times d$ matrix A provides a linear operator via the specification $T(x) := Ax$, $x \in \mathbb{R}^d$. Here, x is to be understood as a column vector. Every such operator is bounded and compact, and it is self-adjoint, if A is a symmetric matrix. Under the latter condition, T is positive if A is positive-semidefinite. However, $\langle Ax, x \rangle > 0$ for every $x \neq 0$ can apply without A being symmetric. As an example, consider the rotation matrix

$$A = \begin{pmatrix} \cos \varphi & -\sin \varphi \\ \sin \varphi & \cos \varphi \end{pmatrix}$$

with $-\frac{\pi}{2} < \varphi < \frac{\pi}{2}$. The notion of trace obviously generalizes the corresponding notion for a symmetric square matrix.

17.4 Remark We will later need that an operator T is uniquely determined by specifying all inner products $\langle Tx, y \rangle$ with $x, y \in \mathbb{H}$. Namely, suppose that for another operator $S : \mathbb{H} \to \mathbb{H}$ we have $\langle Tx, y \rangle = \langle Sx, y \rangle$ for every $x, y \in \mathbb{H}$. Putting $y := (T - S)x$, it follows that $\|(T - S)x\|^2 = 0$ for each $x \in \mathbb{H}$ and thus $S = T$.

A linear mapping $\ell : \mathbb{H} \to \mathbb{R}$ is called a *linear functional*. A linear functional ℓ is said to be *bounded*, if $\|\ell\|^* := \sup\{|\ell(x)| : x \in \mathbb{H}, \|x\| = 1\} < \infty$. For bounded linear functionals, the following result, which goes back to F. Riesz[3], applies. For a proof see, e.g., [HSE], p. 66.

17.5 Theorem (Riesz Representation Theorem)
Suppose ℓ is a bounded linear functional. Then there is exactly one z in $\mathbb{H}$ with

$$\ell(x) = \langle z, x \rangle, \quad x \in \mathbb{H}.$$

Furthermore, $\|\ell\|^* = \|z\|$.

---

[3] Frigyes Riesz (1880–1956), Hungarian mathematician, appointed to a chair at the University of Klausenburg in 1911. Riesz was one of the founders of functional analysis. Together with Alfréd Haar, he founded the János Bolyai Institute for Mathematics in 1922. From 1945 he was a professor in Budapest.

We now turn to $\mathbb{H}$-valued random elements and their distributions. For this purpose, we denote $\mathcal{O}$ the collection of open subsets of $\mathbb{H}$ and $\mathcal{B} := \mathcal{B}(\mathbb{H}) := \sigma(\mathcal{O})$ the σ-field of Borel sets. According to Problem 13.3, $\mathcal{B}$ is generated by the collection of all open balls and thus also by the collection of all closed balls, because $\mathbb{H}$ is separable. The following result provides another generating class of sets, defined using the inner product.

17.6 Theorem Let $\mathcal{M} := \big\{\{x \in \mathbb{H} : \langle x, y \rangle \in B\}\big| y \in \mathbb{H},\ B \in \mathcal{O}^1\big\}$. Then $\sigma(\mathcal{M}) = \mathcal{B}$.

Proof Note that $\mathcal{M} \subset \mathcal{O}$ and thus $\sigma(\mathcal{M}) \subset \sigma(\mathcal{O}) = \mathcal{B}$.

Self-question 3: Why does $\mathcal{M} \subset \mathcal{O}$ hold?

To show $\sigma(\mathcal{M}) \supset \mathcal{B}$, let $\mathcal{G} := \big\{\{x \in \mathbb{H} : \|x - y\| > \varepsilon\}\big| y \in \mathbb{H},\ \varepsilon > 0\big\}$ be the collection of complements of closed balls with respect to $\|\cdot\|$. With Problem 13.3, we have $\sigma(\mathcal{G}) = \mathcal{B}$, and therefore only $\mathcal{G} \subset \sigma(\mathcal{M})$ needs to be shown. To this end, let $\{e_1, e_2, \ldots\}$ be any ONB of $\mathbb{H}$. Due to Parseval's equation (17.1), it follows that

$$\{x : \|x\| > \varepsilon\} = \bigcup_{j=1}^{\infty}\{x : \textstyle\sum_{k=1}^{j}\langle x, e_k\rangle^2 > \varepsilon^2\}.$$

Each of the sets $A_j := \{x : \sum_{k=1}^{j}\langle x, e_k\rangle^2 > \varepsilon^2\}$, $j \geq 1$, belongs to $\sigma(\mathcal{M})$, because initially $A_1 \in \mathcal{M}$, and due to

$$\{x : \langle x, e_1\rangle^2 + \langle x, e_2\rangle^2 > \varepsilon^2\} = \bigcup_{q\in\mathbb{Q}}\big(\{x : \langle x, e_1\rangle^2 > q\} \cap \{x : \langle x, e_2\rangle^2 > \varepsilon^2 - q\}\big),$$

it follows that $A_2 \in \sigma(\mathcal{M})$. Inductively, we get $A_j \in \sigma(\mathcal{M})$ for each $j \geq 1$ and thus $\{x : \|x\| > \varepsilon\} \in \sigma(\mathcal{M})$. For each $y \in \mathbb{H}$ we have

$$\{x : \|x - y\| > \varepsilon\} = \bigcup_{q\in\mathbb{Q}}\big(\{x : \|x\|^2 > q\} \cap \{x : 2\langle x, y\rangle < q - \varepsilon^2 + \|y\|^2\}\big).$$

Consequently, $\{x : \|x - y\| > \varepsilon\} \subset \sigma(\mathcal{M})$, because $\{x : \|x\|^2 > q\} \in \sigma(\mathcal{M})$ as already shown, and the set to the right of the intersection symbol belongs to $\mathcal{M}$. Thus, the complements of closed balls lie in $\sigma(\mathcal{M})$, and the claim follows.   □

17.7 Corollary (Measurability of $\mathbb{H}$-Valued Random Elements)
Suppose $(\Omega, \mathcal{A}, \mathbb{P})$ is a probability space, and $X : \Omega \to \mathbb{H}$ is a mapping. Then the following holds:

$$X \text{ is } (\mathcal{A}, \mathcal{B})\text{-measurable} \iff \langle X, y \rangle \text{ is } (\mathcal{A}, \mathcal{B}^1)\text{-measurable for every } y \in \mathbb{H}.$$

Proof The proof is left to you as Problem 17.4.

If $\mathcal{M}_\cap$ denotes the collection of all finite intersections of sets from the class of sets $\mathcal{M} := \{\{x \in \mathbb{H} : \langle x, y \rangle \in B\} | y \in \mathbb{H}, B \in \mathcal{O}^1\}$ introduced in Theorem 17.6, then the distribution $\mathbb{P}^X$ of an $\mathbb{H}$-valued random element X is uniquely determined by the values $\mathbb{P}(X \in M)$, where $M \in \mathcal{M}_\cap$.   □

> **Self-question 4**: Why does this statement hold?

17.8 Theorem (The Distributions of $\langle X, y \rangle$, $y \in \mathbb{H}$, Determine $\mathbb{P}^X$)

The distribution $\mathbb{P}^X$ of an $\mathbb{H}$-valued random element on a probability space $(\Omega, \mathcal{A}, \mathbb{P})$ is uniquely determined by the totality of the distributions of $\langle X, y \rangle$, where $y \in \mathbb{H}$.

Proof Every set $M \in \mathcal{M}_\cap$ has the form $M = \{x \in \mathbb{H} : \langle x, y_1 \rangle \in B_1, \ldots, \langle x, y_k \rangle \in B_k\}$ for a $k \geq 1$ as well as $y_1, \ldots, y_k \in \mathbb{H}$ and open sets $B_1, \ldots, B_k \subset \mathbb{R}$. This results in

$$\mathbb{P}^X(M) = \mathbb{P}\left(\langle X, y_1 \rangle \in B_1, \ldots, \langle X, y_k \rangle \in B_k\right)$$

$$= \mathbb{P}^{(\langle X, y_1 \rangle, \ldots, \langle X, y_k \rangle)}(B_1 \times \ldots \times B_k).$$

By the Herglotz-Radon-Cramér-Wold device (Theorem 1.17) the distribution of the k-dimensional random vector $(\langle X, y_1 \rangle, \ldots, \langle X, y_k \rangle)$ is determined by the distributions of

$$\sum_{j=1}^k c_j \langle X, y_j \rangle = \left\langle X, \sum_{j=1}^k c_j y_j \right\rangle, \quad (c_1, \ldots, c_k) \in \mathbb{R}^k,$$

from which the assertion follows.   □

Like real-valued random variables or d-dimensional random vectors, also a Hilbert space-valued random element has an expectation under certain conditions.

17.9 Definition (Expectation of an $\mathbb{H}$-Valued Random Element)

Suppose X is a $\mathbb{H}$-valued random element with $\mathbb{E}|\langle X, y \rangle| < \infty$ for each $y \in \mathbb{H}$. If there is an m in $\mathbb{H}$ such that

$$\langle m, y \rangle = \mathbb{E}\langle X, y \rangle \quad \text{for each } y \in \mathbb{H}, \tag{17.6}$$

then m is called the *expectation of* X, and we write $\mathbb{E}(X) := m$. The random element X is called *centered* if $\mathbb{E}(X) = \mathbf{0}$, where $\mathbf{0}$ denotes the zero element of $\mathbb{H}$.

Since m in (17.6) is uniquely determined, the expectation of X—if it exists at all—is characterized by the equations

$$\langle \mathbb{E}X, y \rangle = \mathbb{E}\langle X, y \rangle, \quad y \in \mathbb{H}. \tag{17.7}$$

Here, we have briefly written $\mathbb{E}X = \mathbb{E}(X)$ and will often do so in the sequel.

Self-question 5: Why is m in (17.6) uniquely determined?

The message of (17.7) is that, by definition, the expectation *commutes with the inner product*. If $X = (X_1, \ldots, X_d)^\top$ is a d-dimensional random vector, we defined the expectation of X according to $\mathbb{E}(X) := (\mathbb{E}X_1, \ldots, \mathbb{E}X_d)^\top$. Since taking expectations is a linear operator, it follows that

$$\langle \mathbb{E}X, y \rangle = \sum_{j=1}^{d} \mathbb{E}X_j \, y_j = \mathbb{E}\left(\sum_{j=1}^{d} X_j \, y_j \right) = \mathbb{E}\langle X, y \rangle, \quad y = (y_1, \ldots, y_d)^\top \in \mathbb{R}^d.$$

Hence, Definition 5.1 (a) is compatible with Definition 17.9.

17.10 Theorem *If $\mathbb{E}\|X\| < \infty$, then $\mathbb{E}X$ exists.*

Proof For any $y \in \mathbb{H}$, we have $\mathbb{E}|\langle X, y \rangle| \leq \mathbb{E}(\|X\| \|y\|) = \|y\| \mathbb{E}\|X\| < \infty$. Thus, $\mathbb{E}\langle X, y \rangle$ exists for each $y \in \mathbb{H}$. If one sets $\ell(y) := \mathbb{E}\langle X, y \rangle$, $y \in \mathbb{H}$, then $\ell : \mathbb{H} \to \mathbb{R}$ is a linear functional on $\mathbb{H}$. Furthermore, $|\ell(y)| \leq \mathbb{E}\|X\| \|y\|$, $y \in \mathbb{H}$. Consequently, ℓ is bounded, and therefore, by the Riesz representation theorem (Theorem 17.5), there is exactly one m in $\mathbb{H}$ with $\ell(y) = \langle m, y \rangle$ for each $y \in \mathbb{H}$. □

In connection with the above statement, we emphasize that $\mathbb{E}\|X\| = \infty$ can hold, although $\mathbb{E}|\langle X, y \rangle| < \infty$ for each $y \in \mathbb{H}$ (cf. Problem 17.5).

17.11 Example Suppose the random element X takes on finitely many values $x_1, \ldots, x_k$ in $\mathbb{H}$, i.e., $X(\Omega) = \{x_1, \ldots, x_k\}$. Then $\|X\| \leq \max(\|x_1\|, \ldots, \|x_k\|)$ and thus $\mathbb{E}\|X\| < \infty$. Since for each $y \in \mathbb{H}$ the inner product $\langle X, y \rangle$ takes on the values $\langle x_j, y \rangle$, $j \in \{1, \ldots, k\}$, we get

$$\mathbb{E}\langle X, y \rangle = \sum_{j=1}^{k} \langle x_j, y \rangle \mathbb{P}(X = x_j) = \left\langle \sum_{j=1}^{k} x_j \mathbb{P}(X = x_j), y \right\rangle.$$

As a result of the second equals sign, a comparison with (17.7) shows that in this case we obtain the expectation of X in the form

$$\mathbb{E}(X) = \sum_{j=1}^{k} x_j \, \mathbb{P}(X = x_j), \tag{17.8}$$

which is familiar from real-valued random variables, namely as "sum of value times probability" and thus as a weighted average of the realizations of X.

Representation (17.8) suggests that, for general $\mathbb{H}$-valued random elements X with $\mathbb{E}\|X\| < \infty$, the expectation of X can be defined as a suitable integral of the form

$$\mathbb{E}(X) = \int_{\Omega} X \, d\mathbb{P}. \tag{17.9}$$

Before we pursue this question, it should be pointed out that the following result about expectations can be obtained solely with the help of the characterizing equations (see Problem 17.7).

17.12 Theorem (The Functional $\mathbb{E}(\cdot)$ is Linear)
Suppose $(\Omega, \mathcal{A}, \mathbb{P})$ is a probability space, and denote L^1 the set of all $\mathbb{H}$-valued random elements $X : \Omega \to \mathbb{H}$ with $\mathbb{E}\|X\| < \infty$. The set L^1 is a vector space over $\mathbb{R}$, and we have

$$\mathbb{E}(\alpha X + \beta Y) = \alpha \, \mathbb{E} X + \beta \, \mathbb{E} Y, \quad \alpha, \beta \in \mathbb{R}, \; X, Y \in L^1.$$

The following concept, going back to S. Bochner,[4] allows to define a measure integral for functions that take on values in a Banach space. As a special case, this results in an expectation of the form (17.9) for Hilbert space-valued random elements. We will see that this expectation is identical to the one obtained from Definition 17.9.

17.13 The Bochner Integral (1933)
Let $(\Omega, \mathcal{A}, \mu)$ be a measure space and $(\mathbb{B}, \|\cdot\|_{\mathbb{B}})$ a Banach space with Borel σ-field $\mathcal{B}$. Suppose we want to assign an integral

$$\int_{\Omega} f \, d\mu$$

---

[4] Salomon Bochner (1899–1982), grew up near Krakow, 1921 PhD (Univ. Berlin), 1927 Habilitation (Univ. Munich); after the seizure of power by the Nazis emigration to the USA, 1934–1968 Professor at Princeton Univ., after retirement Professor at Rice Univ. in Houston. Main areas of work: Almost periodic functions, Fourier theory, complex analysis of several variables, differential geometry.

to as many $(\mathcal{A}, \mathcal{B})$-measurable functions $f : \Omega \to \mathbb{B}$ as possible. As in the construction of the measure integral $\int_\Omega f \, d\mu$ for real-valued or $[-\infty, \infty]$-valued functions f, it seems natural to start with simple functions. With regard to the construction of a $\mathbb{B}$-valued integral, a measurable function $f : \Omega \to \mathbb{B}$ is called *simple* if there are an integer k and sets $A_1, \ldots, A_k$ in $\mathcal{A}$ as well as $b_1, \ldots, b_k$ in $\mathbb{B}$, such that f is of the form

$$f = \sum_{j=1}^k \mathbf{1}\{A_j\} \, b_j \tag{17.10}$$

and thus has a finite range. W.l.o.g. the sets $A_1, \ldots, A_k$ can be assumed to be pairwise disjoint with $\Omega = A_1 \uplus \ldots \uplus A_k$.

Self-question 6: Why can the latter assumption be made?

A simple function of the form (17.10) is said to be *Bochner integrable* (short: B-*integrable*), if $\mu(A_j) < \infty$ for each $j \in \{1, \ldots, k\}$ with $b_j \neq \mathbf{0}$. Here, $\mathbf{0}$ denotes the zero element of $\mathbb{B}$. With the definition $\infty \cdot \mathbf{0} := \mathbf{0}$,

$$\int_\Omega f \, d\mu := \sum_{j=1}^k \mu(A_j) \, b_j \tag{17.11}$$

is called the *Bochner integral* (in short: B-*integral*) of f (over Ω). Since the measure μ is additive, this definition does not depend on the specific representation (17.10) of f. If $\mathcal{E}$ denotes the vector space of all B-integrable simple functions, the mapping that assigns the value $\int_\Omega f \, d\mu$ to f in $\mathcal{E}$ is linear, that is, we have

$$\int_\Omega (\alpha f + \beta g) \, d\mu = \alpha \int_\Omega f \, d\mu + \beta \int_\Omega g \, d\mu, \quad f, g \in \mathcal{E}; \alpha, \beta \in \mathbb{R}. \tag{17.12}$$

Furthermore,

$$\left\| \int_\Omega f \, d\mu \right\|_{\mathbb{B}} \leq \int_\Omega \|f\|_{\mathbb{B}} \, d\mu, \quad f \in \mathcal{E}. \tag{17.13}$$

Self-question 7: Can you prove (17.13)?

A measurable function $f : \Omega \to \mathbb{B}$ is called *Bochner integrable* (short: B-*integrable*), if there is a sequence (f_n) of simple B-integrable functions such that

$$\lim_{n \to \infty} \int_\Omega \| f_n - f \|_\mathbb{B} \, d\mu = 0. \tag{17.14}$$

In this case, the *Bochner integral* of f over Ω is defined as

$$\int_\Omega f \, d\mu := \lim_{n \to \infty} \int_\Omega f_n \, d\mu. \tag{17.15}$$

Since

$$\left\| \int_\Omega f_n \, d\mu - \int_\Omega f_m \, d\mu \right\|_\mathbb{B} \leq \int_\Omega \| f_n - f_m \|_\mathbb{B} \, d\mu \leq \int_\Omega \| f - f_n \|_\mathbb{B} \, d\mu + \int_\Omega \| f - f_m \|_\mathbb{B} \, d\mu,$$

the sequence $\left(\int_\Omega f_n d\mu \right)_{n \geq 1}$ is a Cauchy sequence in $\mathbb{B}$. Because $\mathbb{B}$ is complete, it thus follows that the limit in (17.14) exists. Note that the first "$\leq$" in the above chain of inequalities follows if one applies (17.13) to the simple function $f_n - f_m$. The limit in (17.15) does not depend on the specific sequence (f_n) with (17.14). To see this claim, suppose (g_n) is a sequence of B-integrable simple functions with $\int_\Omega \| g_n - f \|_\mathbb{B} \, d\mu \to 0$. In view of

$$\int_\Omega \| f_n - g_n \|_\mathbb{B} \, d\mu \leq \int_\Omega \| f_n - f \|_\mathbb{B} \, d\mu + \int_\Omega \| f - g_n \|_\mathbb{B} \, d\mu$$

and (17.13), applied to $f_n - g_n$, it follows that $\int_\Omega f \, d\mu = \lim_{n \to \infty} \int_\Omega g_n \, d\mu$.

When extending the Bochner integral from simple B-integrable functions to general B-integrable functions, the linearity (17.12) and property (17.13) are preserved (Problem 17.3).

Without proof (see, e.g., [HSE], pp. 42–44), we highlight the following: Suppose $f : \Omega \to \mathbb{B}$ is a measurable function with the property $\int_\Omega \| f \|_\mathbb{B} \, d\mu < \infty$, and assume that there is for each integer n a finite-dimensional subspace $\mathbb{B}_n$ of $\mathbb{B}$ and a measurable function $g_n : \Omega \to \mathbb{B}_n$ such that

$$\lim_{n \to \infty} \int_\Omega \| f - g_n \|_\mathbb{B} \, d\mu = 0.$$

Then there exist simple B-integrable functions f_n, $n \geq 1$, with (17.14), so that f is B-integrable.

If the Banach space $\mathbb{B}$ is even a separable Hilbert space $\mathbb{H}$, then this criterion implies that every measurable function $f : \Omega \to \mathbb{H}$ with $\int_\Omega \| f \| \, d\mu < \infty$ is Bochner integrable.

To understand this assertion, let $\{e_1, e_2, \ldots\}$ be an ONB of $\mathbb{H}$. If one sets

$$g_n(\omega) := \sum_{j=1}^{n} \langle f(\omega), e_j \rangle e_j, \quad \omega \in \Omega,$$

and $\mathbb{B}_n := \left\{ \sum_{j=1}^{n} \alpha_j e_j : \alpha_1, \ldots, \alpha_n \in \mathbb{R} \right\}$, then $g_n : \Omega \to \mathbb{B}_n$ is a measurable mapping. Moreover, we have

$$\| f - g_n \|^2 = \sum_{j=n+1}^{\infty} \langle f, e_j \rangle^2 \to 0 \text{ as } n \to \infty,$$

elementwise on Ω. Because $\| g_n \| \le \| f \|$, it follows that $\int_\Omega \| f - g_n \| \, d\mu \to 0$, which was to be shown.

Self-question 8: Why do $\| g_n \| \le \| f \|$ and $\int_\Omega \| f - g_n \| \, d\mu \to 0$ hold?

If $\mathbb{H}$ is a Hilbert space then, using (17.11) and (17.14), we obtain

$$\left\langle \int_\Omega f \, d\mu, \, y \right\rangle = \int_\Omega \langle f, y \rangle \, d\mu$$

for every Bochner integrable function f and every $y \in \mathbb{H}$ (Problem 17.8). In the special case $f = X$ and $\mu = \mathbb{P}$, we therefore have

$$\langle \mathbb{E}X, \, y \rangle = \left\langle \int_\Omega X \, d\mathbb{P}, \, y \right\rangle = \int_\Omega \langle X, y \rangle \, d\mathbb{P} = \mathbb{E}\langle X, y \rangle, \quad y \in \mathbb{H}.$$

According to (17.7), the Bochner integral $\int_\Omega X \, d\mathbb{P}$ is thus equal to the expectation of X introduced according to Definition 17.9.

Before we turn to the notion of a covariance operator, we state two results related to expectations for Hilbert space-valued random elements. The first generalizes the multiplication rule $\mathbb{E}(XY) = \mathbb{E}X \, \mathbb{E}Y$ for the expectation of the product of independent real-valued random variables. Note that with $\mathbb{H}$-valued random elements X and Y, the pair (X, Y) is an $(\mathbb{H} \times \mathbb{H})$-valued random element due to the separability of $\mathbb{H}$. Thus, $\langle X, Y \rangle$ is a real-valued random variable (cf. the discussion before Definition 14.27 and Problem 13.7).

17.14 Theorem (Expectation and Inner Product)
Let X and Y be independent $\mathbb{H}$-valued random elements with $\mathbb{E}\|X\| < \infty$ and $\mathbb{E}\|Y\| < \infty$. Then the expectation of $\langle X, Y \rangle$ exists, and we have

$$\mathbb{E}\langle X, Y \rangle = \langle \mathbb{E}X, \mathbb{E}Y \rangle.$$

Proof Problem 17.9.
    The second result is a strong law of large numbers.   □

17.15 Theorem (Strong Law of Large Numbers)
If $X_1, X_2, \ldots$ is an i.i.d. sequence of $\mathbb{H}$-valued random elements with $\mathbb{E}\|X_1\| < \infty$, then

$$\lim_{n\to\infty} \frac{1}{n} \sum_{j=1}^{n} X_j = \mathbb{E}(X_1) \quad \mathbb{P}\text{-almost surely.}$$

Proof This result holds more generally for sequences of random elements with values in a separable Banach space (see, e.g., [LET], p. 189). The proof is relatively simple if we assume $\mathbb{E}\|X_1\|^4 < \infty$ (see Problem 17.12). The assumption can be weakened to the effect that $X_1, X_2, \ldots$ is a sequence of *pairwise independent identically distributed* random elements, see [HSE], p. 204 ff.   □

If $X = (X_1, \ldots, X_d)^{\top}$ is a d-dimensional random vector with $\mathbb{E}X_j^2 < \infty$ for each $j \in \{1, \ldots, d\}$, the covariance matrix of X is the $d \times d$ matrix

$$\mathbb{C}\text{ov}(X) := \Big(\text{Cov}(X_j, X_k)\Big)_{1\le j,k\le d}.$$

If we assign to each y in $\mathbb{R}^d$ the value $\mathbb{C}\text{ov}(X)\, y$, the covariance matrix of X defines a linear operator on $\mathbb{R}^d$. The following considerations generalize this concept to general separable Hilbert spaces.

17.16 Theorem
For an $\mathbb{H}$-valued random element X, the following properties are equivalent:

(a) $\mathbb{E}\|X\|^2 < \infty$.
(b) There exists exactly one linear, self-adjoint, and positive operator $T : \mathbb{H} \to \mathbb{H}$ of trace-class such that

$$\langle Tx, y \rangle = \mathbb{E}\big[\langle X, x \rangle \langle X, y \rangle\big], \quad x, y \in \mathbb{H}. \tag{17.16}$$

If (a) or (b) holds, then $\text{tr}(T) = \mathbb{E}\|X\|^2$.

Proof To prove "(a) $\Longrightarrow$ (b)", fix any x, $y \in \mathbb{H}$. The Cauchy–Schwarz inequality (17.2) yields $\mathbb{E}|\langle X, x\rangle\langle X, y\rangle| \leq \|x\|\,\|y\|\,\mathbb{E}\|X\|^2 < \infty$. Thus for each $x \in \mathbb{H}$, the definition

$$\ell_x(y) := \mathbb{E}\big[\langle X, x\rangle\langle X, y\rangle\big], \quad y \in \mathbb{H},$$

provides a bounded linear functional $\ell_x : \mathbb{H} \to \mathbb{R}$. By the Riesz representation theorem (Theorem 17.5), there exists exactly one element $Tx := T(x)$ in $\mathbb{H}$ with $\langle Tx, y\rangle = \mathbb{E}\big[\langle X, x\rangle\langle X, y\rangle\big]$ for each $y \in \mathbb{H}$. Because x was arbitrary, we get (17.16). Since both expectation and inner product are linear, T is a linear operator, which is obviously self-adjoint and positive. If $\{e_1, e_2, \ldots\}$ is any ONB of $\mathbb{H}$, then

$$\mathbb{E}\|X\|^2 = \mathbb{E}\left[\sum_{k=1}^{\infty}\langle X, e_k\rangle^2\right] = \sum_{k=1}^{\infty}\mathbb{E}\big[\langle X, e_k\rangle^2\big] = \sum_{k=1}^{\infty}\langle Te_k, e_k\rangle < \infty. \tag{17.17}$$

Consequently, T is a trace-class operator with $\mathrm{tr}(T) = \mathbb{E}\|X\|^2$. According to Remark 17.4, T is uniquely determined by (17.16).

Self-question 9: Why does the second equal sign hold in (17.17)?

Conversely, if (b) holds, and if $\{e_1, e_2, \ldots\}$ is any ONB of $\mathbb{H}$, then

$$\infty > \sum_{k=1}^{\infty}\langle Te_k, e_k\rangle = \sum_{k=1}^{\infty}\mathbb{E}\langle X, e_k\rangle^2 = \mathbb{E}\left(\sum_{k=1}^{\infty}\langle X, e_k\rangle^2\right) = \mathbb{E}\|X\|^2.$$

$\square$

17.17 Theorem and Definition (Covariance Operator)

Let X be a $\mathbb{H}$-valued random element with $\mathbb{E}\|X\|^2 < \infty$. Then there exists exactly one linear, self-adjoint, positive trace-class operator $\Sigma : \mathbb{H} \to \mathbb{H}$ with the property

$$\langle \Sigma x, y\rangle = \mathbb{E}\big[\langle X - \mathbb{E}X, x\rangle\langle X - \mathbb{E}X, y\rangle\big], \quad x, y \in \mathbb{H}. \tag{17.18}$$

Σ is called the covariance operator (of the distribution) of X.

Proof By Theorem 17.16, there exists a linear, self-adjoint, positive trace-class operator $T : \mathbb{H} \to \mathbb{H}$ with

$$\langle Tx, y\rangle = \mathbb{E}\big[\langle X, x\rangle\langle X, y\rangle\big], \quad x, y \in \mathbb{H}.$$

If we set $\Sigma x := \Sigma(x) := Tx - \langle \mathbb{E}X, x \rangle \mathbb{E}X$, $x \in \mathbb{H}$, then

$$\langle \Sigma x, y \rangle = \mathbb{E}\big[\langle X, x \rangle \langle X, y \rangle\big] - \langle \mathbb{E}X, x \rangle \langle \mathbb{E}X, y \rangle, \quad x, y \in \mathbb{H}. \tag{17.19}$$

Direct calculation (see Problem 17.13) shows that Σ has all the claimed properties. $\square$

Self-question 10: Why are (17.18) and (17.19) equivalent?

Due to the bilinearity of taking covariances for real-valued random variables, and in view of (17.18), the covariance operator of X provides the covariances $\mathrm{Cov}(\langle X, x \rangle, \langle X, y \rangle)$ for all pairs $(x, y) \in \mathbb{H} \times \mathbb{H}$. We briefly consider that, in the special case $\mathbb{H} = \mathbb{R}^d$, this operator is indeed equal to the covariance matrix of X, when this matrix is considered as an operator applied to column vectors. Let $X = (X_1, \ldots, X_d)^\top$, $\mathbb{E}X = (\mathbb{E}X_1, \ldots, \mathbb{E}X_d)^\top$, and $\Sigma = (\sigma_{ij})_{1 \le i, j \le d}$. If $x = (x_1, \ldots, x_d)^\top$ and $y = (y_1, \ldots, y_d)^\top$ are in $\mathbb{R}^d$, then

$$\langle \Sigma x, y \rangle = \sum_{i=1}^{d} \sum_{j=1}^{d} \sigma_{ij} x_i y_j.$$

Since $\sigma_{ij} = \mathrm{Cov}(X_i, X_j) = \mathbb{E}[(X_i - \mathbb{E}X_i)(X_j - \mathbb{E}X_j)]$ and

$$\langle X - \mathbb{E}X, x \rangle \langle X - \mathbb{E}X, y \rangle = \sum_{i=1}^{d} \sum_{j=1}^{d} (X_i - \mathbb{E}X_i) x_i (X_j - \mathbb{E}X_j) y_j,$$

we obtain (17.18) and thus the claim.

Covariance matrices add up when adding independent random vectors (see Problem 5.5). The following result generalizes this fact. To denote the random element associated with a covariance operator, we generally write $\Sigma(Z)$ for the covariance operator of an $\mathbb{H}$-valued random element Z.

17.18 Theorem (Addition Theorem for Covariance Operators)
Let X and Y be independent $\mathbb{H}$-valued random elements with $\mathbb{E}\|X\|^2 < \infty$ and $\mathbb{E}\|Y\|^2 < \infty$. Then $\mathbb{E}\|X + Y\|^2 < \infty$, and we have

$$\Sigma(X + Y) = \Sigma(X) + \Sigma(Y).$$

Proof Since $\|X + Y\|^2 \le 2\|X\|^2 + 2\|Y\|^2$, it follows that $\mathbb{E}\|X + Y\|^2 < \infty$, and thus the covariance operator of $X + Y$ exists. It remains to show

$$\langle \Sigma(X + Y)x, \ y \rangle = \langle (\Sigma(X) + \Sigma(Y))x, \ y \rangle, \quad x, \ y \in \mathbb{H}.$$

To this end, note that $\mathbb{E}(X + Y) = \mathbb{E}X + \mathbb{E}Y$. Putting $\widetilde{X} := X - \mathbb{E}X$ and $\widetilde{Y} := Y - \mathbb{E}Y$, the definition of the expectation yields $\mathbb{E}\langle \widetilde{X}, x \rangle = \mathbb{E}\langle \widetilde{X}, y \rangle = 0$ as well as $\mathbb{E}\langle \widetilde{Y}, x \rangle = \mathbb{E}\langle \widetilde{Y}, y \rangle = 0$. With Theorem 17.14, we get

$$
\begin{aligned}
\langle \Sigma(X{+}Y)x, \ y \rangle &= \mathbb{E}\big[\langle X + Y - \mathbb{E}(X + Y), \ x \rangle \langle X + Y - \mathbb{E}(X + Y), \ y \rangle\big] \\
&= \mathbb{E}\big[\langle \widetilde{X} + \widetilde{Y}, \ x \rangle \langle \widetilde{X} + \widetilde{Y}, \ y \rangle\big] \\
&= \mathbb{E}\big[\langle \widetilde{X}, x \rangle \langle \widetilde{X}, y \rangle\big] + \mathbb{E}\big[\langle \widetilde{Y}, x \rangle \langle \widetilde{X}, y \rangle\big] + \mathbb{E}\big[\langle \widetilde{X}, x \rangle \langle \widetilde{Y}, y \rangle\big] + \mathbb{E}[\langle \widetilde{Y}, x \rangle \langle \widetilde{Y}, y \rangle] \\
&= \langle \Sigma(X)x, \ y \rangle + \langle \Sigma(Y)x, \ y \rangle = \langle (\Sigma(X) + \Sigma(Y))x, \ y \rangle.
\end{aligned}
$$

$\square$

So far, we have considered abstract Hilbert space-valued random elements on a probability space $(\Omega, \mathcal{A}, \mathbb{P})$. In the following, we will concretize these random elements with the help of stochastic processes. The starting point is a compact metric space E with Borel σ-field $\mathcal{B}(E)$ and a family $(X(t))_{t \in E}$ of real-valued random variables on Ω indexed by $t \in E$. The most important special case will be $E = [0, 1]$. We assume that $\mathbb{E}X^2(t) < \infty$ for each $t \in E$. By definition, $(X(t))_{t \in E}$ is then a *second-order stochastic process*. Thus, both the *mean function* $m : E \to \mathbb{R}$, defined by

$$m(t) := \mathbb{E}X(t), \quad t \in E,$$

and the *covariance function* $K : E \times E \to \mathbb{R}$, defined by

$$K(s, t) := \mathrm{Cov}(X(s), X(t)), \quad s, t \in E, \tag{17.20}$$

of the process exist. We make the further assumption that both $m(\cdot)$ and $K(\cdot, \cdot)$ are continuous functions. This is the case if and only if the process $(X(t))_{t \in E}$ is *mean square continuous*, that is, if for every $t \in E$ and every sequence (t_n) in E converging to t, we have

$$\lim_{n \to \infty} \mathbb{E}\left[\left(X(t_n) - X(t)\right)^2\right] = 0$$

(Problem 17.15).

In this situation, the Hilbert space $\mathbb{H} = \mathrm{L}^2(E, \mathcal{B}(E), \mu)$ of all (equivalence classes of almost everywhere equal) $(\mathcal{B}(E), \mathcal{B}^1)$-measurable functions $f : E \to \mathbb{R}$ comes into focus.

Here, μ is a finite measure on $\mathcal{B}(E)$. Note that for each $t \in E$ the random variable $X(t)$ is a $(\mathcal{A}, \mathcal{B}^1)$-measurable mapping defined on Ω. If we add the argument $\omega \in \Omega$ to $X(t)$ and write $X(t, \omega) := (X(t))(\omega)$, we can (while maintaining the notation) consider X as a mapping $X : E \times \Omega \to \mathbb{R}$. For a fixed $\omega \in \Omega$, this results in a mapping $X(\cdot, \omega) : E \to \mathbb{R}$ that assigns to each t in E the value $X(t, \omega)$. We assume that for each $\omega \in \Omega$ this mapping belongs to $\mathbb{H}$ and thus is a $(\mathcal{B}(E), \mathcal{B}^1)$-measurable function with $\int_E X^2(t, \omega) \, \mu(dt) < \infty$. The question then arises whether the mapping

$$\mathbb{X} : \begin{cases} \Omega \to \mathbb{H}, \\ \omega \mapsto \mathbb{X}(\omega), \quad (\mathbb{X}(\omega))(t) := X(t, \omega), \ t \in E, \end{cases} \tag{17.21}$$

is $(\mathcal{A}, \mathcal{B}(\mathbb{H}))$-measurable, which means that $\mathbb{X}$ defines a $\mathbb{H}$-valued random element. The following result provides an answer to this question.

17.19 Theorem (Product-Measurability Yields a $\mathbb{H}$-Valued Random Element)
In the above situation, let the mapping $X : E \times \Omega \to \mathbb{R}$ be measurable with respect to the product σ-field $\mathcal{B}(E) \otimes \mathcal{A}$. If $X(\cdot, \omega) \in \mathbb{H}$ for each fixed $\omega \in \Omega$, then the mapping $\mathbb{X}$ defined in (17.21) is $(\mathcal{A}, \mathcal{B}(\mathbb{H}))$-measurable and thus an $\mathbb{H}$-valued random element.

Proof Fix any $f \in \mathbb{H}$. Due to the product-measurability, the mapping

$$\Omega \ni \omega \mapsto \langle X(\cdot, \omega), f \rangle = \int_E X(t, \omega) f(t) \, \mu(dt)$$

is $(\mathcal{A}, \mathcal{B}^1)$-measurable according to theorems of measure theory (see, e.g., [BI3], Theorem 18.1), so the claim follows from Theorem 17.7.   □

The following theorem provides sufficient conditions for the $(\mathcal{B}(E) \otimes \mathcal{A})$-measurability of the mapping $X : E \times \Omega \to \mathbb{R}$ (see, e.g., [HSE], Theorem 7.4.2).

17.20 Theorem (Criterion for Product-Measurability)
Suppose that the mapping $X(t, \cdot) : \Omega \to \mathbb{R}$ is $(\mathcal{A}, \mathcal{B}^1)$-measurable (that is, a random variable) for each fixed t. Furthermore, assume the mapping $X(\cdot, \omega) : E \to \mathbb{R}$ is continuous for each $\omega \in \Omega$. Then the mapping $X : E \times \Omega \to \mathbb{R}$ is $(\mathcal{B}(E) \otimes \mathcal{A}, \mathcal{B}^1)$-measurable. In this case, the distribution $\mathbb{P}^{\mathbb{X}}$ of the $\mathbb{H}$-valued random element $\mathbb{X}$ (which is a probability measure on $\mathcal{B}(\mathbb{H})$) is uniquely determined by the collection of all finite-dimensional distributions of $(X(t_1), \ldots, X(t_k))$, where $k \geq 1$ and $t_1, \ldots, t_k \in E$.

The covariance function $K(s,t)$ in (17.20) corresponds to the integral operator $\mathbb{K} :$ $\mathbb{H} \to \mathbb{H}$ on $\mathbb{H} = L^2(E, \mathcal{B}(E), \mu)$, which is defined by

$$\mathbb{K}f(s) = \int_E K(s,t)f(t)\,\mu(dt), \quad s \in E, \quad f \in \mathbb{H}.$$

We will now see that, under certain conditions, the mean function $m(t) = \mathbb{E}X(t), t \in E$, of a second-order stochastic process $(X(t))_{t \in E}$ is equal to the expectation $m = \mathbb{E}(\mathbb{X})$ of the random element defined in (17.21), and that the covariance operator $\Sigma(\mathbb{X})$ of $\mathbb{X}$ coincides with the integral operator $\mathbb{K}$.

17.21 Theorem *Let $(X(t))_{t \in E}$ be a mean square continuous second-order stochastic process on a probability space $(\Omega, \mathcal{A}, \mathbb{P})$. For the mapping defined by $X(t, \omega) :=$ $(X(t))(\omega)$, where $(t, \omega) \in E \times \Omega$, $X : E \times \Omega \to \mathbb{R}$, suppose that the prerequisites of Theorem 17.19 are fulfilled. For the random element $\mathbb{X} : \Omega \to \mathbb{H}$ defined in (17.21), we then have:*

(a) $\mathbb{E}(\mathbb{X}) = m(\cdot)$,
(b) $\Sigma(\mathbb{X}) = \mathbb{K}$.

Proof Due to the $(\mathcal{B}(E) \otimes \mathcal{A}, \mathcal{B}^1)$-measurability of $X : E \times \Omega \to \mathbb{R}$, Tonelli's theorem (Theorem 1.33) yields

$$\mathbb{E}\|\mathbb{X}\|^2 = \mathbb{E}\left[\int_E X^2(t)\,\mu(dt)\right] = \int_E \mathbb{E}X^2(t)\,\mu(dt)d = \int_E \left(K(t,t) + m^2(t)\right)\mu(dt)$$

$$(17.22)$$

and thus $\mathbb{E}\|\mathbb{X}\|^2 < \infty$. As a consequence, the covariance operator $\Sigma(\mathbb{X})$ and the expectation $\mathbb{E}(\mathbb{X})$ of $\mathbb{X}$ exist. The last integral is finite, since E is compact and both $K(\cdot, \cdot)$ and $m(\cdot)$ are continuous functions by Problem 17.15.

Self-question 11: Which equals sign in (17.22) follows from Tonelli's theorem?

(a) Fix any f in $\mathbb{H}$. Due to $|\langle \mathbb{X}, f \rangle| \leq \|\mathbb{X}\| \|f\|$, Fubini's theorem (Theorem 1.34) yields

$$\mathbb{E}\langle \mathbb{X}, f \rangle = \mathbb{E}\left(\int_E X(t)f(t)\,\mu(dt)\right) = \int_E m(t)f(t)\,\mu(dt) = \langle m, f \rangle.$$

According to Definition 17.9, it follows that $\mathbb{E}(\mathbb{X}) = m(\cdot)$.

(b) The covariance operator $\Sigma = \Sigma(\mathbb{X})$ of $\mathbb{X}$ is determined by $\langle \Sigma f, g \rangle = \mathbb{E}[\langle \mathbb{X} - \mathbb{E}\mathbb{X}, f \rangle \langle \mathbb{X} - \mathbb{E}\mathbb{X}, g \rangle]$ for arbitrary $f, g \in \mathbb{H}$. Assuming w.l.o.g. $m = \mathbb{E}(\mathbb{X}) = \mathbf{0}$, Fubini's theorem (Theorem 1.34) implies

$$
\begin{aligned}
\langle \Sigma f, g \rangle = \mathbb{E}[\langle \mathbb{X}, f \rangle \langle \mathbb{X}, g \rangle] &= \mathbb{E}\left[\int_E X(s) f(s)\, \mu(ds) \int_E X(t) g(t)\, \mu(dt) \right] \\
&= \mathbb{E}\left[\int_E \int_E X(s) X(t) f(s) g(t)\, \mu(ds) \mu(dt) \right] \\
&= \int_E \int_E \mathbb{E}[X(s) X(t)] f(s) g(t)\, \mu(ds) \mu(dt) \\
&= \int_E \left(\int_E K(s,t) f(s)\, \mu(ds) \right) g(t)\, \mu(dt) = \int_E (\mathbb{K}f)(t) g(t)\, \mu(dt) \\
&= \langle \mathbb{K}f, g \rangle,
\end{aligned}
$$

which was to be shown.

□

The following notion generalizes the concept of a characteristic function.

17.22 Definition (Characteristic Functional)
Let $\mathbb{H}$ be a separable Hilbert space and X an $\mathbb{H}$-valued random element. The function $\varphi_X : \mathbb{H} \to \mathbb{C}$, defined by

$$
\varphi_X(h) := \mathbb{E}\left[e^{i\langle X, h \rangle} \right] = \mathbb{E}\left[\cos(\langle X, h \rangle) \right] + i\, \mathbb{E}\left[\sin(\langle X, h \rangle) \right],
$$

is called the *characteristic functional (of the distribution) of* X.

The characteristic functional of an $\mathbb{H}$-valued random element X has the following properties. Here, $\overline{z}$ generally denotes the complex conjugate of a complex number z. Property (f) justifies the attribute *characteristic*.

17.23 Theorem (Properties of the Characteristic Functional)

(a) $\varphi_X(0) = 1$.
(b) $\varphi_X(-h) = \overline{\varphi_X(h)}, \quad h \in \mathbb{H}$.
(c) *The function* φ_X *is continuous.*
(d) *The function* φ_X *is positive-semidefinite, that is, for each* $n \geq 1$ *and each choice of* $\alpha_1, \dots, \alpha_n \in \mathbb{C}$ *and* $h_1, \dots, h_n \in \mathbb{H}$, *we have*

$$
\sum_{k,\ell=1}^n \alpha_k \overline{\alpha_\ell} \varphi_X(h_\ell - h_k) \geq 0.
$$

(e) If X and Y are independent $\mathbb{H}$-valued random elements, then $\varphi_{X+Y} = \varphi_X \, \varphi_Y$.

(f) If X and Y are $\mathbb{H}$-valued random elements, then: $\varphi_X = \varphi_Y \iff X \overset{\mathcal{D}}{=} Y$.

Proof Properties (a) and (b) are obvious, and (c) follows from the dominated convergence theorem. That φ_X is positive-semidefinite results from

$$0 \leq \mathbb{E}\left|\sum_{k=1}^{n} \alpha_k e^{\mathrm{i}\langle X, h_k\rangle}\right|^2 = \mathbb{E}\left[\sum_{k,\ell=1}^{n} \alpha_k \overline{\alpha_\ell} e^{\mathrm{i}\langle X, h_k - h_\ell\rangle}\right] = \sum_{k,\ell=1}^{n} \alpha_k \overline{\alpha_\ell} \varphi_X(h_k - h_\ell).$$

To show the implication "$\Longrightarrow$" in (f), fix any $h \in \mathbb{H}$ and $t \in \mathbb{R}$. Then

$$\varphi_{\langle X, h\rangle}(t) = \mathbb{E}\left[\exp\left(\mathrm{i}t\langle X, h\rangle\right)\right] = \mathbb{E}\left[\exp\left(\mathrm{i}\langle X, th\rangle\right)\right] = \varphi_X(th)$$
$$= \varphi_Y(th) = \mathbb{E}\left[\exp\left(\mathrm{i}\langle Y, th\rangle\right)\right] = \mathbb{E}\left[\exp\left(\mathrm{i}t\langle Y, h\rangle\right)\right]$$
$$= \varphi_{\langle Y, h\rangle}(t).$$

Therefore, $\langle X, h\rangle \overset{\mathcal{D}}{=} \langle Y, h\rangle$ for every $h \in \mathbb{H}$, and the assertion follows from Theorem 17.8.

□

Self-question 12: Why does property (e) hold?

In Chap. 5 we saw that for every $m \in \mathbb{R}^d$ and for every symmetric positive-semidefinite $d \times d$ matrix Σ there is a random vector X that has the d-variate normal distribution $N_d(m, \Sigma)$ (cf. Theorem 5.10). Without further specification, the normal distribution of X is characterized by the fact that for every $c \in \mathbb{R}^d$ the inner product $c^\top X$ has a (possibly degenerate) univariate normal distribution. According to Problem 5.6, a random vector X with the normal distribution $N_d(m, \Sigma)$ has the characteristic function

$$\varphi_X(t) = \mathbb{E}\left[e^{\mathrm{i}t^\top X}\right] = e^{\mathrm{i}t^\top m} \exp\left(-\frac{1}{2}t^\top \Sigma t\right), \quad t \in \mathbb{R}^d. \tag{17.23}$$

We now generalize the concept of a normal distribution to a separable Hilbert space $\mathbb{H}$. The following definition is a direct generalization of Definition 5.5.

17.24 Definition (Normal Distribution on a Separable Hilbert Space)
Let $\mathbb{H}$ be a separable Hilbert space and $(\Omega, \mathcal{A}, \mathbb{P})$ a probability space. A random element $X : \Omega \to \mathbb{H}$ has a *normal distribution*, if for every $h \in \mathbb{H}$ the inner product $\langle X, h\rangle$ has a (possibly degenerate) univariate normal distribution. In this case, X is called a *Gaussian random element*.

If X is such a Gaussian random element, then for each $k \geq 1$ and for each $h_1, \ldots, h_k$ in $\mathbb{H}$ the k-dimensional random vector $(\langle X, h_1 \rangle, \ldots, \langle X, h_k \rangle)$ has a k-variate normal distribution. This fact follows readily, since for every $c = (c_1, \ldots, c_k) \in \mathbb{R}^k$ the random variable

$$\sum_{j=1}^{k} c_j \langle X, h_j \rangle = \left\langle X, \sum_{j=1}^{k} c_j h_j \right\rangle$$

has a univariate normal distribution (see Definition 5.5).

If $\{e_1, e_2, \ldots\}$ is an ONB of $\mathbb{H}$, and $\mathbb{H}_k := \{\sum_{j=1}^{k} \alpha_j e_j : \alpha_1, \ldots, \alpha_k \in \mathbb{R}\}$ denotes the k-dimensional subspace of $\mathbb{H}$ spanned by $e_1, \ldots, e_k$, we can construct a Gaussian random element X as follows: Let $Y_1, \ldots, Y_k$ be independent normally distributed random variables defined on a common probability space $(\Omega, \mathcal{A}, \mathbb{P})$. If we set

$$X(\omega) := \sum_{j=1}^{k} Y_j(\omega) e_j, \quad \omega \in \Omega, \tag{17.24}$$

then $X = \sum_{j=1}^{k} Y_j e_j$ is an $\mathbb{H}$-valued random element by Theorem 17.7. Since for every $h \in \mathbb{H}$ the inner product $\langle X, h \rangle = \sum_{j=1}^{k} Y_j \langle e_j, h \rangle$ has a univariate normal distribution according to the addition theorem for the normal distribution, X has a normal distribution. However, this is comparatively uninteresting, as it conceptually does not go beyond the case treated in Chap. 5. Because $\mathbb{P}^X(\mathbb{H}_k) = 1$, the distribution of X is entirely concentrated on the subspace $\mathbb{H}_k$, and $\mathbb{P}^X$ is the image measure of a normal distribution as in Chap. 5 under the mapping $(\alpha_1, \ldots, \alpha_k) \mapsto \sum_{j=1}^{k} \alpha_j e_j$.

To go beyond the finite-dimensional case, it makes sense to start with a sequence $(Y_n)_{n \geq 1}$ of independent normally distributed random variables on a probability space $(\Omega, \mathcal{A}, \mathbb{P})$. We initially assume that $\mathbb{E}(Y_n) = 0$ for each $n \geq 1$ and denote the variance of Y_n with σ_n^2, $n \geq 1$. If we set

$$X(\omega) := \sum_{j=1}^{\infty} Y_j(\omega) e_j, \quad \omega \in \Omega, \tag{17.25}$$

in analogy to (17.24), the immediate question arises for which $\omega \in \Omega$ this series converges, and inevitably the variances $\sigma_1^2, \sigma_2^2, \ldots$ come into play. If these do not converge to zero fast enough, no convergence in (17.25) (in whatever sense) is to be expected. If we require

$$\sum_{j=1}^{\infty} \sigma_j^2 < \infty, \tag{17.26}$$

then $\mathbb{E}\left(\sum_{j=1}^{\infty} Y_j^2\right) = \sum_{j=1}^{\infty} \sigma_j^2 < \infty$ by the monotone convergence theorem, and consequently $\mathbb{P}\left(\sum_{j=1}^{\infty} Y_j^2 < \infty\right) = 1$. In the probability space $(\Omega, \mathcal{A}, \mathbb{P})$ there is therefore a set $\Omega_0 \in \mathcal{A}$ with $\mathbb{P}(\Omega_0) = 1$ such that the restriction of X to Ω_0 is a mapping with range $\mathbb{H}$. Since the complement $\Omega \setminus \Omega_0$ of Ω_0 plays no role, we can assume w.l.o.g. that the series in (17.25) converges pointwise on Ω and that $X : \Omega \to \mathbb{H}$ is therefore an $\mathbb{H}$-valued random element.

Self-question 13: Why is the mapping X defined in (17.25) $(\mathcal{A}, \mathcal{B}(\mathbb{H}))$-measurable?

Now fix any $h \in \mathbb{H}$. In view of (17.25) we have (elementwise on Ω)

$$\langle X, h \rangle = \lim_{n \to \infty} \sum_{j=1}^{n} Y_j \langle e_j, h \rangle. \tag{17.27}$$

Due to the addition theorem for the normal distribution, it follows that

$$\sum_{j=1}^{n} Y_j \langle e_j, h \rangle \sim \mathrm{N}\left(0, \sum_{j=1}^{n} \sigma_j^2 \langle e_j, h \rangle^2\right). \tag{17.28}$$

Since due to (17.26) and $\langle e_j, h \rangle^2 \leq \|h\|^2$ the variance of the above normal distribution converges as $n \to \infty$, $\langle X, h \rangle$ has the normal distribution $\mathrm{N}(0, \sum_{j=1}^{\infty} \sigma_j^2 \langle e_j, h \rangle^2)$. By means of (17.25) and the condition (17.26), we have thus obtained an extensive arsenal of Gaussian $\mathbb{H}$-valued random elements, whose distributions—provided infinitely many of the variances σ_j^2 are positive—are not concentrated on any finite-dimensional subspace of $\mathbb{H}$.

In the following, let $\mathcal{L}_{tr}^{+}(\mathbb{H})$ denote the set of all linear, bounded, self-adjoint positive trace-class operators $T : \mathbb{H} \to \mathbb{H}$. We will now see that for every $m \in \mathbb{H}$ and every $\Sigma \in \mathcal{L}_{tr}^{+}(\mathbb{H})$ there exists an $\mathbb{H}$-valued Gaussian random element X, such that X has expectation m and covariance operator Σ. This fact generalizes Theorem 5.10.

17.25 Theorem (Existence of Normal Distributions)
For arbitrary $m \in \mathbb{H}$ and $\Sigma \in \mathcal{L}_{tr}^{+}(\mathbb{H})$ there exists an $\mathbb{H}$-valued Gaussian random element X with $m = \mathbb{E}(X)$ and $\Sigma = \Sigma(X)$.

Proof The starting point is the general fact that for the operator Σ there is a set $\{e_1, e_2, \ldots\}$ of orthonormal elements in $\mathbb{H}$ and a sequence $\lambda_1, \lambda_2, \ldots$ of non-negative numbers such that $\Sigma e_k = \lambda_k e_k$ for each $k \geq 1$ (see, e.g., [HEI], Theorem 7.8.1). For each j, λ_j is thus an eigenvalue of Σ associated with the normalized eigenvector e_j. Since Σ is a trace-class operator, we have

$$\operatorname{tr}(\Sigma) = \sum_{k=1}^{\infty} \langle \Sigma e_k, e_k \rangle = \sum_{k=1}^{\infty} \lambda_k < \infty.$$

In principle, it cannot be ruled out that only finitely many of the λ_j are greater than zero. In this case, however, the following convergence considerations are superfluous. We set $m_k := \langle m, e_k \rangle$, $k \geq 1$. In generalization of the approach leading to (17.24), let $Y_1, Y_2, \ldots$ be independent random variables defined on a common probability space $(\Omega, \mathcal{A}, \mathbb{P})$, where Y_k has the normal distribution $N(m_k, \lambda_k)$, $k \geq 1$. Analogous to (17.24), we define

$$X(\omega) := \sum_{j=1}^{\infty} Y_j(\omega) e_j, \quad \omega \in \Omega. \tag{17.29}$$

Because

$$\mathbb{E}\left(\sum_{j=1}^{\infty} Y_j^2 \right) = \sum_{j=1}^{\infty} \mathbb{E}(Y_j^2) = \sum_{j=1}^{\infty} \left(m_j^2 + \lambda_j \right) = \sum_{j=1}^{\infty} m_j^2 + \sum_{j=1}^{\infty} \lambda_j < \infty, \tag{17.30}$$

it follows that $\mathbb{P}\left(\sum_{j=1}^{\infty} Y_j^2 < \infty \right) = 1$. Therefore, the series in (17.29) converges on a set of probability one in Ω and thus w.l.o.g. on all of Ω (cf. the reasoning after (17.26)).

Self-question 14: Why does $\sum_{j=1}^{\infty} m_j^2 < \infty$ hold?

Consequently, X is an $\mathbb{H}$-valued random element, and due to (17.30) we have $\mathbb{E}\|X\|^2 < \infty$. Thus, both the expectation and the covariance operator of X exist.

In generalization of (17.28) we obtain for each $n \geq 1$ the equality in distribution

$$\sum_{j=1}^{n} Y_j \langle e_j, h \rangle \sim N\left(\sum_{j=1}^{n} m_j \langle e_j, h \rangle, \sum_{j=1}^{n} \lambda_j \langle e_j, h \rangle^2 \right).$$

Because $m_j = \langle m, e_j \rangle$, the generalized Parseval's equation (17.3) entails

$$\lim_{n \to \infty} \sum_{j=1}^{n} m_j \langle e_j, h \rangle = \langle m, h \rangle. \tag{17.31}$$

Since Σ is self-adjoint, using $\Sigma e_j = \lambda_j e_j$ we get

$$\lambda_j \langle e_j, h \rangle^2 = \langle h, \lambda_j e_j \rangle \langle e_j, h \rangle = \langle h, \Sigma e_j \rangle \langle e_j, h \rangle = \langle \Sigma h, e_j \rangle \langle e_j, h \rangle.$$

A repeated application of the generalized Parseval's equation now yields

$$\lim_{n\to\infty} \sum_{j=1}^{n} \lambda_j \langle e_j, h \rangle^2 = \langle \Sigma h, h \rangle. \tag{17.32}$$

From (17.27), (17.31) and (17.32), it follows that $\langle X, h \rangle$ has a normal distribution, where

$$\mathbb{E}\langle X, h \rangle = \langle m, h \rangle, \qquad \mathbb{V}\langle X, h \rangle = \langle \Sigma h, h \rangle, \qquad h \in \mathbb{H}. \tag{17.33}$$

By definition of the expectation, we obtain $\mathbb{E}(X) = m$, and we now show that $\Sigma = \Sigma(X)$. To this end, we use that Σ is linear and self-adjoint to get

$$\langle \Sigma g, h \rangle = \frac{1}{2}\Big(\langle \Sigma(g+h), g+h \rangle - \langle \Sigma g, g \rangle - \langle \Sigma h, h \rangle \Big), \qquad g, h \in \mathbb{H}.$$

In view of (17.33) the right-hand side equals $\frac{1}{2}\big(\mathbb{V}\langle X, g+h \rangle - \mathbb{V}\langle X, g \rangle - \mathbb{V}\langle X, h \rangle\big)$, which, due to $\mathbb{V}\langle X, g+h \rangle = \mathbb{V}\langle X, g \rangle + \mathbb{V}\langle X, h \rangle + 2\mathrm{Cov}(\langle X, g \rangle, \langle X, h \rangle)$, results in the equation $\langle \Sigma g, h \rangle = \mathrm{Cov}(\langle X, g \rangle, \langle X, h \rangle)$. By definition of the covariance operator, we obtain $\Sigma = \Sigma(X)$, which completes the proof.   □

By Theorem 17.8, the distribution of X is uniquely determined by the distributions of $\langle X, h \rangle$ with $h \in \mathbb{H}$. Consequently, the distribution of the random element obtained in Theorem 17.25 depends only on m and Σ. It is then said that X is *normally distributed* (or X is a *Gaussian random element*) *with expectation m and covariance operator Σ*. For short, this is written as

$$X \sim N(m, \Sigma).$$

A direct calculation shows that the random element X has the characteristic functional

$$\varphi_X(h) = e^{i\langle m, h \rangle} \exp\left(-\frac{1}{2} \langle \Sigma h, h \rangle \right), \qquad h \in \mathbb{H}, \tag{17.34}$$

which generalizes (17.23) (Problem 17.17).

Without proof, it should be noted that a random element X, which is normally distributed according to Definition 17.24, *necessarily* satisfies the inequality $\mathbb{E}\|X\|^2 < \infty$ and therefore has an expectation $m := \mathbb{E}(X)$ and a covariance operator $\Sigma = \Sigma(X)$ (see, e.g., [KUK], Chapter 5). Thus, we could have introduced the normal distribution $N(m, \Sigma)$ alternatively via its characteristic functional given in (17.34).

The following theorem shows that the square of the distance of a Gaussian random element from its expectation is distributed like a sum of independent weighted χ_1^2-

distributed random variables. In the special case $\mathbb{H} = \mathbb{R}^d$, this result follows directly from the principal component representation (5.4) of X.

17.26 Theorem (The Distribution of $\|X - m\|^2$)
If X is an $\mathbb{H}$-valued Gaussian random element with $X \sim N(m, \Sigma)$, then

$$\|X - m\|^2 \overset{\mathcal{D}}{=} \sum_{j=1}^{\infty} \lambda_j N_j^2.$$

Here, $\lambda_1, \lambda_2, \ldots$ are the positive eigenvalues of the covariance operator Σ of X, listed according to their geometric multiplicity, and $N_1, N_2, \ldots$ is a sequence of i.i.d. random variables with a standard normal distribution.

Proof W.l.o.g., we can assume $m = 0$. By Theorem 17.25, there is a set $\{e_1, e_2, \ldots\}$ of orthonormal elements of $\mathbb{H}$ with $\Sigma e_j = \lambda_j e_j$ for each $j \geq 1$. If we set $\widetilde{N}_j := \langle X, e_j \rangle$, $j \geq 1$, then for each $k \geq 1$ the random vector $(\widetilde{N}_1, \ldots, \widetilde{N}_k)$ has a k-variate normal distribution. Because $m = 0$, we have $\mathbb{E}(\widetilde{N}_j) = 0$, $j \in \{1, \ldots, k\}$. Moreover, we get

$$\mathbb{E}(\widetilde{N}_i \widetilde{N}_j) = \mathbb{E}\left[\langle X, e_i \rangle \langle X, e_j \rangle\right] = \langle \Sigma e_i, e_j \rangle = \lambda_i \langle e_i, e_j \rangle, \quad i, j \geq 1.$$

Thus, $\widetilde{N}_1, \widetilde{N}_2, \ldots$ are independent random variables with $\widetilde{N}_j \sim N(0, \lambda_j)$, $j \geq 1$.

Self-question 15: Why are $\widetilde{N}_1, \widetilde{N}_2, \ldots$ independent?

If we set $N_j := \widetilde{N}_j / \sqrt{\lambda_j}$, $j \geq 1$, then $N_1, N_2, \ldots$ is a sequence of i.i.d. random variables with a standard normal distribution, and we have

$$\|X\|^2 = \sum_{j=1}^{\infty} \langle X, e_j \rangle^2 = \sum_{j=1}^{\infty} \widetilde{N}_j^2 = \sum_{j=1}^{\infty} \lambda_j N_j^2.$$

$\square$

Problem 17.20 shows that, in generalization of Theorem 5.7, there is a reproduction theorem for the normal distribution in Hilbert spaces: If X is a Gaussian random element in $\mathbb{H}$ with $X \sim N(m, \Sigma)$, and if $T : \mathbb{H} \to \mathbb{H}$ is a bounded linear mapping, then $Y := T(X)$ is a Gaussian random element with $Y \sim N(Tm, T\Sigma T^*)$. Here, T^* is the adjoint mapping to T.

We now turn to convergence in distribution of $\mathbb{H}$-valued random elements. For this purpose, we assume that $\mathbb{H}$ is a separable *infinite-dimensional* Hilbert space, because otherwise the theory presented in Chap. 6 applies. For a fixed ONB $\{e_k : k \geq 1\}$ of $\mathbb{H}$ and each $\ell \geq 1$, let

$$\Pi_\ell : \begin{cases} \mathbb{H} \to \mathbb{H}, \\ x \mapsto \Pi_\ell(x) := \sum_{k=1}^{\ell} \langle x, e_k \rangle e_k \end{cases}$$

denote the orthogonal projection onto the subspace

$$\mathbb{H}_\ell := \left\{ \sum_{j=1}^{\ell} \alpha_j e_j : \alpha_1, \dots, \alpha_\ell \in \mathbb{R} \right\} \tag{17.35}$$

of $\mathbb{H}$ spanned by $\{e_1, \dots, e_\ell\}$.

17.27 Theorem (Convergence in Distribution in $\mathbb{H}$)
Let $X, X_1, X_2, \dots$ be $\mathbb{H}$-valued random elements on a probability space $(\Omega, \mathcal{A}, \mathbb{P})$. If

$$\Pi_\ell(X_n) \xrightarrow{\mathcal{D}} \Pi_\ell(X) \ \text{as } n \to \infty \ \text{for every } \ell \geq 1, \tag{17.36}$$

$$\lim_{\ell \to \infty} \limsup_{n \to \infty} \mathbb{P}\left(\|X_n - \Pi_\ell(X_n)\| \geq \delta \right) = 0 \ \text{for every } \delta > 0, \tag{17.37}$$

then $X_n \xrightarrow{\mathcal{D}} X$.

17.28 Remark Before we prove this theorem, a discussion of the conditions (17.36) and (17.37) is appropriate. Since $\Pi_\ell(X_n)$ and $\Pi_\ell(X)$ are random elements that take on values in the ℓ-dimensional subspace $\mathbb{H}_\ell$ of $\mathbb{H}$ defined in (17.35), condition (17.36) is an analogue of the convergence of all finite-dimensional distributions (fidi-convergence) in the function space $C[0, 1]$. Together with (17.36), condition (17.37) guarantees the tightness of the sequence $(X_n)_{n \geq 1}$. According to Prokhorov's theorem (Theorem 14.22), this property is a necessary condition for convergence in distribution. If, for fixed $\delta > 0$, $\mathbb{H}_\ell^\delta := \{ y \in \mathbb{H} : \|y - \mathbb{H}_\ell\| < \delta \}$ denotes the δ-neighborhood of $\mathbb{H}_\ell$, then the probability in (17.37) is equal to $1 - \mathbb{P}^{X_n}(\mathbb{H}_\ell^\delta)$. After switching to complementary events, we can restate (17.37) as follows: For every $\varepsilon > 0$ and every $\delta > 0$ there exists an ℓ_0 such that for every $\ell \geq \ell_0$:

$$\mathbb{P}^{X_n}\left(\mathbb{H}_\ell^\delta \right) > 1 - \varepsilon \quad \text{for every } n \geq 1.$$

Therefore, uniformly in n, the distributions $\mathbb{P}^{X_n}$ are concentrated with arbitrarily large probability on neighborhoods of finite-dimensional subspaces.

The following considerations show that we can replace condition (17.36) by

$$\langle X_n, h \rangle \xrightarrow{\mathcal{D}} \langle X, h \rangle \quad \text{for every } h \in \mathbb{H}. \tag{17.38}$$

Why? If (17.38) holds, we choose for each $\ell \geq 1$ any real numbers $\alpha_1, \ldots, \alpha_\ell$ and set $h := \sum_{j=1}^{\ell} \alpha_j e_j$. Due to the linearity of the inner product, (17.38) and the Cramér–Wold device (Theorem 6.18) then yield

$$\left(\langle X_n, e_1 \rangle, \ldots, \langle X_n, e_\ell \rangle \right) \xrightarrow{\mathcal{D}} \left(\langle X, e_1 \rangle, \ldots, \langle X, e_\ell \rangle \right) \quad \text{as } n \to \infty.$$

Using the continuous mapping $\psi : \mathbb{R}^\ell \to \mathbb{H}$ defined by $\psi(\alpha_1, \ldots, \alpha_\ell) := \sum_{j=1}^{\ell} \alpha_j e_j$, the mapping theorem (Theorem 14.4) entails

$$\Pi_\ell(X_n) = \psi\left(\langle X_n, e_1 \rangle, \ldots, \langle X_n, e_\ell \rangle \right) \xrightarrow{\mathcal{D}} \psi\left(\langle X, e_1 \rangle, \ldots, \langle X, e_\ell \rangle \right) = \Pi_\ell(X),$$

which is (17.36).

Proof of Theorem 17.27 Let $f : \mathbb{H} \to \mathbb{R}$ be any bounded and uniformly continuous function. According to part (b) of the Portmanteau theorem (Theorem 14.7) we have to show $\mathbb{E}f(X_n) \to \mathbb{E}f(X)$ as $n \to \infty$. To this end, fix any $\varepsilon > 0$. Due to the uniform continuity of f, there exists a $\delta > 0$ such that for all $x, \, y \in \mathbb{H}$ the implication

$$\|x - y\| < \delta \implies |f(x) - f(y)| < \varepsilon \tag{17.39}$$

holds. For a fixed integer ℓ, the triangle inequality implies

$$|\mathbb{E}f(X_n) - \mathbb{E}f(X)| \leq |\mathbb{E}f(X_n) - \mathbb{E}f(\Pi_\ell(X_n))| + |\mathbb{E}f(\Pi_\ell(X_n)) - \mathbb{E}f(\Pi_\ell(X))|$$

$$+ |\mathbb{E}f(\Pi_\ell(X)) - \mathbb{E}f(X)|$$

$$=: u_{n,\ell} + v_{n,\ell} + w_\ell,$$

say. By the dominated convergence theorem, there exists an integer ℓ_0 (depending on ε) such that $w_\ell \leq \varepsilon$ for each $\ell \geq \ell_0$. With $K := \sup_{x \in \mathbb{H}} |f(x)| < \infty$, it follows from (17.39) that

$$u_{n,\ell} \leq |\mathbb{E}\left[(f(X_n) - f(\Pi_\ell(X_n))) \, \mathbf{1}\{\|X_n - \Pi_\ell(X_n)\| \geq \delta\} \right]|$$

$$+ |\mathbb{E}\left[(f(X_n) - f(\Pi_\ell(X_n))) \, \mathbf{1}\{\|X_n - \Pi_\ell(X_n)\| < \delta\} \right]|$$

$$\leq 2K \, \mathbb{P}(\|X_n - \Pi_\ell(X_n)\| \geq \delta) + \varepsilon.$$

For each $\ell \geq \ell_0$, we thus obtain

$$|\mathbb{E}f(X_n) - \mathbb{E}f(X)| \leq 2K \, \mathbb{P}(\|X_n - \Pi_\ell(X_n)\| \geq \delta) + 2\varepsilon + |\mathbb{E}f(\Pi_\ell(X_n)) - \mathbb{E}f(\Pi_\ell(X))|.$$

By (17.36), the last summand on the right-hand side converges to zero, and we get

$$\limsup_{n\to\infty} |\mathbb{E}f(X_n) - \mathbb{E}f(X)| \leq 2\,K \limsup_{n\to\infty} \mathbb{P}(\|X_n - \Pi_\ell(X_n)\| \geq \delta) + 2\varepsilon.$$

From (17.37), it follows that $\limsup_{n\to\infty} |\mathbb{E}f(X_n) - \mathbb{E}f(X)| \leq 2\varepsilon$. This completes the proof, since ε was arbitrary. $\square$

After these preparations, we are able to state and prove a central limit theorem for Hilbert space-valued random elements. This theorem is a generalization of the multivariate central limit theorem (Theorem 6.19).

17.29 Central Limit Theorem for a Sequence of i.i.d. $\mathbb{H}$-valued Random Elements
Let $X_1, X_2, \ldots$ be a sequence of i.i.d. $\mathbb{H}$-valued random elements with $\mathbb{E}\|X_1\|^2 < \infty$. Denote $m := \mathbb{E}X_1$ the expectation of X_1 and $C := \Sigma(X_1)$ the covariance operator of X_1. Then there exists an $\mathbb{H}$-valued Gaussian random element X in $\mathbb{H}$ with $X \sim N(0, C)$, such that

$$\frac{1}{\sqrt{n}} \sum_{j=1}^{n} (X_j - m) \xrightarrow{\mathcal{D}} X \quad \text{as } n \to \infty.$$

Proof of Theorem 17.27 We assume w.l.o.g. $m = 0$ and set $S_n := n^{-1/2}(X_1 + \ldots + X_n)$, $n \geq 1$. Using Problem 17.14 and Theorem 17.18, it follows that $\Sigma(S_n) = C$, and we thus obtain

$$\langle Cx, y \rangle = \mathbb{E}\left[\langle S_n, x \rangle \langle S_n, y \rangle\right], \quad n \geq 1, \; x, y \in \mathbb{H}.$$

Because of $C \in \mathcal{L}_{tr}^+(\mathbb{H})$, there is a random element X with $X \sim N(0, C)$ by Theorem 17.25. In the following, let $\{e_k : k \geq 1\}$ be a fixed ONB of $\mathbb{H}$. For any $\delta > 0$ Markov's inequality and the monotone convergence theorem imply

$$\mathbb{P}\big(\|S_n - \Pi_\ell(S_n)\| \geq \delta\big) \leq \frac{1}{\delta^2} \mathbb{E}\big[\|S_n - \Pi_\ell(S_n)\|^2\big]$$

$$= \frac{1}{\delta^2} \mathbb{E}\Big[\sum_{k=\ell+1}^{\infty} \langle S_n, e_k \rangle^2\Big]$$

$$= \frac{1}{\delta^2} \sum_{k=\ell+1}^{\infty} \mathbb{E}\big[\langle S_n, e_k \rangle \langle S_n, e_k \rangle\big]$$

$$= \frac{1}{\delta^2} \sum_{k=\ell+1}^{\infty} \langle Ce_k, e_k \rangle.$$

Since C is a trace-class operator, we have $\sum_{k=1}^{\infty} \langle Ce_k, e_k \rangle < \infty$, and thus (17.37) follows.

To prove condition (17.36), it is sufficient to show $\langle S_n, h \rangle \xrightarrow{\mathcal{D}} \langle X, h \rangle$ for any $h \in \mathbb{H}$ according to Remark 17.28. To this end, note that

$$\langle S_n, h \rangle = \frac{1}{\sqrt{n}} \sum_{j=1}^{n} \langle X_j, h \rangle$$

with an i.i.d. sequence $(\langle X_j, h \rangle)_{j \geq 1}$, where $\mathbb{E}\langle X_1, h \rangle = 0$ and $\mathbb{V}(\langle X_1, h \rangle) = \langle Ch, h \rangle$. Since $\langle X, h \rangle \sim N(0, \langle Ch, h \rangle)$, the claim follows from the Lindeberg–Lévy central limit theorem (Theorem 1.18). $\qquad\qquad\square$

As already mentioned for the case of random variables or d-dimensional random vectors, the above "Lindeberg–Lévy version" of a central limit theorem for Hilbert space-valued random elements is often not far-reaching enough. Following [KMM], we state and prove a "Lindeberg–Feller version", that is, a central limit theorem for triangular arrays. Such a version is, among other things, a powerful tool for proving the asymptotic validity of bootstrap methods in the sense of statement (c) of Theorem 12.11 under general conditions. Examples of this are Theorem 3.4 in [HLS], Theorem 3 in [HJI], and Theorem 2 as well as Corollary 1 in [BK].

17.30 Central Limit Theorem for Triangular Arrays of $\mathbb{H}$-valued Random Elements
For each $n \geq 1$, let $X_{n,1}, \ldots, X_{n,r_n}$ be independent $\mathbb{H}$-valued random elements with $\mathbb{E}(X_{n,j}) = \mathbf{0}$ and $\mathbb{E}\|X_{n,j}\|^2 < \infty$ for each n and each $j \in \{1, \ldots, r_n\}$. Further, let C_n be the covariance operator of $S_n := X_{n,1} + \ldots + X_{n,r_n}$ and $\{e_1, e_2, \ldots\}$ be any ONB of $\mathbb{H}$. We make the following assumptions:

(a) For any choice of $k, \ell \in \mathbb{N}$, the limit $a_{k,\ell} := \lim_{n \to \infty} \langle C_n e_k, e_\ell \rangle$ exists,

(b) $\displaystyle\lim_{n \to \infty} \sum_{k=1}^{\infty} \langle C_n e_k, e_k \rangle = \sum_{k=1}^{\infty} a_{k,k} < \infty,$

(c) For each $k \geq 1$ and each $\varepsilon > 0$ we have

$$\lim_{n \to \infty} L_n(\varepsilon, e_k) = 0, \tag{17.40}$$

where

$$L_n(\varepsilon, x) := \sum_{j=1}^{r_n} \mathbb{E}\left[\langle X_{n,j}, x \rangle^2 \mathbf{1}\{|\langle X_{n,j}, x \rangle| > \varepsilon\} \right], \quad x \in \mathbb{H}.$$

Then there exists a $C \in \mathcal{L}_{tr}^{+}(\mathbb{H})$ and a Gaussian random element X with $X \sim N(\mathbf{0}, C)$, such that

$$S_n \xrightarrow{\mathcal{D}} X \text{ as } n \to \infty. \tag{17.41}$$

The covariance operator C is characterized by

$$\langle Cx, e_\ell \rangle = \sum_{k=1}^{\infty} a_{\ell,k} \langle x, e_k \rangle \qquad (x \in \mathbb{H}, \, \ell \geq 1). \tag{17.42}$$

Proof of Theorem 17.27 We first show that C exists. The definition of the covariance operator C_n and the Cauchy–Schwarz inequality yield

$$\begin{aligned}
\left| \langle C_n e_k, e_\ell \rangle \right|^2 &= \left| \mathbb{E}\left[\langle S_n, e_k \rangle \cdot \langle S_n, e_\ell \rangle \right] \right|^2 \\
&\leq \mathbb{E}\left[\langle S_n, e_k \rangle^2 \right] \cdot \mathbb{E}\left[\langle S_n, e_\ell \rangle^2 \right] \\
&= \langle C_n e_k, e_k \rangle \cdot \langle C_n e_\ell, e_\ell \rangle.
\end{aligned}$$

Letting n tend to infinity, it follows from (a) that

$$|a_{k,\ell}|^2 \leq a_{k,k} \cdot a_{\ell,\ell}, \qquad k, \, \ell \geq 1. \tag{17.43}$$

Using (17.43) and the Cauchy–Schwarz inequality, we obtain

$$\sum_{j=1}^{\infty} \left(\sum_{k=1}^{\infty} a_{j,k} \langle x, e_k \rangle \right)^2 < \infty, \quad x \in \mathbb{H}.$$

Self-question 16: Can you prove this inequality?

The assignment $x \mapsto Cx$, where

$$Cx := C(x) := \sum_{j=1}^{\infty} \left(\sum_{k=1}^{\infty} a_{j,k} \langle x, e_k \rangle \right) e_j, \qquad x \in \mathbb{H}, \tag{17.44}$$

defines an operator $C : \mathbb{H} \to \mathbb{H}$. This operator is linear, and due to the symmetry of C_n and condition (a), we have $a_{j,k} = a_{k,j}$ $(k, \, j \geq 1)$, which shows that C is symmetric. Obviously, (17.42) holds, and from (17.44) it follows that

$$\langle Ce_i, e_\ell \rangle = a_{\ell,i} = a_{i,\ell}, \qquad i, \, \ell \geq 1. \tag{17.45}$$

The operator C is bounded, because the Cauchy–Schwarz inequality and Parseval's equation as well as (17.43) yield

$$\|Cx\|^2 = \sum_{j=1}^{\infty}\left(\sum_{k=1}^{\infty} a_{j,k}\langle x, e_k\rangle\right)^2 \le \sum_{j,k=1}^{\infty} a_{j,k}^2 \cdot \|x\|^2 \le \left(\sum_{k=1}^{\infty} a_{k,k}\right)^2 \cdot \|x\|^2, \quad x \in \mathbb{H}.$$

The operator C is also positive, because $0 \le \mathbb{E}\langle S_n, x\rangle^2 = \langle C_n x, x\rangle$, $x \in \mathbb{H}$, by definition of C_n. If $x = \sum_{j=1}^{\ell} b_j e_j$ with $b_1, \dots, b_\ell \in \mathbb{R}$ is a linear combination of *finitely many* e_j, then

$$0 \le \left\langle C_n\left(\sum_{i=1}^{\ell} b_i e_i\right), \sum_{j=1}^{\ell} b_j e_j\right\rangle = \sum_{i=1}^{\ell}\sum_{j=1}^{\ell} b_i b_j \langle C_n e_i, e_j\rangle.$$

Due to (17.45) and condition (a) , the above double sum converges as $n \to \infty$ to $\sum_{i,j=1}^{\ell} b_i b_j \langle C e_i, e_j\rangle = \langle Cx, x\rangle$. Thus, we have $\langle Cx, x\rangle \ge 0$ for every x from a dense set in $\mathbb{H}$, namely the union of all subspaces $\mathbb{H}_\ell$, $\ell \ge 1$, with $\mathbb{H}_\ell$ as in (17.35). Since C is bounded we obtain $\langle Cx, x\rangle \ge 0$ for every $x \in \mathbb{H}$.

Self-question 17: Can you prove the last claim?

From (17.45) and condition (b), it follows that $\sum_{k=1}^{\infty}\langle C e_k, e_k\rangle = \sum_{k=1}^{\infty} a_{k,k} < \infty$. Thus, C is a trace-class operator. All together, we get $C \in \mathcal{L}_{tr}^+(\mathbb{H})$. By Theorem 17.25, there exists a Gaussian random element X with $X \sim \mathrm{N}(\mathbf{0}, C)$.

To prove (17.41), we apply Theorem 17.27 and first show the validity of (17.37). Markov's inequality, the monotone convergence theorem, and the definition of C_n as a covariance operator entail

$$\mathbb{P}\big(\|S_n - \Pi_\ell(S_n)\| \ge \delta\big) \le \frac{1}{\delta^2}\,\mathbb{E}\big[\|S_n - \Pi_\ell(S_n)\|^2\big]$$

$$= \frac{1}{\delta^2}\,\mathbb{E}\Big[\sum_{k=\ell+1}^{\infty}\langle S_n, e_k\rangle^2\Big]$$

$$= \frac{1}{\delta^2}\sum_{k=\ell+1}^{\infty}\langle C_n e_k, e_k\rangle, \qquad \delta > 0.$$

Using condition (a) for $k = \ell$ and (b), it follows that

$$\lim_{n\to\infty}\sum_{k=\ell+1}^{\infty}\langle C_n e_k, e_k\rangle = \sum_{k=\ell+1}^{\infty}\langle C e_k, e_k\rangle = \sum_{k=\ell+1}^{\infty} a_{k,k},$$

and thus

$$\limsup_{n\to\infty} \mathbb{P}\big(\|S_n - \Pi_\ell(S_n)\| \geq \delta\big) \leq \frac{1}{\delta^2} \sum_{k=\ell+1}^{\infty} a_{k,k}.$$

In view of $\sum_{k=1}^{\infty} a_{k,k} < \infty$, (17.37) follows. To prove the still missing condition (17.36), we need to show

$$\big(\langle S_n, \mathsf{e}_1\rangle, \ldots, \langle S_n, \mathsf{e}_\ell\rangle\big) \xrightarrow{\;\mathcal{D}\;} \big(\langle X, \mathsf{e}_1\rangle, \ldots, \langle X, \mathsf{e}_\ell\rangle\big)$$

for each integer ℓ. The limit distribution appearing here is the normal distribution $\mathrm{N}_\ell\big(0_\ell, (a_{i,j})_{1\leq i,j\leq\ell}\big)$. Let $Y_{n,j} := \big(\langle X_{n,j}, \mathsf{e}_1\rangle, \ldots, \langle X_{n,j}, \mathsf{e}_\ell\rangle\big)$, $n \geq 1$, $j \in \{1, \ldots, r_n\}$, and $T_n := Y_{n,1} + \ldots + Y_{n,r_n}$, then due to condition (a) $\mathbb{E}(T_n T_n^\top) = \big(\langle C_n \mathsf{e}_i, \mathsf{e}_j\rangle\big)_{1\leq i,j\leq\ell} \to \Gamma := (a_{i,j})_{1\leq i,j\leq\ell}$. Let B be the set of canonical unit vectors in $\mathbb{R}^\ell$. Condition (c) shows that we can apply Theorem 6.22 (with $d = \ell$, $r_n = n$, $X_{n,j} := Y_{n,j}$ and $S_n := T_n$), from which the assertion follows. □

The results presented in this chapter are used for so-called *weighted L^2-statistics*, among other things. To define this class of random variables in relatively general terms, let $(\Omega, \mathcal{A}, \mathbb{P})$ be a probability space, $M \subset \mathbb{R}^d$ a non-empty Borel set and μ a finite measure on the trace σ-field $\mathcal{B}_M^d := M \cap \mathcal{B}^d$. Furthermore, let $X_1, \ldots, X_n$ be M-valued d-dimensional random vectors on Ω.

17.31 Definition (Weighted L^2-Statistic)
Let $z_n : (\mathbb{R}^d)^n \times M \to \mathbb{R}$ be a measurable function with respect to the product σ-field $(\mathcal{B}^d)^n \otimes \mathcal{B}_M^d$, and set $Z_n(t) := z_n(X_1, \ldots, X_n, t)$, $t \in M$. Then the random variable

$$T_n := \int_M Z_n^2(t)\, \mu(\mathrm{d}t) \tag{17.46}$$

(based on z_n and μ) is called a *weighted L^2-statistic*.

The attribute *weighted* refers to the measure μ. In this respect, $\mu(\mathrm{d}t) = w(t)\,\mathrm{d}t$ often applies with a non-negative measurable function $w : M \to \mathbb{R}$, which is called the *weight function*.

Weighted L^2-statistics are used in particular to test various hypotheses about the unknown distribution of a d-dimensional random vector X (for an overview of the relevant literature, see [BEH]). Suppose we want to test the hypothesis

$$H_0 : \mathbb{P}^X \in \mathcal{Q} := \{Q_\vartheta : \vartheta \in \Theta\}$$

that the distribution of X belongs to a family $\mathcal{Q}$ of distributions on $\mathcal{B}^d$ indexed by a parameter ϑ. Here, for some integer s, the parameter space Θ is a nonempty open subset of $\mathbb{R}^s$. If $X_1, X_2, \ldots$ is a sequence of i.i.d. random vectors with the same distribution as X, the usual procedure is to estimate the parameter ϑ, based on $X_1, \ldots, X_n$, using a suitable estimator $\widehat{\vartheta}_n = \widehat{\vartheta}_n(X_1, \ldots, X_n)$ (see Chaps. 10 and 11). In the following, we will explain why weighted L^2-statistics are useful for this testing problem, deliberately omitting technical details.

Typically, $Z_n(t)$ in (17.46) has the form

$$Z_n(t) = \frac{1}{\sqrt{n}} \sum_{j=1}^{n} H(X_j, \widehat{\vartheta}_n, t), \tag{17.47}$$

where $H : M \times \Theta \times M \to \mathbb{R}$ is a continuous, bounded function, which is continuously differentiable with respect to the second argument. Moreover, we assume

$$\mathbb{E}_\vartheta H(X, \vartheta, t) = 0, \quad t \in M, \tag{17.48}$$

for each $\vartheta \in \Theta$. Thus, if the hypothesis H_0 holds, the mean function $M \ni t \mapsto \mathbb{E}_\vartheta H(X, \vartheta, t)$ is identically zero. As in earlier chapters, the indexing with ϑ emphasizes that X has the distribution Q_ϑ. In the Hilbert space $\mathbb{H} := L^2(M, \mathcal{B}_M^d, \mu)$, (17.46) is the same as

$$T_n = \|Z_n\|^2, \tag{17.49}$$

and thus T_n measures the deviation from the zero mean function in a specific way.

In view of (17.49) and (17.47), one may conjecture that, under H_0, there is a centered Gaussian random element Z in $\mathbb{H}$ such that $Z_n \overset{\mathcal{D}}{\longrightarrow} Z$ as $n \to \infty$. By the mapping theorem, $T_n \overset{\mathcal{D}}{\longrightarrow} \|Z\|^2$ would then apply, and we would have a limit distribution of T_n under H_0, at least in qualitative form. The attribute *qualitative* refers primarily to the fact that, provided that H_0 is a composite hypothesis, the "true parameter" $\vartheta \in \Theta$ is not known. The limit distribution of the test statistic T_n under H_0 thus depends on ϑ, at least in general. How to deal with this problem in the practical implementation of a test of H_0 that rejects H_0 for large values of T_n will be discussed later.

First of all, the question arises how to prove $Z_n \overset{\mathcal{D}}{\longrightarrow} Z$ for some centered Gaussian random element in $\mathbb{H}$ if H_0 holds. The procedure is clearly outlined: If H_0 holds, there is a $\vartheta \in \Theta$ with $\mathbb{P}^X = Q_\vartheta$. If the parameter ϑ were to replace $\widehat{\vartheta}_n = \widehat{\vartheta}_n(X_1, \ldots, X_n)$ in (17.47), we would have the $\mathbb{H}$-valued random element $\widetilde{Z}_n = \widetilde{Z}_n(\cdot)$ defined by

$$\widetilde{Z}_n(t) := \frac{1}{\sqrt{n}} \sum_{j=1}^{n} H(X_j, \vartheta, t) \tag{17.50}$$

instead of $Z_n = Z_n(\cdot)$ in (17.47). Note that $\widetilde{Z}_n$ is centered because of (17.48). For the sequence $(\widetilde{Z}_n)$ the central limit theorem (Theorem 17.29) applies. The idea now is to use the smoothness of the function H appearing in (17.47) in its second argument. If $\frac{\partial}{\partial \vartheta}$ denotes the gradient with respect to ϑ, a Taylor expansion of H around the point ϑ yields

$$H(X_j, \widehat{\vartheta}_n, t) \approx H(X_j, \vartheta, t) + \frac{\partial}{\partial \vartheta} H(X_j, \vartheta, t)^\top (\widehat{\vartheta}_n - \vartheta)$$

and thus

$$Z_n(t) \approx \widetilde{Z}_n(t) + \left[\frac{1}{n} \sum_{j=1}^{n} \frac{\partial}{\partial \vartheta} H(X_j, \vartheta, t) \right]^\top \cdot \sqrt{n}(\widehat{\vartheta}_n - \vartheta). \tag{17.51}$$

Of course, it must be worked out that the error made by this approximation is asymptotically negligible. For the term within squared brackets, we require a strong law of large numbers (cf. Theorem 17.15), that is,

$$\left\| \frac{1}{n} \sum_{j=1}^{n} \frac{\partial}{\partial \vartheta} H(X_j, \vartheta, \cdot) - L(\vartheta, \cdot) \right\| \to 0 \quad \mathbb{P}_\vartheta\text{-a.s.,} \tag{17.52}$$

where $L(\vartheta, t) := \mathbb{E}_\vartheta \left[\frac{\partial}{\partial \vartheta} H(X, \vartheta, t) \right]$. For the estimation error $\widehat{\vartheta}_n - \vartheta$, we require a representation of the form

$$\sqrt{n}(\widehat{\vartheta}_n - \vartheta) = \frac{1}{\sqrt{n}} \sum_{j=1}^{n} \ell(X_j, \vartheta) + o_{\mathbb{P}_\vartheta}(1) \quad \text{as } n \to \infty \tag{17.53}$$

(cf.(10.30)). Here, $\mathbb{E}_\vartheta[\ell(X, \vartheta)] = 0_s$ and $\mathbb{E}_\vartheta \|\ell(X, \vartheta)\|_2^2 < \infty$ apply, and $\| \cdot \|_2$ denotes the Euclidean norm in $\mathbb{R}^s$.

Together with (17.47) and (17.51)–(17.53) we now get $Z_n(\cdot) \approx Z_n^*(\cdot)$, where $Z_n^*(\cdot)$ is defined by

$$Z_n^*(t) := \frac{1}{\sqrt{n}} \sum_{j=1}^{n} \left(H(X_j, \vartheta, t) + L(\vartheta, t)^\top \ell(X_j, \vartheta) \right), \quad t \in M.$$

The technical conditions to be specified must ensure that $\|Z_n - Z_n^*\| = o_{\mathbb{P}_\vartheta}(1)$ as $n \to \infty$. Due to (17.48) and $\mathbb{E}_\vartheta \ell(X, \vartheta) = 0$, the summands in Z_n^* are i.i.d. centered $\mathbb{H}$-valued random elements, to which the central limit Theorem 17.29 applies. Due to $\|Z_n - Z_n^*\| = o_{\mathbb{P}_\vartheta}(1)$, this limit theorem and Slutsky's lemma guarantee that there is a centered $\mathbb{H}$-valued Gaussian random element Z with covariance function

$$K(s, t) = \mathbb{E}_\vartheta \left[\left(H(X, \vartheta, s) + L(\vartheta, s)^\top \ell(X, \vartheta) \right) \left(H(X, \vartheta, t) + L(\vartheta, t)^\top \ell(X, \vartheta) \right) \right]$$

such that $Z_n \xrightarrow{\mathcal{D}} Z$, and thus $T_n = \|Z_n\|^2 \xrightarrow{\mathcal{D}} \|Z\|^2$.

By Theorem 17.26, the distribution of $\|Z\|^2$ is that of $\sum_{j \geq 1} \lambda_j N_j^2$, where $\lambda_1, \lambda_2, \ldots$ denote the non-zero eigenvalues of the integral operator associated with $K(\cdot, \cdot)$, and $N_1, N_2, \ldots$ is a sequence of i.i.d. standard normally distributed random variables.

With regard to practical applications, this result is purely theoretical in nature, as the covariance function and thus the eigenvalues depend on the unknown parameter ϑ under H_0. However, as outlined in Sect. 12.10, there is the possibility of performing a parametric bootstrap procedure. In this procedure, given $\widehat{\vartheta}_n$, a computer generates realizations of T_n using pseudorandom numbers, which simulate the distribution of X under the parameter $\widehat{\vartheta}_n$, that is, according to the distribution $\mathbb{P}_{\widehat{\vartheta}_n}$. The hypothesis H_0 is then rejected at level α if the observed realization of T_n is among the largest $\alpha \cdot 100\%$ in relation to all simulated realizations. For a result analogous to Theorem 12.11 (c), that this bootstrap procedure asymptotically maintains a given test level, one needs the strong consistency $\widehat{\vartheta}_n \to \vartheta$ $\mathbb{P}_\vartheta$-a.s. for each $\vartheta \in \Theta$ and the central limit Theorem 17.30 for triangular arrays of independent $\mathbb{H}$-valued random elements.

We finally discuss the asymptotic behavior of T_n as $n \to \infty$ when the hypothesis H_0 does not hold. Then there is usually a $z \in \mathbb{H}$ with $z \neq 0$ and $\frac{1}{n} \sum_{j=1}^n H(X_j, \widehat{\vartheta}_n, \cdot) \xrightarrow{\mathbb{P}} z(\cdot)$ in $\mathbb{H}$, that is,

$$\left\| \frac{1}{n} \sum_{j=1}^n H(X_j, \widehat{\vartheta}_n, \cdot) - z(\cdot) \right\| \xrightarrow{\mathbb{P}} 0 \quad \text{as } n \to \infty.$$

This implies

$$\frac{T_n}{n} = \left\| \frac{1}{n} \sum_{j=1}^n H(X_j, \widehat{\vartheta}_n, \cdot) \right\|^2 \xrightarrow{\mathbb{P}} \Delta := \|z\|^2 = \int_M z^2(t) \, \mu(dt).$$

In the case $\Delta > 0$, it follows that $T_n \xrightarrow{\mathbb{P}} \infty$, and thus the test based on T_n is consistent against each such alternative. Using the central limit Theorem 17.30 and the mapping theorem, this result can even be sharpened to the extent that T_n is asymptotically normally distributed under fixed alternatives to H_0. The details are presented in [BEH], but the idea of proof is quickly explained: If we set $\overline{Z}_n := n^{-1/2} Z_n$ with Z_n as in (17.47), then

$$\sqrt{n} \left(\frac{T_n}{n} - \Delta \right) = \sqrt{n} \left(\|\overline{Z}_n\|^2 - \|z\|^2 \right) = \sqrt{n} \langle \overline{Z}_n - z, \overline{Z}_n + z \rangle$$

$$= \sqrt{n} \langle \overline{Z}_n - z, 2z + \overline{Z}_n - z \rangle$$

$$= 2 \langle V_n, z \rangle + \frac{1}{\sqrt{n}} \|V_n\|^2, \tag{17.54}$$

where $V_n := \sqrt{n}(\overline{Z}_n - z)$. The main step is now to prove $V_n \overset{\mathcal{D}}{\longrightarrow} V$ for some centered $\mathbb{H}$-valued Gaussian random element V. Since the sequence $(\|V_n\|^2)$ is tight, (17.54) together with the mapping theorem and Slutsky's lemma entail

$$\sqrt{n}\left(\frac{T_n}{n} - \Delta\right) \overset{\mathcal{D}}{\longrightarrow} 2\langle V, z\rangle.$$

By definition of the normal distribution in a Hilbert space, $2\langle V, z\rangle$ has the normal distribution $N(0, \sigma^2)$, where $\sigma^2 = 4\mathbb{E}\left[\langle V, z\rangle^2\right]$. In principle, it is possible to construct a consistent estimator $\widehat{\sigma}_n^2 = \widehat{\sigma}_n^2(X_1, \ldots, X_n)$ for σ^2 and thus provide an asymptotic confidence interval for Δ (see [BEH]).

17.32 Example (BHEP Tests of Multivariate Normality)

Tests of multivariate normality have long enjoyed unbroken interest, both from a theoretical and a practical point of view (for an overview see, e.g., [H02], and [EH1]). The hypothesis states that X has *any* non-degenerate d-variate normal distribution, without specifying the parameters of this distribution. In this testing problem, the basic assumption is that the distribution of X has a density with respect to the Borel–Lebesgue measure λ^d. The hypothesis H_0 thus formally reads

$$H_0 : \mathbb{P}^X \in \mathcal{N}_d := \left\{N_d(m, \Sigma) : m \in \mathbb{R}^d, \ \Sigma \in \mathbb{R}^{d \times d} \text{ symmetric and positive definite}\right\}.$$

In this case, the unknown parameter(vector) ϑ consists of the d components of m as well as the diagonal and the part above the diagonal of Σ; thus $s = d + d + \binom{d}{2} = d(d+3)/2$. It is common to estimate the expectation m by the *sample mean* $\overline{X}_n := n^{-1} \sum_{j=1}^n X_j$ and the covariance matrix Σ by the *sample covariance matrix* $S_n := n^{-1} \sum_{j=1}^n (X_j - \overline{X}_n)(X_j - \overline{X}_n)^\top$ of $X_1, \ldots, X_n$. Writing $S_n^{-1/2}$ for the symmetric positive definite square root of S_n^{-1}, the next step is to form the so-called *scaled residuals*

$$X_{n,j} := S_n^{-1/2}(X_j - \overline{X}_n), \quad j \in \{1, \ldots, n\}, \tag{17.55}$$

which represent an empirical standardization of $X_1, \ldots, X_n$. We tacitly assume $n \geq d+1$ in the sequel. Since X has a density with respect to the Borel–Lebesgue measure, it follows that S_n is invertible with probability 1 (see Problem 5.4).

The following weighted L^2-statistic leads to a test of multivariate normality which is very well studied and has many desirable properties. Its practical significance is evident from the fact that it is a part of the freely available and widely used statistical package R (see [CRT]) as function "test.BHEP". The test statistic is motivated by the heuristic that, under H_0 and for large n, the scaled residuals defined in (17.55) should be approximately independent and each nearly $N_d(0_d, I_d)$-distributed. Thus, the *empirical characteristic function*

$$\psi_n(t) := \frac{1}{n} \sum_{j=1}^{n} \exp\left(it^\top X_{n,j}\right), \quad t \in \mathbb{R}^d,$$

of $X_{n,1}, \ldots, X_{n,n}$ should serve as a good approximation for the characteristic function

$$\psi_0(t) := \exp\left(-\frac{\|t\|_2^2}{2}\right), \quad t \in \mathbb{R}^d, \tag{17.56}$$

of the standard normal distribution $N_d(0_d, I_d)$ in $\mathbb{R}^d$. Here, $\|\cdot\|_2$ stands for the Euclidean norm in $\mathbb{R}^d$. The BHEP test of H_0 uses the test statistic

$$T_{n,\beta} := n \int_{\mathbb{R}^d} |\psi_n(t) - \psi_0(t)|^2 w_\beta(t)\, dt, \tag{17.57}$$

where

$$w_\beta(t) := \frac{1}{(2\pi\beta^2)^{d/2}} \exp\left(-\frac{\|t\|_2^2}{2\beta^2}\right), \quad t \in \mathbb{R}^d,$$

and $\beta > 0$ is a so-called *tuning parameter*. The extremely appealing feature to use this weight function is that the integral in (17.57) does not have to be evaluated numerically. In fact, a direct calculation shows that $T_{n,\beta}$ takes the form

$$T_{n,\beta} = \frac{1}{n} \sum_{j,k=1}^{n} \exp\left(-\frac{\beta^2}{2}\|X_{n,j} - X_{n,k}\|_2^2\right)$$

$$- \frac{2}{(1+\beta^2)^{d/2}} \sum_{j=1}^{n} \exp\left(-\frac{\beta^2\|X_{n,j}\|_2^2}{2(1+\beta^2)}\right) + \frac{n}{(1+2\beta^2)^{d/2}}.$$

$$\tag{17.58}$$

From this, it can be seen that the calculation of the square root $S_n^{-1/2}$ of S_n^{-1} does not have to be carried out, and that $T_{n,\beta}$ is an affine-invariant statistic.

Self-question 18: Why is the calculation of $S_n^{-1/2}$ not necessary?

The property of affine invariance means that for every invertible $d \times d$ matrix A and every $b \in \mathbb{R}^d$ we have

$$T_{n,\beta}(AX_1 + b, \ldots, AX_n + b) = T_{n,\beta}(X_1, \ldots, X_n)$$

with probability 1 (cf. Problem 17.22). Affine invariance is a desirable property of any conceivable statistic for testing of multivariate normality, as the class $\mathcal{N}_d$ of all non-degenerate d-variate normal distributions is closed with respect to affine transformations $x \mapsto Ax + b$ with a regular matrix A (cf. the reproduction Theorem 5.7). A consequence of the affine invariance of $T_{n,\beta}$ is that the distribution of $T_{n,\beta}$ under H_0 does not depend on the unknown expectation m and the covariance matrix Σ, and thus we can assume w.l.o.g. $m = 0_d$ and $\Sigma = I_d$.

Note that $T_{n,\beta}$ at first glance is not a weighted L^2-statistic in the sense of Definition 17.31, because *complex-valued* random variables occur within the absolute value of the integrand in (17.57). However, one can write this integrand in the form $\overline{(\psi_n(t) - \psi_0(t))}(\psi_n(t) - \psi_0(t))$ and use that $\int_{\mathbb{R}^d} \sin(t^\top x) w_\beta(t)\, dt = 0$ for each $x \in \mathbb{R}^d$. Putting $M := \mathbb{R}^d$, $\mu(dt) := w_\beta(t)dt$ and

$$Z_n(t) := \frac{1}{\sqrt{n}} \sum_{j=1}^n \left[\cos(t^\top X_{n,j}) + \sin(t^\top X_{n,j}) - \exp\left(-\frac{\|t\|_2^2}{2} \right) \right], \quad t \in \mathbb{R}^d,$$

(17.59)

$T_{n,\beta}$ has the form (17.46) with $X_{n,j}$ as in (17.55).

The statistic $T_{n,\beta}$ was initially proposed by T. Epps and L. Pulley for the case $d = 1$ (see [EPY]) and then generalized to the multivariate case by L. Baringhaus and the author of this book (see [BHE]). The acronym *BHEP test* reflects the first letter of the surnames of these authors. It goes back to S. Csörgő who proved the consistency of a test of multivariate normality that rejects the hypothesis H_0 for large values of $T_{n,\beta}$ against any distribution not belonging to $\mathcal{N}_d$ (see [CSO]). The asymptotic distribution of $T_{n,\beta}$ under H_0 was initially derived using the theory of V-statistics with estimated parameters, in [BHE] for the case $\beta = 1$ and in [HEZ] for general β. The representation $T_{n,\beta} = \int Z_n^2(t)\varphi_\beta(t)\, dt$ with $Z_n(t)$ as in (17.59) goes back to the author and T. Wagner (see [HWA]). In [HWA], $Z_n(\cdot)$ was studied as a random element in the space $\mathcal{C}(\mathbb{R}^d)$ of continuous functions $f : \mathbb{R}^d \to \mathbb{R}$, equipped with the Fréchet metric

$$\rho(f, g) := \sum_{k=1}^\infty 2^{-k} \frac{\rho_k(f, g)}{1 + \rho_k(f, g)}, \qquad \rho_k(f, g) := \max_{\|x\|_2 \le k} |f(x) - g(x)|,$$

The limit distribution of $T_{n,\beta}$ under H_0 is that of

$$T_{\infty,\beta} := \int Z^2(t)\varphi_\beta(t)\, dt. \tag{17.60}$$

Here, $Z(\cdot)$ is a centered Gaussian random element in $\mathcal{C}(\mathbb{R}^d)$ with covariance function

$$K(s,t) := \exp\left(-\frac{\|s-t\|_2^2}{2}\right) - \left\{1 + s^\top t + \frac{1}{2}(s^\top t)^2\right\}\exp\left(-\frac{\|s\|_2^2 + \|t\|_2^2}{2}\right), \qquad (17.61)$$

$s,t \in \mathbb{R}^d$. N. Gürtler (see [GUE]) showed convergence in distribution in the Hilbert space $\mathbb{H} := \mathrm{L}^2(\mathbb{R}^d, \mathcal{B}^d, \varphi_\beta(t)\mathrm{d}t)$ of $Z_n(\cdot)$ in (17.59) to a likewise denoted centered $\mathbb{H}$-valued Gaussian random element with the covariance function $K(\cdot,\cdot)$ defined in (17.61). The mapping theorem then provides $T_{n,\beta} \xrightarrow{\mathcal{D}} T_{\infty,\beta}$ under H_0 in a different way compared to the proof in [HWA]. By Theorem 17.26, we have

$$T_{\infty,\beta} \overset{\mathcal{D}}{=} \sum_{j=1}^{\infty} \lambda_{j,\beta} N_j^2.$$

Here, $N_1, N_2, \ldots$ is a sequence of i.i.d. standard normal random variables, and $\lambda_{1,\beta}, \lambda_{2,\beta}, \ldots$ are the positive eigenvalues of the integral operator $A_\beta : \mathbb{H} \to \mathbb{H}$ defined by

$$A_\beta(f)(s) := \int K(s,t)f(t)\varphi_\beta(t)\,\mathrm{d}t, \quad s \in \mathbb{R}^d.$$

Although the covariance function $K(\cdot,\cdot)$ has a relatively simple form, it seems to be a hopeless endeavor to specify the eigenvalues of A_β in explicit form. A stable numerical method in the case $d = 1$ is provided by [EH]. In the case $d = 1 = \beta$, the first four moments of $T_{\infty,\beta}$ are known ([H90]), and for general d and β, the first three moments of $T_{\infty,\beta}$ were determined in [HWA].

If X has a non-degenerate distribution with $\mathbb{E}\|X\|_2^4 < \infty$, $\mathbb{E}(X) = 0_d$, and $\mathbb{E}(XX^\top) = \mathrm{I}_d$ (the last two properties can be assumed w.l.o.g. because of invariance), then

$$\sqrt{n}\left(\frac{T_{n,\beta}}{n} - \Delta_\beta\right) \xrightarrow{\mathcal{D}} \mathrm{N}(0, \sigma_\beta^2),$$

where

$$\Delta_\beta := \int \left|\mathbb{E}(e^{it^\top X}) - \psi_0(t)\right|^2 \varphi_\beta(t)\,\mathrm{d}t,$$

and $\psi_0(t)$ is given in (17.56), if $\mathbb{P}^X \notin \mathcal{N}_d$.

The variance σ_β^2 depends in a complicated way on the distribution of X. However, there is a sequence of consistent estimators of σ_β^2 (see [GUE], Theorem 5 and Section 1.2).

Although a test of multivariate normality based on $T_{n,\beta}$ is consistent against any fixed alternative distribution to H_0 for each β, the empirical power of the test, that is, the estimated rejection probability of H_0 for a fixed sample size n when choosing such an alternative, depends more or less strongly on β. Recommendations for the choice of β are given, among others, in [TE1]. Recently, attempts have also been made to let β depend on $X_1, \ldots, X_n$ (see [TE2]). Finally, it should be emphasized that the approach underlying the BHEP test also applies to test the hypothesis that a Hilbert space-valued random element has a non-degenerate normal distribution (see [HJI]).

Answers to Self-Questions

Answer 1 (b) follows from (c) if we set $\Omega := \mathbb{N}$ and $\mathcal{A} := \mathcal{P}(\mathbb{N})$ and choose the counting measure for μ on $\mathcal{A}$.

Answer 2 From $T(x) = T(y)$ it follows that $\langle x - y, e_k \rangle = 0$ for each $k \geq 1$ and thus $x = y$, since $\{e_1, e_2, \ldots\}$ is an ONB of $\mathbb{H}$.

Answer 3 Since $\mathbb{H} \ni x \mapsto \langle x, y \rangle$ defines a continuous mapping, the preimage of an open set in $\mathbb{R}$ under this mapping is an open set in $\mathbb{H}$.

Answer 4 The statement follows from the uniqueness theorem for measures (Theorem 1.28), since the collection $\mathcal{M}_\cap$ is a π-system that generates the σ-field $\mathcal{B}$.

Answer 5 If $\langle m_1, y \rangle = \langle m_2, y \rangle$ for each $y \in \mathbb{H}$ then, putting $y := m_1 - m_2$, it follows that $\|m_1 - m_2\|^2 = 0$ and thus $m_1 = m_2$.

Answer 6 Let $\mathbf{0}$ denote the zero element of $\mathbb{B}$. If $k = 2$ then, setting $B_1 := A_1 \setminus A_2$, $B_2 := A_1 \cap A_2$, $B_3 := A_2 \setminus A_1$ and $B_4 := \Omega \setminus (A_1 \cup A_2)$, we have $f = \mathbf{1}\{B_1\}b_1 + \mathbf{1}\{B_2\}(b_1 + b_2) + \mathbf{1}\{B_3\}b_2 + \mathbf{1}\{B_4\}\mathbf{0}$. The conclusion from k to $k + 1$ is done inductively.

Answer 7 W.l.o.g. we assume that $A_1, \ldots, A_k$ in (17.10) are pairwise disjoint sets. Using (17.10), (17.11) and the triangle inequality for the norm $\| \cdot \|_{\mathbb{B}}$, we obtain

$$\left\| \int_\Omega f \, d\mu \right\|_{\mathbb{B}} \leq \sum_{j=1}^{k} \mu(A_j) \|b_j\|_{\mathbb{B}} = \int_\Omega \|f\|_{\mathbb{B}} \, d\mu.$$

The equal sign holds because $\|f(\omega)\|_{\mathbb{B}} = \sum_{j=1}^{k} \mathbf{1}\{A_j\}(\omega)\|b_j\|_{\mathbb{B}}$, $\omega \in \Omega$.

Answer 8 The inequality follows from Bessel's inequality (17.4), and the convergence holds in view of the dominated convergence theorem, because $\|f - g_n\|_{\mathbb{H}} \leq 2\|f\|_{\mathbb{H}}$.

Answer 9 The second equal sign follows with the monotone convergence theorem.

Answer 10 Expanding the inner products within the square brackets in (17.18) yields $\langle X, x \rangle \langle X, y \rangle - \langle \mathbb{E}X, x \rangle \langle X, y \rangle - \langle X, x \rangle \langle \mathbb{E}X, y \rangle + \langle \mathbb{E}X, x \rangle \langle \mathbb{E}X, y \rangle$. Taking expectations, the assertion follows due to $\mathbb{E}\langle X, x \rangle = \langle \mathbb{E}X, x \rangle$ for each $x \in \mathbb{H}$.

Answer 11 Theorem 1.33 is used in the second equality, which reads

$$\int_{\Omega} \left(\int_E X^2(t, \omega) \, \mu(dt) \right) \mathbb{P}(d\omega) = \int_E \left(\int_{\Omega} X^2(t, \omega) \, \mathbb{P}(d\omega) \right) \mu(dt).$$

Answer 12 With X and Y also $e^{i\langle X,h \rangle}$ and $e^{i\langle Y,h \rangle}$ are independent. The assertion then follows from $e^{i\langle X+Y,h \rangle} = e^{i\langle X,h \rangle} e^{i\langle Y,h \rangle}$ using the multiplication rule for expectations of independent random variables.

Answer 13 Due to the continuity of the inner product, $\langle X, h \rangle = \lim_{n \to \infty} \sum_{j=1}^{n} Y_j \langle e_j, h \rangle$, $h \in \mathbb{H}$. Being the limit of $(\mathcal{A}, \mathcal{B}^1)$-measurable functions, $\langle X, h \rangle$ is also $(\mathcal{A}, \mathcal{B}^1)$-measurable. Therefore, X is $(\mathcal{A}, \mathcal{B}(\mathbb{H}))$-measurable according to Theorem 17.7.

Answer 14 Since $m_k = \langle m, e_k \rangle$, $k \geq 1$, the convergence follows from Parseval's equation.

Answer 15 By definition, $\tilde{N}_1, \tilde{N}_2, \ldots$ are independent if for every $k \geq 2$ the random variables $\tilde{N}_1, \ldots, \tilde{N}_k$ are independent. The latter independence follows from $\mathrm{Cov}(\tilde{N}_i, \tilde{N}_j) = 0$ for $i \neq j$ according to Corollary 5.12.

Answer 16 Using the Cauchy–Schwarz inequality as well as (17.43) and Parseval's equation, we get

$$\left(\sum_{k=1}^{\infty} a_{j,k} \langle x, e_k \rangle \right)^2 \leq \sum_{k=1}^{\infty} a_{j,k}^2 \cdot \sum_{k=1}^{\infty} \langle x, e_k \rangle^2 \leq a_{j,j} \cdot \sum_{k=1}^{\infty} a_{k,k} \cdot \|x\|^2.$$

In view of condition (b), we have $\sum_{k=1}^{\infty} a_{k,k} < \infty$, and the assertion follows.

Answer 17 Let $0 \leq \langle Cy, y \rangle$ for every $y \in D$, where $\overline{D} = \mathbb{H}$. For each $x \in \mathbb{H}$ there is a sequence (y_n) in D with $y_n \to x$. The triangle inequality and the Cauchy–Schwarz inequality yield

$$|\langle Cx, x \rangle - \langle Cy_n, y_n \rangle| \leq |\langle Cx, x \rangle - \langle Cx, y_n \rangle| + |\langle Cx, y_n \rangle - \langle Cy_n, y_n \rangle|$$
$$\leq \|Cx\| \cdot \|x - y_n\| + \|C(x - y_n)\| \cdot \|y_n\|.$$

Since $\|x - y_n\| \to 0$ and $\|C(x - y_n)\| \to 0$ (because C is a bounded linear operator and thus continuous) and the sequence $(\|y_n\|)$ is bounded, we have $\langle Cy_n, y_n \rangle \to \langle Cx, x \rangle$. Since $\langle Cy_n, y_n \rangle \geq 0$ for each $n \geq 1$, it follows that $\langle Cx, x \rangle \geq 0$.

Answer 18 By definition of $X_{n,j}$, $\|X_{n,j}\|_2^2 = (X_j - \overline{X}_n)^\top S_n^{-1}(X_j - \overline{X}_n)$ and $\|X_{n,j} - X_{n,k}\|_2^2 = (X_j - X_k)^\top S_n^{-1}(X_j - X_k)$.

Problems

17.1 Problem Prove the generalized Parseval's equation (17.3).

Hint If $\|x_n - x\| \to 0$ then $\langle x_n, y \rangle \to \langle x, y \rangle$.

17.2 Problem Let $\mathbb{H}$ be an infinite-dimensional separable Hilbert space. A measure μ on $\mathcal{B}(\mathbb{H})$ is called *invariant* if $\mu(A) = \mu(A + x)$ for every $A \in \mathcal{B}(\mathbb{H})$ and every $x \in \mathbb{H}$. Here, we set $A + x := \{y + x : y \in A\}$. Let $B(x, \varepsilon) := \{y \in \mathbb{H} : \|x - y\| < \varepsilon\}$ $(x \in \mathbb{H}, \varepsilon > 0)$. Show: If μ is an invariant measure such that $\mu(B(x_0, \varepsilon)) < \infty$ for some $x_0 \in \mathbb{H}$ and every $\varepsilon > 0$, then $\mu \equiv 0$, i.e., $\mu(B) = 0$ for each $B \in \mathcal{B}(\mathbb{H})$.

17.3 Problem Show: If $f : \Omega \to \mathbb{B}$ is a Bochner integrable function, then

$$\left\| \int_\Omega f \, d\mu \right\|_{\mathbb{B}} \leq \int_\Omega \|f\|_{\mathbb{B}} \, d\mu.$$

17.4 Problem Prove Corollary 17.7.

17.5 Problem Let $\Omega := \{\omega_1, \omega_2, \ldots\}$ be a countable set, $\mathcal{A} := \mathcal{P}(\Omega)$ and

$$\mathbb{P}(\{\omega_k\}) := \frac{C}{k(\ln(k+1))^2}, \quad k \geq 1.$$

Here, C is a constant that ensures $\sum_{k=1}^\infty \mathbb{P}(\{\omega_k\}) = 1$. Further, let $\mathbb{H} := \ell_2$ be the Hilbert space of square summable sequences from Example 17.2 (b) and $X : \Omega \to \mathbb{H}$ defined by $X(\omega_k) := x_k$, where $x_k := (x_{k,j})_{j \geq 1}$ and $x_{k,k} := \ln(k+1)$ and $x_{k,j} := 0$, if $j \neq k$, $k \in \mathbb{N}$. Show:

(a) $\mathbb{E}\|X\| = \infty$.
(b) $\mathbb{E}|\langle X, y \rangle| < \infty$ for each $y \in \mathbb{H}$.

Hint $\sum_{k=1}^\infty (k \ln(k+1))^{-1} = \infty$.

17.6 Problem Let X be an $\mathbb{H}$-valued random element with $\mathbb{E}\|X\| < \infty$. Show using (17.7):

$$\|\mathbb{E}X\| \leq \mathbb{E}\|X\|.$$

17.7 Problem Prove Theorem 17.12 using (17.7).

17.8 Problem Let $(\Omega, \mathcal{A}, \mu)$ be a measure space and $\mathbb{H}$ a separable Hilbert space. Using (17.11) and (17.14), show that

$$\left\langle \int_{\Omega} f \, d\mu, y \right\rangle = \int_{\Omega} \langle f, y \rangle \, d\mu$$

for every Bochner integrable function $f : \Omega \to \mathbb{H}$ and every $y \in \mathbb{H}$.

17.9 Problem Prove Theorem 17.14.

Hint $\mathbb{E}\langle X, Y \rangle = \int_{\mathbb{H}} \int_{\mathbb{H}} \langle x, y \rangle \mathbb{P}^X \otimes \mathbb{P}^Y (dx, dy)$. Use Fubini's theorem.

17.10 Problem Let X be an $\mathbb{H}$-valued random element with $\mathbb{E}\|X\|^2 < \infty$. Show that

$$\mathbb{E}\|X - h\|^2 = \mathbb{E}\|X - \mathbb{E}X\|^2 + \|\mathbb{E}X - h\|^2$$

for each $h \in \mathbb{H}$. In particular, as with real random variables, it follows that

$$\mathbb{E}\|X - \mathbb{E}X\|^2 = \min_{h \in \mathbb{H}} \mathbb{E}\|X - h\|^2.$$

17.11 Problem Let $X_1, \ldots, X_n$ be pairwise independent $\mathbb{H}$-valued random elements with $\mathbb{E}\|X_j\|^2 < \infty$ and $\mathbb{E}X_j = 0$, $j \in \{1, \ldots, n\}$. Further, let $S_n := X_1 + \ldots + X_n$. Show:

$$\mathbb{E}\|S_n\|^2 = \sum_{j=1}^{n} \mathbb{E}\|X_j\|^2.$$

17.12 Problem Let $X_1, X_2, \ldots$ be i.i.d. $\mathbb{H}$-valued random elements with $\mathbb{E}\|X_1\|^4 < \infty$. Prove that

$$\lim_{n \to \infty} \frac{1}{n} \sum_{j=1}^{n} X_j = \mathbb{E}(X_1) \quad \mathbb{P}\text{-almost surely.}$$

Hint Show $\mathbb{E}\left\| \frac{1}{n} \sum_{j=1}^{n} X_j - \mathbb{E}X_1 \right\|^4 < \infty$.

17.13 Problem Let X be an $\mathbb{H}$-valued random element with $\mathbb{E}\|X\|^2 < \infty$ and $\Sigma : \mathbb{H} \to \mathbb{H}$ be defined by

$$\Sigma x := \Sigma(x) := Tx - \langle \mathbb{E}X, x \rangle \mathbb{E}X, \quad x \in \mathbb{H},$$

where $\langle Tx, y \rangle = \mathbb{E}\big[\langle X, x \rangle \langle X, y \rangle\big]$, $x, y \in \mathbb{H}$. Show that Σ is a linear, self-adjoint, positive trace-class operator for which (17.18) holds.

17.14 Problem Suppose X is an $\mathbb{H}$-valued random element with $\mathbb{E}\|X\|^2 < \infty$. Show: If $a \in \mathbb{R}$ and $h \in \mathbb{H}$ then

$$\Sigma(aX + h) = a^2 \Sigma(X).$$

17.15 Problem Let E be a compact metric space and $(X(t))_{t \in E}$ a second-order stochastic process with mean function $m(t) = \mathbb{E}X(t)$, $t \in E$, and covariance function $K(s, t) = \mathrm{Cov}((X(s), X(t))$, $s, t \in E$. Prove that the following statements are equivalent:

(a) $(X(t))_{t \in E}$ is a mean square continuous stochastic process.
(b) The functions $m(\cdot)$ and $K(\cdot, \cdot)$ are continuous.

Hint for "(b) $\implies$ **(a)":** Write $\mathbb{E}(X(s) - X(t))^2$ in terms of $m(\cdot)$ and $K(\cdot, \cdot)$.

Hint for "(a) $\implies$ **(b)":** We have $K(s, t) - K(u, v) = (K(s, t) - K(s, v)) + (K(s, v) - K(u, v))$. Apply the Cauchy–Schwarz inequality to each of the terms in brackets.

17.16 Problem Let X be an $\mathbb{H}$-valued random element with characteristic functional φ_X. Show: The distribution of X is symmetric to $\mathbf{0}$ (that is, $X \overset{\mathcal{D}}{=} -X$) if and only if φ_X is real-valued.

17.17 Problem Show that an $\mathbb{H}$-valued random element X with the normal distribution $N(m, \Sigma)$ has the characteristic functional

$$\varphi_X(h) = e^{i\langle m, h \rangle} \exp\left(-\frac{1}{2} \langle \Sigma h, h \rangle\right), \quad h \in \mathbb{H}.$$

17.18 Problem Let $X, X_1, X_2, \ldots$ be $\mathbb{H}$-valued random elements with characteristic functionals $\varphi, \varphi_1, \varphi_2, \ldots$, respectively. Show: If $\lim_{n \to \infty} \varphi_n(h) = \varphi(h)$ for every $h \in \mathbb{H}$, and if the sequence $(X_n)_{n \geq 1}$ is relatively compact, then $X_n \overset{\mathcal{D}}{\longrightarrow} X$.

17.19 Problem Let X and Y be independent $\mathbb{H}$-valued random elements, where $X \sim N(a, \Sigma)$ and $Y \sim N(b, T)$. Show that $X + Y$ has the normal distribution $N(a + b, \Sigma + T)$.

17.20 Problem Let $\mathbb{H}$ and $\mathbb{L}$ be Hilbert spaces, X an $\mathbb{H}$-valued Gaussian random element with $X \sim N(m, \Sigma)$ and $T : \mathbb{H} \to \mathbb{L}$ a linear bounded mapping. Show that $Y := T(X) = T \circ X$ is a Gaussian random element in $\mathbb{L}$. What are the expectation and the covariance operator of Y?

Hint For the adjoint mapping $T^* : \mathbb{L} \to \mathbb{H}$ to T, we have $\langle Tx, y \rangle_{\mathbb{L}} = \langle x, T^*y \rangle_{\mathbb{H}}$ ($x \in \mathbb{H}$, $y \in \mathbb{L}$). The indexing highlights which space the respective inner product refers to.

17.21 Problem Let X be an $\mathbb{H}$-valued Gaussian random element with $X \sim N(m, \Sigma)$. Show:

$$\mathbb{E}\left(\|X - m\|^4\right) = 2 \sum_{j=1}^{\infty} \lambda_j^2 + \left(\sum_{j=1}^{\infty} \lambda_j \right)^2.$$

Here, $\lambda_1, \lambda_2, \ldots$ are the non-zero eigenvalues of the covariance operator Σ appearing according to their geometric multiplicity (the sums can also be finite).

17.22 Problem Let $T_{n,\beta} = T_{n,\beta}(X_1, \ldots, X_n)$ be the test statistic defined in (17.57). Show: If A is a regular $d \times d$ matrix and $b \in \mathbb{R}^d$, then

$$T_{n,\beta}(X_1, \ldots, X_n) = T_{n,\beta}\left(AX_1 + b, \ldots, AX_n + b\right) \quad \mathbb{P}\text{-almost surely.}$$

Hint Use representation (17.58).

Afterword

As emphasized in the preface, this book is designed to give you a quick insight into asymptotic stochastics for self-study. Particular attention is paid to topics at the interface between probability theory and mathematical statistics. Only basic knowledge of stochastics and measure and integration theory is required, as is usually taught in introductory courses. I would be delighted if this introduction to asymptotic stochastics has whetted your appetite for more. There are a plethora of further books on this subject, some of which have already been mentioned in the course of the individual chapters and some of which are added here.

If you are interested in advanced topics from probability theory (including the law of the iterated logarithm, infinitely divisible and especially stable distributions, the theory of large deviations and convergence theorems for martingales), the books [DUR], [GSI], [KAL], and [KLE] should be mentioned in particular. The book [BOG] deals with normal distributions (Gaussian probability measures) in a general framework.

If you want to delve deeper into asymptotic statistics, you will find what you are looking for with [VW]. This book covers many areas of asymptotic statistics that are not addressed in the present work, such as nonparametric density estimation, semiparametric models, Bayes methods, or the asymptotic efficiency of tests. Relevant monographs on nonparametric density estimation include [SIL], [DEV], and [SCO]. The books [EUB], [GRS], and [HAE] provide more or less extensive insights into the field of nonparametric regression estimation.

A standard work on the asymptotic efficiency of nonparametric tests is [NIK]. The book [HOP] particularly addresses the concept of local asymptotic normality with a view towards stochastic processes.

© The Author(s), under exclusive license to Springer-Verlag GmbH, DE, part of Springer Nature 2024
N. Henze, *Asymptotic Stochastics*, Mathematics Study Resources 10,
https://doi.org/10.1007/978-3-662-68923-3

Solutions to the Problems

1.1 Solution W.l.o.g., let $\|\cdot\|$ be the maximum norm $\|x\|_\infty := \max_{k=1,\ldots,d} |x_k|$, where $x = (x_1, \ldots, x_d)$. For each $j \in \{1, \ldots, d\}$ and each $\varepsilon > 0$, we have

$$\left\{ \left|X_n^{(j)} - X^{(j)}\right| > \varepsilon \right\} \subset \left\{ \|X_n - X\|_\infty > \varepsilon \right\} \subset \bigcup_{k=1}^{d} \left\{ \left|X_n^{(k)} - X^{(k)}\right| > \varepsilon \right\}.$$

The first "$\subset$" yields "$\Longrightarrow$". The reverse "$\Longleftarrow$" follows from the second "$\subset$".

1.2 Solution

(a) Let $(\Omega, \mathcal{A}, \mathbb{P})$ be as in the hint. Each integer n can be written uniquely as $n = 2^k + j$ with $k \in \mathbb{N}_0$ and $j \in \{0, \ldots, 2^k - 1\}$. We set $X(\omega) := 0, \omega \in \Omega$, and

$$X_n(\omega) := 1, \quad \text{if } \frac{j}{2^k} \le \omega \le \frac{j+1}{2^k},$$

as well as $X_n(\omega) := 0$, otherwise, so $0 = \liminf_{n \to \infty} X_n(\omega) < \limsup_{n \to \infty} X_n(\omega) = 1$. Thus, the sequence $(X_n(\omega))$ does not converge for any $\omega \in \Omega$. On the other hand, we get $X_n \xrightarrow{\mathbb{P}} X$, because for every $\varepsilon \in (0, 1)$ $\mathbb{P}(|X_n - X| > \varepsilon) = \mathbb{P}(X_n = 1) = 2^{-k}$, if $2^k \le n < 2^{k+1}$. Below are the graphs of $X_2, \ldots, X_6$.

(b) For any $p \in (0, 1)$, with X_n and X as in part (a), $\mathbb{E}|X_n - X|^p = \mathbb{P}(X_n = 1)$. We thus get $X_n \xrightarrow{\mathcal{L}^p} X$. According to (a), however, $(X_n(\omega))$ does not converge for any ω.

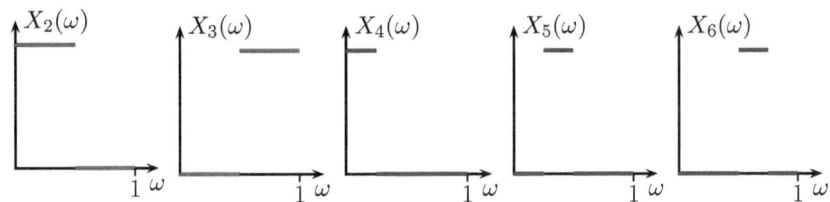

N. Henze, *Asymptotic Stochastics*, Mathematics Study Resources 10, https://doi.org/10.1007/978-3-662-68923-3

1.3 Solution One possible wording is: Suppose $(X_n)_{n \geq 1}$ is a sequence of i.i.d. d-dimensional random vectors on a probability space $(\Omega, \mathcal{A}, \mathbb{P})$. Let $S_n := \sum_{j=1}^{n} X_j, n \geq 1$, and let $\| \cdot \|$ be any norm on $\mathbb{R}^d$. Then the following statements are equivalent:

(a) There exists a d-dimensional random vector X with $\frac{1}{n} S_n \xrightarrow{\text{a.s.}} X$.
(b) $\mathbb{E}\|X_1\| < \infty$.

If (a) or (b) holds, then $X = \mathbb{E}(X_1)$ $\mathbb{P}$-a.s.

Proof Let $S_n^{(k)}$, $X_n^{(k)}$, and $X^{(k)}$ be the k-th component of S_n, X_n, and X, respectively, where $k \in \{1, \dots, d\}$. If (a) holds, then $\frac{1}{n} S_n^{(k)} \xrightarrow{\text{a.s.}} X^{(k)}$. With Theorem 1.2, it follows that $\mathbb{E}|X_1^{(k)}| < \infty$ for each k and thus $\mathbb{E}\|X_1\| < \infty$, since all norms on $\mathbb{R}^d$ are equivalent. From (b) it follows that $\mathbb{E}|X_1^{(k)}| < \infty$ for each $k \in \{1, \dots, d\}$. By Theorem 1.2, there is a random variable $X^{(k)}$ with $\frac{1}{n} S_n^{(k)} \xrightarrow{\text{a.s.}} X^{(k)}, k \in \{1, \dots, d\}$. Then $X := (X^{(1)}, \dots, X^{(k)})$ is a random vector with $\frac{1}{n} S_n \xrightarrow{\text{a.s.}} X$. If (a) or (b) is fulfilled, then according to Theorem 1.2, $X^{(k)} = \mathbb{E}(X_1^{(k)})$ $\mathbb{P}$-a.s. for each $k \in \{1, \dots, d\}$ and thus $X = \mathbb{E}(X_1)$ $\mathbb{P}$-a.s. Note in these considerations that the intersections of finitely many sets of probability one also has probability one. $\qquad \square$

1.4 Solution Let $\varepsilon > 0$ be such that $t + \varepsilon \in \mathcal{C}(F)$ and $t - \varepsilon \in \mathcal{C}(F)$. For sufficiently large n we have $t - \varepsilon \leq t_n \leq t + \varepsilon$. For such n it follows that $\mathbb{P}(X_n \leq t - \varepsilon) \leq \mathbb{P}(X_n \leq t_n) \leq \mathbb{P}(X_n \leq t + \varepsilon)$. Now $X_n \xrightarrow{D} X$ implies $F(t - \varepsilon) \leq \liminf_{n \to \infty} \mathbb{P}(X_n \leq t_n) \leq \limsup_{n \to \infty} \mathbb{P}(X_n \leq t_n) \leq F(t + \varepsilon)$. Letting ε tend to zero under the condition $t \pm \varepsilon \in \mathcal{C}(F)$, the assertion follows.

1.5 Solution In view of (1.3), and since X has the same distribution as $\sigma Y + \mu$ with $Y \sim N(0, 1)$, we can assume $\mu = 0$ and $\sigma^2 = 1$ without loss of generality. Due to the symmetry of the density f of the standard normal distribution, we get

$$\varphi_X(t) = \mathbb{E}[\cos(tX)] = \int_{-\infty}^{\infty} \cos(tx) f(x) \, dx.$$

Differentiation and the equation $f'(x) = -xf(x)$ as well as integration by parts yield

$$\varphi_X'(t) = \int_{-\infty}^{\infty} \sin(tx) f'(x) \, dx = -t \int_{-\infty}^{\infty} \cos(tx) f(x) \, dx = -t\varphi_X(t).$$

Here, differentiation under the integral sign is allowed because of the dominated convergence theorem. The above differential equation together with the constraint $\varphi_X(0) = 1$ has the unique solution $\varphi_X(t) = \exp(-t^2/2)$, $t \in \mathbb{R}$.

1.6 Solution We have $Z_n \sim \mathbf{1}\{A_{n,1}\} + \ldots + \mathbf{1}\{A_{n,n}\}$, where $A_{n,1}, \ldots, A_{n,n}$ are independent events with the same probability p_n/n. With $X_{n,k} := \mathbf{1}\{A_{n,k}\}$ for $k = 1, \ldots, n$, the Lyapunov condition (1.11) with $\delta = 2$ and $r_n = n$ holds, since $\sigma_n^4 = n^2 p_n^2 (1 - p_n)^2$ as well as $\mathbb{E}(X_{n,k}) = p_n/n$ and

$$\sum_{j=1}^{n} \mathbb{E}\left(\mathbf{1}\{A_{n,k}\} - \frac{p_n}{n}\right)^4 = n\mathbb{E}\left(\mathbf{1}\{A_{n,1}\} - \frac{p_n}{n}\right)^4 \leq n.$$

1.7 Solution

(a) Because $X_n \xrightarrow{\mathcal{D}} X$ the sequence (X_n) is tight. If the sequence (μ_n) were unbounded, there would be a subsequence (μ_{n_k}) with $\mu_{n_k} \to \infty$ or $\mu_{n_k} \to -\infty$ as $k \to \infty$. Since $\mathbb{P}(X_{n_k} \geq \mu_{n_k}) = \frac{1}{2}$ and $\mathbb{P}(X_{n_k} \leq \mu_{n_k}) = \frac{1}{2}$, this would contradict the assumption of tightness. Therefore, there is an $M > 0$ with $|\mu_n| \leq M$ for every n. If there were a subsequence $(\sigma_{n_k}^2)$ with $\sigma_{n_k}^2 \to \infty$ as $k \to \infty$, then the triangle inequality would yield

$$\mathbb{P}\left(\left|\frac{X_{n_k} - \mu_{n_k}}{\sigma_{n_k}}\right| \geq M\right) = \mathbb{P}\left(\left|X_{n_k} - \mu_{n_k}\right| \geq \sigma_{n_k} M\right) \leq \mathbb{P}\left(\left|X_{n_k} - M\right| \geq (\sigma_{n_k} - 1)M\right)$$

for each k with $\sigma_{n_k} > 1$. Due to $(X_{n_k} - \mu_{n_k})/\sigma_{n_k} \sim N(0, 1)$, the probability on the left-hand side does not depend on k. We thus would also have a contradiction to the tightness of the sequence (X_n). Hence, the sequence (σ_n^2) is also bounded.

(b) According to (a) and the Bolzano–Weierstrass theorem, there is a subsequence (n_k) of integers such that the limits $\mu := \lim_{k \to \infty} \mu_{n_k}$ and $\sigma^2 := \lim_{k \to \infty} \sigma_{n_k}^2$ exist. Thus, there is a random variable Y with $X_{n_k} \xrightarrow{\mathcal{D}} Y$ as $k \to \infty$, and $Y \sim N(\mu, \sigma^2)$. Since $X_n \xrightarrow{\mathcal{D}} X$, we have $X \sim Y$. Because the distribution of X is not degenerate, we also obtain $\sigma^2 > 0$.

1.8 Solution Let $h_n := g - f_n$, $n \geq 1$. Since $h_n \geq 0$, Fatou's lemma gives

$$\int \liminf_{n \to \infty} h_n \, d\mu \leq \liminf_{n \to \infty} \int h_n \, d\mu.$$

Since g is μ-integrable, we obtain

$$\int \liminf_{n\to\infty} h_n \, d\mu = \int g \, d\mu - \int \limsup_{n\to\infty} f_n \, d\mu,$$

$$\liminf_{n\to\infty} \int h_n \, d\mu = \int g \, d\mu - \limsup_{n\to\infty} \int f_n \, d\mu,$$

and the assertion follows.

2.1 Solution Due to $\mathbb{P}(|X_{n,k_n}| > \varepsilon) \leq \max_{1\leq j\leq n} \mathbb{P}(|X_{n,j}| > \varepsilon)$, (2.4) follows from (2.3). For a proof (by contraposition) of the reverse direction, we assume (2.3) does not hold. Then there is an $\varepsilon > 0$ and a $\delta > 0$ as well as a subsequence (ℓ_n) with $\max_{1\leq j\leq \ell_n} \mathbb{P}(|X_{\ell_n,j}| > \varepsilon) \geq \delta$ for each $n \geq 1$. Thus, there exists a subsequence (m_n) with $m_n \in \{1, \ldots, \ell_n\}$ for each $n \geq 1$ and $\mathbb{P}(|X_{\ell_n,m_n}| > \varepsilon) \geq \delta$ for each $n \geq 1$. This fact contradicts (2.4).

2.2 Solution Let $X_{n,j} := 1\{A_{n,j}\}$ and $p_{n,j} := \mathbb{P}(A_{n,j}) = \mathbb{E}(X_{n,j})$, $1 \leq j \leq n$.

(a) By assumption, $\lim_{n\to\infty} \sum_{j=1}^{n} p_{n,j} = \lim_{n\to\infty} \sum_{j=1}^{n} p_{n,j}(1 - p_{n,j}) = \lambda$. It follows that

$$0 \leq \left(\max_{1\leq j\leq n} p_{n,j} \right)^2 \leq \sum_{j=1}^{n} p_{n,j}^2 = \sum_{j=1}^{n} p_{n,j} - \sum_{j=1}^{n} p_{n,j}(1 - p_{n,j}) \to 0 \text{ as } n \to \infty$$

and thus $\max_{1\leq j\leq n} p_{n,j} = \max_{1\leq j\leq n} \mathbb{E}[|X_{n,j}| \wedge 1] \to 0$ as $n \to \infty$. According to Proposition 2.2, we have a null array, and the assertion follows from Theorem 2.3, since $\mathbb{P}(X_{n,j} > 1) = 0$ and $\sum_{j=1}^{n} \mathbb{P}(X_{n,j} = 1) = \mathbb{E}(S_n)$.

(b) By assumption, $\sigma_n^2 := \mathbb{V}(S_n) = \sum_{j=1}^{n} p_{n,j}(1 - p_{n,j}) \to \infty$. Since $(X_{n,j} - p_{n,j})^2 \leq 1$, the Lyapunov condition (1.11) holds with $\delta = 2$, because

$$\frac{1}{\sigma_n^4} \sum_{j=1}^{n} \mathbb{E}(X_{n,j} - p_{n,j})^4 \leq \frac{1}{\sigma_n^4} \sum_{j=1}^{n} \mathbb{V}(X_{n,j}) = \frac{1}{\sigma_n^2} \to 0 \text{ as } n \to \infty.$$

2.3 Solution The first condition means that the random variables defined by $X_{n,j} := 1\{A_{n,j}\}$ are a null array, see Solution 2.2 (a). The assertion then follows from Theorem 2.3.

2.4 Solution

"(a) $\Longrightarrow$ (b)": Let F_j be the distribution function of X_j, $j \geq 0$. For each $k \geq 0$, $k \pm \frac{1}{2}$ are continuity points of F_0, and we have $\mathbb{P}(X_n = k) = F_n(k + \frac{1}{2}) - F_n(k - \frac{1}{2})$, q.e.d.

"(b) $\Longrightarrow$ (a) ": For each continuity point t of F_0, i.e., for each $t \in \mathbb{R} \setminus N_0$, we have $F_n(t) = \sum_{k \in N_0, k < t} \mathbb{P}(X_n = k)$ for each $n \geq 0$, which implies the assertion.

"(b) $\Longrightarrow$ (c)": W.l.o.g. let $s < 1$. Fix any $\varepsilon > 0$, and put $\Delta_{n,k} := |\mathbb{P}(X_n = k) - \mathbb{P}(X_0 = k)|$. By assumption, $\lim_{n \to \infty} \max_{0 \leq k \leq m} \Delta_{n,k} = 0$ for each $m \geq 0$. Due to

$$|g_n(s) - g_0(s)| \leq \sum_{k=0}^{\infty} \Delta_{n,k} s^k \leq \max_{0 \leq k \leq m} \Delta_{n,k} \sum_{k=0}^{m} s^k + \sum_{k=m+1}^{\infty} s^k$$

$$= \max_{0 \leq k \leq m} \Delta_{n,k} \frac{1 - s^{m+1}}{1 - s} + \frac{s^{m+1}}{1 - s}$$

and the fact that $s^{m+1}/(1 - s) < \varepsilon$ for sufficiently large m, we obtain $\limsup_{n \to \infty} |g_n(s) - g_0(s)| \leq \varepsilon$.

"(c) $\Longrightarrow$ (b)": For each $k \geq 0$, $(\mathbb{P}(X_n = k))_{n \in \mathbb{N}}$ is a bounded sequence. By the Bolzano-Weierstrass theorem and Cantor's diagonal argument, there is a subsequence $(X_{n'})_{n' \in \mathbb{N}}$ such that the limit $q_k := \lim_{n' \to \infty} \mathbb{P}(X_{n'} = k)$ exists for each $k \geq 0$. If we set $f(s) := \sum_{k=0}^{\infty} q_k s^k$, $0 \leq s \leq 1$, part "(b) $\Rightarrow$ (c)" of the proof yields $\lim_{n' \to \infty} g_{n'}(s) = f(s)$, $0 \leq s \leq 1$, and thus, according to condition (c), $f(s) = g_0(s)$, $0 \leq s \leq 1$. This implies $q_k = \mathbb{P}(X_0 = k)$, $k \in N_0$, as well as (b), because there can be no $k \geq 0$ and no subsequence (X_{n*}) with $\lim_{n* \to \infty} \mathbb{P}(X_{n*} = k) \neq \mathbb{P}(X_0 = k)$.

2.5 Solution

"$\Longrightarrow$": Let $X \sim \mathrm{Po}(\lambda)$ and $f : N_0 \to [0, \infty)$. Then

$$\mathbb{E}[Xf(X)] = \sum_{k=0}^{\infty} k f(k) e^{-\lambda} \frac{\lambda^k}{k!} = e^{-\lambda} \sum_{k=1}^{\infty} f(k) \frac{\lambda^k}{(k-1)!}$$

$$= e^{-\lambda} \lambda \sum_{k=0}^{\infty} f(k+1) \frac{\lambda^k}{k!} = \lambda \mathbb{E}[f(X+1)].$$

"$\Longleftarrow$": For the specific choice $f(x) := \mathbf{1}\{x = k\}$, $k \in \mathbb{N}$, $\mathbb{E}[Xf(X)] = \lambda \mathbb{E}[f(X+1)]$ is equivalent to $k\mathbb{P}(X = k) = \lambda \mathbb{P}(X = k - 1)$. This implies recursively $\mathbb{P}(X = k) = \frac{\lambda^k}{k!} \mathbb{P}(X = 0)$, $k \in N_0$, and thus the assertion, since $1 = \sum_{k=0}^{\infty} \mathbb{P}(X = k)$.

2.6 Solution Let $x > 0$ be arbitrary. Using the binomial series, for each integer n with $x/n \leq 1$, we have

$$\mathbb{E}(U_n(x)) = n\left(1 - F\left(1 - \frac{x}{n}\right)\right) = n\left(1 - \left(1 - \frac{x}{n}\right)^{\vartheta}\right) \to \vartheta x \quad \text{as } n \to \infty.$$

Theorem 2.3 yields $U_n(x) \xrightarrow{D} \mathrm{Po}(\vartheta x)$. For $M_n := \max(X_1, \ldots, X_n)$, we thus obtain $\mathbb{P}(n(1 - M_n) \leq x) = 1 - \mathbb{P}(M_n < 1 - \frac{x}{n}) = 1 - \mathbb{P}(U_n(x) = 0) \to 1 - e^{-\vartheta x}$, $x > 0$. The limit distribution is thus an exponential distribution.

2.7 Solution Due to the assumption about F, we have $\mathbb{E}(V_n(x)) = nF(x/n^{1/\alpha}) \to \lambda x^\alpha$ as $n \to \infty$. In view of Theorem 2.3, we get $V_n(x) \xrightarrow{D} \mathrm{Po}(\lambda x^\alpha)$, from which (2.15) follows.

3.1 Solution The assertion follows from the inequality (valid for any $a > 0$)

$$|X_n + Y_n| \mathbf{1}\{|X_n + Y_n| \geq a\} \leq 2|X_n| \mathbf{1}\left\{|X_n| \geq \frac{a}{2}\right\} + 2|Y_n| \mathbf{1}\left\{|Y_n| \geq \frac{a}{2}\right\}.$$

3.2 Solution

"(a) $\Longrightarrow$ (b)": Jensen's inequality (Theorem 1.7), applied to the absolute value function, yields $|\mathbb{E}|X_n| - \mathbb{E}|X|| \leq \mathbb{E}||X_n| - |X|| \leq \mathbb{E}|X_n - X|$, and thus $\mathbb{E}|X_n| \to \mathbb{E}|X|$. Fix any $\varepsilon > 0$. With the function Ψ_C given in the hint, according to the dominated convergence theorem, there is a $C_0(\varepsilon)$ with $\mathbb{E}|X| - \mathbb{E}\Psi_C(|X|) \leq \varepsilon/2$, if $C \geq C_0(\varepsilon)$. Since $|X_n| \xrightarrow{P} |X|$ and the function Ψ_C is continuous and bounded, we also have $\mathbb{E}\Psi_C(|X_n|) \to \mathbb{E}\Psi_C(|X|)$ and hence in particular $\mathbb{E}\Psi_C(|X|) - \mathbb{E}\Psi_C(|X_n|) \leq \varepsilon/4$, if $n \geq n_0(\varepsilon)$. After possibly increasing n_0, we also get $\mathbb{E}|X_n| \leq \mathbb{E}|X| + \varepsilon/4$ for each $n \geq n_0(\varepsilon)$. We therefore obtain

$$\mathbb{E}\left[|X_n| \mathbf{1}\{|X_n| > C\}\right] \leq \mathbb{E}|X_n| - \mathbb{E}\Psi_C(|X_n|) \leq \mathbb{E}|X| - \mathbb{E}\Psi_C(|X|) + \frac{\varepsilon}{2} \leq \varepsilon$$

for each $n \geq n_0(\varepsilon)$ and each $C \geq C_0(\varepsilon)$. By possibly increasing $C_0(\varepsilon)$, this inequality also holds for each $n \in \{1, \ldots, n_0(\varepsilon) - 1\}$, and thus (X_n) is uniformly integrable.

"(b) $\Longrightarrow$ (a) ": $X_n \xrightarrow{P} X$ implies $X_n \xrightarrow{D} X$ which, due to the uniform integrability of (X_n), implies $\mathbb{E}|X| < \infty$ in view of Theorem 3.1. The $\mathcal{L}^1$-convergence was shown in the proof of Theorem 3.4 (see inequality (3.3)).

3.3 Solution Let t with $|t| < \delta$ be arbitrary. With the hint, $e^{|tX|} \leq e^{tX} + e^{-tX}$, and thus $\mathbb{E}[e^{|tX|}] \leq \mathbb{E}[e^{tX}] + \mathbb{E}[e^{-tX}] < \infty$. Setting

$$Y_n := \sum_{k=0}^{n} \frac{|X|^k}{k!} |t|^k, \quad n \geq 0,$$

the sequence (Y_n) converges element-wise on the underlying probability space $(\Omega, \mathcal{A}, \mathbb{P})$ to $e^{|tX|}$. Since $|Y_n| \leq e^{|tX|}$ for each $n \geq 0$ and $e^{|tX|}$ is integrable, the dominated convergence theorem yields

$$\lim_{n \to \infty} \mathbb{E}(Y_n) = \lim_{n \to \infty} \sum_{k=0}^{n} \frac{\mathbb{E}[|X|^k]}{k!} |t|^k = \sum_{k=0}^{\infty} \frac{\mathbb{E}[|X|^k]}{k!} |t|^k = \mathbb{E}[e^{|tX|}] < \infty.$$

Because $|\mathbb{E}(X^k)| \leq \mathbb{E}[|X|^k]$, the assertion follows.

3.4 Solution Note that $\mathbb{E}(S_n/\sqrt{n}) = 0$ and $\mathbb{V}(S_n/\sqrt{n}) = 1$. According to Example 3.8 and Theorem 3.6, we thus have to prove

$$\lim_{n \to \infty} \mathbb{E}\left[\left(\frac{S_n}{\sqrt{n}}\right)^{2k+1}\right] = 0, \qquad \lim_{n \to \infty} \mathbb{E}\left[\left(\frac{S_n}{\sqrt{n}}\right)^{2k}\right] = \prod_{j=1}^{k}(2j-1)$$

for each $k \geq 1$. Since taking expections is a linear operation, we have

$$\mathbb{E}\left[\left(\frac{S_n}{\sqrt{n}}\right)^r\right] = \frac{1}{n^{r/2}} \sum_{j_1=1}^{n} \cdots \sum_{j_r=1}^{n} \mathbb{E}(X_{j_1} X_{j_2} \ldots X_{j_r}) \qquad (*)$$

for each $r \geq 3$. We split the r-fold sum over all r-tuples $(j_1, \ldots, j_r)$ according to the number $\ell := |\{j_1, \ldots, j_r\}|$ of *different occurring indices* and consider the multiplication rule $\mathbb{E}(UV) = \mathbb{E}(U)\mathbb{E}(V)$ for the expectation of the product of independent random variables as well as $\mathbb{E}(X_j) = 0$ and $\mathbb{P}(|X_j| \leq M) = 1$ $(j = 1, \ldots, n)$. Thus, for example, each term in $(*)$ vanishes in which an index occurs only once. If $\ell < \frac{r}{2}$, the contribution to the sum in $(*)$ is at most equal to $M^r n^\ell$ and thus asymptotically negligible due to the factor $n^{-r/2}$. If $\ell > \frac{r}{2}$, then at least one isolated index occurs. Due to the multiplication rule for expectations and $\mathbb{E}(X_j) = 0$ these contributions to the sum are thus all equal to zero. Since the case $\ell = \frac{r}{2}$ cannot occur for odd r, we have obtained the first limit relation to be shown. It thus only remains to tackle the case that $r = 2k$ is an even number and $\ell = k$, which means that each of the $2k$ indices appears exactly twice. Without considering the factor $n^{-r/2}$, under the above constraint on the indices each summand in $(*)$ is equal to one. According to the multiplication rule of combinatorics, the number of these summand is equal to

$$\binom{n}{k}\binom{2k}{2}\binom{2k-2}{2}\cdot \ldots \cdot \binom{2}{2} = \frac{n!}{(n-k)!} \cdot \frac{(2k)!}{2^k \cdot k!}.$$

Summarizing, we thus obtain

$$\lim_{n\to\infty} \mathbb{E}\left[\left(\frac{S_n}{\sqrt{n}}\right)^{2k}\right] = \lim_{n\to\infty} \frac{1}{n^k} \cdot \frac{n!}{(n-k)!} \cdot \frac{(2k)!}{2^k \cdot k!} = \frac{(2k)!}{2^k \cdot k!} = \prod_{j=1}^{k}(2j-1),$$

as desired.

3.5 Solution

(a) Due to $|\mathbb{E}(X^k)| \leq M^k$, the power series in (3.6) has a positive radius of convergence.

(b) Equation (3.14) follows using the substitution $\log t = s + k$, $t = e^{s+k}$, $dt = e^{s+k}\,ds$. If we set $g(t) := f(t)(1 + \sin(2\pi \log t)$ for $t > 0$, and $g(t) := 0$, otherwise, then g is a non-negative measurable function with $\int_{-\infty}^{\infty} g(t)\,dt = 1$ and thus the density of a random variable, denoted Y. We have $\mathbb{P}^X \neq \mathbb{P}^Y$. On the other hand, (3.14) yields the equality $\mathbb{E}(X^k) = \mathbb{E}(Y^k)$, $k \in \mathbb{N}$, of all moments.

3.6 Solution

(a) The proof is done by induction over k. The base case $k = 1$ holds due to $\mathbb{E}(X^2) = \lambda(1 + \lambda)$. The induction step $k \mapsto k + 1$ is obtained using the binomial theorem, because

$$\mathbb{E}(X^{k+1}) = \sum_{j=0}^{\infty} j^{k+1}\frac{\lambda^j}{j!}e^{-\lambda} = \lambda\sum_{j=1}^{\infty} j^k\frac{\lambda^{j-1}}{(j-1)!}e^{-\lambda} = \lambda\sum_{j=0}^{\infty}(j+1)^k\frac{\lambda^j}{j!}e^{-\lambda}$$

$$= \lambda\sum_{j=0}^{\infty}\sum_{\ell=0}^{k}\binom{k}{\ell}j^\ell\frac{\lambda^j}{j!}e^{-\lambda} = \lambda\sum_{\ell=0}^{k}\binom{k}{\ell}\sum_{j=0}^{\infty}j^\ell\frac{\lambda^j}{j!}e^{-\lambda} = \lambda\sum_{\ell=0}^{k}\binom{k}{\ell}\mathbb{E}(X^\ell).$$

(b) In view of part (a), we have

$$\mathbb{E}(X^k) = \lambda\sum_{\ell=0}^{k-1}\frac{k-\ell}{k}\binom{k}{\ell}\mathbb{E}(X^\ell) \geq \frac{\lambda}{k}\left(\sum_{\ell=0}^{k}\binom{k}{\ell}\mathbb{E}(X^\ell) - \mathbb{E}(X^k)\right)$$

$$= \frac{1}{k}\mathbb{E}(X^{k+1}) - \frac{\lambda}{k}\mathbb{E}(X^k),$$

and the assertion follows.

3.7 Solution For each $\omega \in \Omega$, we have

$$\binom{S_n(\omega)}{k} = \sum_{1\leq i_1 < \ldots < i_k \leq n} \mathbf{1}\{A_{i_1} \cap \ldots \cap A_{i_k}\}(\omega),$$

since in the case $S_n(\omega) < k$ both sides of this equation are equal to zero, because ω is an element of less than k of the $A_1, \ldots, A_n$. If $S_n(\omega) \geq k$ then the left-hand side is the number of ways to choose k from among $S_n(\omega)$ objects. However, the same number of summands are equal to one on the right-hand side. The number of these summands is equal to the number of ways to choose k from the $S_n(\omega)$ occurring events. If we take expectations on both sides, the assertion follows.

3.8 Solution Note that $|M| = k^\ell$ and $|A_i| = (k-1)^\ell$ for $i = 1, \ldots, k$. If $j \in \{2, \ldots, k-1\}$ and $1 \leq i_1 < \ldots < i_j \leq k$, then $|A_{i_1} \cap \ldots \cap A_{i_j}| = (k-j)^\ell$. The claim now follows with the principle of inclusion and exclusion.

3.9 Solution Let $F_n := \sum_{j=1}^{n} 1\{A_{n,j}\}$, where $A_{n,j}$ denotes the event that j is a fixed point in a purely random permutation of $1, 2, \ldots, n$. Due to $\mathbb{P}(A_{i_1} \cap \ldots \cap A_{i_k}) = (n-k)!/n!$ for $1 \leq i_1 < \ldots < i_k \leq n$, it follows that

$$\lim_{n \to \infty} \mathbb{E}\binom{F_n}{k} = \frac{1}{k!} \lim_{n \to \infty} \mathbb{E}[F_n(F_n - 1) \ldots (F_n - k + 1)] = \lim_{n \to \infty} \binom{n}{k} \cdot \frac{(n-k)!}{n!} = \frac{1}{k!}$$

for every $k \in \mathbb{N}$. According to (3.13), this was to be shown.

3.10 Solution Let $S_n := \sum_{i=1}^{n} 1\{A_{n,i}\}$. By Problem 3.7, and due to the exchangeability of the events $A_{n,1}, \ldots, A_{n,n}$, we have for each n with $n \geq k$

$$\mathbb{E}\binom{S_n}{k} = \frac{1}{k!} \mathbb{E}[S_n(S_n - 1) \cdot \ldots \cdot (S_n - k + 1)] = \binom{n}{k} \mathbb{P}(A_{n,1} \cap \ldots \cap A_{n,k}).$$

The convergence $n^k \mathbb{P}(A_{n,1} \cap \ldots \cap A_{n,k}) \to \lambda^k$ yields $\mathbb{E}[S_n(S_n - 1) \cdot \ldots \cdot (S_n - k + 1)] \to \lambda^k$, $k \geq 1$, and thus the sufficient condition (3.13) for convergence in distribution to the Poisson distribution Po(λ) holds.

4.1 Solution By definition, $Y_1, Y_2, \ldots$ are independent if the random variables $Y_1, \ldots, Y_k$ are independent for each $k \geq 2$. The independence of $Y_1, \ldots, Y_k$ is equivalent to the fact that the joint distribution of $Y_1, \ldots, Y_k$ equals the product of the individual marginal distributions, i.e., to the equation

$$\mathbb{P}^{(Y_1, \ldots, Y_k)} = \mathbb{P}^{Y_1} \otimes \ldots \otimes \mathbb{P}^{Y_k}. \tag{*}$$

By definition, 0-dependence holds if the sigma-fields $\sigma(Y_1, \ldots, Y_s)$ and $\sigma(Y_{s+1}, Y_{s+2}, \ldots)$ are independent for each $s \geq 1$. Since independence is preserved when reducing independent classes of sets, $\sigma(Y_1)$ and $\sigma(Y_2)$ are independent (set $s = 1$), and thus (*) holds for $k = 2$ (base case). For the induction step $k \mapsto k + 1$, we use that

$\sigma(Y_1, \ldots, Y_k)$ and $\sigma(Y_{k+1}, Y_{k+2}, \ldots)$ and consequently also $\sigma(Y_1, \ldots, Y_k)$ and $\sigma(Y_{k+1})$ are independent, which is equivalent to $\mathbb{P}^{(Y_1, \ldots, Y_k, Y_{k+1})} = \mathbb{P}^{(Y_1, \ldots, Y_k)} \otimes \mathbb{P}^{Y_{k+1}}$. Together with the induction hypothesis $(*)$, the claim follows.

4.2 Solution With the hint, let $Y_j := X_j - X_{j+1}$, $j \geq 1$. Then $\sigma_{0,0} = \mathbb{V}(Y_1) = 2\mathbb{V}(X_1) = \frac{1}{2}$ and $\sigma_{0,1} = \mathrm{Cov}(Y_1, Y_2) = \mathrm{Cov}(X_1 - X_2, X_2 - X_3) = -\mathbb{V}(X_2) = -\frac{1}{4}$ and thus $\sigma^2 = \sigma_{0,0} + 2\sigma_{0,1} = 0$. And indeed, $\sum_{j=1}^{n} Y_j = X_1 - X_{n+1}$ is not asymptotically normally distributed.

4.3 Solution The sequence (Y_j) is 2-dependent and stationary. Since F is continuous, we have $\mathbb{P}\left(\bigcap_{i \neq j}\{X_i \neq X_j\}\right) = 1$. From reasons of symmetry, $\mathbb{E}(Y_1) = \mathbb{P}(X_0 > X_1 < X_2) = \mathbb{P}(X_1 = \min(X_0, X_1, X_2)) = \frac{1}{3}$. Since $Y_1 \sim \mathrm{Bin}(1, \frac{1}{3})$, $\sigma_{0,0} = \mathbb{V}(Y_1) = \frac{1}{3} \cdot \frac{2}{3} = \frac{2}{9}$. Further, $\mathbb{E}(Y_1 Y_2) = \mathbb{P}(X_0 > X_1 < X_2, X_1 > X_2 < X_3) = 0$ and thus $\mathrm{Cov}(Y_1, Y_2) = \mathbb{E}(Y_1 Y_2) - \mathbb{E}(Y_1)\mathbb{E}(Y_2) = -\frac{1}{9}$. Finally, $\mathbb{E}(Y_1 Y_3) = \mathbb{P}(X_0 > X_1 < X_2, X_2 > X_3 < X_4) = \frac{16}{120} = \frac{2}{15}$, because in exactly 16 out of 120 $(= 5!)$ equally probable permutations of five different numbers, the second number is smaller than the first and the third and the fourth number is smaller than the third and the fifth. It follows that $\sigma_{0,2} = \mathrm{Cov}(Y_1, Y_3) = \mathbb{E}[Y_1 Y_3] - \mathbb{E}[Y_1]\mathbb{E}[Y_3] = \frac{2}{15} - \frac{1}{9} = \frac{1}{45}$, and thus $\sigma^2 = \sigma_{0,0} + 2\sum_{j=1}^{2} \sigma_{0,j} = \mathbb{V}(Y_1) + 2\mathrm{Cov}(Y_1, Y_2) + 2\mathrm{Cov}(Y_1, Y_3) = \frac{2}{9} - \frac{2}{9} + \frac{2}{45} = \frac{2}{45}$. Theorem 4.6 now yields the assertion.

4.4 Solution The sequence $(Y_j)_{j\geq 1}$ defined by $Y_j := X_{j-1} X_j$ is stationary and 1-dependent. Since $Y_1 \sim \mathrm{Bin}(1, p^2)$, we have $\mathbb{E}(Y_1) = p^2$ and $\mathbb{V}(Y_1) = p^2(1 - p^2)$. Furthermore, $\mathbb{E}(Y_1 Y_2) = p^3$ and thus $\mathrm{Cov}(Y_1, Y_2) = p^3 - p^4$. For σ^2 in (4.2), we have $\sigma^2 = \mathbb{V}(Y_1) + 2\mathrm{Cov}(Y_1, Y_2) = p^2(1 + 2p - 3p^2)$. The assertion now follows from Theorem 4.6.

4.5 Solution We have $\mathbb{E}Y_1 = q^2 p^r = \mathbb{E}Y_1^2$ and thus $\mathbb{V}(Y_1) = q^2 p^r - q^4 p^{2r} = \sigma_{00}$. Furthermore, $\mathbb{E}(Y_1 Y_{1+r+1}) = q^3 p^{2r}$ and (considering $X_j(1 - X_j) = 0$) $\mathbb{E}(Y_j Y_{j+k}) = 0$, if $k \in \{1, \ldots, r\}$. Thus, $\sigma_{0k} = \mathrm{Cov}(Y_j, Y_{j+k}) = -q^4 p^{2r}$, if $k \in \{1, \ldots, r\}$. Finally, we get $\sigma_{0,r+1} = \mathrm{Cov}(Y_1, Y_{1+r+1}) = q^3 p^{2r} - q^4 p^{2r}$. Since all other covariances vanish, direct calculations entail $\sigma^2 = \sigma_{00} + 2\sum_{j=1}^{r+1} \sigma_{0j} = q^2 p^r + 2q^3 p^{2r} - (2r+3)q^4 p^{2r}$. The assertion now follows from Theorem 4.6.

4.6 Solution This is a special case of 4.3 with a stationary and q-dependent sequence. We have $\mathbb{E}(X_1) = 0$ and $\mathbb{V}(X_1) = \tau^2 \sum_{\ell=0}^{q} \vartheta_\ell^2$. Using the bilinearity of taking covariances and $\mathrm{Cov}(\varepsilon_i, \varepsilon_j) = 0$ for $i \neq j$, it follows that $\mathrm{Cov}(X_1, X_{1+k}) = \tau^2 \sum_{\ell=0}^{q-k} \vartheta_\ell \vartheta_{\ell+k}$. Putting $\sigma^2 := \mathbb{V}(X_1) + 2\sum_{k=1}^{q} \mathrm{Cov}(X_1, X_{1+k})$, we thus get

$$\sigma^2 = \tau^2 \left(\sum_{\ell=0}^{q} \vartheta_\ell^2 + 2\sum_{k=1}^{q} \sum_{\ell=0}^{q-k} \vartheta_\ell \vartheta_{\ell+k} \right) = \tau^2 \left(\sum_{\ell=0}^{q} \vartheta_\ell \right)^2 .$$

The assertion then follows with Theorem 4.6.

5.1 Solution Suppose $(X_1, Y_1)^\top$ has density f_ϱ, $(X_2, Y_2)^\top$ has density $f_{-\varrho}$, and Z has the binomial distribution $\mathrm{Bin}(1, \alpha)$. Moreover, assume $(X_1, Y_1)^\top$, $(X_2, Y_2)^\top$ and Z are independent. Then

$$\begin{pmatrix} U \\ V \end{pmatrix} \overset{\mathcal{D}}{=} Z \begin{pmatrix} X_1 \\ X_2 \end{pmatrix} + (1 - Z) \begin{pmatrix} X_2 \\ Y_2 \end{pmatrix}.$$

Since the conditional distributions of X_i, Y_j are standard normal both under the condition $Z = 1$ and under the condition $Z = 0$, it follows that $U \sim ZX_1 + (1 - Z)X_2 \sim N(0, 1)$ and $V \sim ZY_1 + (1 - Z)Y_2 \sim N(0, 1)$. Because of the assumption of independence, we further obtain $\mathbb{E}(UV) = \mathbb{E}\big[(ZX_1 + (1 - Z)X_2)(ZY_1 + (1 - Z)Y_2)\big] = 0$. Thus, U and V are uncorrelated random variables that both have a standard normal distribution. However, the vector $(U, V)^\top$ does not follow a bivariate normal distribution.

5.2 Solution For any $c \in \mathbb{R}^d$ we have $c^\top(X + Y) = c^\top X + c^\top Y$, where the summands on the right-hand side are independent and have the normal distributions $N(c^\top\mu, c^\top\Sigma c)$ and $N(c^\top\nu, c^\top Tc)$, respectively. Due to the addition theorem for the univariate normal distribution, the sum on the right-hand side has the normal distribution $N(c^\top(\mu + \nu), c^\top(\Sigma + T)c)$, The assertion now follows from the definition of the normal distribution $N_d(\mu + \nu, \Sigma + T)$. An alternative proof uses characteristic functions (see Problem 5.6).

5.3 Solution

(a) If Z is a real-valued random variable that takes on the value z_j with probability p_j ($j \in \{1, \ldots, k\}$), then $\mathbb{E}(Z) = \sum_{j=1}^k p_j z_j$. According to the definition of the expectation of a random vector, we have $\mathbb{E}(Y) = \frac{1}{n}\sum_{j=1}^n x_j = \bar{x}_n$, and by the definition of a covariance matrix, it follows that

$$\mathbb{Cov}(Y) = \mathbb{E}\big[(Y - \bar{x}_n)(Y - \bar{x}_n)^\top\big] = \frac{1}{n}\sum_{j=1}^n (x_j - \bar{x}_n)(x_j - \bar{x}_n)^\top = s_n.$$

(b) Suppose $n \leq d$. Then $x_1, \ldots, x_n$ lie in a $(d - 1)$-dimensional hyperplane of $\mathbb{R}^d$. According to part (a) and Theorem 5.3 (b), s_n is singular, i.e., not invertible.

(c) The condition states that Y cannot take on values in a suitable hyperplane of $\mathbb{R}^d$ with probability one (which implies $n \geq d + 1$). The assertion follows from Theorem 5.3 (see also the remark following that theorem).

5.4 Solution We can use part (c) of Problem 5.3. For this purpose, let Y be a random vector with the following property: For each n-tuple $(x_1, \ldots, x_n) \in (\mathbb{R}^d)^n$, $n^{-1}\sum_{j=1}^n \delta_{x_j}$ is the conditional distribution of Y given $X_1 = x_1, \ldots, X_n = x_n$. Here, δ_x generally denotes the one-point distribution at $x \in \mathbb{R}^d$. In the case that $x_1, \ldots, x_n$ are pairwise distinct, this conditional distribution is therefore the uniform distribution on $x_1, \ldots, x_n$. Note that s_n from Problem 5.3 is the covariance matrix of Y under the condition $X_1 = x_1, \ldots, X_n = x_n$. Because X_1 has a density with respect to the Borel–Lebesgue measure, $X_1, \ldots, X_n$ are pairwise distinct with probability one, and for each hyperplane $\mathcal{H} \subset \mathbb{R}^d$ we have $\mathbb{P}(\mathcal{H}^c \cap \{X_1, \ldots, X_n\} \neq \emptyset) = 1$. In view of Problem 5.3 (c), S_n is then invertible. If f denotes the density of X_1, it follows that

$$\mathbb{P}(S_n \text{ is invertible})$$
$$= \int_{\mathbb{R}^d} \cdots \int_{\mathbb{R}^d} \mathbb{P}(S_n \text{ is invertible} | X_1 = x_1, \ldots, X_n = x_n) \prod_{j=1}^{n} f(x_j) \, dx_1 \ldots dx_n$$
$$= 1.$$

5.5 Solution Due to $\|X + Y\|^2 \leq 2\|X\|^2 + 2\|Y\|^2$, the covariance matrix of $X + Y$, denoted by S, exists. We have $S = \mathbb{E}\big[(X + Y - \mathbb{E}(X + Y))(X + Y - \mathbb{E}(X + Y))^\top\big]$. With $\tilde{X} := X - \mathbb{E}(X)$ and $\tilde{Y} := Y - \mathbb{E}(Y)$, it follows that $S = \mathbb{E}[(\tilde{X} + \tilde{Y})(\tilde{X} + \tilde{Y})^\top] = \mathbb{E}[\tilde{X}\tilde{X}^\top] + \mathbb{E}[\tilde{Y}\tilde{Y}^\top] + 2\mathbb{E}[\tilde{X}\tilde{Y}^\top] = \Sigma + T$, since the matrix $\mathbb{E}[\tilde{X}\tilde{Y}^\top]$ is the $d \times d$ zero matrix.

5.6 Solution In view of the proof of Theorem 5.10, let $X = AY + \mu$ with $\Sigma = AA^\top$ and $Y \sim N_d(0_d, I_d)$. Since $t^\top Y \sim N(0, \|t\|^2) \sim \|t\| N(0, 1)$, it follows that $\varphi_Y(t) = \exp(-\frac{1}{2}\|t\|^2)$. We thus obtain

$$\varphi_X(t) = \mathbb{E}\big[e^{it^\top(AY+\mu)}\big] = e^{it^\top \mu} \varphi_Y(A^\top t) = e^{it^\top \mu} \exp\left(-\tfrac{1}{2} t^\top AA^\top t\right)$$
$$= e^{it^\top \mu} \exp\left(-\tfrac{1}{2} t^\top \Sigma t\right).$$

5.7 Solution First assume $X \sim N_d(0_d, I_d)$, and let $Q := X^\top A X \sim \chi_d^2$, so $\mathbb{E}(Q) = d$ and $\mathbb{E}(Q^2) = d^2 + 2d$. With $A =: (a_{i,j})_{i,j=1,\ldots,d}$, it follows that $d = \mathbb{E}(Q) = \sum_{i=1}^{d} a_{i,i}$, and

$$d^2 + 2d = \mathbb{E}(Q^2) = \sum_{i,j,k,\ell=1}^{d} a_{i,j} a_{k,\ell} \mathbb{E}[X_i X_j X_k X_\ell]$$
$$= 3 \sum_{i=1}^{d} a_{i,i}^2 + \sum_{i \neq k} a_{i,i} a_{k,k} + 2 \sum_{i \neq j} a_{i,j}^2$$
$$= \left(\sum_{i=1}^{d} a_{i,i}\right)^2 + 2 \sum_{i,j=1}^{d} a_{i,j}^2,$$

so $\sum_{i,j=1}^{d} a_{i,j}^2 = d$. Since $d^2 = \left(\sum_{i=1}^{d} a_{i,i}\right)^2 \le d \sum_{i=1}^{d} a_{i,i}^2$, the equal sign holds in the Cauchy–Schwarz inequality. Thus $a_{i,i} = 1$ for each $i = 1, \ldots, d$. Furthermore, $a_{i,j} = 0$ if $i \ne j$. Therefore, $A = I_d^{-1} = I_d$. In the general case $X \sim N_d(0_d, \Sigma)$, there is a regular matrix B with $\Sigma = BB^\top$. Then $Y := B^{-1}X \sim N_d(0_d, I_d)$ and $X^\top A X = Y^\top B^\top A B Y \sim \chi_d^2$. After what has been shown so far, we get $B^\top A B = I_d$ and thus $A = (B^\top)^{-1} B^{-1} = (BB^\top)^{-1} = \Sigma^{-1}$.

5.8 Solution (a) Let H be an orthogonal $d \times d$ matrix with $Ha = \|a\| e_1$, where $e_1 = (1, 0, \ldots, 0)^\top \in \mathbb{R}^d$, and let $Z =: (Z_1, \ldots, Z_d)^\top = HX$. Then $Z \sim N_d(0_d, I_d)$, and setting $\delta := \|a\|$ yields

$$Y = \|H(X + a)\|^2 = \|HX + \delta e_1\|^2 = \|Z + \delta e_1\|^2 = (Z_1 + \delta)^2 + \sum_{j=2}^{d} Z_j^2,$$

from which (b) follows. (c) Let $S(\delta e_1, \sqrt{t}) := \{x \in \mathbb{R}^d : \|x - \delta e_1\| \le \sqrt{t}\}$ be the d-dimensional sphere centered at δe_1 with radius $\sqrt{t}$. We have

$$G_{d,\delta^2}(t) = \int_{S(\delta e_1, \sqrt{t})} \frac{1}{(2\pi)^{d/2}} \exp\left(-\frac{\|x\|^2}{2}\right) dx.$$

Since the integrand is a strictly decreasing function of $\|x\|$, assertion (c) follows.

5.9 Solution We have $(X, Y)^\top \overset{\mathcal{D}}{=} (\sigma U, \tau(\rho U + \sqrt{1 - \rho^2} V))^\top$, where U and V are independent standard normal random variables. Since $\mathbb{E}(UV) = 0$, it follows that $\mathbb{E}(XY) = \mathbb{E}[\sigma U \tau(\rho U + \sqrt{1 - \rho^2} V)] = \sigma \tau \rho$. Due to $\mathbb{E}(U^4) = 3$ and $\mathbb{E}(U^3 V) = 0$, we further obtain

$$\mathbb{E}(X^2 Y^2) = \sigma^2 \tau^2 \mathbb{E}[U^2(\rho U + \sqrt{1 - \rho^2} V)^2] = \sigma^2 \tau^2 (3\rho^2 + 1 - \rho^2) = \sigma^2 \tau^2 (1 + 2\rho^2).$$

With the addition theorem for the cosine function and $\mathbb{E}[\sin(tV)] = 0$, $t \in \mathbb{R}$, it follows that

$$\mathbb{E}(X^2 \cos Y) = \sigma^2 \mathbb{E}[U^2 \cos(\tau \rho U)] \cdot \mathbb{E}[\cos(\tau \sqrt{1 - \rho^2} V)].$$

We have $\mathbb{E}[\cos(tU)] = \exp(-t^2/2)$, $t \in \mathbb{R}$ (characteristic function of $N(0, 1)$!), and differentiating twice yields $\mathbb{E}[U^2 \cos(tU)] = (1 - t^2) \exp(-\frac{1}{2}t^2)$. The assertion now follows by direct calculation.

5.10 Solution

(a) We have $\mathbb{P}(X/\|X\| = 1) = 1$ and $\|HX\| = \|X\|$ for every orthogonal matrix H. The assertion now follows with Corollary 5.7.

(b) In view of (a), let w.l.o.g. $Y = \frac{X}{\|X\|}$ with $X \sim N_d(0_d, I_d)$. Since $-X/\|X\| = -X/\| - X\| \stackrel{\mathcal{D}}{=} X/\|X\|$ it follows that $\mathbb{E}(Y) = 0_d$. Setting $X := (X_1, \ldots, X_d)^\top$ and $Y := (Y_1, \ldots, Y_d)^\top$, $X_1^2 + \ldots + X_d^2 = \|X\|^2$ and symmetry arguments yield $\mathbb{E}(Y_j^2) = \frac{1}{d}$, $j = 1, \ldots, d$. Because of

$$\frac{(-X_1, X_2, \ldots, X_d)^\top}{\|(-X_1, X_2, \ldots, X_d)\|^\top} \stackrel{\mathcal{D}}{=} \frac{X}{\|X\|}$$

we obtain $\mathbb{E}(Y_1 Y_2) = 0$ and analogously $\mathbb{E}(Y_i Y_j) = 0$ for all $i \neq j$.

5.11 Solution (a) Define $X := (X_1, \ldots, X_n)^\top$, $Y := (Y_1, \ldots, Y_n)^\top$, and $Y := HX$, where H has the property given in the hint. We have $Y \sim N_n(H\mu, H\sigma^2 I_n H^\top) = N_n(H\mu, \sigma^2 I_n)$, and thus $Y_1, \ldots, Y_n$ are independent. In view of $\sum_{j=1}^n X_j^2 = \|X\|^2 = \|HX\|^2 = \|Y\|^2 = \sum_{j=1}^n Y_j^2$, $\sum_{j=1}^n (X_j - \overline{X}_n)^2 = \sum_{j=1}^n X_j^2 - n\overline{X}_n^2$ and $Y_n = \sqrt{n}\overline{X}_n$, it follows that $(n-1)S_n^2 = \sum_{j=1}^{n-1} Y_j^2$. Being a function of (only) Y_n, $\overline{X}_n$ is independent of $(n-1)S_n^2$. Regarding part (b), we can assume $\mu = 0$ without loss of generality. If we set $N_j := Y_j/\sigma \sim N(0, 1)$, $j = 1, \ldots, n$, then $\frac{n-1}{\sigma^2} S_n^2 \sim \sum_{j=1}^{n-1} N_j^2$, which was to be shown.

6.1 Solution

(a) With the hint, we have $0 \leq \mathbb{P}(X \in (x, y]) = F(y) - \mathbb{P}(\bigcup_{j=1}^d A_j)$, and the inclusion-exclusion formula gives $\mathbb{P}(\bigcup_{j=1}^d A_j) = \sum_{r=1}^d (-1)^{r-1} \sum_{1 \leq i_1 < \ldots < i_r \leq d} \mathbb{P}(A_{i_1} \cap \ldots \cap A_{i_r})$. Here, $\{A_{i_1} \cap \ldots \cap A_{i_r}\} = \{X_1 \leq y_1^{\varepsilon_1} x_1^{1-\varepsilon_1}, \ldots, X_d \leq y_d^{\varepsilon_d} x_d^{1-\varepsilon_d}\}$ with $\varepsilon := (\varepsilon_1, \ldots, \varepsilon_d) \in \{0, 1\}^d$, where $\varepsilon_v := 0$ for $v \in \{i_1, \ldots, i_r\}$ and $\varepsilon_v := 1$, otherwise. The definition of the distribution function yields $\mathbb{P}(A_{i_1} \cap \ldots \cap A_{i_r}) = F(y_1^{\varepsilon_1} x_1^{1-\varepsilon_1}, \ldots, y_d^{\varepsilon_d} x_d^{1-\varepsilon_d})$, and with $s(\varepsilon) := \varepsilon_1 + \ldots + \varepsilon_d$ we get $\sum_{1 \leq i_1 < \ldots < i_r \leq d} \mathbb{P}(A_{i_1} \cap \ldots \cap A_{i_r}) = \sum_{\varepsilon \in \{0,1\}^d, s(\varepsilon) = d-r} F(y_1^{\varepsilon_1} x_1^{1-\varepsilon_1}, \ldots, y_d^{\varepsilon_d} x_d^{1-\varepsilon_d})$. We therefore obtain

$$\mathbb{P}(X \in (x, y]) = F(y_1, \ldots, y_d) - \sum_{r=1}^d (-1)^{r-1}$$

$$\times \sum_{\varepsilon \in \{0,1\}^d, s(\varepsilon) = d-r} F(y_1^{\varepsilon_1} x_1^{1-\varepsilon_1}, \ldots, y_d^{\varepsilon_d} x_d^{1-\varepsilon_d})$$

$$= \sum_{\varepsilon \in \{0,1\}^d} (-1)^{d-s(\varepsilon)} F(y_1^{\varepsilon_1} x_1^{1-\varepsilon_1}, \ldots, y_d^{\varepsilon_d} x_d^{1-\varepsilon_d}) = \Delta_x^y F.$$

(b) The assertion follows since $(-\infty, x^{(n)}]$ is a decreasing sequence of sets with limit $(-\infty, x]$ and $\mathbb{P}^X$ is continuous from above.

(c) In the first case, $(-\infty, x^{(n)}] \downarrow \varnothing$, in the second $(-\infty, x^{(n)}] \uparrow \mathbb{R}^d$. Due to the continuity from above and from below of the probability measure $\mathbb{P}^X$, the assertion follows.

6.2 Solution In the case $d = 2$, let $X := (X_1, 0)$, where X_1 has the normal distribution $N(0, 1)$. Then each point $(x_1, 0)$ with $x_1 \in \mathbb{R}$ is a discontinuity point of the distribution function of X.

6.3 Solution Let $U \sim N(0, 1)$, and let $U = U_1 = U_2 = \dots$. Then trivially $U_n \xrightarrow{D} U$. If we set $V_n := U, n \geq 1$ and $V := -U$, then $V_n \xrightarrow{D} V$. But $U_n + V_n = 2U \sim N(0, 4)$, $n \geq 1$, and $U + V = 0$. This means that $U_n + V_n$ does not converge in distribution to $U + V$.

6.4 Solution Let G_n, H_n, G, and H be the distribution functions (df's) of Y_n, Z_n, Y, and Z, respectively. Further, set $d := k + \ell$ and $X_n := (Y_n, Z_n)$, and write F_n for the df of X_n. With $x = (x_1, \dots, x_d) =: (y_1, \dots, y_k, z_1, \dots, z_\ell)$, $y = (y_1, \dots, y_k)$ and $z = (z_1, \dots, z_\ell)$, we have $F_n(x) = G_n(y)H_n(z)$ due to the independence of Y_n and Z_n. If each component of x is a continuity point of the corresponding marginal df, then $Y_n \xrightarrow{D} Y$ and $Z_n \xrightarrow{D} Z$ imply $G_n(y) \to G(y)$ and $H_n(z) \to H(z)$, and it follows that $F_n(x) \to G(y)H(z)$. According to Remark 6.4, we have $X_n \xrightarrow{D} (Y, Z)$, where Y and Z are independent, since the df of (X, Y) is the product $G(y)H(z)$ of the distribution function of Y and Z.

6.5 Solution

"$\Longrightarrow$": For each $\varepsilon > 0$, there is a compact set $K = \times_{j=1}^{d}[a_j, b_j]$ with $-\infty < a_j < b_j < \infty$, $j = 1, \dots, d$, and $\mathbb{P}(X_n \in K) \geq 1 - \varepsilon$ for each $n \geq 1$. Then $\mathbb{P}(a_j \leq X_n^{(j)} \leq b_j) \geq 1 - \varepsilon, n \geq 1$, for each $j \in \{1, \dots, d\}$.

"$\Longleftarrow$": Suppose $\mathbb{P}(a_j \leq X_n^{(j)} \leq b_j) \geq 1 - \frac{\varepsilon}{d}$ for each $n \geq 1$ and each $j \in \{1, \dots, d\}$. Setting $K = \times_{j=1}^{d}[a_j, b_j]$ then yields $\mathbb{P}(X \in K) \geq 1 - \varepsilon$ for each $n \geq 1$.

6.6 Solution We have $\|X_n - \mu_n\|^2 \sim \chi_d^2$. Fix any $\varepsilon \in (0, 1)$, and let a be the $(1 - \varepsilon)$-quantile of the χ_d^2-distribution. Then $\mathbb{P}(\|X_n - \mu_n\| \leq a) = 1 - \varepsilon, n \geq 1$. If the sequence (μ_n) is bounded, there is a $K > 0$ with $\{\mu_n : n \geq 1\} \subset \{x : \|x\| \leq K\}$. It follows that $\mathbb{P}(\|X_n\| \leq K + a) \geq 1 - \varepsilon$ for each $n \geq 1$. Thus, (X_n) is tight. Conversely, if (X_n) is tight, then for every $\varepsilon \in (0, 1)$ there is a $K > 0$ with $\mathbb{P}(\|X_n\| \leq K) > 1 - \varepsilon, n \geq 1$. If the sequence $(\mu_n)_{n \geq 1}$ were unbounded, we would have $\|\mu_n\| > K + a$ for infinitely many n. Since $\mathbb{P}(\|X_n - \mu_n\| \leq a) = 1 - \varepsilon)$ for each such n according to the first part of the proof, we obtain a contradiction to $\mathbb{P}(\|X_n\| \leq K) > 1 - \varepsilon$.

6.7 Solution (b) follows from the inequality $\mathbb{P}(\|X_n + Y_n\| > \varepsilon) \leq \mathbb{P}(\|X_n\| > \frac{\varepsilon}{2}) + \mathbb{P}(\|Y_n\| > \frac{\varepsilon}{2})$.

(c): According to the assumption, there are K, $L > 0$ with $\mathbb{P}(\|X_n\| \leq K) \geq 1 - \frac{\varepsilon}{2})$ and $\mathbb{P}(|Z_n| \leq L) \geq 1 - \frac{\varepsilon}{2})$ for each $n \geq 1$. Since $\|Z_n X_n\| \leq |Z_n| \|X_n\|$, it follows that $\mathbb{P}(\|Z_n X_n\| \leq KL) \geq 1 - \varepsilon$ for each $n \geq 1$.

(d): For each $\eta > 0$ there is a $K > 0$ with $\mathbb{P}(\|X_n\| \leq K) \geq 1 - \eta$ for every $n \geq 1$. Let $\varepsilon > 0$ be arbitrary. We have $\mathbb{P}(\|Z_n X_n\| > \varepsilon) = \mathbb{P}(\|Z_n X_n\| > \varepsilon, \|X_n\| > K) + \mathbb{P}(\|Z_n X_n\| > \varepsilon, \|X_n\| > K) \leq \mathbb{P}(\|X_n\| > K) + \mathbb{P}(|Z_n| > \frac{\varepsilon}{K})$ and thus $\limsup_{n \to \infty} \mathbb{P}(\|Z_n X_n\| > \varepsilon) \leq \eta$. Since η was arbitrary, the claim follows. (e) results from the fact that the image of a compact set under a continuous mapping is a compact set.

6.8 Solution We have $A_n X_n = AX_n + (A_n - A)X_n$. The mapping theorem yields $AX_n \xrightarrow{\mathcal{D}} AX$. Since $X_n = O_{\mathbb{P}}(1)$, for each $\delta > 0$ there is a $K > 0$ with $\mathbb{P}(\|X_n\| > K) \leq \delta$ for every $n \geq 1$. For each $\varepsilon > 0$ we thus have $\mathbb{P}(\|(A_n - A)X_n\| > \varepsilon) \leq \delta + \mathbb{P}(\|(A_n - A)X_n\| > \varepsilon, \|X_n\| \leq K)$. For sufficiently large n, the last probability is zero due to $\|(A_n - A)X_n\| \leq \|A_n - A\|_{\mathrm{sp}} \|X_n\| \leq \|A_n - A\|_{\mathrm{sp}} K$ and $\|A_n - A\|_{\mathrm{sp}} \to 0$. Since δ and ε are arbitrary, we get $(A_n - A)X_n \xrightarrow{\mathbb{P}} 0_s$, and thus the claim follows from Slutzky's lemma.

6.9 Solution Assume the setting of Theorem 6.19. We set $r_n := n$ and $X_{n,j} := (X_j - \mu)/\sqrt{n}$, $j \in \{1, \ldots, n\}$. Then $X_{n,1}, \ldots, X_{n,n}$ are independent d-dimensional random vectors with $\mathbb{E}(X_{n,j}) = 0_d$ and $\mathbb{E}\|X_{n,j}\|^2 < \infty$ $(n \geq 1, j \in \{1, \ldots, r_n\})$. Furthermore,

$$S_n = \frac{1}{\sqrt{n}} \sum_{j=1}^{n} (X_j - \mu) = X_{n,1} + \ldots + X_{n,n}$$

and $\mathbb{E}(S_n S_n^{\top}) = \mathbb{E}[(X_1 - \mu)(X_1 - \mu)^{\top}] = \Sigma$. Since $X_{n,1}, \ldots, X_{n,n}$ are identically distributed, we have $L_n(\varepsilon, b) = \mathbb{E}[(b^{\top}(X_1 - \mu)^2 \mathbf{1}\{|b^{\top}(X_1 - \mu)| > \varepsilon\sqrt{n}\}]$, $b \in \mathbb{R}^d$. The condition $\mathbb{E}\|X_1\|^2 < \infty$ and the dominated convergence theorem then yield $\lim_{n \to \infty} L_n(\varepsilon, b) = 0$, and the claim follows.

6.10 Solution W.l.o.g. let $\mathbb{E}(X_{n,j}) = 0_d$ $(n \geq 1, 1 \leq j \leq n)$. Further, let $c \in \mathbb{R}^d$ with $c \neq 0_d$ be arbitrary, and put $Z_{n,j} := c^{\top} X_{n,j}$. According to the Cramér–Wold device, we must show that $n^{-1/2} \sum_{j=1}^{n} Z_{n,j} \xrightarrow{\mathcal{D}} N(0, c^{\top}\Sigma c)$. Since $\Sigma_{n,j}$ is positive definite, we have $\mathbb{V}(Z_{n,j}) = c^{\top}\Sigma_{n,j} c > 0$. Let $\sigma_n^2 := \sum_{j=1}^{n} \mathbb{V}(Z_{n,j})$. Putting $\overline{\Sigma}_n := \frac{1}{n}\sum_{j=1}^{n} \Sigma_{n,j}$ gives $\sigma_n^2 = nc^{\top}\overline{\Sigma}_n c$. Since $\overline{\Sigma}_n \to \Sigma$, the inequality $\sigma_n^2 \geq \frac{n}{2}c^{\top}\Sigma c$ holds for sufficiently large n. In view of Theorem 1.19, for every $\varepsilon > 0$ we have to show $\lim_{n\to\infty} L_n(\varepsilon) = 0$, where

$$L_n(\varepsilon) = \frac{1}{\sigma_n^2} \sum_{j=1}^{n} \mathbb{E}\left(Z_{n,j}^2 \mathbf{1}\{|Z_{n,j}| > \varepsilon\sigma_n\}\right).$$

By the Cauchy–Schwarz inequality, $|c^\top Z_{n,j}| \le \|c\| \, \|Z_{n,j}\|$. For sufficiently large n we thus have

$$L_n(\varepsilon) \le \frac{2\|c\|^2}{c^\top \Sigma c} \cdot \frac{1}{n} \sum_{j=1}^{n} \mathbb{E}\left(\|X_{n,j}\|^2 \mathbf{1}\left\{\|X_{n,j}\| > \sqrt{n} \cdot \frac{\varepsilon\sqrt{c^\top \Sigma c}}{\sqrt{2}\|c\|}\right\}\right),$$

so that (6.25) provides the assertion.

6.11 Solution If $(X_1 = (X_{1,1}, \ldots, X_{1,s})^\top$ has a uniform distribution on all $\binom{s}{r}$ s-tuples of ones and zeros that have exactly r ones, then $\mathbb{P}(X_{1,j} = 1) = \binom{s-1}{r-1}/\binom{s}{r} = \frac{r}{s}$, and we obtain the claim made for $\mathbb{E}(X_1)$. Since $X_{1,j} \sim \text{Bin}(1, \frac{r}{s})$, it follows that $\mathbb{V}(X_{1,j}) = \sigma_{jj} = \frac{r}{s}(1 - \frac{r}{s})$. If $i \ne j$ then $\mathbb{P}(X_{1,i} = 1, X_{1,j} = 1) = \binom{s-2}{r-2}/\binom{s}{r} = \frac{r(r-1)}{s(s-1)}$ and thus $\text{Cov}(X_{1,i}, X_{1,j}) = \frac{r(r-1)}{s(s-1)} - \frac{r}{s} \cdot \frac{r}{s}$. Thus also the claim made for $\text{Cov}(X_1)$ has been proved.

6.12 Solution

(a) Since $X_{1,1} + \ldots + X_{1,s} = r$, $\text{Cov}(X_1)$ is singular according to Theorem 5.3 (b). Thus the matrix Σ in (6.10) is also singular, since it is a multiple of $\text{Cov}(X_1)$.
(b) follows by direct calculation (matrix multiplication).
(c) Starting from (6.13), everything proceeds in exactly the same way as after (6.9). The difference to (6.9) lies only in the correction factor $s/\sqrt{r(s-r)}$ occurring in (6.13). This is due to the fact that r different numbers are drawn in a lottery draw, and thus there are no independent repetitions of a trial with s different outcomes.
(d) The test statistic T_n yields the value 60.9. Since the 0.95 quantile of the χ_{48}^2-distribution is 65.17, the observed fluctuation of the winning frequencies is still compatible with the model assumptions made.

6.13 Solution The random vector $X_n/\sqrt{n}$ has a uniform distribution on the surface of the unit sphere in $\mathbb{R}^n$. According to Problem 5.10, $X_n/\sqrt{n} \overset{\mathcal{D}}{=} Y_n/\|Y_n\|$, where $Y_n =:$ $(Y_{n,1}, \ldots, Y_{n,n})^\top \sim N_n(0_n, I_n)$. Thus, $(X_{n,1}, \ldots, X_{n,d})^\top \overset{\mathcal{D}}{=} (Y_{n,1}, \ldots, Y_{n,d})^\top/U_n$, where $U_n = +\sqrt{U_n^2}$ and $U_n^2 = (Y_{n,1}^2 + \ldots + Y_{n,n}^2)/n$. In view of $U_n^2 \overset{\mathbb{P}}{\longrightarrow} 1$ the assertion follows, since $(Y_{n,1}, \ldots, Y_{n,d})^\top \sim N_d(0_d, I_d)$.

6.14 Solution

"(b) $\Longrightarrow$ (a)" follows by setting $\delta = 1$.

"(a) $\Longrightarrow$ (c)": Since the sum in (c) decreases monotonically in ε, let w.l.o.g. $0 < \varepsilon \le 1$. For such ε, $\mathbf{1}\{W_{n,j} > \varepsilon\} \le \frac{1}{\varepsilon}\min(1, W_{n,j})$, and the claim follows.

"(c) $\Longrightarrow$ (b)": Fix any $\delta > 0$. For each $\varepsilon > 0$, the hint gives

$$W_{n,j}\min\left(1, W_{n,j}^{\delta}\right) \le W_{n,j}\varepsilon^{\delta} + W_{n,j}\mathbf{1}\{W_{n,j} > \varepsilon\}, \qquad j \in \{1, \dots, r_n\}.$$

If we abbreviate the sum in (b) with $a_n(\delta)$, it follows that

$$a_n(\delta) \le \varepsilon^{\delta}\left(\sum_{j=1}^{r_n}\mathbb{E}[W_{n,j}]\right) + \sum_{j=1}^{r_n}\mathbb{E}[W_{n,j}\mathbf{1}\{W_{n,j} > \varepsilon\}].$$

By assumption, the second sum on the right-hand side converges to zero, and therefore (6.27) yields $\limsup_{n\to\infty} a_n(\delta) \le \varepsilon^{\delta}K$. Since $\varepsilon > 0$ was arbitrary, the claim follows.

6.15 Solution Let $V := (Z_1, \dots, Z_{s-1})^{\top}$. Further, let B be an $(s-1) \times (s-1)$ matrix with $A = BB^{\top}$. Then $V \overset{\mathcal{D}}{=} BU$, where $U \sim N_{s-1}(B^{-1}\Delta_{s-1}, I_{s-1})$. It follows that

$$V^{\top}A^{-1}V \overset{\mathcal{D}}{=} U^{\top}B^{\top}(BB^{\top})^{-1}BU = U^{\top}U.$$

According to Problem 5.8, $U^{\top}U$ has a non-central χ_{s-1}^2 distribution with non-centrality parameter

$$\delta^2 = \|B^{-1}\Delta_{s-1}\|^2 = \Delta_{s-1}^{\top}(B^{-1})^{\top}B^{-1}\Delta_{s-1} = \Delta_{s-1}^{\top}A^{-1}\Delta_{s-1}$$

$$= \sum_{j=1}^{s-1}\delta_j^2/p_j + (\delta_1 + \dots + \delta_{s-1})^2/p_s = \sum_{j=1}^{s}\delta_j^2/p_j.$$

6.16 Solution According to (6.24), g must satisfy the equation $p(1-p)g'(p)^2 = 1$, which is true for the function defined by $g(p) := 2\arcsin\sqrt{p}$.

7.1 Solution Let $x \ge x_{m,m-1}$. Using (7.7) it follows that

$$F_n^{\omega}(x) \ge F_n^{\omega}(x_{m,m-1}) \ge F(x_{m,m-1}) - D_{m,n}^{\omega} \ge 1 - \frac{1}{m} - D_{m,n}^{\omega} \ge F(x) - \frac{1}{m} - D_{m,n}^{\omega},$$

$$F(x) \ge F(x_{m,m-1}) \ge 1 - \frac{1}{m} \ge F_n^{\omega}(x) - \frac{1}{m} - D_{m,n}^{\omega}$$

and thus the assertion.

7.2 Solution Let $U_1, \ldots, U_n$ be independent random variables with a uniform distribution on $[0, 1]$, and let $\widehat{G}_n(t) := \frac{1}{n} \sum_{j=1}^{n} \mathbb{1}\{U_j \leq t\}$, $0 \leq t \leq 1$, be the empirical distribution function of $U_1, \ldots, U_n$. Suppose that $\mathbb{P}(\sup_{0 \leq u \leq 1} |\widehat{G}_n(u) - u| > t) \leq 2 \exp(-2nt^2)$. Set $X_j := F^{-1}(U_j)$, $j = 1, \ldots, n$. Then $X_1, \ldots, X_n$ are independent with the same distribution function F. Since $X_j \leq x \iff U_j \leq F(x)$, we obtain $\widehat{F}_n(x) = \widehat{G}_n(F(x))$, $x \in \mathbb{R}$, and the assertion follows from $\sup_{x \in \mathbb{R}} |\widehat{F}_n(x) - F(x)| \leq \sup_{0 \leq u \leq 1} |\widehat{G}_n(u) - u|$.

7.3 Solution Let $D_n := \sup_{x \in \mathbb{R}} |\widehat{F}_n(x) - F(x)|$ and $\varepsilon > 0$ be arbitrary. From (7.11) it follows that

$$\sum_{n=1}^{\infty} \mathbb{P}(D_n > \varepsilon) \leq 2 \sum_{n=1}^{\infty} \exp\left(-2n\varepsilon^2\right) < \infty.$$

The Borel–Cantelli lemma yields $\mathbb{P}\left(\limsup_{n \to \infty}\{D_n > \varepsilon\}\right) = 0$ for every $\varepsilon > 0$ and thus the Glivenko–Cantelli theorem.

7.4 Solution Let $Z_\ell := (Z_{\ell,1}, \ldots, Z_{\ell,k})^\top$, where $Z_{\ell,i} = \mathbb{1}\{X_\ell \leq x_i\} - F(x_i)$ $(1 \leq \ell \leq n, 1 \leq i \leq k)$. $Z_1, \ldots, Z_n$ are i.i.d. random vectors with $\mathbb{E}(Z_1) = 0_k$ and $\mathbb{E}(Z_{1,i}^2) = F(x_i)(1 - F(x_i))$. If $i \neq j$ then $\mathbb{1}\{X_1 \leq x_i\}\mathbb{1}\{X_1 \leq x_j\} = \mathbb{1}\{X_1 \leq \min(x_i, x_j)\}$ entails

$$\mathbb{E}(Z_{1,i} Z_{1,j}) = \mathbb{E}[\mathbb{1}\{X_1 \leq \min(x_i, x_j)\}] - F(x_i)F(x_j) = F\left(\min(x_i, x_j)\right) - F(x_i)F(x_j).$$

Since $B_n(x_\ell) = n^{-1/2} \sum_{j=1}^{n} Z_{j,\ell}$, $\ell \in \{1, \ldots, k\}$, the assertion follows from the multivariate central limit theorem.

8.1 Solution For a k-element subset $A = \{i_1, \ldots, i_k\}$ of $\{1, \ldots, n\}$, we set $h_A := h(X_{i_1}, \ldots, X_{i_k})$. It follows that $\mathbb{V}(U_n) = \text{Cov}(U_n, U_n) = \binom{n}{k}^{-2} \sum_A \sum_B \text{Cov}(h_A, h_B)$. Here, the sums run over all k-element subsets A and B of $\{1, \ldots, n\}$. We have $\text{Cov}(h_A, h_B) = 0$, if $A \cap B = \emptyset$ and $\text{Cov}(h_A, h_B) = \sigma_c^2$, if $|A \cap B| = c$, where $c \in \{1, \ldots, k\}$. If we split the double sum over A and B according to the number $c = |A \cap B|$, and note that after choosing A (which can be done in $\binom{n}{k}$ ways) there are $\binom{k}{c}\binom{n-k}{k-c}$ subsets B with $|A \cap B| = c$, the assertion follows.

8.2 Solution We have $\vartheta = \mathbb{E}(X_1 X_2) = \mu^2$, $h_1(x_1) = \mathbb{E}[h(x_1, X_2)] = x_1 \mu$ and $\sigma_1^2 = \mathbb{V}(h_1(X_1)) = \mu^2 \sigma^2$. If $\mu \neq 0$ then $\sigma_1^2 > 0$, and Theorem 8.9 provides the assertion. In the case $\mu = 0$, we are in the situation of Example 8.12 with $s = 1$, $\lambda_1 = \sigma^2$ and $\varphi_1(x) = x/\sigma$. It follows that $nU_n \xrightarrow{D} \sigma^2(N^2 - 1)$, where N has a standard normal distribution.

8.3 Solution By Theorem 8.3, $\mathbb{V}(\tau_n) = \binom{n}{2}^{-1}(2(n-2)\sigma_1^2 + \sigma_2^2)$, where $\sigma_1^2 = \mathbb{V}(h_1((X_1, Y_1)))$ and $\sigma_2^2 = 1 - \tau^2$, since $\mathrm{sgn}^2(t) = 1$ for $t \neq 0$. Putting $(X, Y) := (X_1, Y_1)$, the assumed independence and a distinction between cases yields

$$h_1(x, y) = \mathbb{E}\big[h((x, y), (X, Y))\big] = \mathbb{P}\left((x-X)(y-Y) > 0\right) - \mathbb{P}\left((x-X)(y-Y) < 0\right)$$

$$= \mathbb{P}(X > x)\,\mathbb{P}(Y > y) + \mathbb{P}(X < x)\,\mathbb{P}(Y < y) - \mathbb{P}(X > x)\,\mathbb{P}(Y < y)$$

$$-\mathbb{P}(X < x)\,\mathbb{P}(Y > y)$$

$$= (1 - G(x))(1 - H(y)) + G(x)H(y) - (1 - G(x))H(y) - G(x)(1 - H(y))$$

$$= (1 - 2G(x))(1 - 2H(y)).$$

The random variables $U := 1 - 2G(X)$ and $V := 1 - 2H(Y)$ are independent and each uniformly distributed on $[-1, 1]$. It follows that $\sigma_1^2 = \mathbb{V}(UV) = \mathbb{E}(U^2)\mathbb{E}(V^2) = \frac{1}{9}$. Since $\tau = 0$, the assertion in (a) follows. Part (b) is a direct consequence of Theorem 8.9.

8.4 Solution The equation $\mathbb{E}(S) = \mathbb{E}(\widehat{S})$ follows by taking iterated conditional expectations. We then assume $\mathbb{E}(S) = 0$ and briefly set $f(X_i) := \mathbb{E}[S|X_i] - \ell_i(X_i)$, $i = 1, \ldots, n$. Due to the independence of $X_1, \ldots, X_n$, $\mathbb{E}\big[\mathbb{E}[S|X_j]\big|X_i\big]$ is equal to $\mathbb{E}(S)$ and thus equal to zero, if $j \neq i$ and equal to $\mathbb{E}[S|X_i]$, if $j = i$. This implies $\mathbb{E}[S - \widehat{S}|X_i] = 0$ for each i and thus

$$\mathbb{E}\big[(S - \widehat{S})(\widehat{S} - L)\big] = \sum_{i=1}^{n} \mathbb{E}\big[(S - \widehat{S})f(X_i)\big] = \sum_{i=1}^{n} \mathbb{E}\big[f(X_i)\mathbb{E}[S - \widehat{S}|X_i]\big] = 0.$$

Here, equations involving conditional expectations are always to be understood almost surely with respect to $\mathbb{P}$.

8.5 Solution We assume that U_n and V_n are non-degenerate and consider the Hájek projections $\widetilde{U}_n = \frac{k}{n}\sum_{j=1}^{n}(f_1(X_j) - \vartheta) + \vartheta$ and $\widetilde{V}_n = \frac{\ell}{n}\sum_{j=1}^{n}(g_1(X_j) - \eta) + \eta$ of U_n and V_n, respectively. We have $\sigma_1^2 := \mathbb{V}(f_1(X_1)) > 0$, $\tau_1^2 := \mathbb{V}(g_1(X_1)) > 0$, and

$$\sqrt{n}\begin{pmatrix} U_n - \vartheta \\ V_n - \eta \end{pmatrix} = \sqrt{n}\begin{pmatrix} \widetilde{U}_n - \vartheta \\ \widetilde{V}_n - \eta \end{pmatrix} + \sqrt{n}\begin{pmatrix} U_n - \widetilde{U}_n \\ V_n - \widetilde{V}_n \end{pmatrix} = \frac{1}{\sqrt{n}}\sum_{j=1}^{n}\begin{pmatrix} k(f_1(X_j) - \vartheta) \\ \ell(g_1(X_j) - \eta) \end{pmatrix} + R_n,$$

where $R_n \overset{\mathbb{P}}{\longrightarrow} 0_2$. With Theorem 6.19 and the Slutsky's lemma,

$$\sqrt{n}\begin{pmatrix} U_n - \vartheta \\ V_n - \eta \end{pmatrix} \overset{D}{\longrightarrow} N_2\left(\begin{pmatrix} 0 \\ 0 \end{pmatrix}, \begin{pmatrix} k^2\sigma_1^2 & \sigma_{1,2} \\ \sigma_{1,2} & \ell^2\tau_1^2 \end{pmatrix}\right),$$

where $\sigma_{1,2} = \mathrm{Cov}\big(k(f_1(X_1) - \vartheta), \ell(g_1(X_1) - \eta)\big) = k\ell\big(\mathbb{E}[f_1(X_1)g_1(X_1)] - \vartheta\eta\big)$.

8.6 Solution The kernel has the representation (8.16) from Example 8.12. We only need to fulfill the conditions (8.17) and (8.18). Setting

$$\varphi_1(x) := \frac{x}{\sigma}, \quad \varphi_2(x) := \frac{x^2 - \sigma^2}{\sqrt{m_4 - \sigma^4}}$$

and using $X_1 \overset{\mathcal{D}}{=} -X_1$, we obtain $\mathbb{E}[\varphi_1(X_1)] = 0$. This implies $\sigma^2 = \mathbb{E}(X_1^2)$, and thus we get $\mathbb{E}[\varphi_2(X_1)] = 0$. Further, $\mathbb{E}[\varphi_1^2(X_1)] = 1$, and $\mathbb{E}[(X_1 - \sigma^2)^2] = m_4 - \sigma^4$ results in $\mathbb{E}[\varphi_2^2(X_1)] = 1$. Additionally, we obtain $\mathbb{E}[\varphi_1(X_1)\varphi_2(X_1)] = 0$, because $\mathbb{E}(X_1^3) = 0$ due to $X_1 \overset{\mathcal{D}}{=} -X_1$. It follows that $h(x_1, x_2) = \sigma^2 \varphi_1(x_1)\varphi_1(x_2) + (m_4 - \sigma^4)\varphi_2(x_1)\varphi_2(x_2)$, and (8.19) yields the claim.

8.7 Solution We have $U_n = \binom{n}{3}^{-1} \sum_{1 \le i < j < k \le n} X_i X_j X_k = \binom{n}{3}^{-1} \frac{1}{6} \sum_{1 \ne j \ne k \ne i} X_i X_j X_k$. Putting $S_n := \sum_{j=1}^n X_j$, the last triple sum is equal to

$$\sum_{i,j,k=1}^n X_i X_j X_k - \sum_{i=1}^n X_i^3 - 3\sum_{i=1}^n X_i^2 \left(\sum_{j:j\ne i} X_j\right) = S_n^3 - \sum_{i=1}^n X_i^3 - 3\sum_{i=1}^n X_i^2(S_n - X_i).$$

If we set $Z_n := \frac{1}{\sqrt{n}} S_n$, $\overline{X_n^r} := \frac{1}{n}\sum_{j=1}^n X_j^r$, $r \in \{2, 3, 4\}$, as well as $a_n := \frac{n^2}{(n-1)(n-2)}$, $b_n := \frac{a_n}{\sqrt{n}}$, then a direct calculation yields $n^{3/2} U_n = a_n Z_n^3 - b_n \overline{X_n^3} - 3a_n Z_n \overline{X_n^2} + 3b_n \overline{X_n^3}$. By the law of large numbers, $\overline{X_n^3} = O_\mathbb{P}(1)$ and $\overline{X_n^2} = 1 + o_\mathbb{P}(1)$. Since $a_n \to 1$, $b_n \to 0$, and $Z_n \overset{\mathcal{D}}{\to} Z$, the claim follows from the mapping theorem and Slutzky's lemma.

8.8 Solution We have $U_n = \binom{n}{4}^{-1} \sum_{1 \le i < j < k < \ell \le n} X_i X_j X_k X_\ell = \binom{n}{4}^{-1} \frac{1}{24} \sum_{\ne} X_i X_j X_k X_\ell$. Here, the last sum runs over all quadruples (i, j, k, ℓ) with pairwise distinct components. We decompose this sum according to the number of different indices occurring. Using the notation from Solution 8.7 and symmetry arguments, it follows that

$$S_n^4 - \sum_{i=1}^n X_i^4 - 3\sum_{i=1}^n X_i^2(n\overline{X_n^2} - X_i^2) - 4\sum_{i=1}^n X_i^3(S_n - X_i) - 6\sum_{i=1}^n X_i^2 \left(\sum_{i \ne k \ne \ell \ne i} X_k X_\ell\right),$$

The last sum runs over k and ℓ. Since

$$\sum_{i=1}^n X_i^2 \left(\sum_{i \ne k \ne \ell \ne i} X_k X_\ell\right) = \sum_{i=1}^n X_i^2 \sum_{k:k \ne i} X_k(S_n - X_k - X_i)$$

$$= \sum_{i=1}^n X_i^2 \left(S_n^2 - 2S_n X_i + 2X_i^2 - n\overline{X_n^2}\right)$$

straightforward calculations (with $a_n := \frac{n^3}{(n-1)(n-2)(n-3)}$ and $b_n := \frac{a_n}{n}$) yield

$$n^2 U_n = a_n \overline{Z_n^4} - 6a_n \overline{Z_n^2 X_n^2} + 3a_n (\overline{X_n^2})^2 - 4a_n \overline{X_n} \overline{X_n^3} - 6b_n \overline{X_n^4} + 12\frac{b_n}{\sqrt{n}} \overline{Z_n X_n^3}.$$

As in Problem 8.7, the assertion now follows with the mapping theorem and Slutsky's lemma, because $a_n \to 1$, $b_n \to 0$, $Z_n \xrightarrow{\mathcal{D}} Z$, $\overline{X_n^2} = 1 + o_{\mathbb{P}}(1)$, as well as $\overline{X}_n = o_{\mathbb{P}}(1)$, $\overline{X_n^3} = O_{\mathbb{P}}(1)$, and $\overline{X_n^4} = O_{\mathbb{P}}(1)$.

8.9 Solution

(a) From $X \overset{\mathcal{D}}{=} -X$ and $\mathbb{E}(X^6) < \infty$, it follows that $\mathbb{E}(X) = 0 = \mathbb{E}(X^3)$, and thus $\mathbb{E}h(X_1, X_2) = 0$ and $\mathbb{E}(U_n) = 0$. Furthermore, $h_1(x_1) = \mathbb{E}h(x_1, X_2) = x_1 \mathbb{E}(X_2) + x_1^3 \mathbb{E}(X_2^3) = 0$, $x_1 \in \mathbb{R}$, and therefore $\sigma_1^2 = \mathbb{V}(h_1(X_1)) = 0$. Since $\sigma_2^2 = \mathbb{V}(h(X_1, X_2)) = \mathbb{E}h^2(X_1, X_2) > 0$, U_n is simply-degenerate.

(b) We have

$$(Ag)(x) = \int xy(1 + x^2 y^2)g(y)\,dF(y) = a(g)x + b(g)x^3, \quad x \in \mathbb{R}, \qquad (*)$$

where $a(g) := \int yg(y)\,dF(y)$, $b(g) := \int y^3 g(y)\,dF(y)$. The eigenvalue equation $(Ag)(x) = \lambda g(x)$, $x \in \mathbb{R}$, with $\lambda \neq 0$ thus leads to $g(x) = ax + bx^3$, $x \in \mathbb{R}$, where $a = a(g)/\lambda$ and $b = b(g)/\lambda$. If one inserts the function $y \mapsto g_0(y) := ay + by^3$ into $(*)$, it follows that

$$(Ag_0)(x) = (am_2 + bm_4)x + (am_4 + bm_6)x^3, \quad x \in \mathbb{R}.$$

Equating with $\lambda g_0(x)$ yields

$$\lambda a = am_2 + bm_4, \qquad \lambda b = am_4 + bm_6. \qquad (**)$$

Since $\lambda \neq 0$, we obtain $a \neq 0$ and $b \neq 0$.

(c) If one divides both sides of the first and the second equation in $(**)$ by a and b, respectively, and equates the two resulting expressions for λ, one obtains the quadratic equation $z^2 m_4 + z(m_2 - m_6) - m_4 = 0$, where $z := b/a$. This equation has the solutions

$$z_{1,2} = \frac{1}{2m_4}\left(m_6 - m_2 \pm \sqrt{(m_6 - m_2)^2 + 4m_4^2}\right).$$

From the first equation in $(**)$ it follows that $\lambda_{1,2} = m_2 + z_{1,2}m_4$, which provides (8.51) and (8.52). The convergence in distributione follows from Theorem 8.17. In the special case $X \sim N(0, 1)$, we have $m_2 = 1$, $m_4 = 3$, and $m_6 = 15$. A direct calculation yields $\lambda_1 = 8 + \sqrt{58}$, $\lambda_2 = 8 - \sqrt{58}$.

8.10 Solution For $s < m$, $(Y_m - Y_s)^2 = \sum_{i,j=s+1}^{m} \lambda_i \lambda_j (N_i^2 - 1)(N_j^2 - 1)$. Since $\mathbb{E}(N_i^2 - 1) = 0$, it follows that $\mathbb{E}[(N_i^2 - 1)(N_j^2 - 1)] = 0$ for $i \neq j$. Moreover, $\mathbb{E}[(N_1^2 - 1)^2] = \mathbb{E}(N_1^4) - 1 = 2$ yields $\mathbb{E}[(Y_m - Y_s)^2] = 2 \sum_{i=s+1}^{m} \lambda_i^2 \leq 2 \sum_{i=s+1}^{\infty} \lambda_i^2$, and the assertion follows.

8.11 Solution According to the definition of the empirical distribution function, we have

$$\omega_n^2 = \frac{1}{n} \sum_{i=1}^{n} \sum_{j=1}^{n} \int_0^1 \big(1\{X_i \leq t\} - t\big)\big(1\{X_j \leq t\} - t\big) dt$$

$$= \frac{1}{n} \sum_{i=1}^{n} \sum_{j=1}^{n} \int_0^1 \Big(1\{\max(X_i, X_j) \leq t\} - t1\{X_i \leq t\} - t1\{X_j \leq t\} + t^2\Big) dt$$

$$= \frac{1}{n} \sum_{i=1}^{n} \sum_{j=1}^{n} \Big(1 - \max(X_i, X_j) - \frac{1}{2}\big(1 - X_i^2\big) - \frac{1}{2}\big(1 - X_j^2\big) + \frac{1}{3}\Big)$$

$$= \frac{1}{n} \sum_{i=1}^{n} \sum_{j=1}^{n} h(X_i, X_j) = (n-1)U_n + \frac{1}{n} \sum_{j=1}^{n} h(X_j, X_j) = (n-1)U_n + \frac{1}{6} + o_{\mathbb{P}}(1).$$

8.12 Solution (a) $U_{m,n} := \binom{m}{2}^{-1} \frac{1}{n} \sum_{1 \leq i < j \leq m} \sum_{\ell=1}^{n} 1\{X_i < Y_\ell, X_j < Y_\ell\}$ is an unbiased estimator of ϑ. (b) With $h(x_1, x_2, y) := 1\{x_1 < y, x_2 < y\}$, we have $h_{1,0}(x_1) = \mathbb{P}(x_1 < Y_1, X_2 < Y_1) = \int_{x_1}^{1} \int_0^y \frac{1}{2} dx dy = \frac{1}{4}(1 - x_1^2)$, $0 \leq x_1 \leq 1$. Thus, $\sigma_{1,0}^2 = \mathbb{V}(h_{1,0}(X_1)) = \frac{1}{16}\mathbb{V}(X_1^2) = \frac{1}{180}$. Further, $h_{0,1}(y_1) = \mathbb{P}(X_1 < y_1, X_2 < y_1) = \mathbb{P}(X_1 < y_1)^2 = y_1^2$ for $0 \leq y_1 \leq 1$ and $h_{0,1}(y_1) = 0$, if $y_1 < 0$. Hence, $\sigma_{0,1}^2 = \mathbb{V}(h_{0,1}(Y_1)) = \mathbb{V}(Y_1^2 1\{Y_1 \geq 0\}) = \mathbb{E}[Y_1^4 1\{Y_1 \geq 0\}] - \big(\mathbb{E}[Y_1^2 1\{Y_1 \geq 0\}]\big)^2 = \frac{1}{10} - \frac{1}{36} = \frac{13}{180}$. Since $\vartheta = \frac{1}{6}$ under the specific assumptions $X_1 \sim U(0, 1)$ and $Y_1 \sim U(-1, 1)$, Theorem 8.9 gives

$$\sqrt{m+n}\Big(U_{m,n} - \frac{1}{6}\Big) \xrightarrow{\mathcal{D}} N\Big(0, \frac{1}{45\tau} + \frac{13}{(1-\tau)180}\Big).$$

8.13 Solution According to (8.48), we have $\sqrt{n}(V_n - \vartheta) = \sqrt{n}(U_n^{(2)} - \vartheta) + \sqrt{n}(V_n - U_n^{(2)})$. In the case $\sigma_1^2 > 0$, Theorem 8.9 implies $\sqrt{n}(U_n^{(2)} - \vartheta) \xrightarrow{\mathcal{D}} N(0, 4\sigma_1^2)$. Further, $\sqrt{n}(V_n - U_n^{(2)}) = n^{-1/2}(U_n^{(1)} - U_n^{(2)}) = o_{\mathbb{P}}(1)$, so (8.49) follows from Slutsky's lemma. In the case $0 = \sigma_1^2 < \sigma_2^2$, we use the representation $n(V_n - \vartheta) = n(U_n^{(2)} - \vartheta) + U_n^{(1)} - U_n^{(2)}$ that results from (8.48). With Theorem 8.17, $n(U_n^{(2)} - \vartheta) \xrightarrow{\mathcal{D}} \sum_{j=1}^{\infty} \lambda_j (N_j^2 - 1)$, which entails $U_n^{(2)} = \vartheta + o_\mathbb{P}(1)$. By the law of large numbers, $U_n^{(1)} = \mu + o_\mathbb{P}(1)$. This leads to (8.50).

9.1 Solution Let $T : \{0, 1, \ldots, \ell\} \to \mathbb{R}$ be an arbitrary function. If T is unbiased for $\gamma(\vartheta) = 1/\vartheta$ then

$$\mathbb{E}_\vartheta\big(T(X)\big) = \sum_{k=0}^{\ell} T(k) \binom{\ell}{k} \vartheta^k (1 - \vartheta)^{\ell-k} = \frac{1}{\vartheta}$$

for each $\vartheta \in (0, 1)$. This equation cannot hold for each such ϑ since the sum converges to $T(0)$ as $\vartheta \to 0$, whereas $\frac{1}{\vartheta}$ tends to infinity.

9.2 Solution Let $\gamma(\vartheta) =: (\gamma_1(\vartheta), \ldots, \gamma_k(\vartheta))$. We have to show $T_{n,j} \xrightarrow{\mathbb{P}_\vartheta} \gamma_j(\vartheta)$ for each $j \in \{1, \ldots, k\}$ and each $\vartheta \in \Theta$. Let $\varepsilon > 0$, $j \in \{1, \ldots, k\}$, and $\vartheta \in \Theta$ be arbitrary. Since $\mathbb{E}_\vartheta(T_{n,j}) \to \gamma_j(\vartheta)$, we have $|\mathbb{E}(T_{n,j}) - \gamma_j(\vartheta)| \le \frac{\varepsilon}{2}$ for sufficiently large n. For every such n, the triangle inequality and Chebyshev's inequality yield

$$\mathbb{P}_\vartheta\big(|T_{n,j} - \gamma_j(\vartheta)| > \varepsilon\big) \le \mathbb{P}_\vartheta\left(|T_{n,j} - \mathbb{E}(T_{n,j})| > \frac{\varepsilon}{2}\right) \le \frac{4\mathbb{V}_\vartheta(T_{n,j})}{\varepsilon^2}$$

and thus the claim, since $\mathbb{V}_\vartheta(T_{n,j}) \to 0$ as $n \to \infty$.

9.3 Solution No. Suppose that $\mathbb{E}_\vartheta(T_n^2) = \vartheta^2$ for each $\vartheta \in \Theta$. Then $\mathbb{V}_\vartheta(T_n) = \mathbb{E}_\vartheta(T_n^2) - (\mathbb{E}_\vartheta(T_n))^2 = 0$, and thus there would be a $c \in \mathbb{R}$ with $\mathbb{P}_\vartheta(T_n = c) = 1$ for each $\vartheta \in \Theta$. This contradicts the assumption that T_n is an unbiased estimator of ϑ if Θ has at least two elements. If, in generalization of the square function, $g : \mathbb{R} \to \mathbb{R}$ is a strictly convex function, then according to Jensen's inequality, $\mathbb{E}g(X) > g(\mathbb{E}X)$ (Theorem 1.7), we would have $\mathbb{E}_\vartheta\big(g(T_n)\big) > g\big(\mathbb{E}_\vartheta(T_n)\big) = g(\vartheta)$. In all these considerations, we have tacitly assumed that all expectations exist.

9.4 Solution (a) and (b): Let $M_n := \max(X_1, \ldots, X_n)$. Then $M_n \sim \vartheta \tilde{M}_n$ with $\tilde{M}_n := \max(U_1, \ldots, U_n)$, where $U_1, \ldots, U_n$ are i.i.d. and $U_1 \sim U(0, 1)$. It follows that $\mathbb{E}_\vartheta(S_n^k) = ((n + 1)/n)^k \vartheta^k \mathbb{E}(\tilde{M}_n^k)$. The distribution function of $\tilde{M}_n$ is $F_n(t) = t^n$, $0 \le t \le 1$. Thus, the associated density f_n is given by $f_n(t) = nt^{n-1}$, $0 \le t \le 1$. It follows that $\mathbb{E}(\tilde{M}_n^k) = \int_0^1 nt^{k+n-1} \, dt = \frac{n}{n+k}$, and we thus obtain (a) and (b) by direct calculation.

(c): Let $t \in \mathbb{R}$. We have $\mathbb{P}_\vartheta(n(S_n - \vartheta) \le t) = \mathbb{P}_\vartheta\big(M_n \le \frac{n}{n+1}(\vartheta + \frac{t}{n})\big) = \mathbb{P}\big(\tilde{M}_n \le \frac{n}{n+1}(1 + \frac{t}{n\vartheta})\big)$. The last probability is independent of n and equals one if $t \ge \vartheta$. Otherwise, it follows that

$$\lim_{n \to \infty} \mathbb{P}\left(\tilde{M}_n \le \frac{n}{n+1}\left(1 + \frac{t}{n\vartheta}\right)\right) = \lim_{n \to \infty}\left(\left(1 - \frac{1}{n+1}\right)\left(1 + \frac{t}{\vartheta n}\right)\right)^n = e^{-1}e^{t/\vartheta}.$$

9.5 Solution Since trivially $\mathbb{P}_\vartheta(M_n \leq \vartheta) = 1$, we only need to show $\mathbb{P}_\vartheta(\vartheta \leq M_n \alpha^{-1/n}) = 1 - \alpha$. Due to $\mathbb{P}_\vartheta(M_n \leq t) = \mathbb{P}_\vartheta(X_1 \leq t)^n = \left(\frac{t}{\vartheta}\right)^n$, it follows that $\mathbb{P}_\vartheta(\vartheta \leq M_n \alpha^{-1/n}) = 1 - \mathbb{P}_\vartheta(M_n \leq \vartheta \alpha^{1/n}) = 1 - (\alpha^{1/n})^n = 1 - \alpha$.

9.6 Solution According to Problem 6.16, $\sqrt{n}(g(\overline{X}_n) - g(\vartheta)) \xrightarrow{\mathcal{D}_\vartheta} N(0, 1)$, where $g(x) = 2\arcsin\sqrt{x}$ and thus $\lim_{n\to\infty} \mathbb{P}_\vartheta(\sqrt{n}|g(\overline{X}_n) - g(\vartheta)| \leq z_\alpha) = 1 - \alpha$. The inequality $\sqrt{n}|g(\overline{X}_n) - g(\vartheta)| \leq z_\alpha$ is equivalent to

$$\sin^2\left(\arcsin\sqrt{\overline{X}_n} - \frac{z_\alpha}{2\sqrt{n}}\right) \leq \vartheta \leq \sin^2\left(\arcsin\sqrt{\overline{X}_n} + \frac{z_\alpha}{2\sqrt{n}}\right),$$

from which the claim follows.

9.7 Solution

(a) We assume w.l.o.g. $\mathbb{E}(X_1) = 0_d$. Thus, $\Sigma = \mathbb{E}[X_1 X_1^\top]$. Further, $\mathbb{E}[X_i X_j^\top] = 0_{d\times d}$ for every pair (i, j) with $i \neq j$. Since taking expectations is a linear operator, symmetry arguments yield

$$\mathbb{E}(S_n) = \mathbb{E}[(X_1 - \overline{X}_n)(X_1 - \overline{X}_n)^\top] = \mathbb{E}[X_1 X_1^\top] - \mathbb{E}[\overline{X}_n X_1^\top]$$

$$-\mathbb{E}[X_1 \overline{X}_n^\top] + \mathbb{E}[\overline{X}_n \overline{X}_n^\top]$$

$$= \Sigma - \frac{1}{n}\sum_{j=1}^n \mathbb{E}[X_j X_1^\top] - \frac{1}{n}\sum_{j=1}^n \mathbb{E}[X_1 X_j^\top] + \frac{1}{n^2}\sum_{i,j=1}^n \mathbb{E}[X_i X_j^\top]$$

$$= \Sigma - \frac{1}{n}\Sigma - \frac{1}{n}\Sigma + \frac{1}{n}\Sigma = \frac{n-1}{n}\Sigma.$$

(b) A direct calculation yields $S_n = n^{-1}\sum_{j=1}^n X_j X_j^\top - \overline{X}_n \overline{X}_n^\top$. According to the strong law of large numbers for random vectors (cf. Problem 1.3), $n^{-1}\sum_{j=1}^n X_j X_j^\top \xrightarrow{\text{a.s.}} \mathbb{E}[X_1 X_1^\top]$ and $\overline{X}_n \xrightarrow{\text{a.s.}} \mathbb{E}(X_1)$, which implies the claim.

10.1 Solution Due to $\nu \ll \tilde{\nu}$, according to the Radon–Nikodým theorem (Theorem 1.32) there is a non-negative measurable function $g : \mathcal{X}_0 \to \mathbb{R}$ with $\mu(B) = \int_B g(x)\tilde{\nu}(dx)$ for each $B \in \mathcal{B}_0$. This implies

$$Q_\vartheta(B) = \int_B f(x, \vartheta)\,\nu(dx) = \int_B f(x, \vartheta)g(x)\,\tilde{\nu}(dx).$$

Furthermore, $Q_\vartheta(B) = \int_B \tilde{f}(x, \vartheta)\,\tilde{\nu}(dx)$, $B \in \mathcal{B}_0$. According to a result of integration theory (see, e.g., [KLE], Th. 7.29), it follows that $f(x, \vartheta)g(x) = \tilde{f}(x, \vartheta)$ $\tilde{\nu}$-a.e. In view of $\nu \ll \tilde{\nu}$, this last equality also holds ν-a.e. Therefore,

$$\prod_{j=1}^{n} f(x_j, \vartheta) \prod_{j=1}^{n} g(x_j) = \prod_{j=1}^{n} \tilde{f}(x_j, \vartheta) \quad \text{for } v \otimes \ldots \otimes v\text{-almost all } (x_1, \ldots, x_n) \in \mathcal{X}_0^n.$$

Since $g(x_1) \ldots g(x_n)$ does not depend on ϑ, the claim follows.

10.2 Solution The likelihood function is

$$L_{x_1, \ldots, x_n}(\vartheta) = \frac{\vartheta^{n\alpha}}{\Gamma(\alpha)^n} \left(\prod_{j=1}^{n} x_j \right)^{\alpha-1} \exp\left(-\vartheta \sum_{j=1}^{n} x_j \right), \quad x_1, \ldots, x_n \in (0, \infty).$$

Thus, the likelihood equation reads

$$0 = \frac{\partial}{\partial \vartheta} \log L_{x_1, \ldots, x_n}(\vartheta) = \frac{n\alpha}{\vartheta} - \sum_{j=1}^{n} x_j,$$

and it follows that $\widehat{\vartheta}_n(x_1, \ldots, x_n) = n\alpha / \left(\sum_{j=1}^{n} x_j \right)$. Since the sign of the derivative of the log-likelihood function changes from positive to negative at the point $\widehat{\vartheta}_n(x_1, \ldots, x_n)$, there is a maximum of the likelihood function at this point.

10.3 Solution The log-likelihood function is

$$\log \left(\prod_{j=1}^{n} f(x_j, \vartheta) \right) = n\alpha \log \lambda - n \log \Gamma(\alpha) + (\alpha - 1) \sum_{j=1}^{n} \log x_j - \lambda \sum_{j=1}^{n} x_j,$$

where $x_1 > 0, \ldots, x_n > 0$. Setting the partial derivatives with respect to α and λ to zero yields the claim.

10.4 Solution Problem 9.4 gives the limit distribution of $n(S_n - \vartheta)$, where $S_n = \frac{n+1}{n} \widehat{\vartheta}_n$. Since $n(\widehat{\vartheta}_n - \vartheta) = n(S_n - \vartheta) - \frac{n}{n+1} S_n$ and $\frac{n}{n+1} S_n \xrightarrow{\mathbb{P}_\vartheta} \vartheta$, $n(\widehat{\vartheta}_n - \vartheta)$ has a limit distribution shifted to the left by ϑ compared to the limit distribution in the above problem.

10.5 Solution

(a) Since the density is equal to $1/(\vartheta_2 - \vartheta_1)$ on the interval $[\vartheta_1, \vartheta_2]$, we have

$$L_{n,\mathbf{x}}(\vartheta) = \prod_{j=1}^{n} \left(\frac{\mathbf{1}_{[\vartheta_1, \vartheta_2]}(x_j)}{\vartheta_2 - \vartheta_1} \right) = \left(\frac{1}{\vartheta_2 - \vartheta_1} \right)^n \mathbf{1}\{\vartheta_1 \le u_n, v_n \le \vartheta_2\},$$

where $u_n := \min(x_1, \ldots, x_n)$, $v_n := \max(x_1, \ldots, x_n)$. Hence the claim made for $\widehat{\vartheta}_n$ follows.

(b) Fix any $s, t > 0$, and let n be so large that $\vartheta_1 + \frac{s}{n} < \vartheta_2 - \frac{t}{n}$. With $A_{n,j} := \{X_j \le \vartheta_1 + \frac{s}{n}\}$ and $B_{n,j} := \{X_j \ge \vartheta_2 - \frac{t}{n}\}$ as well as $A_n := \bigcup_{j=1}^n A_{n,j}$ and $B_n := \bigcup_{j=1}^n B_{n,j}$, we then get

$$\mathbb{P}_\vartheta\big(n(U_n - \vartheta_1) \le s, n(\vartheta_2 - V_n) \le t\big) = \mathbb{P}_\vartheta(A_n \cap B_n) = \mathbb{P}_\vartheta(A_n)$$
$$+ \mathbb{P}_\vartheta(B_n) - \mathbb{P}_\vartheta(A_n \cup B_n).$$

Since $X_1, \ldots, X_n$ are independent and each uniformly distributed on $[\vartheta_1, \vartheta_2]$, we have

$$\mathbb{P}_\vartheta(A_n) = 1 - \left(\mathbb{P}_\vartheta\left(X_1 > \vartheta_1 + \frac{s}{n}\right)\right)^n = 1 - \left(\frac{\Delta - \frac{s}{n}}{\Delta}\right)^n$$
$$= 1 - \left(1 - \frac{s}{\Delta n}\right)^n \to 1 - e^{-s/\Delta}.$$

Similarly, $\mathbb{P}_\vartheta(B_n) \to e^{-t/\Delta}$ and $\mathbb{P}_\vartheta(A_n \cup B_n) \to e^{-(s+t)/\Delta}$, which proves the assertion.

10.6 Solution According to (5.3), the density of X_1 is

$$f(x, \vartheta) = \frac{1}{(2\pi)^{d/2}\sqrt{\det(\Sigma)}} \exp\left(-\frac{1}{2}(x - \vartheta)^\top \Sigma^{-1}(x - \vartheta)\right), \qquad x \in \mathbb{R}^d,$$

in which the dependence of $f(x, \vartheta)$ on Σ has been suppressed in the notation. Since the dependence on ϑ of the joint density $\prod_{j=1}^n f(x_j, \vartheta)$ of $X_1, \ldots, X_n$ is only in the exponential term

$$\prod_{j=1}^n \exp\left(-\frac{1}{2}(x_j - \vartheta)^\top \Sigma^{-1}(x_j - \vartheta)\right) = \exp\left(-\frac{1}{2}\sum_{j=1}^n (x_j - \vartheta)^\top \Sigma^{-1}(x_j - \vartheta)\right),$$

we have to *minimize* the sum $\sum_{j=1}^n (x_j - \vartheta)^\top \Sigma^{-1}(x_j - \vartheta)$ with respect to ϑ due to the *negative* sign. With $\bar{x}_n := n^{-1}\sum_{j=1}^n x_j$ and $\sum_{j=1}^n (x_j - \bar{x}_n) = 0_d$ we use $x_j - \vartheta = x_j - \bar{x}_n + \bar{x}_n - \vartheta$ to obtain

$$\sum_{j=1}^n (x_j - \vartheta)^\top \Sigma^{-1}(x_j - \vartheta) = \sum_{j=1}^n (x_j - \bar{x}_n)^\top \Sigma^{-1}(x_j - \bar{x}_n) + n(\bar{x}_n - \vartheta)^\top \Sigma^{-1}(\bar{x}_n - \vartheta).$$

Since Σ^{-1} is positive definite, $(\bar{x}_n - \vartheta)^\top \Sigma^{-1}(\bar{x}_n - \vartheta)$ attains its minimum value if $\vartheta := \bar{x}_n$. The ML estimator of ϑ based on $X_1, \ldots, X_n$ is therefore $\overline{X}_n = n^{-1}\sum_{j=1}^n X_j$.

According to the addition theorem (Theorem 5.13) for the multivariate normal distribution, $\sum_{j=1}^{n} X_j \sim N_d(n\vartheta, n\Sigma)$, and from the reproduction theorem (Corollary 5.7), it follows that $\sqrt{n}(\overline{X}_n - \vartheta) \sim N_d(0_d, \Sigma)$.

10.7 Solution Since $\int_0^\infty e^{-u}\, du = 1 = \int_0^\infty u e^{-u}\, du$, we have

$$I_{KL}(\vartheta : \vartheta') = \int_0^\infty \log\left(\frac{\vartheta e^{-\vartheta x}}{\vartheta' e^{-\vartheta' x}}\right) \vartheta e^{-\vartheta x}\, dx = \log\frac{\vartheta}{\vartheta'} - (\vartheta - \vartheta')\frac{1}{\vartheta}.$$

10.8 Solution With the hint we get

$$
\begin{aligned}
I_{KL}(\vartheta + \varepsilon : \vartheta) &= \int_{\mathcal{X}_0} \frac{f(x, \vartheta + \varepsilon)}{f(x, \vartheta)} \log \frac{f(x, \vartheta + \varepsilon)}{f(x, \vartheta)} f(x, \vartheta)\, \nu(dx) \\
&= \int_{\mathcal{X}_0} \left(\frac{f(x, \vartheta + \varepsilon)}{f(x, \vartheta)} - 1 + \frac{1}{2}\left(\frac{f(x, \vartheta + \varepsilon)}{f(x, \vartheta)} - 1\right)^2 \right) \\
&\qquad \times f(x, \vartheta)\, \nu(dx) + o(\varepsilon^2) \\
&= \frac{1}{2}\int_{\mathcal{X}_0} \frac{(f(x, \vartheta + \varepsilon) - f(x, \vartheta))^2}{f(x, \vartheta)}\, \nu(dx) + o(\varepsilon^2).
\end{aligned}
$$

Since $f(x, \vartheta + \varepsilon) - f(x, \vartheta) = \varepsilon \frac{d}{d\vartheta} f(x, \vartheta) + o(\varepsilon)$ and $\left(\frac{d}{d\vartheta} f(x, \vartheta)\right)^2 / f(x, \vartheta) = \left(\frac{d}{d\vartheta} \log f(x, \vartheta)\right)^2 f(x, \vartheta)$, the assertion follows.

10.9 Solution Let $\varepsilon := \min\{I_{KL}(\vartheta_0 : \vartheta) : \vartheta \in \Theta, \vartheta \neq \vartheta_0\} > 0$. Furthermore, fix any $\mathbf{x} = (x_1, x_2, \ldots) \in M$. The set M is an intersection of finitely many sets M_ϑ. By definition of M_ϑ, there is an integer n_0 that depends on ε such that for every $n \geq n_0$:

$$\frac{1}{n}\sum_{j=1}^{n} \log \frac{f(x_j, \vartheta)}{f(x_j, \vartheta_0)} \leq -\frac{\varepsilon}{2} \quad \text{for every } \vartheta \in \Theta \text{ with } \vartheta \neq \vartheta_0.$$

This statement is equivalent to

$$\prod_{j=1}^{n} \frac{f(x_j, \vartheta_0)}{f(x_j, \vartheta)} \geq \exp\left(\frac{n\varepsilon}{2}\right) > 1 \quad \text{for every } \vartheta \in \Theta \text{ with } \vartheta \neq \vartheta_0$$

for every $n \geq n_0$. Hence, $\widehat{\vartheta}_n(\mathbf{x}) = \vartheta_0$ for every $n \geq n_0$.

10.10 Solution Condition (a) holds, since Θ is a closed and bounded interval and thus a compact set. To prove (b), we must show that for every $x \in \mathcal{X}_0 := \mathbb{R}$ the function $[1, 2] = \Theta \ni \vartheta \mapsto f(x, \vartheta) = \frac{1}{\vartheta} \mathbf{1}_{[0,\vartheta]}(x)$ is upper semi-continuous. If $x < 0$ or $x > 2$, then $f(x, \vartheta) = 0$ for every $\vartheta \in [1, 2]$. For such x, $f(x, \cdot)$ is therefore continuous and in particular upper semi-continuous. In the case $0 \le x \le 1$, one argues in the same way, because then $f(x, \vartheta) = \frac{1}{\vartheta}$, $\vartheta \in \Theta$. In the remaining case $1 < x \le 2$, we have

$$f(x, \vartheta) = \begin{cases} 0, & \text{if } 1 \le \vartheta < x, \\ \frac{1}{\vartheta}, & \text{if } x \le \vartheta \le 2. \end{cases}$$

This function is upper semi-continuous. If one were to set the density on the open interval $(0, 1)$ to $\frac{1}{\vartheta}$ and outside to 0, then in the case $1 < x \le 2$ with 0, if $1 \le \vartheta \le x$ and $\frac{1}{\vartheta}$, if $x < \vartheta \le 2$, there would be *no* upper semi-continuous function of ϑ. To show (c), a direct calculation yields

$$M(x) := \max_{1 \le \vartheta \le 2} \frac{f(x, \vartheta)}{f(x, \vartheta_0)} = \begin{cases} \vartheta_0, & \text{if } x \le 1, \\ \frac{\vartheta_0}{x}, & \text{if } 1 < x \le \vartheta_0, \\ \infty, & \text{if } x > \vartheta_0. \end{cases}$$

With $K(x) := \log M(x)$ we have $V(x, \vartheta) \le K(x)$ for every x and every ϑ. Furthermore, $\mathbb{E}_{\vartheta_0}|K(X)| < \infty$, because $\infty \cdot \mathbb{P}_{\vartheta_0}(X > \vartheta_0) = \infty \cdot 0 = 0$.

It remains to provide the proof of (d). If $1 < \vartheta < 2$, and if r is so small that the inequalities $\vartheta - r > 1$ and $\vartheta + r < 2$ are satisfied, then a straightforward algebra yields

$$\overline{f}_{\vartheta, r}(x) = \begin{cases} 0, & \text{if } x < 0 \text{ or } x > \vartheta + r, \\ \frac{1}{\vartheta - r}, & \text{if } 0 \le x < \vartheta - r, \\ \frac{1}{x}, & \text{if } |\vartheta - x| \le r. \end{cases}$$

As a piecewise continuous function, $\overline{f}_{\vartheta, r}$ is measurable and integrable with respect to $U(0, \vartheta_0)$. In the case $\vartheta = 1$, the supremum in (10.19) runs over all ϑ' with $1 \le \vartheta' < 1 + r$. Here, a direct calculation yields $\overline{f}_{1, r}(x) = 0$, if $x < 0$ or $x > 1 + r$, $\overline{f}_{1, r}(x) = 1$, if $0 \le x < 1$, and $\overline{f}_{1, r}(x) = \frac{1}{x}$, if $1 \le x \le 1 + r$. Also in this case, $\overline{f}_{1, r}$ is measurable and integrable with respect to $U(0, \vartheta_0)$. In the remaining case $\vartheta = 2$, the supremum in (10.19) runs over all ϑ' with $2 - r < \vartheta' \le 2$. We have $\overline{f}_{2, r}(x) = 0$, if $x < 0$ or $x > 2$, $\overline{f}_{2, r}(x) = \frac{1}{2 - r}$, if $0 \le x \le 2 - r$, and $\overline{f}_{2, r}(x) = \frac{1}{x}$, if $2 - r < x \le 2$. Also the function $\overline{f}_{2, r}$ is measurable and integrable with respect to $U(0, \vartheta_0)$.

10.11 Solution Since $f_n(X_1, \ldots, X_n, \vartheta) = \prod_{j=1}^{n} f(X_j, \vartheta)$, and due to the additivity when taking expectations as well as $\mathbb{E}_\vartheta[\frac{d}{d\vartheta} \log f(X_j, \vartheta)] = 0_k$, we have $\mathbb{E}_\vartheta[U_n(\vartheta)] = 0_k$. Since $U_n(\vartheta) = \sum_{j=1}^{n} \frac{d}{d\vartheta} \log f(X_j, \vartheta)$ is a sum of independent random vectors, the covariance matrix of $U_n(\vartheta)$ is equal to $n I_1(\vartheta)$ according to Problem 5.5.

10.12 Solution In the situation of Example 10.3, $\mathcal{X}_0 = \{0, 1\}$, $\nu = \delta_0 + \delta_1$, $f(x, \vartheta) = \vartheta^x(1 - \vartheta)^{1-x}$ and $\Theta^\circ = (0, 1)$. We have $\log f(x, \vartheta) = x \log \vartheta + (1 - x) \log(1 - \vartheta)$ and $\frac{d}{d\vartheta} \log f(x, \vartheta) = \frac{x}{\vartheta} - \frac{1-x}{1-\vartheta}$. If $X \sim \text{Bin}(1, \vartheta)$, then $X^2 = X$, $(1 - X)^2 = 1 - X$ and $X(1 - X) = 0$, and it follows that

$$I_1(\vartheta) = \mathbb{E}_\vartheta \left(\frac{X}{\vartheta} - \frac{1-X}{1-\vartheta} \right)^2 = \frac{\mathbb{E}_\vartheta(X)}{\vartheta^2} + \frac{1 - \mathbb{E}_\vartheta(X)}{(1-\vartheta)^2} = \frac{1}{\vartheta} + \frac{1}{1-\vartheta} = \frac{1}{\vartheta(1-\vartheta)}.$$

10.13 Solution

(a) We have

$$\frac{\partial \log f(X, \vartheta)}{\partial \vartheta_1} = -\frac{1}{\vartheta_1 + \vartheta_2} + \frac{X \mathbf{1}_{[0,\infty)}(X)}{\vartheta_1^2}, \quad \frac{\partial \log f(X, \vartheta)}{\partial \vartheta_2}$$

$$= -\frac{1}{\vartheta_1 + \vartheta_2} - \frac{X \mathbf{1}_{(-\infty,0)}(X)}{\vartheta_2^2}.$$

With $\mathbb{E}_\vartheta \left(X^2 \mathbf{1}_{[0,\infty)}(X) \right) = (\vartheta_1 + \vartheta_2)^{-1} \int_0^\infty x^2 \exp(-x/\vartheta_1) \, dx = 2\vartheta_1^3/(\vartheta_1 + \vartheta_2)$ and $\vartheta_1^2/(\vartheta_1 + \vartheta_2) = \mathbb{E}_\vartheta \left(X \mathbf{1}_{[0,\infty)}(X) \right)$ it follows that $\mathbb{E}_\vartheta \left[(\frac{\partial}{\partial \vartheta_1} \log f(X, \vartheta))^2 \right] = (\vartheta_1 + 2\vartheta_2)/(\vartheta_1(\vartheta_1 + \vartheta_2)^2)$. In the same way, we get $\mathbb{E}_\vartheta \left[(\frac{\partial}{\partial \vartheta_2} \log f(X, \vartheta))^2 \right] = (2\vartheta_1 + \vartheta_2)/(\vartheta_2(\vartheta_1 + \vartheta_2)^2)$. Because $\mathbf{1}_{[0,\infty)}(X) \mathbf{1}_{(-\infty,0)}(X) = 0$ we obtain $\mathbb{E}_\vartheta \left[\frac{\partial}{\partial \vartheta_1} \log f(X, \vartheta) \cdot \frac{\partial}{\partial \vartheta_2} \log f(X, \vartheta) \right] = -1/(\vartheta_1 + \vartheta_2)^2$.

(b) The likelihood equations are

$$-\frac{n}{\vartheta_1 + \vartheta_2} + \frac{s_n}{\vartheta_1^2} = 0, \quad -\frac{n}{\vartheta_1 + \vartheta_2} + \frac{t_n}{\vartheta_2^2} = 0.$$

This implies that $s_n/\vartheta_1^2 = t_n/\vartheta_2^2$ and thus $\vartheta_2 = \sqrt{t_n/s_n}\vartheta_1$. Substituting this expression for ϑ_2 into the first equation yields the claimed equation for $\widehat{\vartheta}_{n,1}(\mathbf{x})$ after multiplying the denominators. The second equation then follows from $\widehat{\vartheta}_{n,2}(\mathbf{x}) = \sqrt{t_n(\mathbf{x})/s_n(\mathbf{x})}\widehat{\vartheta}_{n,1}(\mathbf{x})$.

(c) Let $a(\vartheta) := \mathbb{E}_\vartheta[X \mathbf{1}_{[0,\infty)}(X)] = \vartheta_1^2/(\vartheta_1 + \vartheta_2)$, $b(\vartheta) := \mathbb{E}_\vartheta[-X \mathbf{1}_{(-\infty,0)}(X)] = \vartheta_2^2/(\vartheta_1 + \vartheta_2)$. By the strong law of large numbers, we have

$$\widehat{\vartheta}_{n,1} = \sqrt{\frac{S_n}{n}} \left(\sqrt{\frac{S_n}{n}} + \sqrt{\frac{T_n}{n}} \right) \to \sqrt{a(\vartheta)} \left(\sqrt{a(\vartheta)} + \sqrt{b(\vartheta)} \right) = \vartheta_1 \quad \mathbb{P}_\vartheta\text{-a.s.}$$

In the same way, $\widehat{\vartheta}_{n,2} \to \vartheta_2$ $\mathbb{P}_\vartheta$-a.s., which shows the claimed consistency.

10.14 Solution According to the main theorem, $\sqrt{n}(\widehat{\vartheta}_n - \vartheta) \xrightarrow{D_\vartheta} N(0, 1/I_1(\vartheta))$ and therefore $\sqrt{I_1(\vartheta)}\sqrt{n}(\widehat{\vartheta}_n - \vartheta) \xrightarrow{D_\vartheta} N(0, 1)$, $\vartheta \in \Theta$. Due to the continuity of $I_1(\cdot)$ and $\widehat{\vartheta}_n \xrightarrow{P_\vartheta} \vartheta$, it follows that $\sqrt{I_1(\widehat{\vartheta}_n)} \xrightarrow{P_\vartheta} \sqrt{I_1(\vartheta)}$, and thus Slutsky's lemma gives $\sqrt{I_1(\widehat{\vartheta}_n)}\sqrt{n}(\widehat{\vartheta}_n - \vartheta) \xrightarrow{D_\vartheta} N(0, 1)$, $\vartheta \in \Theta$. This leads to the assertion.

If $X_1 \sim \text{Bin}(1, \vartheta)$, where $\vartheta \in \Theta := (0, 1)$, then the density of X_1 with respect to the counting measure on $\{0, 1\}$ is given by $f(x, \vartheta) = \vartheta^x(1 - \vartheta^{1-x})$ for $x \in \{0, 1\}$ and $f(x, \vartheta) := 0$, otherwise. For $x \in \{0, 1\}$, we have

$$\frac{\partial}{\partial \vartheta} \log f(x, \vartheta) = \frac{x - \vartheta}{\vartheta(1 - \vartheta)}$$

and thus $I_1(\vartheta) = \mathbb{V}_\vartheta\left(\frac{X_1 - \vartheta}{\vartheta(1-\vartheta)}\right) = \frac{1}{\vartheta(1-\vartheta)}$. It follows that $(\widehat{\vartheta}_n(1 - \widehat{\vartheta}_n))^{-1/2}\sqrt{n}(\widehat{\vartheta}_n - \vartheta) \xrightarrow{D_\vartheta} N(0, 1)$, $\vartheta \in \Theta$, and we obtain

$$\lim_{n\to\infty} P_\vartheta\left(\widehat{\vartheta}_n - \frac{\Phi^{-1}(1-\alpha/2)}{\sqrt{n}}\sqrt{\widehat{\vartheta}_n(1-\widehat{\vartheta}_n)} \leq \vartheta \leq \widehat{\vartheta}_n\right.$$
$$\left. + \frac{\Phi^{-1}(1-\alpha/2)}{\sqrt{n}}\sqrt{\widehat{\vartheta}_n(1-\widehat{\vartheta}_n)}\right) = 1 - \alpha,$$

$\vartheta \in \Theta$. Here, $\widehat{\vartheta}_n = \frac{1}{n}\sum_{j=1}^n X_j$ is the relative frequency of successes.

11.1 Solution We have $f(x, \vartheta) = e^{-\vartheta}\vartheta^x/x!$, $x \in \mathbb{N}_0$, and thus $\left(\frac{\partial}{\partial \vartheta} \log f(x, \vartheta)\right)^2 = \left(\frac{x}{\vartheta} - 1\right)^2$. Therefore, $I_1(\vartheta) = \vartheta^{-2}\mathbb{E}_\vartheta[(X - \vartheta)^2] = \vartheta^{-2}\mathbb{V}_\vartheta(X) = \frac{1}{\vartheta}$. Here, X has the Poisson distribution $\text{Po}(\vartheta)$. The estimator T_n is unbiased for ϑ, and $\mathbb{V}_\vartheta(T_n) = \frac{\vartheta}{n} = \frac{1}{nI_1(\vartheta)}$, $\vartheta \in \Theta$. Since every other unbiased estimator satisfies the information inequality(11.3), the assertion follows.

11.2 Solution We have $f(x, \vartheta) = \vartheta^x(1 - \vartheta)^{1-x}$, $x \in \{0, 1\}$, cf. Example 10.3. If $0 < \vartheta < 1$, then $\frac{\partial}{\partial \vartheta} \log f(x, \vartheta) = (x - \vartheta)/(\vartheta(1 - \vartheta))$ and thus

$$I_1(\vartheta) = \mathbb{E}_\vartheta\left[\left(\frac{X_1 - \vartheta}{\vartheta(1 - \vartheta)}\right)^2\right] = \frac{\mathbb{V}_\vartheta(X_1)}{\vartheta^2(1 - \vartheta)^2} = \frac{1}{\vartheta(1 - \vartheta)}.$$

The estimator $T_n = n^{-1}\sum_{j=1}^n X_j$ is unbiased for ϑ, and $\mathbb{V}_\vartheta(T_n) = \frac{\vartheta(1-\vartheta)}{n} = \frac{1}{nI_1(\vartheta)}$. Since $\mathbb{V}_\vartheta(T_n) = 0$ if $\vartheta = 0$ or $\vartheta = 1$ and any other unbiased estimator of ϑ satisfies the information inequality(11.3), the assertion follows.

11.3 Solution

(a) We have $f(x, \vartheta) = (2\pi\vartheta)^{-1/2} \exp\left(-\frac{(x-\mu)^2}{2\vartheta}\right)$ and thus $\frac{\partial}{\partial\vartheta}\log f(x, \vartheta) = \left((x-\mu)^2 - \vartheta\right)/(2\vartheta^2)$. With $X \sim N(\mu, \vartheta)$, we get $I_1(\vartheta) = \mathbb{E}_\vartheta\left[\left((X-\mu)^2 - \vartheta\right)^2\right]/(4\vartheta^4)$, and due to $\mathbb{E}_\vartheta\left[(X-\mu)^4\right] = 3\vartheta^2$ as well as $\mathbb{E}_\vartheta\left[(X-\mu)^2\right] = \vartheta$, we obtain $I_1(\vartheta) = 1/(2\vartheta^2)$.

(b) Due to $\mathbb{E}_\vartheta\left[(X-\mu)^2\right] = \vartheta$, T_n is unbiased for ϑ, and with $\mathbb{E}_\vartheta\left[(X-\mu)^4\right] = 3\vartheta^2$, we get $\mathbb{V}_\vartheta(T_n) = \frac{1}{n}\mathbb{V}_\vartheta(X-\mu)^2 = \frac{1}{n}\left(3\vartheta^2 - \vartheta^2\right) = (2\vartheta^2)/n = 1/(nI_1(\vartheta))$.

(c) Since $S_n = \frac{n}{n+2}T_n$, we obtain $\mathbb{E}_\vartheta\left[(S_n - \vartheta)^2\right] = \mathbb{E}\left[\left(\frac{n}{n+2}(T_n - \vartheta) - \frac{2}{n+2}\vartheta\right)^2\right] = \left(\frac{n}{n+2}\right)^2\frac{2\vartheta^2}{n} + \frac{4}{(n+2)^2}\vartheta^2$, and thus $\mathbb{E}_\vartheta\left[(S_n - \vartheta)^2\right] = \frac{2\vartheta^2}{n+2}$, $\vartheta \in \Theta$.

11.4 Solution

(a) For each $c \in \mathbb{R}^k$, we have $0 \leq \mathbb{E}\left[\left(c^\top Y\right)^2\right] = \mathbb{E}(c^\top Y Y^\top c) = c^\top\mathbb{E}\left[YY^\top\right]c$.

(b) Let $Y =: (Y_1, \ldots, Y_k)^\top$. From $\mathbb{E}(YY^\top) = 0_{k\times k}$, it follows that $\mathbb{E}(Y_j^2) = 0$ and thus $\mathbb{P}(Y_j = 0) = 1$ for each $j \in \{1, \ldots, k\}$, so $Y = 0_k$ $\mathbb{P}$-almost surely. The reverse implication is trivial.

11.5 Solution

(a) From $f(x, \vartheta) = (2\pi)^{-1/2}\exp\left(-\frac{1}{2}(x - \vartheta)^2\right)$, it follows that $\left(\frac{\partial}{\partial\vartheta}\log f(x, \vartheta)\right)^2 = (x - \vartheta)^2$ and thus $I_1(\vartheta) = \mathbb{E}_\vartheta(X_1 - \vartheta)^2 = \mathbb{V}_\vartheta(X_1) = 1$.

(b) Let $Y_n := \mathbf{1}\{|\overline{X}_n| > n^{-1/4}\} + \frac{1}{2}\mathbf{1}\{|\overline{X}_n| \leq n^{-1/4}\}$. If $\vartheta = 0$ then $\sqrt{n}\overline{X}_n \xrightarrow{D_\vartheta} N(0, 1)$ and thus $Y_n \xrightarrow{\mathbb{P}_\vartheta} \frac{1}{2}$. With Slutsky's lemma, it follows that $\sqrt{n}(T_n - \vartheta) \xrightarrow{D_\vartheta} N(0, \frac{1}{4})$. If $\vartheta \neq 0$, we have $\sqrt{n}(T_n - \vartheta) = \sqrt{n}(\overline{X}_n - \vartheta)Y_n - \frac{\vartheta}{2}\sqrt{n}\mathbf{1}\{|\overline{X}_n| \leq n^{-1/4}\}$. Since $\overline{X}_n \xrightarrow{\mathbb{P}_\vartheta} \vartheta \neq 0$, it follows that $Y_n \xrightarrow{\mathbb{P}_\vartheta} 1$ and $\frac{\vartheta}{2}\sqrt{n}\mathbf{1}\{|\overline{X}_n| \leq n^{-1/4}\} \xrightarrow{\mathbb{P}_\vartheta} 0$. With Slutsky's lemma, we obtain $\sqrt{n}(T_n - \vartheta) \xrightarrow{D_\vartheta} N(0, 1)$.

11.6 Solution With $f(x, \vartheta) = e^{-\vartheta}\vartheta^x/x!$ ($x \in \mathbb{N}_0$), it follows that $\frac{\partial}{\partial\vartheta}\left(\log\prod_{j=1}^n f(x_j, \vartheta)\right) = -n + \frac{1}{\vartheta}\sum_{j=1}^n x_j$. Thus, the log-likelihood equation has the solution $\widehat{\vartheta}_n(x_1, \ldots, x_n) = \frac{1}{n}\sum_{j=1}^n x_j$. With $\widehat{\vartheta}_n := \widehat{\vartheta}_n(X_1, \ldots, X_n)$, we have $\widehat{\vartheta}_n \xrightarrow{\mathbb{P}_\vartheta} \vartheta$, $\vartheta \in \Theta := (0, \infty)$, and the central limit theorem gives $\sqrt{n}(\widehat{\vartheta}_n - \vartheta) \xrightarrow{D_\vartheta} N(0, \vartheta)$. Since $\left(\frac{\partial}{\partial\vartheta}\log f(X_1, \vartheta)\right)^2 = (X_1 - \vartheta)^2/\vartheta^2$, it follows that $I_1(\vartheta) = \frac{1}{\vartheta}$. Therefore, the sequence $(\widehat{\vartheta}_n)$ is asymptotically efficient.

11.7 Solution The derivation of (11.17) is based on the delta method and thus on the Taylor expansion (6.22). With this, we get $\sqrt{n}(\widehat{\vartheta}_n - \vartheta) = n^{-1/2}\sum_{j=1}^n \ell(X_j, \vartheta) + o_{\mathbb{P}_\vartheta}(1)$ as $n \to \infty$, where $\ell(X_j, \vartheta) = g'\left(g^{-1}(\vartheta)\right)\left((X_j, X_j^2, \ldots, X_j^k)^\top - (m_1, m_2, \ldots, m_k)^\top\right)$.

11.8 Solution

(a) Let k be any integer. With the substitution $u := \lambda x$, we have

$$\mathbb{E}(X_1^k) = \frac{\lambda^\alpha}{\Gamma(\alpha)} \int_0^\infty x^{k+\alpha-1} e^{-\lambda x} \, dx = \frac{\lambda^\alpha}{\Gamma(\alpha)} \frac{1}{\lambda^{k+\alpha}} \int_0^\infty u^{k+\alpha-1} e^{-u} \, du = \frac{\Gamma(k+\alpha)}{\Gamma(\alpha)\lambda^k}.$$

Because $\Gamma(x+1) = x\Gamma(x)$ for each $x > 0$, the claim follows.

(b) We set $m_j := \mathbb{E}(X_1^j)$, $j \geq 1$. Solving the equations in (a) for α and λ gives

$$\alpha = \frac{m_1^2}{m_2 - m_1^2}, \qquad \lambda = \frac{m_1}{m_2 - m_1^2}.$$

Hence, the claim follows by the definition of the method of moments estimator.

(c) We have $(\alpha, \lambda) = g(m_1, m_2) = \big(g_1(m_1, m_2), g_2(m_1, m_2)\big)$ with

$$g_1(u, v) = \frac{u^2}{v - u^2}, \qquad g_2(u, v) = \frac{u}{v - u^2}, \qquad u, v > 0, \ u > v^2.$$

The Jacobian matrix of g at the point $a = (m_1, m_2)$ and the matrix T in (11.16) are given by

$$g'(a) = \frac{1}{(m_2 - m_1^2)^2} \begin{pmatrix} 2m_1 m_2 & -m_1^2 \\ m_2 + m_1^2 & -m_1 \end{pmatrix}, \qquad T = \begin{pmatrix} m_2 - m_1^2 & m_3 - m_1 m_2 \\ m_3 - m_1 m_2 & m_4 - m_2^2 \end{pmatrix}.$$

11.9 Solution

(a) Fix any $t \in \mathbb{R}$. With $p_n(t) := F\big(F^{-1}(p) + t/\sqrt{n}\big)$ and $Z_n \sim \text{Bin}(n, p_n(t))$, (11.25) implies $\mathbb{P}\big(\sqrt{n}(X_{n:r_n} - F^{-1}(p)) \leq t\big) = \mathbb{P}\big(X_{n:r_n} \leq F^{-1}(p) + t/\sqrt{n}\big) = \mathbb{P}(Z_n \geq r_n)$. Furthermore, $\mathbb{P}(Z_n \geq r_n) = \mathbb{P}(Z_n^* \geq t_n)$, where

$$Z_n^* = \frac{Z_n - np_n(t)}{\sqrt{np_n(t)(1 - p_n(t))}}, \qquad t_n := \frac{r_n - np_n(t)}{\sqrt{np_n(1)(1 - p_n(t))}}.$$

According to Problem 1.6, $Z_n^* \xrightarrow{D} N(0, 1)$. Because $F\big(F^{-1}(p) + t/\sqrt{n}\big) = p + f\big(F^{-1}(p)\big)t/\sqrt{n} + o(1/\sqrt{n})$ and $r_n = np + o(\sqrt{n})$ it follows that

$$\lim_{n\to\infty} t_n = t^* := -\frac{f\big(F^{-1}(p)\big)t}{\sqrt{p(1-p)}}.$$

In view of Problem 1.5, we have $\mathbb{P}(Z_n^* \geq t_n) \to \mathbb{P}(Z \geq t^*)$, where $Z \sim N(0, 1)$, and thus the assertion follows.

(b) follows from (a) and the hint, since for even n the sequences $r_n := \frac{n}{2}$ and $r_n := \frac{n}{2} + 1$ both fall under the scheme (11.28) with $p = \frac{1}{2}$.

11.10 Solution The distribution function of X_1 is given by $F(x) = \frac{1}{2}e^{(x-a)/\sigma}$, if $x \leq a$, and $F(x) = 1 - \frac{1}{2}e^{-(x-a)/\sigma}$, if $x > a$. The derivative of F at the point $x = a$ is $\frac{1}{2\sigma}$. The variance $\sigma^2(F)$ of X_1 does not depend on a and is

$$\sigma^2(F) = \frac{1}{2\sigma} \int_{-\infty}^{\infty} x^2 e^{-|x|/\sigma}\, dx = \frac{1}{\sigma} \int_0^{\infty} x^2 e^{-x/\sigma}\, dx = 2\sigma^2.$$

As in Example 11.6, let S_n be the sample mean and T_n the sample median, then according to (11.27) $\mathrm{ARE}_F\big((T_n) : (S_n)\big) = 4\left(\frac{1}{2\sigma}\right)^2 2\sigma^2 = 2$. In this case, the sample median is asymptotically twice as efficient as the sample mean, which is perhaps not surprising given the shape of the density, which has a peak at the point a.

12.1 Solution With the abbreviation $\sum_j^* := \sum_{j=1,\ldots,s:N_{n,j}>0}$ and $0 \cdot \log 0 := 0$, we have

$$\log\left(\prod_{j=1}^n f(X_j, \vartheta)\right) = \sum_j^* N_{n,j} \log p_j = \sum_j^* N_{n,j} \log\left(\frac{N_{n,j}}{n} \cdot \frac{np_j}{N_{n,j}}\right)$$

$$= \sum_{j=1}^n N_{n,j} \log\frac{N_{n,j}}{n} + \sum_j^* N_{n,j} \log\left(\frac{np_j}{N_{n,j}}\right)$$

$$\leq \sum_{j=1}^n N_{n,j} \log\frac{N_{n,j}}{n} + \sum_j^* N_{n,j} \left(\frac{np_j}{N_{n,j}} - 1\right).$$

Since the last sum is less than or equal to $n \sum_j^* p_j - (N_{n,1} + \ldots + N_{n,s}) = 0$ and thus, due to $\sum_{j=1}^s N_{n,j} = n$, less than or equal to zero, the assertion follows.

12.2 Solution A Taylor expansion of the 3rd order of the function $g(t) := t \log t$ around $t = 1$ yields $t \log t = g(t) = t - 1 + \frac{1}{2}(t - 1)^2 - \frac{1}{6}(t - 1)^3/\rho^2$, where $|\rho - 1| \leq |t - 1|$. If we insert $t = \frac{N_{n,j}}{nq_j}$ into this expansion, we get

$$\frac{N_{n,j}}{nq_j} \log\left(\frac{N_{n,j}}{nq_j}\right) = \frac{N_{n,j} - nq_j}{nq_j} + \frac{1}{2}\left(\frac{N_{n,j} - nq_j}{nq_j}\right)^2 - \frac{1}{6}\left(\frac{N_{n,j} - nq_j}{nq_j}\right)^3 \frac{1}{R_{n,j}^2},$$

where

$$|R_{n,j} - 1| \le \left| \frac{N_{n,j}}{nq_j} - 1 \right|.$$ (*)

A direct calculation now yields

$$T_n - M_n = -\frac{1}{3} \sum_{j=1}^{s} \frac{(N_{n,j} - nq_j)^3}{R_{n,j}^2 n^2 q_j^2}.$$

It suffices to show that each summand converges to zero in probability as $n \to \infty$. From (*) it follows that $R_{n,j}^{-2} \le \max(1, (nq_j/N_{n,j})^2)$ and thus

$$\frac{|N_{n,j} - nq_j|^3}{R_{n,j}^2 n^2} \le \max\left(1, \left(\frac{nq_j}{N_{n,j}}\right)^2\right) \cdot \left| \frac{N_{n,j} - nq_j}{\sqrt{n}} \right|^3 \cdot \frac{1}{\sqrt{n}}.$$ (**)

With the strong law of large numbers, $\frac{N_{n,j}}{nq_j} \xrightarrow{\text{a.s.}} 1$. Due to the Lindeberg–Lévy central limit theorem and the mapping theorem, the second factor on the right-hand side of (**) is a tight sequence, from which the assertion follows.

12.3 Solution We first assume $\Sigma = I_k$. Then A is idempotent with rank r, and there exists an orthogonal $k \times k$ matrix P with $A = P^\top C P$, where

$$C := \begin{pmatrix} I_r & 0_{r \times (k-r)} \\ 0_{(k-r) \times r} & 0_{(k-r) \times (k-r)} \end{pmatrix}.$$

Let $Z := PX$. Since $P^\top P = I_k$, it follows from Theorem 5.7 that $Z \sim N_k(0_k, I_k)$. Furthermore,

$$X^\top AX = X^\top P^\top C P X = Z^\top C Z = \sum_{j=1}^{r} Z_j^2 \sim \chi_r^2,$$

where $Z =: (Z_1, \ldots, Z_k)^\top$. In the general case, we have $X \sim BY$, where $Y \sim N_k(0_k, I_k)$ and $\Sigma = BB^\top$. It follows that $X^\top AX \sim Y^\top \tilde{A} Y$, where $\tilde{A} := B^\top AB$. With $\Sigma = BB^\top$ and the fact that $A\Sigma$ is an idempotent matrix, it follows that $\tilde{A}^2 = B^\top ABB^\top AB = B^\top A\Sigma AB = B^\top A\Sigma A\Sigma\Sigma^{-1}B = B^\top A\Sigma\Sigma^{-1}B = B^\top AB = \tilde{A}$. The matrix $\tilde{A}$ is thus idempotent. By assumption, $r = \text{rank}(A\Sigma)$, where generally $\text{rank}(M)$ stands for the rank of a matrix M. If we write $\text{tr}(D)$ for the trace of a square matrix D, the assertion follows from $\text{rank}(\tilde{A}) = \text{tr}(\tilde{A}) = \text{tr}(B^\top AB) = \text{tr}(ABB^\top) = \text{tr}(A\Sigma) = \text{rank}(A\Sigma) = r$.

12.4 Solution

(a) Since $I_1(\vartheta)$ and $\tilde{I}_1(u)^{-1}$ are symmetric, the rule $(CD)^\top = D^\top C^\top$ implies

$$A^\top = \left(I_k - I_1(\vartheta) h'(u) \tilde{I}_1(u)^{-1} h'(u)^\top \right) I_1(\vartheta) = A.$$

(b) With $B := h'(u) \tilde{I}_1(u)^{-1} h'(u)^\top$, it follows from $I_1(\vartheta)^{-1} I_1(\vartheta) = I_k$ that

$$(A\Sigma)^2 = I_1(\vartheta) \{ I_k - B I_1(\vartheta) \} \{ I_k - B I_1(\vartheta) \} I_1(\vartheta)^{-1}.$$

Because of (12.12), the product of the curly brackets equals $I_k - B I_1(\vartheta)$, from which the assertion follows.

(c) Since $I_1(\vartheta)$ and $I_1(\vartheta)^{-1}$ each have full rank k, the rank of $A\Sigma$ is that of $I_k - B I_1(\vartheta)$ with B as in part (b). Due to the assumptions on the function h, the matrix $h'(u)$ has rank ℓ. Since $\tilde{I}_1(u)$ and $I_1(\vartheta)$ are invertible, $B I_1(\vartheta)$ also has rank ℓ, and thus $I_k - B I_1(\vartheta)$ has rank $k - \ell$.

12.5 Solution In view of (12.22) and $M_n = -2 \log \Lambda_n$, we have

$$M_n = 2 \left\{ \sum_{i,j} N_{ij} \log \frac{N_{ij}}{n} - \sum_i N_{i+} \log \frac{N_{i+}}{n} - \sum_j N_{+j} \log \frac{N_{+j}}{n} \right\}.$$

Since $1 = \frac{1}{n} \sum_i N_{i+} = \frac{1}{n} \sum_j N_{+j}$ and $\log(ab) = \log a + \log b$, it follows that

$$M_n = 2n \sum_{i,j} \left\{ \frac{N_{ij}}{n} \log \frac{N_{ij}}{n} - \frac{N_{i+}}{n} \cdot \frac{N_{+j}}{n} \log \left(\frac{N_{i+}}{n} \cdot \frac{N_{+j}}{n} \right) \right\}.$$

12.6 Solution Let $\vartheta \in \Theta_0$. Using (12.25), a direct calculation yields

$$M_n - T_n = -\frac{1}{3} \sum_{i=1}^r \sum_{j=1}^s \frac{N_{i+} N_{+j}}{n} \left(\frac{N_{ij}}{n} \cdot \frac{n^2}{N_{i+} N_{+j}} - 1 \right)^3 \cdot \frac{1}{R_{n,i,j}^2}, \qquad (*)$$

where

$$\left| R_{n,i,j} - 1 \right| \le \left| \frac{N_{ij}}{n} \cdot \frac{n^2}{N_{i+} N_{+j}} - 1 \right|. \qquad (**)$$

W.l.o.g. we assume that $N_{ij} \geq 1$ for each pair (i, j), as the probability of this event tends to 1 as $n \to \infty$. It must be shown that each term on the right-hand side of $(*)$ converges to 0 in probability. Due to $(**)$, we first have

$$\frac{1}{R_{n,i,j}^2} \leq \max\left(1, \left(\frac{N_{i+}N_{+j}}{n \cdot n} \cdot \frac{n}{N_{ij}}\right)\right),$$

and it follows that $R_{n,i,j}^{-2} = O_{\mathbb{P}_\vartheta}(1)$, because both $\frac{1}{n^2}N_{i+}N_{+j}$ and $\frac{1}{n}N_{ij}$ converge in probability (to the same value). It remains to show that

$$\frac{N_{i+}N_{+j}}{n^2} \cdot \left\{n\left(\frac{N_{ij}}{n} \cdot \frac{n^2}{N_{i+}N_{+j}} - 1\right)^2\right\} \cdot \left(\frac{N_{ij}}{n} \cdot \frac{n^2}{N_{i+}N_{+j}} - 1\right)$$

converges to zero in probability. The factor before the curly bracket is bounded in probability, and the factor within the round bracket converges to 0 in probability. It remains to show that the factor within the curly bracket is bounded in probability. According to the multivariate central limit theorem, the rs-dimensional random vector

$$\frac{1}{\sqrt{n}}\left(N_{11} - np_{11}, \ldots, N_{1s} - np_{1s}, \ldots, N_{r1} - np_{r1}, \ldots, N_{rs} - np_{rs}\right)$$

converges in distribution to a centered rs-variate normal distribution as $n \to \infty$. With $p_i := \sum_j p_{ij}$ and $q_j := \sum_i p_{ij}$, it follows that $\frac{1}{n}(N_{i+} - np_i)(N_{+j} - nq_j) = O_{\mathbb{P}_\vartheta}(1)$. Further, $\frac{1}{n}N_{i+} = p_i + O_{\mathbb{P}_\vartheta}(n^{-1/2})$ and $\frac{1}{n}N_{+j} = q_j + O_{\mathbb{P}_\vartheta}(n^{-1/2})$, and therefore $n^{-2}N_{i+}N_{+j} = p_i q_j + O_{\mathbb{P}_\vartheta}(n^{-1/2})$. Together with $\frac{1}{n}N_{ij} = p_{ij} + O_{\mathbb{P}_\vartheta}(n^{-1/2})$ and $p_{ij} = p_i q_j$, it follows that the curly bracket is of order $O_{\mathbb{P}_\vartheta}(1)$ and thus bounded in probability. From this, the assertion follows.

12.7 Solution Let $p_{ij} := \mathbb{P}(X = x_i, Y = y_j)$, $p_i := \mathbb{P}(X = x_i)$ and $q_j := \mathbb{P}(Y = y_j)$.

(a) Due to $\frac{1}{n}N_{ij} \xrightarrow{\mathrm{P}} p_{ij}$, $\frac{1}{n}N_{i+} \xrightarrow{\mathrm{P}} p_i$ and $\frac{1}{n}N_{+j} \xrightarrow{\mathrm{P}} q_j$, we get

$$\frac{M_n}{n} \xrightarrow{\mathrm{P}} 2\sum_{i=1}^{r}\sum_{j=1}^{s}\left\{p_{ij}\log p_{ij} - p_i q_j \log\left(p_i q_j\right)\right\}.$$

The asserted representation follows from $\sum_{i,j}\left(p_{ij} - p_i q_j\right)\log(p_i q_j) = 0$, since $\log(ab) = \log a + \log b$.

(b) Let $Q := \mathbb{P}^{(X,Y)}$ be the joint distribution of X and Y and $Q' := \mathbb{P}^X \otimes \mathbb{P}^Y$ the product of the marginal distributions of (X, Y). According to (10.11), the stochastic limit in (a) is equal to $2I_{KL}(Q : Q')$.

(c) If X and Y are not independent then, with the notations from (b), $Q \neq Q'$, and thus the stochastic limit in (a) is (strictly) positive. We then have $M_n \overset{\mathbb{P}}{\longrightarrow} \infty$ (see Self-question 2) and thus the asserted consistency.

12.8 Solution Let $\Delta := N_{1+}N_{+1}N_{2+}N_{+2}$. Since $r = s = 2$, there are four terms in (12.24). If we introduce a common denominator, some algebra yields

$$T_n = \frac{n}{\Delta}\left[N_{2+}N_{+2}\left(N_{11} - \frac{N_{1+}N_{+1}}{n}\right)^2 + N_{+1}N_{2+}\left(N_{12} - \frac{N_{1+}N_{+2}}{n}\right)^2\right.$$
$$\left. + N_{1+}N_{+2}\left(N_{21} - \frac{N_{2+}N_{+1}}{n}\right)^2 + N_{1+}N_{+1}\left(N_{22} - \frac{N_{2+}N_{+2}}{n}\right)^2\right].$$

If we calculate the four brackets using the binomial formula and employ the relations $n = N_{+1}+N_{+2} = N_{11}+N_{12}+N_{21}+N_{22}$, it follows that the expression within the above square brackets is equal to

$$N_{11}^2\left(N_{21}N_{12} + N_{22}N_{12} + N_{21}N_{22} + N_{22}^2\right)$$
$$+N_{12}^2\left(N_{11}N_{21} + N_{21}^2 + N_{11}N_{22} + N_{21}N_{22}\right)$$
$$+N_{21}^2\left(N_{11}N_{22} + N_{12}^2 + N_{11}N_{22} + N_{12}N_{22}\right)$$
$$+N_{22}^2\left(N_{11}^2 + N_{12}N_{11} + N_{11}N_{21} + N_{12}N_{21}\right) - \Delta.$$

Since $\Delta = (N_{11} + N_{12})(N_{11} + N_{21})(N_{21} + N_{22})(N_{12} + N_{22})$, expanding shows that the term within the square brackets simplifies to

$$N_{11}^2 N_{22}^2 + N_{12}^2 N_{21}^2 - 2N_{11}N_{12}N_{21}N_{22} = \left(N_{11}N_{22} - N_{12}N_{21}\right)^2,$$

which was to be shown.

12.9 Solution The density of (X_1, Y_1) is $f(x, y, \vartheta) = (2\pi)^{-1}\exp\left(-\frac{1}{2}((x - \vartheta_1)^2 + (y - \vartheta_2)^2)\right)$. In the numerator and the denominator of (12.5), $(2\pi)^{-n}$ cancels out in each case, and since

$$\inf_{\vartheta_1 \geq 0} \sum_{j=1}^{n}\left(X_j - \vartheta_1\right)^2 = \sum_{j=1}^{n}\left(X_j - \overline{X}_n\right)^2 + n\overline{X}_n^2 \mathbf{1}\{\overline{X}_n < 0\},$$

where $\overline{X}_n = n^{-1}\sum_{j=1}^{n} X_j$, putting $\overline{Y}_n := n^{-1}\sum_{j=1}^{n} Y_j$ gives

$$\Lambda_n = \frac{\exp\left(-\frac{1}{2}\sum_{j=1}^{n}\left(X_j^2 + Y_j^2\right)\right)}{\exp\left(-\frac{1}{2}\sum_{j=1}^{n}\left(Y_j - \overline{Y}_n\right)^2\right)\exp\left(-\frac{1}{2}\left(\sum_{j=1}^{n}\left(X_j - \overline{X}_n\right)^2 + n\overline{X}_n^2 \mathbf{1}\{\overline{X}_n < 0\}\right)\right)}.$$

Using $\sum_{j=1}^{n} \left(X_j - \overline{X}_n \right)^2 = \sum_{j=1}^{n} X_j^2 - n\overline{X}_n^2$, $\sum_{j=1}^{n} \left(Y_j - \overline{Y}_n \right)^2 = \sum_{j=1}^{n} Y_j^2 - n\overline{Y}_n^2$, it thus follows by direct calculation that

$$M_n = -2 \log \Lambda_n = n\overline{Y}_n^2 + n\overline{X}_n^2 \mathbf{1}\{\overline{X}_n \geq 0\}.$$

Under H_0 we have $U_n := \sqrt{n}\overline{Y}_n \sim N(0, 1)$ and $V_n := \sqrt{n}\overline{X}_n \sim N(0, 1)$, and U_n and V_n are independent. Further, $\mathbb{P}(\overline{X}_n \geq 0) = \mathbb{P}(V_n \geq 0) = \frac{1}{2}$, and the condition $V_n \geq 0$ has no influence on the distribution of V_n^2. Since

$$M_n = U_n^2 + V_n^2 \mathbf{1}\{V_n \geq 0\},$$

$U_n^2 \sim \chi_1^2$, and $V_n^2 \sim \chi_1^2$, the assertion follows from the addition law for the chi-square distribution.

12.10 Solution According to Problem 10.6, $\overline{X}_n$ is the MLE of ϑ. Since in the numerator and denominator of (12.4) the factors of the joint density $\prod_{j=1}^{n} f(x_j, \vartheta)$ that are independent of $x_1, \ldots, x_n$ cancel out, we get

$$\Lambda_n = \Lambda_n(X_1, \ldots, X_n) = \frac{\exp\left(-\frac{1}{2} \sum_{j=1}^{n} (X_j - \vartheta_0)^\top \Sigma^{-1} (X_j - \vartheta_0) \right)}{\exp\left(-\frac{1}{2} \sum_{j=1}^{n} (X_j - \overline{X}_n)^\top \Sigma^{-1} (X_j - \overline{X}_n) \right)}.$$

With $X_j - \vartheta_0 = X_j - \overline{X}_n + \overline{X}_n - \vartheta_0$ and $\sum_{j=1}^{n} (X_j - \overline{X}_n) = 0_d$ a direct calculation yields

$$M_n = -2 \log \Lambda_n = n(\overline{X}_n - \vartheta_0)^\top \Sigma^{-1} (\overline{X}_n - \vartheta_0).$$

Under the hypothesis H_0, $\sqrt{n}(\overline{X}_n - \vartheta_0)$ has the normal distribution $N_d(0_d, \Sigma)$. According to Theorem 5.16, M_n has a χ_d^2 distribution.

13.1 Solution If $y \in \partial S(x, r)$, then there exists a sequence (x_n) in $S(x, r)$ and a sequence (z_n) in $S(x, r)^c$ with $\rho(x_n, y) \to 0$ and $\rho(z_n, y) \to 0$. Because $\rho(x_n, x) < r$ and $\rho(z_n, x) \geq r$, the triangle inequality yields $\rho(x, y) \leq \rho(x, x_n) + \rho(x_n, y)$ and $r \leq \rho(x, z_n) \leq \rho(x, y) + \rho(y, z_n)$. In the limit as $n \to \infty$, we get $\rho(x, y) \leq r$ and $r \leq \rho(x, y)$, hence $\rho(x, y) = r$. As the following example shows, strict inclusion can hold. If (S, ρ) is a discrete metric space, then $\{y \in S : \rho(x, y) = 1\} = S \setminus \{x\}$. Since every subset of S is both open and closed, we have $S(x, 1) = \overline{S(x, 1)}$ and thus $\partial S(x, 1) = \emptyset$.

13.2 Solution We assume w.l.o.g. that $M \neq \emptyset$ and first show that M is bounded. We have $M \subset \cup_{x \in M} B(x, 1)$. Because M is compact, this open cover has a finite subcover.

Thus, there exists an $n \geq 1$ and $x_1, \ldots, x_n$ in S such that $M \subset \cup_{j=1}^{n} B(x_j, 1)$. With $r := \max_{1 \leq i, j \leq n} \rho(x_i, x_j)$ the triangle inequality then yields $M \subset B(x_1, r + 1)$.

To show that M is closed, we can assume $M \neq S$, as the set S is closed. We show that $S \setminus M$ is open. Fix any $x \in S \setminus M$, and set $U_k := \{y \in S : \rho(y, x) > \frac{1}{k}\}, k \geq 1$. The set U_k is open, and we have $M \subset S \setminus \{x\} = \cup_{k=1}^{\infty} U_k$. Since $U_1 \subset U_2 \subset U_3 \ldots$, the compactness of M implies that there is an integer n with $M \subset U_n$. By definition of U_n, we have $B(x, \frac{1}{n}) \subset S \setminus M$, and thus $S \setminus M$ is open, i.e., M is closed.

13.3 Solution Let $N \subset S$ be a countable dense subset of S, and put $\mathcal{M}_0 := \{B(x, \varepsilon) : x \in N, \varepsilon \in \mathbb{Q}, \varepsilon > 0\}$. If $O \subset S$ is any non-empty open set, then for each $x \in O$ there exists a $y(x) \in N$ and a positive rational number $\varepsilon(x)$ such that $x \in B(y(x), \varepsilon(x)) \subset O$. Thus, $O = \cup_{x \in O} B(y(x), \varepsilon(x))$. The union extends over countably many balls in $\mathcal{M}_0$. Hence, $O \subset \sigma(\mathcal{M}_0) \subset \sigma(\mathcal{M})$. This implies $\mathcal{B}(S) \subset \sigma(\mathcal{M})$ and thus the assertion.

13.4 Solution If M were not totally bounded, there would exist an $\varepsilon > 0$ and a sequence (x_n) in M with $\rho(x_m, x_n) \geq \varepsilon$ for all m, n with $m \neq n$. Then, however, (x_n) could not have a convergent subsequence, which contradicts the assumed sequential compactness. To prove the completeness of $\overline{M}$, Let (x_j) be a Cauchy sequence in $\overline{M}$. For each j, there is a $y_j \in M$ with $\rho(x_j, y_j) < 2^{-j}$. Then (y_j) is a Cauchy sequence from M. Due to the sequential compactness of M, there exists a subsequence that converges to some $y \in \overline{M}$. Since (y_j) is a Cauchy sequence, it follows that $y_n \to y$, and due to $\rho(x_j, y) \leq \rho(x_j, y_j) + \rho(y_j, y)$, we also have $x_j \to y$. Thus, $\overline{M}$ is complete.

13.5 Solution (a) Clearly, $\tilde{\rho}$ is a symmetric function, and $\tilde{\rho}(x, y) \geq 0$ and $\tilde{\rho}(x, y) = 0 \iff x = y$. The triangle inequality follows from $\min(1, \rho(x, y)) \leq \min(1, \rho(x, z)) + \rho(z, y)) \leq \min(1, \rho(x, z)) + \min(1, \rho(z, y))$.

(b) Since spheres with the same center and the same radius ε coincide if $\varepsilon \leq 1$, the collections of open sets with respect to ρ and $\tilde{\rho}$ are also identical.

(c), (d): Since $\tilde{\rho}(x, y) = \rho(x, y)$, if $\rho(x, y) \leq 1$, a countable subset of S, which is dense in S with respect to ρ, is also dense in S with respect to $\tilde{\rho}$. The same argument applies to completeness, because every Cauchy sequence with respect to ρ is also a Cauchy sequence with respect to $\tilde{\rho}$.

13.6 Solution

(a) ρ is a metric, since $\rho(x, y) = 0$ if and only if $\rho_j(x_j, y_j) = 0$ for each $j = 1, 2, \ldots$. Since ρ_j is a metric for each j, we have $x_j = y_j$ for each j and thus $x = y$. Since $\rho_j(x_j, y_j) = \rho_j(y_j, x_j)$ for each j, it follows that $\rho(x, y) = \rho(y, x)$. The triangle inequality $\rho(x, z) \leq \rho(x, y) + \rho(y, z)$ follows from the convergence of the series $\sum_{j=1}^{\infty} 2^{-j}$ as well as from part (a) of Solution 13.5.

The equivalence still needs to be shown. Here, "$\Longrightarrow$" is obvious. To show "$\Longleftarrow$", fix any $\varepsilon > 0$ and choose k such that $\sum_{j=k+1}^{\infty} 2^{-j} < \varepsilon$. It follows that

$$\rho(x^{(n)}, x) \leq \sum_{j=1}^{k} \frac{\rho_j(x_j^{(n)}, x_j)}{2^j} + \varepsilon.$$

Since each summand $\rho_j(x_j^{(n)}, x_j)/2^j$ converges to zero, we have $\limsup_{n \to \infty} \rho(x^{(n)}, x) \leq \varepsilon$ and thus the assertion.

(b) Let $D_j \subset S_j$ be countable and dense in S_j, $j \geq 1$. Further, let $y = (y_j)_{j \geq 1} \in S$ be arbitrary and $D \subset S$ be the countable set $D := \{(x_1, \ldots, x_k, y_{k+1}, y_{k+2}, \ldots) \in S : k \geq 1, x_1 \in D_1, \ldots, x_k \in D_k\}$. For any $\varepsilon > 0$, choose k such that $\sum_{j=k+1}^{\infty} 2^{-j} < \frac{\varepsilon}{2}$ and $\rho_j(x_j, y_j) < \frac{\varepsilon}{2}$ for $j \in \{1, \ldots, k\}$. With $a \wedge b = \min(a, b)$, it follows that

$$\rho(x, y) = \sum_{j=1}^{k} \frac{1 \wedge \rho_j(x_j, y_j)}{2^j} + \sum_{j=k+1}^{\infty} \frac{1 \wedge \rho_j(x_j, y_j)}{2^j} \leq \sum_{j=1}^{k} \frac{\rho_j(x_j, y_j)}{2^j}$$

$$+ \sum_{j=k+1}^{\infty} \frac{1}{2^j} < \sum_{j=1}^{k} \frac{\varepsilon}{2^{j+1}} + \frac{\varepsilon}{2} < \varepsilon.$$

Hence, D is dense in S, and S is thus separable. Alternatively, one can also construct a countable and dense set with the help of $\times_{j=1}^{\infty} D_j$.

(c) Let $x^{(n)} = (x_1^{(n)}, x_2^{(n)}, \ldots)$ be a Cauchy sequence in S. Then for each $i \geq 1$, the sequence $(x_i^{(n)})_{n \geq 1}$ is a Cauchy sequence in S_i. Since S_i is complete, there is an $x_i \in S_i$ with $\rho_i(x_i^{(n)}, x_i) \to 0$ as $n \to \infty$. If we set $x := (x_i)_{i \geq 1}$, the convergence $\rho(x^{(n)}, x) \to 0$ results analogously to the proof of the triangle inequality in part (a) . Thus, the metric space (S, ρ) is complete.

(d) If $x^{(n)} = (x_1^{(n)}, x_2^{(n)}, \ldots)$ is a sequence in A, then for each $i \geq 1$, $(x_i^{(n)})_{n \geq 1}$ is a sequence in A_i. Since A_1 is compact, there exist a subsequence $(x_1^{(n(1,k))})_{k \geq 1}$ of $(x_1^{(n)})$ and an $x_1 \in A_1$ with $\rho_1(x_1^{(n(1,k))}, x_1) \to 0$ as $k \to \infty$. Since A_2 is compact, there is a subsequence $(n(2, k))_{k \geq 1}$ of $(n(1, k))_{k \geq 1}$ and an $x_2 \in A_2$ with $\rho_2(x_1^{(n(2,k))}, x_2) \to 0$ as $k \to \infty$. Using Cantor's diagonal argument, there is finally a sequence $(n(k, k))_{k \geq 1}$ such that $\rho_i(x_i^{(n(k,k))}, x_i) \to 0$ as $k \to \infty$ for each $i \geq 1$. With $x^{(n(k,k))} := (x_i^{(n(k,k))})_{i \geq 1}$ and $x := (x_i)_{i \geq 1}$, it follows that $\rho(x^{(n(k,k))}, x) \to 0$ as $k \to \infty$. According to Theorem 13.6, A is compact.

13.7 Solution (a) The function ρ is non-negative and symmetric, and from $\rho((x', x''), (y', y'')) = 0$ it follows that $\rho'(x', y') = 0$ and $\rho''(x'', y'') = 0$ and thus $(x', x'') = (y', y'')$. The triangle inequality is obtained using the inequality

$\max(a + b, c + d) \leq \max(a, c) + \max(b, d)$, valid for each choice of real numbers a, b, c, d.

(b1) If (S', ρ') and (S'', ρ'') are separable, then there are countable, dense subsets $A' \subset S'$ and $A'' \subset S''$. The set $A := A' \times A''$ is countable and dense in S, because for $x = (x', x'') \in A$ and fixed $r > 0$, the inequality $\rho(x, y) < r$ holds for $y = (y', y'')$ with $\rho'(x', y') < r$ and $\rho''(x'', y'') < r$. Thus, (S, ρ) is separable. Conversely, if (S, ρ) is separable, then there exists a countable dense set $A = A' \times A''$ in S. Since the projections defined by $\pi' : S \to S'$, $x = (x', x'') \mapsto \pi'(x) = x'$, and $\pi'' : S \to S''$, $x = (x', x'') \mapsto \pi''(x) = x''$, are continuous, the sets $\pi'(A)$ and $\pi''(A)$ are countable and dense in S' and S'', respectively. Therefore, (S', ρ') and (S'', ρ'') are separable.

(b2) The projections π' and π'' from part (b1) are continuous and thus $(\mathcal{B}, \mathcal{B}')$- and $(\mathcal{B}, \mathcal{B}'')$-measurable, respectively. According to the definition of the product-σ-field, $\mathcal{B}' \otimes \mathcal{B}'' = \sigma(\{A' \times A'' : A' \in \mathcal{B}', A'' \in \mathcal{B}''\})$. Since $(\pi')^{-1}(A') \in \mathcal{B}$ and $(\pi'')^{-1}(A'') \in \mathcal{B}$, it follows that $A' \times A'' = (\pi')^{-1}(A') \cap (\pi'')^{-1}(A'') \in \mathcal{B}$ and thus $\mathcal{B}' \otimes \mathcal{B}'' \subset \mathcal{B}$. Since (S, ρ) is separable, every open set $O \in \mathcal{O}$ can be written as a countable union of open balls $B_\rho(x, \varepsilon) = B_{\rho'}(x', \varepsilon) \times B_{\rho''}(x'', \varepsilon) \in \mathcal{B}' \otimes \mathcal{B}''$. Therefore, $O = \bigcup_{x \in O} B_\rho(x, \varepsilon) \in \mathcal{B}' \otimes \mathcal{B}''$ and thus $\mathcal{O} \subset \mathcal{B}' \otimes \mathcal{B}''$, so also $\mathcal{B} = \sigma(\mathcal{O}) \subset \mathcal{B}' \otimes \mathcal{B}''$.

13.8 Solution Since every subset of S is open, the Borel σ-field is equal to the power set of S. For the open balls we have $B(x, \varepsilon) = \{x\}$, if $0 < \varepsilon \leq 1$, and $B(x, \varepsilon) = S$, if $\varepsilon > 1$. The smallest σ-field that contains all open balls thus consists of all subsets A of S, for which either A or the complement A^c is countable.

13.9 Solution Let $\delta \in (0, 1]$ and $x, y \in C$. If $s, t \in [0, 1]$ and $|s - t| \leq \delta$, then

$$|x(s) - x(t)| \leq |x(s) - y(s)| + |y(s) - y(t)| + |y(t) - x(t)|$$
$$\leq \rho(x, y) + w_y(\delta) + \rho(x, y)$$
$$\leq w_y(\delta) + 2\rho(x, y).$$

Therefore, $w_x(\delta) \leq w_y(\delta) + 2\rho(x, y)$. The assertion now follows from reasons of symmetry.

13.10 Solution We have $\mathbb{R}^\infty = \pi_1^{-1}(\mathbb{R}) \in \mathcal{R}_f^\infty$, and if $A = \pi_k^{-1}(H) \in \mathcal{R}_f^\infty$, then $A^c = \pi_k^{-1}(\mathbb{R}^k \setminus H) \in \mathcal{R}_f^\infty$. To show that $\mathcal{R}_f^\infty$ is a π-system, let $A = \pi_k^{-1}(H) \in \mathcal{R}_f^\infty$ and $B = \pi_\ell^{-1}(K) \in \mathcal{R}_f^\infty$, where $H \in \mathcal{B}^k$ and $K \in \mathcal{B}^\ell$. If $k = \ell$ then $A \cap B = \pi_k^{-1}(H \cap K)$, and in the case $k < \ell$ (the case $k > \ell$ follows from symmetry), $A \cap B = \pi_\ell^{-1}((H \times \mathbb{R}^{\ell-k}) \cap K)$.

13.11 Solution

"$\Longrightarrow$": The projection $\pi_k : \mathbb{R}^\infty \to \mathbb{R}^k$ is continuous. Thus, if A is compact then $\pi_k(A) \subset \mathbb{R}^k$ is compact and thus bounded. This implies that the set $\{x_k : x = (x_j)_{j \geq 1} \in A\}$ is bounded.

"$\Longleftarrow$": Let $(x^{(n)})_{n \geq 1}$ with $x^{(n)} = (x_j^{(n)})_{j \geq 1}$ be any sequence in A. According to the assumption, the set $\{x_1^{(n)} : n \geq 1\}$ is bounded. Therefore, there exists a subsequence $(n(1, k))_{k \geq 1}$ and an $x_1 \in \mathbb{R}$ with $x_1^{(n(1,k))} \to x_1$ as $k \to \infty$. Since the set $\{x_2^{(n(1,k))} : k \geq 1\}$ is bounded, there exists an $x_2 \in \mathbb{R}$ as well as a subsequence $(n(2, k))_{k \geq 1}$ of $(n(1, k))_{k \geq 1}$ with $x_2^{(n(2,k))} \to x_2$ as $k \to \infty$. If we continue indefinitely, there exists a sequence $x = (x_j)_{j \geq 1} \in \mathbb{R}^\infty$ with the property that for each $j \geq 1$ the sequence $x_j^{(n(k,k))}$ converges to x_j as $k \to \infty$. According to (13.10), this means $\rho(x^{(n(k,k))}, x) \to 0$ as $k \to \infty$. In view of Theorem 13.6, A is compact.

13.12 Solution Only the properties of $\mathcal{G}$ indicated in the hint need to be shown, because this would imply $\mathcal{B}(S) = \sigma(\mathfrak{A}) \subset \sigma(\mathcal{G})$. We have $\mathfrak{A} \subset \mathcal{G}$, because A^δ is an open set for each $\delta > 0$, and $A^\delta \downarrow A$ as $\delta \downarrow 0$. Since P is continuous from above, it follows that $P(A^\delta \setminus A) < \varepsilon$ for sufficiently small δ. Since S is open and closed, we have $S \in \mathcal{G}$, and with $A \in \mathfrak{A}$, $B \in \mathcal{B}(S)$ and $O \in \mathcal{O}$ with $A \subset B \subset O$ and $P(O \setminus A) < \varepsilon$, it follows that $O^c \subset B^c \subset A^c$ and $P(A^c \setminus O^c) = P(O \setminus A) < \varepsilon$. Thus, $\mathcal{G}$ contains with each set also its complement. If $B_1, B_2, \ldots \in \mathcal{G}$, then there exist $A_1, A_2, \ldots \in \mathfrak{A}$ and $O_1, O_2, \ldots \in \mathcal{O}$ with $A_j \subset B_j \subset O_j$ and $P(O_j \setminus A_j) < \varepsilon/2^j$, $j \geq 1$. With $\tilde{A} := \cup_{j=1}^\infty A_j$, $B := \cup_{j=1}^\infty$, and $O := \cup_{j=1}^\infty$, it follows that $P(O \setminus \tilde{A}) \leq \sum_{j=1}^\infty P(O_j \setminus A_j) < \varepsilon$. The set O is open. If we choose k to be so large that $P(A \setminus (\cup_{j=1}^k A_j)) < \varepsilon$, then $A := \cup_{j=1}^k A_j$ is a closed set with $A \subset B \subset O$ and $P(O \setminus A) < 2\varepsilon$. Thus, $\mathcal{G}$ is a σ-field.

14.1 Solution The collection $\mathcal{C}(P) := \{B \in \mathcal{B}(S) : P(\partial B) = 0\}$ is a field, because due to $\overline{S} = S^\circ = S$, we have $\partial S = \emptyset$ and thus $S \in \mathcal{C}(P)$. Since a subset A of S and its complement have the same boundary, each set in $\mathcal{C}(P)$ also contains its complement. Finally, if $A, B \in \mathcal{C}(P)$, then, due to $\partial(A \cup B) \subset \partial A \cup \partial B$, also $A \cup B \in \mathcal{C}(P)$.

14.2 Solution We first assume that there exists an $x \in S$ and a subsequence (x_{n_k}) with $x_{n_k} \to x$ as $k \to \infty$. According to Example 14.2, then $\delta_{x_{n_k}} \xrightarrow{\mathcal{D}} \delta_x$. Due to the assumption $\delta_{x_n} \xrightarrow{\mathcal{D}} P$ and the uniqueness of the limit under weak convergence (Self-Question 1), it follows that $P = \delta_x$.

We now show that the sequence (x_n) must necessarily contain a convergent subsequence in order for $\delta_{x_n} \xrightarrow{\mathcal{D}} P$ to hold for some $P \in \mathcal{P}$, and we prove this claim by contraposition. To this end, assume that (x_n) contains *no* convergent subsequence. In particular, for each $k \geq 1$ there is *no* convergent subsequence of (x_n) that converges to

x_k. Consequently, there exists an $\varepsilon_k > 0$ with $x_n \notin B(x_k, \varepsilon_k)$ for infinitely many n. With part (d) of the Portmanteau Theorem 14.3, it follows that

$$P\big(B(x_k, \varepsilon_k)\big) \leq \liminf_{n\to\infty} \delta_{x_n}\big(B(x_k, \varepsilon_k)\big) = 0.$$

Since the set $A := \{x_n : n \geq 1\}$ contains no convergent subsequence, it is closed and thus $O := A^c$ is open. Again with part (d) of Theorem 14.3, we get $P(O) \leq \liminf_{n\to\infty} \delta_{x_n}(O) = 0$. Because $S = O \cup \bigcup_{k=1}^\infty B(x_k, \varepsilon_k)$, the equation $P(S) = 0$ would then follow from the σ-subadditivity of P, which is a contradiction.

14.3 Solution

(a) By construction, $\varrho(P, Q) \geq 0$ and $\varrho(Q, P) = \varrho(P, Q)$. From $\varrho(P, Q) = 0$ it follows in particular that $P(A) \leq Q(A^\varepsilon) + \varepsilon$ for each $A \in \mathfrak{A}$ and each $\varepsilon > 0$. As $\varepsilon \to 0$ we obtain $P(A) \leq Q(A)$. By symmetry, $Q(A) \leq P(A)$. It now follows that $P = Q$ (see the discussion after 13.7). If $P(B) \leq Q(B^\varepsilon) + \varepsilon$ and $Q(B) \leq P(B^\varepsilon) + \varepsilon$ as well as $Q(B) \leq R(B^\eta) + \eta$ and $R(B) \leq Q(B^\eta) + \eta$ $(B \in \mathcal{B}(S))$ for $\varepsilon > 0$ and $\eta > 0$, it follows that

$$P(B) \leq Q(B^\varepsilon) + \varepsilon \leq R\big((B^\varepsilon)^\eta\big) + \varepsilon + \eta \leq R(B^{\varepsilon+\eta}) + \varepsilon + \eta$$

and analogously $R(B) \leq P(B^{\varepsilon+\eta}) + \varepsilon + \eta$ $(B \in \mathcal{B}(S))$. These results imply the still missing triangle inequality $\varrho(P, R) \leq \varrho(P, Q) + \varrho(Q, R)$.

(b) For every $B \in \mathcal{B}(S)$, let $P(B) \leq Q(B^\varepsilon) + \varepsilon$. Due to $\big((B^\varepsilon)^c\big)^\varepsilon \subset B^c$, it follows that

$$1 - P(B^\varepsilon) = P\big(B^\varepsilon\big)^c \leq Q\big(((B^\varepsilon)^c)^\varepsilon\big) + \varepsilon \leq Q(B^c) + \varepsilon = 1 - Q(B) + \varepsilon$$

and thus $Q(B) \leq P(B^\varepsilon) + \varepsilon$.

(c) Let B be any P-continuity set, and fix any $\varepsilon > 0$. Since $B \in C(P)$, there exists a δ with $0 < \delta < \varepsilon$ and $P(B^\delta \setminus B) < \varepsilon$ as well as $P\big((B^c)^\delta \setminus B^c\big) < \varepsilon$. Since $\varrho(P_n, P) \to 0$, for sufficiently large n we have both $P_n(B) \leq P(B^\delta) + \delta \leq P(B) + 2\varepsilon$ and $P_n(B^c) \leq P\big((B^c)^\delta\big) + \delta \leq P(B^c) + 2\varepsilon$. For each such n, it follows that $|P_n(B) - P(B)| \leq 2\varepsilon$ and thus $P_n(B) \to P(B)$. According to the Portmanteau theorem, we obtain $P_n \xrightarrow{D} P$.

14.4 Solution If X is $(\mathcal{A}, \mathcal{B}(C))$-measurable then $X(t) = \pi_t \circ X$, as a composition of X with the continuous (and thus $(\mathcal{B}(C), \mathcal{B}^1)$-measurable) projection $\pi_t : C \to \mathbb{R}$, is $(\mathcal{A}, \mathcal{B}^1)$-measurable.

We now assume that $X(t) = \pi_t \circ X$ is a $(\mathcal{A}, \mathcal{B}^1)$-measurable mapping for each $t \in [0, 1]$. Let $k \geq 1$ and $t_1, \ldots, t_k \in [0, 1]$ with $0 \leq t_1 < \ldots < t_k \leq 1$ and $B_1, \ldots, B_k \in \mathcal{B}^1$ be arbitrary. Since $\bigcap_{j=1}^k X(t_j)^{-1}(B_j) = \big(\pi_{t_1,\ldots,t_k} \circ X\big)^{-1}\big(\times_{j=1}^k B_j\big)$ and $\mathcal{B}^k = \sigma\big(\{B_1 \times \ldots \times B_k : B_1, B_2, \ldots, B_k \in \mathcal{B}^1\}\big)$, it follows that $(X(t_1), \ldots, X(t_k)) = \pi_{t_1,\ldots,t_k} \circ X$ is a $(\mathcal{A}, \mathcal{B}^k)$-

measurable mapping. Since $C_f = \{\pi_{t_1,\dots,t_k}^{-1}(H) : k \geq 1, 0 \leq t_1 < \dots < t_k \leq 1, H \in \mathcal{B}^k\}$, it follows that $X^{-1}(B) \in \mathcal{A}$ for each $B \in C_f$. Since C_f is a generating system for $\mathcal{B}(C)$ (see the end of Sect. 13.12), the $(\mathcal{A}, \mathcal{B})$-measurability of X follows.

14.5 Solution Fix any $\varepsilon > 0$. The triangle inequality yields $\mathbb{P}(\rho(X, Y) \geq 2\varepsilon) \leq \mathbb{P}(\rho(X_n, X) \geq \varepsilon) + \mathbb{P}(\rho(X_n, Y) \geq \varepsilon)$. Letting $n \to \infty$ gives $\mathbb{P}(\rho(X, Y) \geq 2\varepsilon) = 0$ for each $\varepsilon > 0$. We thus obtain $\mathbb{P}(\rho(X, Y) = 0) = 1$ and therefore $\mathbb{P}(X = Y) = 1$.

14.6 Solution According to Theorem 6.3, $P_n \xrightarrow{\mathcal{D}} P$ for probability measures $P, P_1, P_2, \dots$ on $\mathcal{B}^d$ is equivalent to $Q_n((-\infty, x]) \to Q((-\infty, x])$ for every point $x \in \mathbb{R}^d$, at which the function defined by $\mathbb{R}^d \ni y \mapsto P((-\infty, y])$ (the distribution function of P) is continuous. This continuity is equivalent to $P(\partial(-\infty, x]) = 0$. According to Remark 6.4, the collection of sets $\mathcal{M}$ can even be made smaller.

14.7 Solution

(a) Let $\alpha := \inf\{\varepsilon \geq 0 : \mathbb{P}(\rho(X, Y) > \varepsilon) \leq \varepsilon\}$. By definition of the infimum, there is a decreasing sequence (ε_n) with $\varepsilon_n \downarrow \alpha$ and $\mathbb{P}(\rho(X, Y) > \varepsilon_n) \leq \varepsilon_n \leq \varepsilon_k$ for every $n \geq k$. As $n \to \infty$, we have $\mathbf{1}\{\rho(X, Y) > \varepsilon_n\} \uparrow \mathbf{1}\{\rho(X, Y) > \alpha\}$. The monotone convergence theorem now yields $\mathbb{P}(\rho(X, Y) > \alpha) \leq \varepsilon_k$ for every k and thus $\mathbb{P}(\rho(X, Y) > \alpha) \leq \alpha$.

(b) The implication "$\Longleftarrow$" is trivial. In view of (a), $d_K(X, Y) = 0$ implies $\mathbb{P}(\rho(X, Y) > 0) = 0$ and thus $\mathbb{P}(\rho(X, Y) = 0) = 1$, so $\mathbb{P}(X = Y) = 1$.

(c) Let $\alpha := d_K(X, Y)$, $\beta := d_K(Y, Z)$. According to (a), we have $\mathbb{P}(\rho(X, Y) > \alpha) \leq \alpha$ and $\mathbb{P}(\rho(Y, Z) > \beta) \leq \beta$. Since $\rho(X, Z) \leq \rho(X, Y) + \rho(Y, Z)$, it follows that $\mathbb{P}(\rho(X, Z) > \alpha + \beta) \leq \alpha + \beta$ and thus the assertion.

(d) Let $X_n \xrightarrow{\mathbb{P}} X$, and fix any integer m. For sufficiently large n, $\mathbb{P}(\rho(X_n, X) > \frac{1}{m}) \leq \frac{1}{m}$ and thus $d_K(X_n, X) \leq \frac{1}{m}$. So $\limsup_{n\to\infty} d_K(X_n, X) \leq \frac{1}{m}$ for every $m \geq 1$, which implies $d_K(X_n, X) \to 0$. Conversely, if $d_K(X_n, X) \to 0$, and $\delta > 0$ is arbitrary, then $d_K(X_n, X) < \delta$ for sufficiently large n. For such n we therefore get $\mathbb{P}(\rho(X_n, X) > \delta) < \delta$ and thus $X_n \xrightarrow{\mathbb{P}} X$.

14.8 Solution According to Problem 7.4 (in which X_j is to be replaced by Z_j and only values in $\mathbb{N}_0$ are considered for x), we have for every $k \in \mathbb{N}_0$

$$\left(X_0^{(n)}, X_1^{(n)}, \dots, X_k^{(n)}\right) \xrightarrow{\mathcal{D}} N_{k+1}\left(0_{k+1}, \Sigma_k\right).$$

Here, the $(k+1)$-by-$(k+1)$ matrix Σ_k is given by the entries $\sigma(i, j) := F(i \wedge j) - F(i)F(j)$ for $i, j \in \{0, \dots, k\}$. According to (14.8), $X_n \xrightarrow{\mathcal{D}} X$, where $X = (X_j)_{j \geq 0}$ is a random element in $\mathbb{R}^{\mathbb{N}_0}$, whose distribution is characterized by the finite-dimensional distributions $(X_0, \ldots, X_k) \sim \mathrm{N}_{k+1}(0_{k+1}, \Sigma_k)$, $k \geq 0$.

14.9 Solution By definition, a sequence $(X_n)_{n \geq 1}$ is uniformly integrable if

$$\lim_{a \to \infty} \sup_{n \geq 1} \mathbb{E}\big[|X_n| 1\{|X_n| \geq a\}\big] = 0.$$

Thus, for every $\varepsilon > 0$ there exists an $a \geq 1$ such that $\sup_{n \geq 1} \mathbb{E}\big[|X_n| 1\{|X_n| \geq a\}\big] < \varepsilon$. It follows that

$$\sup_{n \geq 1} \mathbb{P}(|X_n| > a) = \sup_{n \geq 1} \mathbb{E}\big[1\{|X_n| > a\}\big] \leq \sup_{n \geq 1} \mathbb{E}\left[\frac{|X_n|}{a} 1\{|X_n| \geq a\}\right] \leq \frac{\varepsilon}{a} \leq \varepsilon.$$

Putting $K := [-a, a]$, the set K is compact, and we have $\mathbb{P}^{X_n}(K) \geq 1 - \varepsilon$ for every $n \geq 1$.

14.10 Solution The assertion follows from the equivalence $\delta_{x_n} \xrightarrow{D} \delta_x \iff x_n \to x$ for every sequence (x_n) in A and every $x \in S$ (cf. Example 14.2).

14.11 Solution Since the projections $\pi' : S \to S'$, $\pi'((x', x'')) = x'$, and $\pi'' : S \to S''$, $\pi''((x', x'')) = x''$, are continuous mappings, (a) follows from the mapping theorem.

(b) Let $S' = S'' = \mathbb{R}$ and P be the uniform distribution on the square $[0, 1]^2$. Moreover, denote Q the uniform distribution on the diagonal $\{(x, x) : 0 \leq x \leq 1\}$. Each of the marginal distributions of P and Q is the uniform distribution on $[0, 1]$. Therefore the converse in (a) cannot hold.

(c) We apply Theorem 14.13. According to Problem 13.7, the space (S, ρ) is separable. Due to the special choice of the metric ρ, for $x = (x', x'')$ and $y = (y', y'')$ and $\varepsilon > 0$

$$B(x, \varepsilon) = \{y \in S : \rho(x, y) < \varepsilon\} = \{y' \in S' : \rho'(x', y') < \varepsilon\} \cap \{y'' \in S'' : \rho''(x'', y'') < \varepsilon\}.$$

Thus, $\mathcal{M}$ is a π-system, and $B(x, \varepsilon) \in \mathcal{M}$ for every $x \in S$ and every $\varepsilon > 0$. Let $\mathcal{M}_{x,\varepsilon} := \{B \in \mathcal{M} : x \in \mathring{B} \subset B \subset B_\rho(x, \varepsilon)\}$ and $\partial \mathcal{M}_{x,\varepsilon} := \{\partial B : B \in \mathcal{M}_{x,\varepsilon}\}$, $x \in S$, $\varepsilon > 0$. Since the sets $\partial B(x, \varepsilon)$ are pairwise disjoint for different values of ε, Theorem 14.13 is applicable, and thus $\mathcal{M}$ is a convergence-determining class.

(d) "$\Longrightarrow$" follows from part (e) of Theorem 14.3 (Portmanteau Theorem), since for $B' \in \mathcal{B}'$ and $B'' \in \mathcal{B}''$

$$\partial(B' \times B'') \subset (\partial B' \times S'') \cup (S' \times \partial B'') \tag{*}$$

and thus $P(\partial(B' \times B'')) \leq P'(\partial B') + P''(\partial B'')$.

"$\Longleftarrow$": Let $\mathcal{M}_P := \{A' \times A'' \in \mathcal{M} : P'(\partial A') = 0 = P''(\partial A'')\}$. Due to $\partial(A' \cap B') \subset \partial A' \cup \partial B'$ (and similarly for A'' and B''), $\mathcal{M}_P$ is a π-system. Because of (*), every set in $\mathcal{M}_P$ is a P-continuity set. Let $x \in S$ and $\varepsilon > 0$ be arbitrary. Since the sets in the representation of $S(x, \varepsilon)$ in part (c) on the right-hand side of the equality have disjoint boundaries for different values of ε, there exists an r with $0 < r < \varepsilon$ and $B(x, r) \in \mathcal{M}_P$. Therefore, $\mathcal{M}_P$ satisfies the conditions of Theorem 14.12, and the claim follows.

14.12 Solution

(a) Let $x \in A_{\varepsilon,\delta}$ and $y, z \in S$ with $\rho(x, y) < \delta$, $\rho(x, z) < \delta$ and $\rho'(h(y), h(z)) \geq \varepsilon$. Further, let $\eta := \min(\delta - \rho(x, y), \delta - \rho(x, z))$. For each $\tilde{x}$ with $\rho(\tilde{x}, x) < \eta$, the triangle inequality implies $\rho(\tilde{x}, y) < \delta$ and $\rho(\tilde{x}, z) < \delta$. Thus, the set $A_{\varepsilon,\delta}$ is open.

(b) If $\mathbb{Q}_+$ denotes the set of positive rational numbers, then

$$C_h = S \setminus \left(\bigcup_{\varepsilon \in \mathbb{Q}_+} \bigcap_{\delta \in \mathbb{Q}_+} A_{\varepsilon,\delta} \right),$$

which implies the claim.

15.1 Solution With $\Delta_1 := t_1$ and $\Delta_j := t_j - t_{j-1}$ for $j \in \{2, \dots, k\}$, AD is given by

$$
\begin{pmatrix}
1 & 0 & 0 & 0 & \cdots & 0 \\
1 & 1 & 0 & 0 & \cdots & 0 \\
1 & 1 & 1 & 0 & \cdots & 0 \\
\vdots & \vdots & \vdots & 1 & \cdots & 0 \\
\vdots & \vdots & \vdots & \vdots & \ddots & 0 \\
1 & 1 & 1 & 1 & \cdots & 1
\end{pmatrix}
\cdot
\begin{pmatrix}
\Delta_1 & 0 & 0 & 0 & \cdots & 0 \\
0 & \Delta_2 & 0 & 0 & \cdots & 0 \\
0 & 0 & \Delta_3 & 0 & \cdots & 0 \\
\vdots & \vdots & \vdots & \ddots & \cdots & 0 \\
\vdots & \vdots & \vdots & \vdots & \ddots & 0 \\
0 & 0 & 0 & 0 & \cdots & \Delta_k
\end{pmatrix}
=
\begin{pmatrix}
\Delta_1 & 0 & 0 & 0 & \cdots & 0 \\
\Delta_1 & \Delta_2 & 0 & 0 & \cdots & 0 \\
\Delta_1 & \Delta_2 & \Delta_3 & 0 & \cdots & 0 \\
\vdots & \vdots & \vdots & \ddots & \cdots & 0 \\
\vdots & \vdots & \vdots & \vdots & \ddots & 0 \\
\Delta_1 & \Delta_2 & \Delta_3 & \Delta_4 & \cdots & \Delta_k
\end{pmatrix}.
$$

It follows that

$$
\begin{pmatrix}
\Delta_1 & 0 & 0 & 0 & \cdots & 0 \\
\Delta_1 & \Delta_2 & 0 & 0 & \cdots & 0 \\
\Delta_1 & \Delta_2 & \Delta_3 & 0 & \cdots & 0 \\
\vdots & \vdots & \vdots & \ddots & \cdots & 0 \\
\vdots & \vdots & \vdots & \vdots & \ddots & 0 \\
\Delta_1 & \Delta_2 & \Delta_3 & \Delta_4 & \cdots & \Delta_k
\end{pmatrix}
\cdot
\begin{pmatrix}
1 & 1 & 1 & \cdots & \cdots & 1 \\
0 & 1 & 1 & \cdots & \cdots & 1 \\
0 & 0 & 1 & \cdots & \cdots & 1 \\
0 & 0 & 0 & 1 & \cdots & 1 \\
\vdots & \vdots & \vdots & \vdots & \ddots & \vdots \\
0 & 0 & 0 & 0 & \cdots & 1
\end{pmatrix}
=
\begin{pmatrix}
t_1 & t_1 & t_1 & \cdots & \cdots & t_1 \\
t_1 & t_2 & t_2 & \cdots & \cdots & t_2 \\
t_1 & t_2 & t_3 & \cdots & \cdots & t_3 \\
\vdots & \vdots & \vdots & t_4 & \cdots & t_4 \\
\vdots & \vdots & \vdots & \vdots & \ddots & \vdots \\
t_1 & t_2 & t_3 & t_4 & \cdots & t_k
\end{pmatrix},
$$

which was to be shown.

15.2 Solution For each $\ell \in \{2, \ldots, k\}$, we have

$$\frac{S_{\lfloor nt_\ell \rfloor} - S_{\lfloor nt_{\ell-1} \rfloor}}{\sigma \sqrt{\lfloor nt_\ell \rfloor - \lfloor nt_{\ell-1} \rfloor}} = \frac{1}{\sigma \sqrt{\lfloor nt_\ell \rfloor - \lfloor nt_{\ell-1} \rfloor}} \sum_{j=\lfloor nt_{\ell-1} \rfloor+1}^{\lfloor nt_\ell \rfloor} Z_j \overset{D}{\longrightarrow} N(0, 1).$$

The convergence in distribution follows from the Lindeberg–Lévy central limit theorem. Since the factor before the sum is asymptotically equal to $1/(\sqrt{n}\sqrt{t_\ell - t_{\ell-1}})$, Slutsky's lemma and the mapping theorem yield

$$\frac{1}{\sigma \sqrt{n}} \left(S_{\lfloor nt_\ell \rfloor} - S_{\lfloor nt_{\ell-1} \rfloor} \right) \overset{D}{\longrightarrow} N(0, t_\ell - t_{\ell-1}).$$

This result also applies to the case $\ell = 1$ if we set $t_0 := 0$. Thus, the random vector in (15.28) converges in distribution to a random vector $N =: (N_1, \ldots, N_k)$ having independent components, where $N_1 \sim N(0, t_1)$ and $N_j \sim N(0, t_j - t_{j-1})$ for $j = 2, \ldots, k$. In other words, we have $N \sim N_k(0_k, D)$.

15.3 Solution Let $M := \max_{1 \le j \le k} \sup_{t_{j-1} \le s \le t_j} |x(s) - x(t_{j-1})|$ and $I_j := [t_{j-1}, t_j]$, $j \in \{1, \ldots, k\}$. If $s, t \in [0, 1]$ with $|s - t| \le \delta$, then due to (15.29) we have to distinguish between only two cases. The first is that s and t both lie in one of the intervals $I_1, \ldots, I_k$, and the second means that s and t lie in two adjacent of these intervals. In the first case, there is a j with $s, t \in I_j$, and we have $|x(s) - x(t)| \le |x(s) - x(t_{j-1})| + |x(t) - x(t_{j-1})| \le 2M$. In the second case, we assume w.l.o.g. that $s < t$, $s \in I_j$, and $t \in I_{j+1}$ with $j \in \{0, \ldots, k-1\}$. We thus get

$$|x(s) - x(t)| \le |x(s) - x(t_{j-1})| + |x(t_j) - x(t_{j-1})| + |x(t) - x(t_j)| \le 3M.$$

To derive the statement of Lemma 15.4 from the inequality (15.30), note that the assumption $w(x, \delta) \ge 3\varepsilon$ together with (15.30) implies $\sup_{t_{j-1} \le s \le t_j} |x(s) - x(t_{j-1})|$ for at least one $j \in \{1, \ldots, k\}$. It thus follows that

$$\{w(X, \delta) \ge 3\varepsilon\} \subset \bigcup_{j=1}^{k} \left\{ \sup_{t_{j-1} \le s \le t_j} |X(s) - X(t_{j-1})| \ge \varepsilon \right\},$$

which entails the statement of Lemma 15.4.

15.4 Solution

(a) We consider the polygonal line that connects the points $(0, 0)$, $(1, S_1)$, $\ldots$, (n, S_n) in a Cartesian coordinate system. Each such polygonal line (path) with the property $S_n > k$ attains the level k at some abszissa t (say) for the first time. If we reflect the part of the path that starts at (t, k) across the line $y = k$, we obtain $\mathbb{P}(M_n \geq k, S_n > k) = \mathbb{P}(M_n \geq k, S_n < k)$. Because $\{S_n = k\} \subset \{M_n \geq k\}$ and $\{S_n > k\} \subset \{M_n \geq k\}$, it follows that

$$\mathbb{P}(M_n \geq k) = \mathbb{P}(M_n \geq k, S_n > k) + \mathbb{P}(M_n \geq k, S_n < k) + \mathbb{P}(M_n \geq k, S_n = k)$$
$$= 2\mathbb{P}(M_n \geq k, S_n > k) + \mathbb{P}(S_n = k)$$
$$= 2\mathbb{P}(S_n > k) + \mathbb{P}(S_n = k).$$

(b) The Lindeberg-Lévy central limit theorem yields $S_n/\sqrt{n} \xrightarrow{D} N(0, 1)$. For fixed $t > 0$ we set $k = k_n = \lceil t\sqrt{n} \rceil$. Since M_n takes on integer values, we have

$$\mathbb{P}\left(\frac{M_n}{\sqrt{n}} \geq t\right) = \mathbb{P}(M_n \geq t\sqrt{n}) = \mathbb{P}(M_n \geq k_n) = 2\mathbb{P}(S_n > k_n) + \mathbb{P}(S_n = k_n)$$
$$= 2\mathbb{P}\left(\frac{S_n}{\sqrt{n}} > \frac{k_n}{\sqrt{n}}\right) + o(1) \to 2(1 - \Phi(t)) = \mathbb{P}(|N| \geq t),$$

where $N \sim N(0, 1)$. This proves the assertion.

15.5 Solution Due to the continuity of the paths of W, the intersection of uncountably many sets is indeed a countable intersection over all pairs (s, t) with rational numbers s and t. For fixed $n \geq 2$, let $u_{n,j} := a + j(b - a)/2^n$, $j \in \{0, \ldots, 2^n - 1\}$. Then

$$\bigcap_{\{(s,t):a \leq s < t \leq b\}} \{W(s) \leq W(t)\} \subset \bigcap_{j=0}^{2^n-1} \{W(u_{n,j+1}) - W(u_{n,j}) \geq 0\}.$$

The increments of W on the right-hand side are independent, and due to the stationarity of these increments, they all have the normal distribution $N(0, (b - a)/2^n)$. Thus, the probability of the event on the left-hand side of "$\subset$" is at most $1/2^n$. Since n is arbitrary, the assertion follows.

15.6 Solution

(a) Let $k \geq 1$ and $t_1, \ldots, t_k$ with $0 \leq t_1 < \ldots < t_k$. We have $\mathbb{P}(U \notin \{t_1, \ldots, t_k\}) = 1$, and thus $(W(t_1), \ldots, W(t_k))$ and $(Y(t_1), \ldots, Y(t_k))$ are equal $\mathbb{P}$-almost surely and thus also in distribution.

(b) If we condition on the values t of the random variable U, the probability that Y has discontinuous paths is given by

$$\int_0^1 \mathbb{P}\big(W(U) \neq 0 | U = t\big) \, dt = \int_0^1 \mathbb{P}(W(t) \neq 0) \, dt = 1.$$

The first equality holds due to the independence of W and U.

15.7 Solution Let $\widetilde{W}(t) := -W(t)$, $0 \leq t \leq 1$. Then $(\widetilde{W}(t))_{0 \leq t \leq 1}$ is also a Wiener process. With Theorem 15.11, we have

$$\max_{0 \leq t \leq 1} \widetilde{W}(t) \sim |N|.$$

Since $\max_{0 \leq t \leq 1} \widetilde{W}(t) = -\min_{0 \leq t \leq 1} W(t)$, the assertion follows.

15.8 Solution If $B_1, \ldots, B_n$ are independent Brownian bridges, and $a_1, \ldots, a_n$ are real numbers with $a_1^2 + \ldots + a_n^2 = 1$, then $B := a_1 B_1 + \ldots + a_n B_n$ is also a Brownian bridge.

The proof is completely analogous to the proof of Theorem 15.21. First, $\mathbb{P}(B(0) = 0) = 1 = \mathbb{P}(B(1) = 0)$. For $k \geq 1$ and $t_1, \ldots, t_k$ with $1 \leq t_1 < \ldots < t_k \leq 1$, the addition theorem for the multivariate normal distribution implies that $(B(t_1), \ldots, B(t_k))$ has a k-variate normal distribution. Furthermore, $\mathbb{E}(B(t)) = 0$, $0 \leq t \leq 1$, and for s, $t \in [0, 1]$

$$\mathbb{E}\big[B(s)B(t)\big] = \mathbb{E}\left[\left(\sum_{i=1}^{n} a_i B_i(s)\right)\left(\sum_{j=1}^{n} a_j B_j(t)\right)\right] = \sum_{i=1}^{n}\sum_{j=1}^{n} a_i a_j \mathbb{E}\big[B_i(s)B_j(t)\big]$$

$$= \sum_{i=1}^{n} a_i^2 \mathbb{E}\big[B_1(s)B_1(t)\big] = \min(s, t) - st.$$

$\square$

15.9 Solution With the triangle inequality for integrals and Fubini's theorem, we have

$$\mathbb{E}\left|\int_0^1 \left(\widetilde{W}^2(t) - \widetilde{W}_n^2(t)\right)dt\right| \leq \mathbb{E}\int_0^1 \left|\left(\widetilde{W}^2(t) - \widetilde{W}_n^2(t)\right)\right|dt = \int_0^1 \mathbb{E}\left|\left(\widetilde{W}^2(t) - \widetilde{W}_n^2(t)\right)\right|dt.$$

Since $a^2 - b^2 = (a - b)(a + b)$, the last integral equals

$$\int_0^1 \mathbb{E}\left|\left(\widetilde{W}(t) - \widetilde{W}_n(t)\right)\left(\widetilde{W}(t) + \widetilde{W}_n(t)\right)\right|dt,$$

and due to the Cauchy–Schwarz inequality $\mathbb{E}|UV| \leq \big(\mathbb{E}(U^2)\big)^{1/2}\big(\mathbb{E}(V^2)\big)^{1/2}$, it has the asserted upper bound

$$\int_0^1 \sqrt{\mathbb{E}\big[(\widetilde{W}(t) - \widetilde{W}_n(t))^2\big]}\sqrt{\mathbb{E}\big[(\widetilde{W}(t) + \widetilde{W}_n(t))^2\big]}\,dt.$$

15.10 Solution

(a) From

$$Af(s) = \int_0^1 K(s,t)f(t)\,dt = \int_0^s tf(t)\,dt + s \int_s^1 f(t)\,dt \overset{!}{=} \lambda f(s), \quad \lambda \neq 0,$$

it follows that $f(0) = 0$. Differentiating twice yields

$$\lambda f'(s) = sf(s) + \int_s^1 f(t)\,dt - sf(s) = \int_s^1 f(t)\,dt, \qquad \lambda f''(s) = -f(s).$$

We guess $\varphi(s) = \sin(as)$ for some $a \neq 0$, which leads to $\varphi''(s) = -a^2\varphi(s)$ and $\frac{1}{a^2}\varphi''(s) = -\varphi(s)$, i.e., $\lambda = \frac{1}{a^2}$. A direct calculation using integration by parts gives

$$\lambda\varphi(s) = \frac{1}{a^2}\sin(as) \overset{!}{=} \int_0^1 \min(s,t)\cdot\sin(at)\,dt = \frac{1}{a^2}\sin(as) - \frac{s}{a}\cos(a).$$

We therefore have $\cos(a) = 0$ must hold, which leads to $a \in \left\{ \left(\ell - \frac{1}{2}\right)\pi : \ell \in \mathbb{N} \right\} =:$
L. The eigenfunctions are thus given by $\{c\sin(as) : c \neq 0,\ a \in L\}$, and since these eigenfunctions are normalized we have $1 = c^2 \int_0^1 \sin^2(at)\,dt$ and thus $c = \sqrt{2}$.

(b) The proof is completed if we can show $\sum_{j=1}^\infty \lambda_j^2 = \int_0^1 \int_0^1 K(s,t)^2\,ds\,dt$ (compare the reasoning after (8.42)). A direct calculation shows that the double integral equals $\frac{1}{6}$. Since $\sum_{j=1}^\infty j^{-4} = \pi^4/90$, it follows that

$$\sum_{j=1}^\infty \lambda_j^2 = \frac{1}{\pi^4} \sum_{j=1}^\infty \frac{2^4}{(2j-1)^4} = \frac{16}{\pi^4} \left(\sum_{j=1}^\infty \frac{1}{j^4} - \sum_{j=1}^\infty \frac{1}{(2j)^4} \right)$$

$$= \frac{16}{\pi^4} \left(\frac{\pi^4}{90} - \frac{\pi^4}{16\cdot 90} \right) = \frac{1}{6}.$$

15.11 Solution We have $\mathbb{P}(W^*(0) = 0) = 1$. The fidis of W^* are centered normal distributions with the same covariance function as that of W, because $\mathrm{Cov}\big(W^*(s), W^*(t)\big) = \frac{1}{a}\mathrm{Cov}\big(W(as), W(at)\big) = \min(s,t)$. The assertion now follows, since $\widetilde{W}$ has continuous paths. Note the equation $W^* = h \circ W$, where $h : C[0,\infty) \to C[0,\infty)$ is defined by $h(x)(t) := a^{-1/2}x(at), t \geq 0$.

15.12 Solution We have $\mathbb{P}(\widetilde{W}(0) = 0) = 1$. All fidis of $\widetilde{W}$ are centered normal distributions. Since taking covariances is a bilinear operator, it follows for s, t with $s \leq t$ that

$$
\begin{aligned}
\mathrm{Cov}\big(\widetilde{W}(s), \widetilde{W}(t)\big) &= \mathrm{Cov}\big(W(s+r) - W(r), W(t+r) - W(r)\big) \\
&= \mathrm{Cov}\big(W(s+r), W(t+r)\big) - \mathrm{Cov}\big(W(s+r), W(r)\big) \\
&\quad -\mathrm{Cov}\big(W(r), W(t+r)\big) + \mathrm{Cov}\big(W(r), W(r)\big) \\
&= s + r - r - r + r = s = \min(s, t).
\end{aligned}
$$

Thus, the finite-dimensional distributions of W and $\widetilde{W}$ coincide. Since with probability one $\widetilde{W}$ has continuous paths, the assertion follows.

15.13 Solution Let W be defined on the probability space $(\Omega, \mathcal{A}, \mathbb{P})$. For any $k \geq 1$ and $t_1, \ldots, t_k$ with $0 < t_1 < \ldots < t_k$, the distribution of $\big(\widetilde{W}(t_1), \ldots, \widetilde{W}(t_k)\big)$ is a centered k-variate normal distribution, where in general

$$
\mathrm{Cov}\big(\widetilde{W}(s), \widetilde{W}(t)\big) = st \min \left(\frac{1}{s}, \frac{1}{t}\right) = st \min \left(\frac{t}{st}, \frac{s}{st}\right) = \min(s, t), \quad s, t > 0.
$$

The paths $\Omega \ni \omega \mapsto \widetilde{W}(\omega)$ are continuous on $(0, \infty)$. According to Theorem 15.23, we have

$$
\lim_{s \to \infty} \frac{W(s)}{s} = 0 \ \mathbb{P}\text{-almost surely, thus} \ \lim_{t \to 0} t W\left(\frac{1}{t}\right) = 0 \ \mathbb{P}\text{-almost surely.}
$$

The set $\Omega_0 := \{\omega \in \Omega : \lim_{t \to 0} t W(1/t, \omega) = 0\}$ has probability 1, and for each ω in Ω_0 the path $\Omega_0 \ni \omega \mapsto \widetilde{W}(\omega)$ is a continuous function on $[0, \infty)$. Thus $\widetilde{W}$ is a Wiener process on the set Ω_0, endowed with the trace-σ-field, and the probability measure $\mathbb{P}$ is restricted to $\Omega_0 \cap \mathcal{A}$.

15.14 Solution Let W be defined on the probability space $(\Omega, \mathcal{A}, \mathbb{P})$. On the half-open interval $[0, 1)$, all fidis of B are centered normal distributions with covariances

$$
\begin{aligned}
\mathrm{Cov}\big(B(s), B(t)\big) &= (1 - s)(1 - t)\mathrm{Cov}\left(W\left(\frac{s(1 - t)}{(1 - s)(1 - t)}\right), W\left(\frac{t(1 - s)}{(1 - s)(1 - t)}\right)\right) \\
&= \min(s, t) - st,
\end{aligned}
$$

$0 \leq s, t < 1$. The paths $\Omega \ni \omega \mapsto B(\omega)$ are continuous on $[0, 1)$. According to Theorem 15.23, we have

$$
\lim_{t \to 1}(1 - t)W\left(\frac{t}{1 - t}\right) = \lim_{s \to \infty} \frac{1}{s + 1}W(s) = 0 \ \mathbb{P}\text{-almost surely.}
$$

The set $\Omega_0 := \{\omega \in \Omega : \lim_{t \to 1}(1-t)W(\frac{t}{1-t}, \omega) = 0\}$ has probability 1, and for each ω in Ω_0 the path $\Omega_0 \ni \omega \mapsto B(\omega)$ is a continuous function on $[0, 1]$. Thus B is a Brownian bridge on the set Ω_0, equipped with the trace-σ-field $\Omega_0 \cap \mathcal{A}$, and the probability measure $\mathbb{P}$ is restricted to $\Omega_0 \cap \mathcal{A}$.

15.15 Solution Let $I : [0, 1] \to [0, 1]$, $t \mapsto t$, be the identity map on $[0, 1]$. If B and Z are defined on the probability space $(\Omega, \mathcal{A}, \mathbb{P})$, then $W := B + IZ : \Omega \to C$ is a C-valued random element on Ω. We have $\mathbb{P}(W(0) = 0) = 1$, and due to the independence of B and Z, all finite-dimensional distributions of W are centered normal distributions. The independence of B and Z and the bilinearity of taking covariances yield

$$\mathrm{Cov}\big(W(s), W(t)\big) = \mathrm{Cov}\big(B(s) + sZ, B(t) + tZ\big) = \min(s, t) - st + st = \min(s, t),$$

$0 \le s, t \le 1$, which was to be shown.

15.16 Solution Let $\mathcal{Z}_n := \big[t_{n,0}, t_{n,1}, \ldots, t_{n,k_n}\big]$ with $0 =: t_{n,0} < t_{n,1} < \ldots < t_{n,k_n} := a$, $n \ge 1$, $1 \le k_n \le n$, be partitions of the interval $[0, a]$ with $\Delta_n := \max_{0 \le j < k_n} \big(t_{n,j+1} - t_{n,j}\big) \to 0$ as $n \to \infty$. Since W is a Gaussian process, the Riemann sum

$$R_n := \sum_{j=0}^{k_n-1} W\big(t_{n,j}\big)\big(t_{n,j+1} - t_{n,j}\big)$$

associated with the partition $\mathcal{Z}_n$ has the normal distribution $N(0, \sigma_n^2)$, where

$$\sigma_n^2 = \mathrm{Cov}\left(\sum_{i=0}^{k_n-1} W\big(t_{n,i}\big)\big(t_{n,i+1} - t_{n,i}\big), \sum_{j=0}^{k_n-1} W\big(t_{n,j}\big)\big(t_{n,j+1} - t_{n,j}\big)\right)$$

$$= \sum_{i=0}^{k_n-1}\sum_{j=0}^{k_n-1} \big(t_{n,i+1} - t_{n,i}\big)\big(t_{n,j+1} - t_{n,j}\big) \min\big(t_{n,i}, t_{n,j}\big).$$

Since $\Delta_n \to 0$, the case $i = j$ in this double sum is asymptotically negligible, and symmetry arguments yield

$$\sigma_n^2 = 2 \sum_{i=0}^{k_n-2} t_{n,i}\big(t_{n,i+1} - t_{n,i}\big)\left(\sum_{j=i+1}^{k_n-1} \big(t_{n,j+1} - t_{n,j}\big)\right) + o(1).$$

The sum over j is equal to $a - t_{n,i+1}$, and we thus obtain

$$\sigma_n^2 = 2a \int_0^a t\, dt - 2 \int_0^a t^2\, dt + o(1).$$

Hence $\sigma_n^2 \to a^3/3$ and thus $R_n \xrightarrow{\mathcal{D}} N(0, a^3/3)$.

16.1 Solution Fix any x in D, and let $A_n := \{t \in (0, 1] : |x(t) - x(t-)| > \frac{1}{n}\}, n \geq 1$. According to Lemma 16.2, the set A_n is finite. Since every point of discontinuity belongs to the set $\bigcup_{n=1}^{\infty} A_n$, the assertion follows.

16.2 Solution The proof is by contraposition. Assume that $x \notin$ D. Then there is either a $t \in [0, 1)$ with $x(t) \neq x(t+)$ or a $t \in (0, 1]$ with $\lim \sup_{s<t,s\to t} x(t) > \lim \inf_{s<t,s\to t} x(t)$. In each of these cases, the condition $\lim_{n\to\infty} w_x'(\delta) = 0$ is violated.

16.3 Solution

(a) If $\delta < \frac{1}{2}$, there is a δ-thin partition of $[0, 1]$ with $t_j - t_{j-1} \leq 2\delta$ for each $j \in \{1, \ldots, k\}$. It follows that

$$w_x'(\delta) \leq \max_{1 \leq j \leq k} w_x([t_j, t_{j-1})) \leq \sup_{|s-t| \leq 2\delta} |x(s) - x(t)| \leq w_x(2\delta).$$

(b) For every $\varepsilon > 0$ there is a δ-thin partition $0 = t_0 < \ldots < t_k = 1$ of $[0, 1]$ with $w([t_j, t_{j-1})) < w_x'(\delta) + \varepsilon$, $j \in \{1, \ldots, k\}$. If $s, t \in [0, 1]$ with $|s - t| \leq \delta$, then s and t either lie both in one of the intervals $[t_j, t_{j-1})$ or in two directly adjacent intervals. In the first case, $w_x(\delta) \leq w_x'(\delta) + \varepsilon$, and in the second case we get the estimate $w_x(\delta) \leq 2w_x'(\delta) + \varepsilon + H(x)$. Since $\varepsilon > 0$ was arbitrary, the assertion follows.

16.4 Solution Since every function in D[0, 1] is bounded, we have $d_S(x, y) < \infty$ (choose g as the identity map on [0, 1]). By definition, $d_S(x, y) \geq 0$. If $d_S(x, y) = 0$, then letting $g(t) = t$ we obtain $\sup_{s\in[0,1]} |x(s) - y(g(s))| = 0$. We thus conlude that $x(t) = y(t)$ or $x(t) = \lim_{t_n \uparrow t} y(t_n)$ for each $t \in [0, 1]$ and thus $x = y$. To show the symmetry of d_S, note that if $g \in \mathcal{G}$ then also $g^{-1} \in \mathcal{G}$. Thus, for $x, y \in$ D[0, 1] we have

$$\sup_{t\in[0,1]} |g^{-1}(t) - t| = \sup_{t\in[0,1]} |g(t) - t| \quad \text{and} \quad \sup_{t\in[0,1]} \left| x\left(g^{-1}(t)\right) - y(t) \right|$$

$$= \sup_{t\in[0,1]} |x(t) - y(g(t))|,$$

and it follows that $d_S(x, y) = d_S(y, x)$. For the proof of the triangle inequality, we use that with g_1 and g_2 also the composition $g_2 \circ g_1$ belongs to $\mathcal{G}$. We thus obtain

$$\sup_{t\in[0,1]} |g_2 \circ g_1(t) - t| \leq \sup_{t\in[0,1]} |g_2(t) - t| + \sup_{t\in[0,1]} |g_1(t) - t|,$$

and for $x, y, z \in D[0, 1]$ we get

$$\sup_{t\in[0,1]} |x(t) - z(g_2 \circ g_1(t))| \leq \sup_{t\in[0,1]} |x(t) - y(g_1(t))| + \sup_{t\in[0,1]} |y(t) - z(g_2(t))|.$$

16.5 Solution

(a) With the hint, the assertion follows from

$$\|g - \mathrm{I}\| = \sup_{0<t\leq 1} t \left| \frac{g(t) - g(0)}{t - 0} - 1 \right| \leq \sup_{0\leq s<t\leq 1} \left| \frac{g(t) - g(s)}{t - s} - 1 \right|.$$

(b) With the hint, $\|x \circ g - y\| \leq \exp(\|x \circ g - y\|) - 1$, which yields the assertion.

16.6 Solution

(a) With $v = g^{-1}(t)$ and $u = g^{-1}(s)$ we get

$$\|g^{-1}\|^{\circ} = \sup_{s<t} \left| \log \frac{g^{-1}(t) - g^{-1}(s)}{t - s} \right| = \sup_{u<v} \left| \log \frac{v - u}{g(v) - g(u)} \right|$$

$$= \sup_{u<v} \left| -\log \frac{g(v) - g(u)}{v - u} \right| = \|g\|^{\circ}.$$

(b) Putting $v = g_2(t)$ and $u = g_2(s)$ we obtain

$$\|g_1 \circ g_2\|^{\circ} = \sup_{s<t} \left| \log \frac{g_1 \circ g_2(t) - g_1 \circ g_2(s)}{t - s} \right|$$

$$= \sup_{u<v} \left| \log \left(\frac{g_1(v) - g_1(u)}{v - u} \cdot \frac{v - u}{g_2^{-1}(v) - g_2^{-1}(u)} \right) \right|.$$

The assertion now follows from $\log(ab) = \log a + \log b$, $a, b > 0$, as well as from part (a).

16.7 Solution We refer to the hint. In the case $t_2 - t_1 \leq \delta$ there are only the two cases that t_1 and t_2 lie either both in one of the intervals $I_j := [s_j, s_{j+1})$, $j \in \{0, \ldots, k-1\}$, or in adjacent intervals I_j and I_{j+1}, where $j \in \{0, \ldots, k-1\}$. If $t_1 \leq t \leq t_2$, then in the first case t is in the same interval as t_1 and t_2, and in the second case either t_1 and t or t_2 and t are in the same interval. In each of the cases at least one of the inequalities $|x(t) - x(t_1)| < \eta$ or $|x(t) - x(t_2)| < \eta$ holds. If we let η converge to $w_x'(\delta)$ from above, the assertion follows.

16.8 Solution Fix any $\varepsilon > 0$. For each $n \geq n_0(\varepsilon)$, we have $d_S(x_n, x) \leq \varepsilon$, and thus according to the definition of d_S for each such n there is a $g_n \in \mathcal{G}$ with $\|x_n - x \circ g_n\| \leq \varepsilon$ and $\|g_n - I\| \leq \varepsilon$. Using the triangle inequality, we get

$$\int_0^1 |x_n(t) - x(t)| \, dt \leq \int_0^1 |x_n(t) - x(g_n(t))| \, dt + \int_0^1 |x(g_n(t)) - x(t)| \, dt.$$

Since $\|x_n - x \circ g_n\| \leq \varepsilon$, the first integral on the right-hand side is at most ε. For the second integral, the integrand converges to zero almost everywhere as $n \to \infty$, and it is bounded in absolute value by the integrable function $2\|x\|$. According to the dominated convergence theorem, it follows that $\limsup_{n\to\infty} \int_0^1 |x_n(t) - x(t)| \, dt \leq \varepsilon$ and thus the assertion, since ε was arbitrary.

16.9 Solution

(a) Let $H(x) = |x(t_0) - x(t_0-)|$, where $0 < t_0 \leq 1$. The triangle inequality yields

$$H(x) \leq |x(t_0) - y(t_0)| + |y(t_0) - y(t_0-)| + |y(t_0-) - x(t_0-)|.$$

Since $\|x - y\| < \varepsilon$, it follows that $H(x) < 2\varepsilon + H(y)$. The inequality $H(y) < 2\varepsilon + H(x)$ follows by symmetry.

(b) If $d_S(x, y) < \varepsilon$, then there exists a $g \in \mathcal{G}$ with $\|x - y \circ g\| < \varepsilon$. Since $H(y) = H(y \circ g)$ for each $g \in \mathcal{G}$, it follows from part (a) that $|H(x) - H(y)| = |H(x) - H(y \circ g)| < 2\varepsilon$.

16.10 Solution Let (x_n) be any sequence in C and $x \in$ D with $d_S(x_n, x) \to 0$. According to Problem 16.9 (b), $\lim_{n\to\infty} H(x_n) = H(x)$. Since $H(x_n) = 0$ for each n, it follows that $H(x) = 0$ and consequently $x \in$ C. The set C is therefore closed with respect to the Skorokhod topology. For $x \in$ C and $\varepsilon > 0$, let $y(t) := x(t)$, if $t < 0.5$, and $y(t) := x(t) + 0.9\varepsilon$, if $0.5 \leq t \leq 1$. Then $y \in$ D $\setminus$ C, and we have $\|x - y\| = 0.9\varepsilon$ and thus $d_S(x, y) < \varepsilon$. Consequently, C contains no interior point and is therefore nowhere dense in D.

16.11 Solution (a) Since $\mathbb{E}[1_A] = \mathbb{P}(A)$ and $1_A^2 = 1_A$ for each indicator function 1_A with $A \in \mathcal{A}$, it follows that $\mathbb{E}[\alpha_1^2] = s - r - 2(s-r)^2 + (s-r)^2 = (s-r)(1 - (s-r)) \leq t - r$ and analogously $\mathbb{E}[\beta_2^2] = (t-s)(1 - (t-s)) \leq t - r$. (b) Since $1_A 1_B = 0$ if $A \cap B = \emptyset$, it follows that $\mathbb{E}[\alpha_1 \beta_1] = \mathbb{E}[\alpha_2 \beta_2] = -(s-r)(t-s)$.

16.12 Solution Since $Y = \psi(B) = B \circ F$, using $\mathrm{Cov}(B(s), B(t)) = s \wedge t - st$ for the Brownian bridge B, we obtain for each $k \geq 1$ and any choice of $t_1, \ldots, t_k$ with $1 \leq t_1 < \ldots < t_k \leq 1$:

$$\big(Y(t_1), \ldots, Y(t_k)\big) = \big(B(F(t_1)), \ldots, B(F(t_k))\big) \sim \mathrm{N}_k\Big(0_k, \big(\sigma_{i,j}\big)_{1 \leq i,j, \leq k}\Big).$$

Here, $\sigma_{i,j} = \mathrm{Cov}\big(B(F(t_i)), B(F(t_j))\big) = F(t_i) \wedge F(t_j) - F(t_i)F(t_j)$. The stochastic process Y thus has the asserted properties.

16.13 Solution

(a) Due to right-continuity of x, we have $\{x \in D : \|x\| < a\} = \bigcap_{t\in[0,1]\cap\mathbb{Q}}\{x \in D : |x(t)| < a\}$. As a countable intersection of Borel sets in D, $\{x \in D : \|x\| < a\}$ belongs to $\mathcal{B}(D)$.

(b) From $d_S(x_n, x) \to 0$ and $x \in C$ it follows that $\|x_n - x\| \to 0$ and thus $\|x_n\| \to \|x\|$.

16.14 Solution

(a) If $x \in D$ then also the function defined by $x^2(t) := (x(t))^2$, $t \in [0, 1]$, belongs to D. Since x^2 is bounded and continuous almost everywhere, $h(x) = \int_0^1 x^2(t)\,dt$ is a Riemann integral. Since h is the limit of Riemann sums that are $(\mathcal{B}(D), \mathcal{B}^1)$-measurable, also h is $(\mathcal{B}(D), \mathcal{B}^1)$-measurable. Each such Riemann sum is $(\mathcal{B}(D), \mathcal{B}^1)$-measurable, because this is true for the mappings $D \ni x \mapsto (\pi_t \circ x)^2 = x^2(t), 0 \le t \le 1$.

(b) If $d_S(x_n, x) \to 0$ and $x \in C$, then $\|x_n - x\| \to 0$ and $\|x_n\| \to \|x\|$ (cf. Problem 16.13 (b)). Due to

$$\int_0^1 \left|x_n^2(t) - x^2(t)\right| dt = \int_0^1 \left|x_n(t) - x(t)\right| \cdot \left|x_n(t) + x(t)\right| dt$$

and $\left|x_n(t)+x(t)\right| \le K\|x\|$ for a $K > 0$ and Problem 16.8, it follows that $\int_0^1 x_n^2(t)dt \to \int_0^1 x^2(t)dt$ and thus the assertion.

16.15 Solution We set $V_j := F_0(X_{(j)})$, $j \in \{1, \ldots, n\}$, $V_0 := 0$, and $V_{n+1} := 1$. Furthermore, let $G_n(t) := \widehat{F}_n(F_0^{-1}(t))$. Since $G_n(V_j) = \frac{j}{n}$, $j \in \{0, \ldots, n\}$, and the fact that G_n is constant on each of the intervals $[V_j, V_{j+1})$, $j \in \{0, \ldots, n\}$, it follows that

$$\omega_n^2 = n \int_0^1 \left(\widehat{F}_n(F_0^{-1}(t)) - t\right)^2 dt = n \sum_{j=0}^n \int_{V_j}^{V_{j+1}} \left(\frac{j}{n} - t\right)^2 dt.$$

Calculations give $\omega_n^2 = \frac{1}{n}\sum_{j=0}^n j^2(V_{j+1} - V_j) - \sum_{j=0}^n j(V_{j+1}^2 - V_j^2) + \frac{n}{3}\sum_{j=0}^n(V_{j+1}^3 - V_j^3)$. The last, telescoping sum is equal to 1, and for the other two sums one can also use telescoping effects by setting $j^2 = (j+1)^2 - (2j+1)$ or $j = (j+1) - 1$, respectively. After straightforward algebra, we get $\omega_n^2 = \frac{n}{3} - \frac{2}{n}\sum_{j=1}^n jV_j + \frac{1}{n}\sum_{j=1}^n V_j + \sum_{j=1}^n V_j^2$. The same result is obtained if one calculates the right-hand side of (16.34).

16.16 Solution If $F \neq F_0$, then there is a $t \in \mathbb{R}$ with $F(t) \neq F_0(t)$. Because $\sup_{x\in\mathbb{R}} |\widehat{F}_n(x) - F_0(x)| \ge |\widehat{F}_n(t) - F_0(t)|$ and $|\widehat{F}_n(t) - F_0(t)| \to |F(t) - F_0(t)|$

almost surely, it follows that $\sqrt{n}K_n \to \infty$ almost surely. We can assume w.l.o.g. that $F_0(t) > F(t)$ for the above t. Since F and F_0 are continuous, there exists an $\varepsilon > 0$ and an interval $[a, b]$ with $a < b$, such that $F_0(x) - F(x) \geq \varepsilon$ holds for every $x \in [a, b]$. Here, we can assume $F_0(a) < F_0(b)$. Due to $\omega_n^2 \geq n \int_a^b (\widehat{F}_n(x) - F_0(x))^2 dF_0(x)$ and $\sup_{x \in [a,b]} |\widehat{F}_n(x) - F_1(x)| \to 0$ $\mathbb{P}$-a.s. (Glivenko–Cantelli!) there exists a set $\Omega_0 \subset \Omega$ with $\mathbb{P}(\Omega_0) = 1$, such that for every $\omega \in \Omega_0$ there is an n_0 with $\sup_{x \in [a,b]} |\widehat{F}_n(\omega, x) - F_1(x)| < \frac{\varepsilon}{2}$ for every $n \geq n_0$. For such n we have

$$\int_a^b (\widehat{F}_n(\omega, x) - F_0(x))^2 dF_0(x) \geq \left(F_0(b) - F_0(a) \right) \frac{\varepsilon^2}{4} > 0.$$

Since the integral is multiplied by n, it follows that $\omega_n^2 \to \infty$ $\mathbb{P}$-a.s.

17.1 Solution The hint states that for a fixed $y \in \mathbb{H}$, the mapping $\mathbb{H} \ni x \mapsto \langle x, y \rangle$ is continuous. With $x_n := \sum_{j=1}^n \langle x, e_j \rangle e_j$, we have $x_n \to x$ and thus

$$\langle x, y \rangle = \lim_{n \to \infty} \left\langle \sum_{j=1}^n \langle x, e_j \rangle e_j, y \right\rangle = \lim_{n \to \infty} \sum_{j=1}^n \langle x, e_j \rangle \langle e_j, y \rangle = \sum_{j=1}^\infty \langle x, e_j \rangle \langle e_j, y \rangle.$$

17.2 Solution For an orthonormal system $\{e_1, e_2, \ldots\}$ of $\mathbb{H}$, we have $\|e_i - e_j\| = \sqrt{2}$ if $i \neq j$. Thus, the sets $B(e_i, \frac{1}{2})$, $i \in \mathbb{N}$, are pairwise disjoint, and the triangle inequality yields $\uplus_{i=1}^\infty B(e_i, \frac{1}{2}) \subset B(e_1, 2)$. Since $\mu(B(x_0, \varepsilon)) < \infty$ for each $\varepsilon > 0$, the translation invariance of μ yields $\mu(B(e_1, 2)) < \infty$. Thus, translation invariance and the σ-additivity of μ imply $\mu(B(e_i, \frac{1}{2})) = 0$ for each $i \geq 1$. Using once more the property of translation invariance, we obtain $\mu(B(x, \frac{1}{2})) = 0$ for every $x \in \mathbb{H}$. If $\{x_1, x_2, \ldots\}$ is a countable dense subset of $\mathbb{H}$, we have $\mathbb{H} = \cup_{j=1}^\infty B(x_j, \frac{1}{2})$ and thus $\mu(\mathbb{H}) \leq \sum_{j=1}^\infty \mu(B(x_j, \frac{1}{2}))$. Hence, $\mu(\mathbb{H}) = 0$ and thus $\mu(B) = 0$ for each $B \in \mathcal{B}(\mathbb{H})$.

17.3 Solution We briefly write $\| \cdot \| := \| \cdot \|_\mathbb{B}$ and $\int := \int_\Omega$. Let (f_n) be a sequence of simple B-integrable functions with $\int \|f_n - f\| d\mu \to 0$. Since the inequality to be shown holds for each f_n, it follows that

$$\left\| \int f \, d\mu \right\| \leq \left\| \int f \, d\mu - \int f_n \, d\mu \right\| + \left\| \int f_n \, d\mu \right\| \leq \left\| \int f \, d\mu - \int f_n \, d\mu \right\| + \int \|f_n\| \, d\mu$$

$$\leq \left\| \int f \, d\mu - \int f_n \, d\mu \right\| + \int \|f_n - f\| \, d\mu + \int \|f\| \, d\mu.$$

Because $\| \int f \, d\mu - \int f_n \, d\mu \| \to 0$, the claim follows.

17.4 Solution For $y \in \mathbb{H}$, let $s_y : \mathbb{H} \to \mathbb{R}$ denote the mapping defined by $s_y(x) := \langle x, y \rangle$, $x \in \mathbb{H}$. If $X : \Omega \to \mathbb{H}$ is $(\mathcal{A}, \mathcal{B})$-measurable, then the composition $s_y \circ X = \langle X, y \rangle$ is $(\mathcal{A}, \mathcal{B}^1)$-measurable, because s_y is continuous and thus $(\mathcal{B}, \mathcal{B}^1)$-measurable. For the reverse implication "$\Longleftarrow$", only $X^{-1}(\mathcal{M}) \subset \mathcal{A}$ needs to be shown, since $\mathcal{B} = \sigma(\mathcal{M})$. According to the definition of $\mathcal{M}$, every set in $\mathcal{M}$ is of the type $B = s_y^{-1}(O) = \{x \in \mathbb{H} : \langle x, y \rangle \in O\}$ for some y in $\mathbb{H}$ and an open set $O \subset \mathbb{R}$. It follows that $X^{-1}(B) = X^{-1}\big(s_y^{-1}(O)\big) = (s_y \circ X)^{-1}(O) \in \mathcal{A}$, since $s_y \circ X$ is $(\mathcal{A}, \mathcal{B}^1)$-measurable.

17.5 Solution

(a) With the definition of the norm in $\mathbb{H} = \ell_2$, we have

$$\mathbb{E}\|X\| = \sum_{k=1}^{\infty} \|X(\omega_k)\| \, \mathbb{P}(\{\omega_k\}) = \sum_{k=1}^{\infty} \ln(k+1) \, \frac{C}{k(\ln(k+1))^2} = \infty.$$

(b) For $y := (y_k)_{k \geq 1} \in \mathbb{H}$, the definition of the inner product in ℓ_2 yields

$$\mathbb{E}|\langle X, y \rangle| = \sum_{k=1}^{\infty} |\langle X(\omega_k), y \rangle| \, \mathbb{P}(\{\omega_k\}) = \sum_{k=1}^{\infty} \ln(k+1)|y_k| \, \frac{C}{k(\ln(k+1))^2}$$

$$= C \sum_{k=1}^{\infty} \frac{1}{k \ln(k)} \, |y_k| \leq C \sqrt{\sum_{k=1}^{\infty} \frac{1}{k^2(\ln(k+1))^2}} \sqrt{\sum_{k=1}^{\infty} y_k^2} < \infty.$$

17.6 Solution If we insert $y := \mathbb{E}X$ into the equation $\langle \mathbb{E}X, y \rangle = \mathbb{E}\langle X, y \rangle$, it follows that

$$\|\mathbb{E}X\|^2 = \langle \mathbb{E}X, \mathbb{E}X \rangle = \mathbb{E}\langle X, \mathbb{E}X \rangle \leq \mathbb{E}|\langle X, \mathbb{E}X \rangle| \leq \mathbb{E}\|X\| \, \|\mathbb{E}X\|$$

and thus the assertion.

17.7 Solution Because $\|\alpha X + \beta Y\| \leq |\alpha| \|X\| + |\beta| \|Y\|$, L^1 is a vector space over $\mathbb{R}$. Fix any y in $\mathbb{H}$. Using (17.7) and the linearity of both the inner product and the expectation operator, we obtain

$$\langle \mathbb{E}(\alpha X + \beta Y), y \rangle = \mathbb{E}\langle \alpha X + \beta Y, y \rangle = \alpha \mathbb{E}\langle X, y \rangle + \beta \mathbb{E}\langle Y, y \rangle = \mathbb{E}\langle \alpha X + \beta Y, y \rangle$$

and thus the assertion.

17.8 Solution Fix any x, y in $\mathbb{H}$ and $A \in \mathcal{A}$ with $\mu(A) < \infty$. For the B-integrable simple function $f := 1\{A\}x$ we have $\int_{\Omega} f \, d\mu = \mu(A)x$, and it follows that

$$\left\langle \int_{\Omega} f \, d\mu, y \right\rangle = \mu(A)\langle x, y \rangle = \int_{\Omega} \langle f, y \rangle \, d\mu.$$

Because of (17.11) and the additivity of both the integral and the inner product, this equation holds for every B-integrable *simple* function f. If f is a general B-integrable function, then there is a sequence (f_n) of B-integrable simple functions such that $\int_{\Omega} \| f_n - f \| \, d\mu \to 0$ as $n \to \infty$. The definition of the B-integral and the continuity of the inner product then yield

$$\left\langle \int_{\Omega} f \, d\mu, y \right\rangle = \left\langle \lim_{n \to \infty} \int_{\Omega} f_n \, d\mu, y \right\rangle = \lim_{n \to \infty} \left\langle \int_{\Omega} f_n \, d\mu, y \right\rangle = \lim_{n \to \infty} \int_{\Omega} \langle f_n, y \rangle \, d\mu.$$

Due to $\left| \int_{\Omega} \langle f_n, y \rangle d\mu - \int_{\Omega} \langle f, y \rangle d\mu \right| \leq \int_{\Omega} |\langle f_n - f, y \rangle| d\mu \leq \|y\| \int \| f_n - f \| d\mu \to 0$, the assertion follows.

17.9 Solution We have $|\langle X, Y \rangle| \leq \|X\| \|Y\|$, and due to the independence of $\|X\|$ and $\|Y\|$, the expectation of $\langle X, Y \rangle$ exists. Since X and Y are independent, the joint distribution of X and Y (i.e., the image measure of $\mathbb{P}$ under (X, Y)) equals the product measure $\mathbb{P}^X \otimes \mathbb{P}^Y$. With the hint and Fubini's theorem, we get

$$\mathbb{E}\langle X, Y \rangle = \int_{\mathbb{H}} \left[\int_{\mathbb{H}} \langle x, y \rangle \mathbb{P}^Y (\mathrm{d}y) \right] \mathbb{P}^X (\mathrm{d}x) = \int_{\mathbb{H}} \mathbb{E}\langle x, Y \rangle \mathbb{P}^X (\mathrm{d}x) = \int_{\mathbb{H}} \langle x, \mathbb{E}Y \rangle \mathbb{P}^X (\mathrm{d}x)$$

$$= \langle \mathbb{E}X, \mathbb{E}Y \rangle.$$

17.10 Solution We set $m := \mathbb{E}X$ and fix any $h \in \mathbb{H}$. Then

$$\mathbb{E}\|X - h\|^2 = \mathbb{E}\|X - m + m - h\|^2 = \mathbb{E}\|X - m\|^2 + \|m - h\|^2 + 2\mathbb{E}\langle X - m, m - h \rangle.$$

Since $\mathbb{E}\langle X - m, m - h \rangle = \langle \mathbb{E}(X - m), m - h \rangle = \langle \mathbf{0}, m - h \rangle = 0$, the assertion follows. Here, $\mathbf{0}$ is the zero element of $\mathbb{H}$.

17.11 Solution Due to

$$\|S_n\|^2 = \left\langle \sum_{i=1}^{n} X_i, \sum_{j=1}^{n} X_j \right\rangle = \sum_{i,j=1}^{n} \langle X_i, X_j \rangle = \sum_{i=1}^{n} \|X_i\|^2 + 2 \sum_{i<j} \langle X_i, X_j \rangle,$$

the assertion follows from the linearity of taking expectations and $\mathbb{E}\langle X_i, X_j \rangle = \langle \mathbb{E}X_i, \mathbb{E}X_j \rangle = \langle \mathbf{0}, \mathbf{0} \rangle = 0$ if $i \neq j$ (cf. Theorem 17.14).

17.12 Solution We assume w.l.o.g. $\mathbb{E}(X_1) = \mathbf{0}$. With $\overline{X}_n := n^{-1} \sum_{j=1}^{n} X_j$, we have $\|\overline{X}_n\|^4 = \langle \overline{X}_n, \overline{X}_n \rangle \langle \overline{X}_n, \overline{X}_n \rangle = n^{-4} \sum_{i,j,k,\ell=1}^{n} \langle X_i, X_j \rangle \langle X_k, X_\ell \rangle$. When taking expectations, we differentiate the quadruple sum according to how many different indices i, j, k and ℓ occur. By Theorem 17.14, $\mathbb{E}(\langle X_1, X_2 \rangle \langle X_3, X_4 \rangle) = 0$. Thus, terms with four different indices do not contribute to $\mathbb{E}\|\overline{X}_n\|^4$. Since $\mathbb{E}(\langle X_1, X_1 \rangle \langle X_2, X_3 \rangle) = \mathbb{E}\|X_1\|^2 \mathbb{E}\langle X_2, X_3 \rangle =$

$\mathbb{E}\|X_1\|^2\langle\mathbb{E}X_2,\mathbb{E}X_3\rangle = 0$ as well as

$$\begin{aligned}
\mathbb{E}(\langle X_1, X_2\rangle\langle X_1, X_3\rangle) &= \mathbb{E}\big[\mathbb{E}\langle X_1, X_2\rangle\langle X_1, X_3\rangle|X_1\big] \\
&= \mathbb{E}\big[\mathbb{E}[\langle X_1, X_2\rangle|X_1]\cdot\mathbb{E}[\langle X_1, X_3\rangle|X_1]\big] \\
&= \mathbb{E}\big[\langle X_1, \mathbb{E}X_2\rangle\cdot\langle X_1, \mathbb{E}X_3\rangle\big] = 0,
\end{aligned}$$

there is also no contribution to $\mathbb{E}\|\overline{X}_n\|^4$ from terms with three different indices. Because each term of the quadruple sum in which only two different indices i, j, k and ℓ occur, is bounded by $(\mathbb{E}\|X_1\|^2)^2$, and since $\mathbb{E}\|X_1\|^4 < \infty$, there is a positive constant C with $\mathbb{E}\|\overline{X}_n\|^4 \le C/n^2$. The monotone convergence theorem now yields $\mathbb{E}(\sum_{n=1}^\infty \|\overline{X}_n\|^4) < \infty$, and thus $\sum_{n=1}^\infty \|\overline{X}_n\|^4$ is finite $\mathbb{P}$-almost surely. Therefore, it follows that $\|\overline{X}_n\| \to 0$ $\mathbb{P}$-almost surely.

17.13 Solution Due to the linearity of T and the linearity of the inner product, Σ is linear. Display (17.19) shows that Σ is self-adjoint and, due to $\langle \Sigma x, x\rangle = \mathbb{E}(\langle X, x\rangle^2) - \langle\mathbb{E}X, x\rangle^2 = \mathbb{V}\langle X, x\rangle \ge 0$, also positive. To show that Σ is a trace class operator, let $\{e_1, e_2, \ldots\}$ be any orthonormal basis of $\mathbb{H}$. With (17.19) and Parseval's equation it follows that

$$\sum_{k=1}^\infty |\langle\Sigma e_k, e_k\rangle| \le \sum_{k=1}^\infty |\langle Te_k, e_k\rangle| + \sum_{k=1}^\infty\langle\mathbb{E}X, e_k\rangle^2 = \sum_{k=1}^\infty |\langle Te_k, e_k\rangle| + \|\mathbb{E}X\|^2 < \infty,$$

since T is a trace class operator.

17.14 Solution Fix any, x, y in $\mathbb{H}$. Due to $\mathbb{E}(aX + h) = a\mathbb{E}(X) + h$, it follows that

$$\begin{aligned}
\langle\Sigma(aX + h)x, y\rangle &= \mathbb{E}\big[\langle(aX + h) - \mathbb{E}(aX + h), x\rangle\langle(aX + h) - \mathbb{E}(aX + h), y\rangle\big] \\
&= \mathbb{E}\big[a\langle X - \mathbb{E}X, x\rangle a\langle X - \mathbb{E}X, y\rangle\big] = \langle a^2\Sigma(X)x, y\rangle.
\end{aligned}$$

17.15 Solution The implication "(b) $\Longrightarrow$ (a)" follows from $\mathbb{E}\big[(X(s) - X(t))^2\big] = K(s, s) + K(t, t) - 2K(s, t) + (m(s) - m(t))^2$.

"(a) $\Longrightarrow$ (b)": Due to $|m(s) - m(t)| \le (\mathbb{E}(X(s) - X(t))^2)^{1/2}$, $m(\cdot)$ is continuous. To show the continuity of $K(\cdot, \cdot)$, assume w.l.o.g. $m(t) = 0$ for each $t \in E$. Due to the Cauchy–Schwarz inequality, $|K(s, t) - K(s, v)| = |\mathbb{E}[X(s)(X(t) - X(v))]| \le K(s, s)^{1/2}(\mathbb{E}(X(t) - X(v))^2)^{1/2}$ and analogously $|K(s, v) - K(u, v)| \le K(v, v)^{1/2}(\mathbb{E}(X(s) - X(u))^2)^{1/2}$. This implies the continuity of the covariance function.

17.16 Solution We have to show $X \overset{\mathcal{D}}{=} -X \Longleftrightarrow \varphi_X(h) = \overline{\varphi_X(h)}$ for every $h \in \mathbb{H}$. From $X \overset{\mathcal{D}}{=} -X$ it follows that $\varphi_X = \varphi_{-X}$. Due to $\varphi_{-X}(h) = \mathbb{E}\big[e^{i\langle X, -h\rangle}\big] = \varphi_X(-h) = \overline{\varphi_X(h)}$,

$h \in \mathbb{H}$, the function φ_X is thus real-valued. Conversely, $\varphi_X(h) = \overline{\varphi_X(h)}$, $h \in \mathbb{H}$, yields $\varphi_X = \varphi_{-X}$ and consequently $X \overset{\mathcal{D}}{=} -X$.

17.17 Solution According to (1.5), the characteristic function of a random variable Y with the normal distribution $\mathrm{N}(a, \sigma^2)$ is given by

$$\varphi_Y(t) = \mathrm{e}^{\mathrm{i}at} \cdot \exp\left(-\frac{\sigma^2 t^2}{2}\right), \quad t \in \mathbb{R}.$$

Because of $\varphi_X(h) = \varphi_{\langle X, h \rangle}(1)$ and $\langle X, h \rangle \sim \mathrm{N}(\langle m, h \rangle, \langle \Sigma h, h \rangle)$ the assertion follows.

17.18 Solution If X_n did not converge in distribution to X, there would be a continuous bounded function $f : \mathbb{H} \to \mathbb{R}$ and an $\varepsilon > 0$, such that for a subsequence $(n_k)_{k \geq 1}$

$$\left| \mathbb{E} f(X_{n_k}) - \mathbb{E} f(X) \right| \geq \varepsilon, \quad k \geq 1. \quad (*)$$

Because $(X_{n_k})_{k \geq 1}$ is relatively compact, there is a sub-subsequence $(X_{n'_k})_{k \geq 1}$ and a $\mathbb{H}$-valued random element Y with $X_{n'_k} \overset{\mathcal{D}}{\longrightarrow} Y$ as $k \to \infty$. Let $\psi(h) := \mathbb{E} \exp(\mathrm{i} \langle Y, h \rangle)$, $h \in \mathbb{H}$, be the characteristic functional of Y. Since for fixed $h \in \mathbb{H}$ the functions $\mathbb{H} \ni x \mapsto \cos\langle x, h \rangle$ and $\mathbb{H} \ni x \mapsto \sin\langle x, h \rangle$ are continuous and bounded, it follows that $\lim_{k \to \infty} \varphi_{n'_k}(h) = \psi(h)$, $h \in \mathbb{H}$. By assumption, $\lim_{n \to \infty} \varphi_n(h) = \varphi(h)$, $h \in \mathbb{H}$, and thus $\psi = \varphi$. With Theorem 17.23 f) it follows that $X \sim Y$. This equality in distribution contradicts the fact that inequality $(*)$ holds for infinitely many k.

17.19 Solution Since for each $h \in \mathbb{H}$ the random variables $\langle X, h \rangle$ and $\langle Y, h \rangle$ are independent, it follows from the addition theorem for the normal distribution that $\langle X + Y, h \rangle$ is normally distributed if both $\langle X, h \rangle$ and $\langle Y, h \rangle$ have a normal distribution. According to Theorem 17.18, the covariance operator of $X + Y$ exists, and it is equal to $\Sigma + T$. Since $\mathbb{E}(X + Y) = \mathbb{E}(X) + \mathbb{E}(Y)$, the assertion follows. Alternatively, one can also argue using characteristic functionals, because $\varphi_{X+Y} = \varphi_X \varphi_Y$.

17.20 Solution For each $\ell \in \mathbb{L}$, we have $\langle Y, \ell \rangle_{\mathbb{L}} = \langle TX, \ell \rangle_{\mathbb{L}} = \langle X, T^* \ell \rangle_{\mathbb{H}}$. Since X has a normal distribution, putting $h := T^* \ell$ the inner product $\langle X, h \rangle_{\mathbb{H}}$ has a univariate normal distribution. This shows that Y is a Gaussian random element in $\mathbb{L}$ (note the measurability of the mapping $T \circ X$ on the underlying probability space). Because $\langle X, h \rangle_{\mathbb{H}} \sim \mathrm{N}(\langle m, h \rangle_{\mathbb{H}}, \langle \Sigma h, h \rangle_{\mathbb{H}})$ (cf. (17.33)), $\langle m, h \rangle_{\mathbb{H}} = \langle m, T^* \ell \rangle_{\mathbb{H}} = \langle Tm, \ell \rangle_{\mathbb{L}}$, and $\langle \Sigma h, h \rangle_{\mathbb{H}} = \langle \Sigma T^* \ell, T^* \ell \rangle_{\mathbb{L}} = \langle T \Sigma T^* \ell, \ell \rangle_{\mathbb{L}}$, it follows that $\mathbb{E}(Y) = Tm$, and Y has the covariance operator $\Sigma(Y) = T \Sigma T^*$.

17.21 Solution By Theorem 17.26, $\|X - m\|^2$ has the same distribution as $\sum_{j=1}^{\infty} \lambda_j N_j^2$, with an i.i.d. sequence $N_1, N_2, \ldots$ of standard normally distributed random variables. We write the square of the above series as a double series. From the monotone convergence theorem, the independence of the N_j^2 and $\mathbb{E}(N_1^4) = 3$ as well as $\mathbb{E}(N_1^2) = 1$, it then follows that

$$\mathbb{E}(\|X - m\|^4) = \mathbb{E}\left(\sum_{j=1}^{\infty}\sum_{k=1}^{\infty}\lambda_j\lambda_k N_j^2 N_k^2\right) = \sum_{j=1}^{\infty}\sum_{k=1}^{\infty}\lambda_j\lambda_k\mathbb{E}(N_j^2 N_k^2)$$

$$= 3\sum_{j=1}^{\infty}\lambda_j^2 + \sum_{j\neq k}\lambda_j\lambda_k = 2\sum_{j=1}^{\infty}\lambda_j^2 + \left(\sum_{j=1}^{\infty}\lambda_j\right)^2.$$

17.22 Solution Due to (17.58), T_n depends on $X_1, \ldots, X_n$ only via

$$\|X_{n,j}\|_2^2 = (X_j - \overline{X}_n)^\top S_n^{-1}(X_j - \overline{X}_n), \quad \|X_{n,j} - X_{n,k}\|_2^2 = (X_j - X_k)^\top S_n^{-1}(X_j - X_k),$$

where $j, k \in \{1, \ldots, n\}$. Here, S_n^{-1} exists $\mathbb{P}$-almost surely. If for each $j \in \{1, \ldots, n\}$, we switch from X_j to $X_j' := AX_j + b$, the sample mean and the sample covariance matrix of $X_1', \ldots, X_n'$ become $\overline{X}_n' = A\overline{X}_n + b$ and $S_n' = AS_n A^\top$, respectively. It follows that

$$(X_j' - \overline{X}_n')^\top S_n'^{-1}(X_j' - \overline{X}_n') = \left(A(X_j - \overline{X}_n)\right)^\top \left(AS_n A^\top\right)^{-1} A(X_j - \overline{X}_n)$$

$$= (X_j - \overline{X}_n)^\top A^\top (A^\top)^{-1} S_n^{-1} A^{-1} A(X_j - \overline{X}_n)$$

$$= (X_j - \overline{X}_n)^\top S_n^{-1}(X_j - \overline{X}_n)^\top.$$

In the same way, it can be seen that $\|X_{n,j} - X_{n,k}\|_2^2 = (X_j - X_k)^\top S_n^{-1}(X_j - X_k)$ remains invariant to the transformation $X_j \mapsto AX_j + b$.

Bibliography

[ABS] *Abramowitz, M., and Stegun, I.A. (eds.)* (1972): Handbook of Mathematical Functions with Formulas, Graphs, and Mathematical Tables. National Bureau of Standards, Applied Mathematics Series 55, 10th Printing, December 1972.

[ALA] *Alexandrov, A.D.* (1943): Additive Set-Functions in Abstract Spaces. Matematicheskii Sbornik 13, 169–238.

[ALB] *Aliprantis, C., and Border K.* (2006): Infinite Dimensional Analysis. A Hitchhiker's Guide. 3rd Edition. Springer, Berlin.

[BAH] *Bahadur, R. R.* (1964): On Fisher's Bound for Asymptotic Variances. Annals of Mathematical Statistics 35(4), 1545–1552.

[BHJ] *Barbour, A.D., Holst, L., and Janson, S.* (1992): Poisson Approximation. Oxford: Clarendon Press

[BHE] *Baringhaus, L., and Henze, N.* (1988): A consistent Test for Multivariate Normality based on the Empirical Characteristic Function. Metrika 35, 339–348.

[BEH] *Baringhaus, L., Ebner, B., and Henze, N.* (2017): The Limit Distribution of weighted L^2-Goodness-of-Fit Statistics under Fixed Alternatives, with applications. Annals of the Institute of Statistical Mathematics 69, 969–995.

[BK] *Baringhaus, L., and Kolbe, D.* (2017): Two-sample Tests based on empirical Hankel Transforms. Statistical Papers 56, 597–617.

[BGO] *Bentkus, V., and Götze, F.* (1999): Optimal Bounds in Non-Gaussian Limit Theorems for U-Statistics. The Annals of Probability 27, 454–521.

[BJZ] *Bentkus, V., Jing, B.-J., and Zhou, W.* (2009): On Normal Approximations to U-Statistics. The Annals of Probability 37, 2174–2199.

[BER] *Berry, A.* (1941): The Accuracy of the Gaussian Approximation to the Sum of Independent Variates. Transactions of the American Mathematical Society 49, 122–136.

[BRR] *Bhattacharya, R.N., and Ranga Rao, R.* (2010). Normal Approximation and Asymptotic Expansions. Updated Republication of the 1986 Reprint Published by Robert E. Krieger Publ. Co. Classics in Applied Mathematics, 64. Society for Industrial and Applied Mathematics (SIAM), Philadelphia, PA.

[BD1] *Bickel, P.J., and Doksum, K.A.* (2015): Mathematical Statistics. Basic Ideas and Related Topics. Vol. 1, 2nd Edition. Boca Raton, London.

[BD2] *Bickel, P.J., and Doksum, K.A.* (2016): Mathematical Statistics. Basic Ideas and Related Topics. Vol. 2. Boca Raton, London.

[BI1] *Billingsley, P.* (1968): Convergence of Probability Measures, 1st Edition. J. Wiley & Sons, New York.

© The Author(s), under exclusive license to Springer-Verlag GmbH, DE, part of Springer Nature 2024
N. Henze, *Asymptotic Stochastics*, Mathematics Study Resources 10,
https://doi.org/10.1007/978-3-662-68923-3

[BI2] *Billingsley, P.* (1999): Convergence of Probability Measures. 2nd Edition. J. Wiley & Sons, New York.

[BI3] *Billingsley, P.* (1995): Probability and Measure. 3rd Edition. J. Wiley & Sons, New York.

[BIK] *Bingham, N.H., and Kiesel, R.* (2004): Risk-Neutral Valuation. Pricing and Hedging of Financial Derivatives. 2nd Edition, Springer, London.

[BOG] *Bogachev, V. I.* (1998): Gaussian Measures. Mathematical Surveys and Monographs, 62. American Mathematical Society, Providence, RI.

[CAN] *Cantelli, F.* (1933): Sulla Determinazione Empirica delle Leggi di Probabilità. Giornale dell' Istituto Italiano degli Attuari 4, 421–424.

[CLH] *Chen, L.H.Y.* (1975): Poisson Approximation for Dependent Trials. The Annals of Probability 3, 534–545.

[CGS] *Chen, L.H.Y., Goldstein, L., and Shao, Q.-M.* (2011): Normal Approximations by Stein's Method. Springer, Berlin.

[CRA] *Cramér, H.* (1945), Mathematical Methods of Statistics. Almquist and Wiksells Boktryckeri AB, Uppsala.

[CSO] *Csörgő, S.* (1989): Consistency of some Tests for Multivariate Normality. Metrika 36, 107–116.

[DEU] *Deuflhard, P.* (2011): Newton Methods for Nonlinear Problems. Affine Invariance and Adaptive Algorithms. Springer, Berlin.

[DEV] *Devroye, L.* (1987): A Course in Density Estimation. Progress in Probability and Statistics 14. Birkhäuser, Boston, MA.

[DJS] *Döring, H., Jansen, S., and Schubert, K.* (2022): The Method of Cumulants for the Normal Approximation. Probability Surveys 19, 185–270.

[DE1] *Dörr, Ph., Ebner, B., and Henze, N.* (2021): Testing Multivariate Normality by Zeros of the Harmonic Oscillator in Characteristic Function Spaces. Scandinavian Journal of Statistics 48, 456–501.

[DE2] *Dörr, Ph., Ebner, B., and Henze, N.* (2021): A new Test of Multivariate Normality by a Double Estimation in a Characterizing PDE. Metrika 84, 401–427.

[DUD] *Dudley, R.M.* (2002): Real Analysis and Probability. Cambridge Studies in Advanced Mathematics. 74. Cambridge University Press, Cambridge.

[DSE] *Durante, F., and Sempi, C.* (2015): Principles of Copula Theory. Taylor & Francis, London.

[DUR] *Durrett, R.* (2010): Probability. Theory and Examples. 4th Edition. Cambridge Series in Statistical and Probabilistic Mathematics, 49. Cambridge University Press, Cambridge.

[DKW] *Dvoretsky, A., Kiefer, J., and Wolfowitz, J.* (1956): Asymptotic Minimax Character of the Sample Distribution Function and of the Classical Multinomial Estimator. Annals of Mathematical Statistics 27, 642–669.

[EDW] *Edwards, A.W.F.* (1974): The History of Likelihood. International Statistical Review 42(1), 9–15.

[EH1] *Ebner, B., and Henze, N.* (2020): Tests for Multivariate Normality – a Critical Review with Emphasis on Weighted L^2-Statistics. TEST 29, 845–892.

[EH] *Ebner, B., and Henze, N.* (2023): On the Eigenvalues Associated with the Limit Null Distribution of the Epps–Pulley Test for Normality. Statistical Papers 64, 739–752.

[EKM] *Embrechts, P., Klüppelberg, C., and Mikosch, Th.* (1997): Modelling Extremal Events. Springer, Berlin.

[EPY] *Epps, T., and Pulley, L.* (1983): A Test for Normality based on the Empirical Characteristic Function. Biometrika 70, 723–726.

[ESS] *Esseen, C.* (1942): On the Liapounoff Limit of Error in the Theory of Probability. Arkiv för Matematik, Astronomi och Fysik, 1–19.

[EUB] *Eubank, R.* (1999): Nonparametric Regression and Spline Smoothing. Second Edition. Statistics: Textbooks and Monographs, 157. Marcel Dekker, New York.

[FH] *Fabian, V., and Hannan, J.* (1985): Introduction to Probability and Mathematical Statistics. J. Wiley & Sons, New York.

[FEL] *Feller, W.* (1970): An Introduction to Probability Theory and Its Applications Vol. 1, 3rd Edition. Wiley, New York.

[FEL] *Feller, W.* (1970): An Introduction to Probability Theory and Its Applications Vol. 2, 2nd Edition. Wiley, New York.

[FER] *Ferguson, Th. S.* (1996): A Course in Large Sample Theory. Chapmann & Hall, London.

[FRE] *Fréchet, M.* (1943): Sur l'extension de certaines évaluations statistiques au cas de petits echantillons. Revue de l'Institute International de Statistique 11, No. 3/4, 182–205.

[GST] *Gänssler, P., and Stute, W.* (1976): On uniform convergence of measures with applications to uniform convergence of empirical distributions. Lecture Notes in Math., Vol. 566, Springer, Berlin.

[GLI] *Glivenko, W.I.* (1933): Sulla Determinazione empirica della Legge di Probabilità. Giornale dell' Istituto Italiano degli Attuari 4, 92–99.

[GKP] *Graham, R.L., Knuth, D.E., and Patashnik, O.* (1994): Concrete Mathematics, 2nd Edition. Addison-Wesley, Reading, Massachusetts.

[GRS] *Green, P.J., and Silverman, B.W.* (1994): Nonparametric Regression and Generalized Linear Models. A Roughness Penalty Approach. Monographs on Statistics and Applied Probability. Chapman & Hall, London.

[GRG] *Gregory, G.G.* (1977): Large Sample Theory for U-Statistics and Tests of Fit. The Annals of Statistics 5, 110–123.

[GSI] *Grimmett, G.R., and Stirzaker, D.R.* (2020): Probability and Random Processes. 4th Edition. Oxford University Press, Oxford.

[GUE] *Gürtler, N.* (2000): Asymptotische Untersuchungen zur Klasse der BHEP-Tests auf multivariate Normalverteilung mit festem und variablem Glättungsparameter. Doctoral dissertation, University of Karlsruhe, Germany.

[GUT] *Guttman, L.* (1948): A Distribution-Free Confidence Interval for the Mean. Annals of Mathematical Statistics 19, 410–413.

[HAA] *Haase, M.* (2014): Functional Analysis. An Elementary Introduction. Graduate Studies in Mathematics Volume 156. American Mathematical Society, Providence, Rhode Island.

[HAE] *Härdle, W.* (1990): Applied Nonparametric Regression. Econometric Society Monographs, 19. Cambridge University Press, Cambridge.

[HAJ] *Hájek, J.* (1968): Asymptotic Normality of Simple Linear Rank Statistics. Annals of Mathematical Statistics 39, 325–346.

[HAL] *Hald, A.* (1998): A History of Mathematical Statistics from 1750 to 1930. J. Wiley & Sons, New York.

[HHE] *Hall, P., and Heyde, C.C.* (1980): Martingale Limit Theory and its Application. Academic Press, New York.

[HEI] *Heil, Ch.* (2018): Metrics, Norms, Inner Products, and Operator Theory. Birkhäuser, Cham, Switzerland.

[H90] *Henze, N.* (1990): An Approximation to the Limit Distribution of the Epps–Pulley Test Statistic for Normality. Metrika 37, 7–18.

[H02] *Henze, N.* (2002): Invariant Tests for Multivariate Normality: A Critical Review. Statistical Papers 43, 467–506.

[HE0] *Henze, N.* (2018): Irrfahrten – Faszination der Random Walks. 2nd Edition. Springer Spektrum, Heidelberg.

[HE1] *Henze, N.* (2019): Stochastik: Eine Einführung mit Grundzügen der Maß- und Integrations-theorie. Springer Spektrum, Heidelberg.

[HJI] *Henze, N., and Jiménez-Gamero, M.D.* (2021): A Test for Gaussianity in Hilbert Spaces via the Empirical Characteristic Functional. Scandinavian Journal of Statistics 48, 406–428.

[HLS] *Henze, N., Lafaye de Micheaux, P., and Meintanis. S.G.* (2022): Tests for circular Symmetry of complex-valued random Vectors. Test 31(2), 488–518.

[HEK] *Henze, N., and Koch, St.* (2020): On a Test of Normality based on the Empirical Moment Generating Function. Statistical Papers 61, 17–29.

[HMA] *Henze, N., and Mayer, C.* (2021): More good News on the HKM Test for Multivariate Reflected Symmetry about an Unknown Centre. Annals of the Institute of Statistical Mathematics 72, 741–770.

[HWA] *Henze, N., and Wagner, Th.* (1997): A new Approach to the BHEP Tests for Multivariate Normality. Journal of Multivariate Analysis 62, 1–23.

[HEZ] *Henze, N., and Zirkler, B.* (1990): A Class of Invariant and Consistent Tests for Multivariate Normality. Communications in Statistics A –Theory Methods – 19, 3595–3617.

[HID] *Hida, T.* (1980): Brownian Motion. Springer, Berlin.

[HOE] *Hoeffding, W.* (1948): A Class of Statistics with Asymptotically Normal Distributions. Annals of Mathematical Statistics 19, 293–325.

[HOP] *Höpfner, R.* (2014): Asymptotic Statistics. With a View Towards Stochastic Processes. De Gruyter Graduate. W. de Gruyter & Co., Berlin.

[HOK] *Horváth, L., and Kokoszka, P.* (2012): Inference for Functional Data Analysis. Springer, New York.

[HSE] *Hsing, T., and Eubank, R.* (2015): Theoretical Foundations of Functional Data Analysis, with an Introduction to Linear Operators. Wiley & Sons, New York.

[KAL] *Kallenberg, O.* (2021): Foundations of Modern Probability. Probability and Stochastic Models Vol. 99. Springer, New York.

[KLE] *Klenke, A.* (2020): Probability Theory. A comprehensive Course. 3rd Edition. Springer Spektrum, Heidelberg.

[KOE] *König, H.* (1986): Eigenvalue Distributions of Compact Operators. Birkhäuser, Basel.

[KOL] *Kolmogorov, A.N.* (1933): Grundbegriffe der Wahrscheinlichkeitsrechnung. Springer, Berlin, Heidelberg, New York, Reprint 1973.

[KOB] *Korolyuk, V.S., and Borovskich, Y.V.* (1994): Theory of U-Statistics. Springer, New York.

[KUK] *Kukush, A.* (2019): Gaussian Measures in Hilbert Space. Construction and Properties. Wiley, New York.

[KUM] *Kumar, P.* (2019): Copula Functions and Applications in Engineering. In: *Deep, K., Jain, M., and Salhi, S. (Eds.)* (2019): Logistics, Supply Chain and Financial Predictive Analysis. Theory and Practices. Springer Nature, Singapore.

[KMM] *Kundu, S., Majumdar, S., and Mukherjee, K.* (2000): Central Limit Theorems revisited. Statistics & Probability Letters 47, 265–275.

[LAP] *Last, G., and Penrose, M.* (2018): Lectures on the Poisson Process. Institute of Mathematical Statistics. Textbooks 7. Cambridge University Press, Cambridge.

[LET] *Ledoux, M., and Talagrand, M.* (2011): Probability in Banach Spaces. Isoperimetry and Processes. Reprint of the 1991 Edition. Classics in Mathematics. Springer, Berlin.

[LEE] *Lee, A.J.* (1990): U-Statistics. Theory and Practice. Marcel Dekker, Inc. New York.

[LEG] *Le Gall, J.-F.* (2013): Brownian Motion, Martingales, and Stochastic Calculus. Graduate Texts in Mathematics 274. Springer, Berlin, Heidelberg.

[LEC] *Lehmann, E.L., and Casella, G.* (2003): Theory of Point Estimation. 2nd Edition. Springer, New York.

[LIM] *Liese, F., and Miescke, K.-J.* (2008): Statistical Decision Theory. Estimation, Testing, and Selection. Springer Series in Statistics. Springer, New York.

[LUK] *Lukacs, E.* (1970): Characteristic functions. 2nd Edition. Griffin, London.

[MAS] *Massart, P.* (1990): The tight constant in the Dvoretsky–Kiefer–Wolfowitz inequality. The Annals of Probability 18, 1269–1283.

[MES] *Mees, A.* (2015): Zur Robustheit von Konfidenzbereichen und Tests für Erwartungswerte. Mit einem Geleitwort von Prof. Dr. Lutz Mattner. Best Masters. Heidelberg: Springer Spektrum; Trier: Universität Trier (Masters Thesis).

[MOP] *Mörters, P., and Peres, Y.* (2010): Brownian Motion. With an Appendix by Oded Schramm and Wendelin Werner. Cambridge Series in Statistical and Probabilistic Mathematics, 30. Cambridge University Press, Cambridge.

[MUS] *Muscat, J.* (2014): Functional Analysis. An Introduction to Metric Spaces, Hilbert Spaces, and Banach Algebras. Springer International Publishing, Switzerland.

[NIK] *Nikitin, Ya. Yu.* (1995): Asymptotic Efficiency of Nonparametric Tests. Cambridge University Press, Cambridge.

[NOV] *Novak, S.Y.* (2019): Poisson Approximation. Probability Surveys 16, 228–276.

[PAR] *Parthasarathy, K.R.* (1967): Probability Measures on Metric Spaces. Wiley, New York.

[PET] *Petrov, V.V.* (1995); Limit Theorems of Probability Theory. Sequences of Independent Random Variables. Oxford Studies in Probability, 4. Oxford Science Publications. Clarendon Press, Oxford University Press, New York.

[PF1] *Pfanzagl, J:* (1994): Parametric Statistical Theory. In Collaboration with Ralf Hamböker. De Gruyter Textbook. W. de Gruyter & Co., Berlin.

[PF2] *Pfanzagl, J.* (2017): Mathematical Statistics. Essays on History and Methodology. Springer Series in Statistics. Perspectives in Statistics. Springer, Berlin.

[PSH] *Pfanzagl, J., and Sheynin, O.* (1996): Studies in the History of Probability and Statistics XLIV. A forerunner of the *t*-distribution. Biometrika 83(4), 891–898.

[PFW] *Pfanzagl, J.* (1985): Asymptotic Expansions for General Statistical Models. With the assistance of W. Wefelmeyer. Lecture Notes in Statistics, 31. Springer, Berlin.

[PIN] *Pinelis, I.* (2017): Optimal-Order Uniform and Nonunform Bounds on the Rate of Convergence to Normality for Maximum Likelihood Estimators. Electronic Journal of Statistics 11, 1160–1179.

[PIM] *Pinelis, I., and Molzon, R.* (2016): Optimal-Order Bounds on the Rate of Convergence to Normality in the Multivariate Delta Method. Electronic Journal of Statistics 10, 1001–1063.

[POL] *Pollard, D.* (1984): Convergence of Stochastic Processes. Springer, New York.

[RAI] *Raič, M.* (2019): A Multivariate Berry–Essen Theorem with Explicit Constants. Bernoulli 25, 2824–2853.

[CRT] *R Core Team* (2022): *R*: A Language and Environment for Statistical Computing. R Foundation for Statistical Computing, Vienna. https://www.R-project.org/.

[REI] *Reiss, R.-D.* (1989): Approximate Distributions of Order Statistics. With Applications to Nonparametric Statistics. Springer Series in Statistics. Springer, New York.

[ROS] *Rosenblatt, M.* (1956): A Central Limit Theorem under a Strong Mixing Condition. Proceedings of the National Academy of Sciences U.S.A. 42, 43-47.

[SCH] *Schilling, R.L.* (2021): Brownian Motion: A Guide to Random Processes and Stochastic Calculus. With a Chapter on Simulation by Björn Böttcher. 3rd Edition. De Gruyter, Berlin.

[SHE] *Schilling, J., and Henze, N.* (2021): Two Poisson Limit Theorems for the Coupon Collector's Problem with Group Drawings. Journal of Applied Probability 58, 966–977.

[SCO] *Scott, D.W.* (2015): Multivariate Density Estimation. Theory, Practice, and Visualization. J. Wiley & Sons, Hoboken, NJ.

[SER] *Serfling, R.* (1980): Approximation Theorems of Mathematical Statistics. J. Wiley & Sons, New York.

[SHA] *Shao. J.* (2003): Mathematical Statistics, 2nd Edition. Springer, New York.

[SHE] *Shevtsova, I.* (2013): On the Absolute Constants in the Berry–Esseen Inequality and its Structural and Nonuniform Improvements. Informatika i ejö Primenenija 7, 124–125.

[SHI] *Shiryayev, A.N.* (1984): Probability. Springer, New York.

[SHW] *Shorack, G.R., and Wellner, J.A.* (2009): Empirical Processes with Applications to Statistics. Society for Industrial and Applied Mathematics (SIAM), Philadelphia. Reprint of the first Edition published by J. Wiley & Sons in 1986.

[SIL] *Silverman, B.W.* (1986): Density Estimation for Statistics and Data Analysis. Monographs on Statistics and Applied Probability. Chapman & Hall, London.

[SPD] *Spokoiny, V., and Dickhaus, Th.* (2015): Basics of Modern Mathematical Statistics. Springer, Berlin.

[STE] *Stein, C.* (1972). A Bound for the Error in the Normal Approximation to the Distribution of a Sum of Dependent Random Variables. In: Proceedings of the Sixth Berkeley Symposium on Mathematical Statistics and Probability, Vol. 2, 586–602. Berkeley: University of California Press.

[STO] *Stone, C.* (1975): Adaptive Maximum Likelihood Estimators of a Location Parameter. The Annals of Statistics 3, 267–284.

[TE1] *Tenreiro, C.* (2009): On the Choice of the Smoothing Parameter for the BHEP Goodness-of-Fit Test. Computational Statistics and Data Analysis 53, 1038–1053.

[TE2] *Tenreiro, C.* (2019): On the Automatic Selection of the Tuning Parameter appearing in Certain Families of Goodness-of-fit Tests. Journal of Statistical Computation and Simulation 89, 1780–1797.

[VBE] *van Beek, P.* (1972): An Application of Fourier Methods to the Problem of Sharpening the Berry–Esseen Inequality. Zeitschrift für Wahrscheinlichkeitstheorie und Verwandte Gebiete 23, 187–196.

[VW] *van der Vaart, A.* (1998): Asymptotic Statistics. Cambridge Series in Statistical and Probabilistic Mathematics, 3. Cambridge: Cambridge University Press.

[VW1] *van der Vaart, A., and Wellner, J.* (2023): Weak Convergence and Empirical Processes: With Applications to Statistics. 2nd Edition. Springer Series in Statistics. Springer, New York.

[VW2] *van der Vaart, A., and Wellner, J.* (2021): Stein 1956: Efficient Nonparametric Testing and Estimation. The Annals of Statistics 49, 1836–1849.

[VOL] *Volkert, K.* (1987): Die Geschichte der pathologischen Funktionen – Ein Betrag zur Entstehung der mathematischen Methodologie. Archive for History of Exact Sciences 37, 193–232.

[WAL] *Wald, A.* (1949): A note on the Consistency of the Maximum Likelihood Estimate. Annals of Mathematical Statistics 20, 595–601.

[WEI] *Weidmann, J.* (1980): Linear Operators in Hilbert Spaces. Springer, New York.

[WIM] *Witting, H., and Müller-Funk, U.* (1995): Mathematische Statistik II. B.G. Teubner, Stuttgart.

[ZEI] *Zeidler, E.* (1986): Nonlinear Functional Analysis and its Applications I: Fixed-Point Theorems. Springer, New York.

[ZO1] *Zolotarev, V.M.* (1967): An Absolute Estimate of the Remainder Term in the Central Limit Theorem. Theory of Probability and its Applications 11, 95–105.

[ZO2] *Zolotarev, V.M.* (1967): A Sharpening of the Inequality of Berry–Esseen. Zeitschrift für Wahrscheinlichkeitstheorie und Verwandte Gebiete 8, 332–342.

[ZR1] *Zorich, V.A.* (2015): Mathematical Analysis I. 2nd Edition. Springer, New York.

[ZR2] *Zorich, V.A.* (2016): Mathematical Analysis II. 2nd Edition. Springer, New York.

Index